T0221052

AN INTRODUCTION TO 3D COMPUTER VISION TECHNIQUES AND ALGORITHMS

AN INTRODUCTION TO 3D COMPUTER VISION TECHNIQUES AND ALGORITHMS

Bogusław Cyganek
Department of Electronics, AGH University of Science and Technology, Poland

J. Paul Siebert
Department of Computing Science, University of Glasgow, Scotland, UK

A John Wiley and Sons, Ltd., Publication

This edition first published 2009

Registered office
John Wiley & Sons Ltd, The Atrium, Southern Gate, Chichester, West Sussex, PO19 8SQ, United Kingdom

For details of our global editorial offices, for customer services and for information about how to apply for permission to reuse the copyright material in this book please see our website at www.wiley.com.

Library of Congress Cataloging-in-Publication Data

Cyganek, Boguslaw.
An introduction to 3D computer vision techniques and algorithms / by Boguslaw Cyganek and J. Paul Siebert.
p. cm.
Includes index.
ISBN 978-0-470-01704-3 (cloth)
1. Computer vision. 2. Three-dimensional imaging. 3. Computer algorithms. I. Siebert, J. Paul. II. Title
TA1634.C94 2008
006.3′7–dc22

2008032205

A catalogue record for this book is available from the British Library.

ISBN 978-0-470-01704-3

Set in 10/12pt Times by Aptara Inc., New Delhi, India.
Printed in Great Britain by CPI Antony Rowe, Chippenham, Wiltshire

To Magda, Nadia and Kamil
From Bogusław

To Sabina, Konrad and Gustav
From Paul

Contents

Preface

Recent decades have seen rapidly growing research in many areas of computer science, including computer vision. This comes from the natural interest of researchers as well as demands from industry and society for qualitatively new features to be afforded by computers. One especially desirable capability would be automatic reconstruction and analysis of the surrounding 3D environment and recognition of objects in that space. Effective 3D computer vision methods and implementations would open new possibilities such as automatic navigation of robots and vehicles, scene surveillance and monitoring (which allows automatic recognition of unexpected behaviour of people or other objects, such as cars in everyday traffic), medical reasoning, remote surgery and many, many more.

This book is a result of our long fascination with computers and vision algorithms. It started many years ago as a set of short notes with the only purpose 'to remember this or that' or to have a kind of 'short reference' just for ourselves. However, as this diary grew with the years we decided to make it available to other people. We hope that it was a good decision! It is our hope that this book facilitates access to this enthralling area, especially for students and young researchers. Our intention is to provide a very concise, though as far as possible complete, overview of the basic concepts of 2D and 3D computer vision. However, the best way to get into the field is to try it oneself! Therefore, in parallel with explaining basic concepts, we provide also a basic programming framework with the hope of making this process easier. We greatly encourage the reader to take the next step and try the techniques in practice.

Bogusław Cyganek, Kraków, Poland
J. Paul Siebert, Glasgow, UK

Acknowledgements

We would like to express our gratitude to all the people who helped in the preparation of this book!

In particular, we are indebted to the whole Wiley team who helped in the preparation of the manuscript. In this group special thanks go to Simone Taylor who believed in this project and made it happen. We would also like to express our gratitude to Sian Andrews, Laura Bell, Liz Benson, Emily Bone, Lucy Bryan, Kate Griffiths, Wendy Hunter, Alex King, Erica Peters, Kathryn Sharples, and Nicky Skinner.

We are also very grateful to the individuals and organizations who agreed to the use of their figures in the book. These are Professor Yuichi Ohta from Tsukuba University, as well as Professor Ryszard Szeliski from Microsoft Research. Likewise we would like to thank Dimensional Imaging Ltd. and Precision 3D Ltd. for use of their images. In this respect we would also like to express our gratitude to Springer Science and Business Media, IEEE Computer Society Press, the IET, Emerald Publishing, the ACM, Maney Publishing and Elsevier Science.

We would also like to thank numerous colleagues from the AGH University of Science and Technology in Kraków. We owe a special debt of gratitude to Professor Ryszard Tadeusiewicz and Professor Kazimierz Wiatr, as well as to Lidia Krawentek for their encouragement and continuous support.

We would also like to thank members of the former Turing Institute in Glasgow (Dr Tim Niblett, Joseph Jin, Dr Peter Mowforth, Dr Colin Urquhart and also Arthur van Hoff) as well as members of the Computer Vision and Graphics Group in the Department of Computing Science, University of Glasgow, for access to and use of their research material (Dr John Patterson, Dr Paul Cockshott, Dr Xiangyang Ju, Dr Yijun Xiao, Dr Zhili Mao, Dr Zhifang Mao (posthumously), Dr J.C Nebel, Dr Tim Boyling, Janet Bowman, Susanne Oehler, Stephen Marshall, Don Whiteford and Colin McLaren). Similarly we would like to thank our collaborators in the Glasgow Dental Hospital and School (Professor Khursheed Moos, Professor Ashraf Ayoub and Dr Balvinder Khambay), Canniesburn Plastic Surgery Unit (Mr Arup Ray), Glasgow, the Department of Statistics (Professor Adrian Bowman and Dr Mitchum Bock), Glasgow University, Professor Donald Hadley, Institute of Neurological Sciences, Southern General Hospital, Glasgow, and also those colleagues formerly at the Silsoe Research Institute (Dr Robin Tillett, Dr Nigel McFarlane and Dr Jerry Wu), Silsoe, UK.

Special thanks are due to Dr Sumitha Balasuriya for use of his Matlab codes and graphs. Particular thanks are due to Professor "Keith" van Rijsbergen and Professor Ray Welland without whose support much of the applied research we report would not have been possible.

We wish to express our special thanks and gratitude to Steve Brett from Pandora Inc. for granting rights to access their software platform.

Some parts of the research for which results are provided in this book were possible due to financial support of the European Commission under RACINE-S (IST-2001-37117) and IP-RACINE (IST-2-511316-IP) as well as Polish funds for scientific research in 2007–2008. Research described in these pages has also been funded by the UK DTI and the EPSRC & BBSRC funding councils, the Chief Scientist Office (Scotland), Wellcome Trust, Smith's Charity, the Cleft Lip and Palate Association, the National Lottery (UK) and the Scottish Office. Their support is greatly appreciated.

Finally, we would like to thank Magda and Sabina for their encouragement, patience and understanding over the three-year period it took to write this book.

Notation and Abbreviations

$I_k(x, y)$	Intensity value of a k-th image at a point with local image coordinates (x, y)
$\overline{I_k(x, y)}$	Average intensity value of a k-th image at a point with local image coordinates (x, y)
$\mathbf{I}$	Identity matrix; image treated as a matrix
$\mathbf{P}$	A vector (a point), matrix, tensor, etc.
$T[\mathbf{I}, \mathbf{P}]$	The Census transformation T for a pixel $\mathbf{P}$ in the image $\mathbf{I}$
i, j	Free coordinates
d_x, d_y	Displacements (offset) in the x and y directions
$D(\mathbf{p}_l, \mathbf{p}_r)$	Disparity between points $\mathbf{p}_l$ and $\mathbf{p}_r$
$\mathbf{D}$	Disparity map (a matrix)
$U(x, y)$	Local neighbourhood of pixels around a point (x, y)
$\mathbf{O}_c$	Optical centre point
$\mathbf{P}_c = [X_c, Y_c, Z_c]^T$	Coordinates of a 3D point in the camera coordinate system
$\boldsymbol{\Pi}$	Camera plane; a projective plane
$\mathbf{o} = (o_x, o_y)^T$	Central point of a camera plane
f	Focus length of a camera
b	Base line in a stereo system (a distance between cameras)
h_x, h_y	Physical horizontal and vertical dimensions of a pixel
$\mathbf{P} = [X, Y, Z]^T$	3D point and its coordinates
$\wp^n$	N-dimensional projective space
$\mathbf{P} = [X_h, Y_h, Z_h, 1]^T$	Homogenous coordinates of a point
$\mathbf{M}$	Camera matrix
$\mathbf{M}_\mathrm{i}$	Intrinsic parameters of a camera
$\mathbf{M}_\mathrm{e}$	Extrinsic parameters of a camera
$\mathbf{E}$	Essential matrix.
$\mathbf{F}$	Fundamental matrix.
$\mathbf{e}_i$	Epipole in an i-th image
SAD	Sum of absolute differences
SSD	Sum of squared differences
ZSAD	Zero-mean sum of absolute differences
ZSSD	Zero-mean sum of squared differences
ZSSD-N	Zero-mean sum of squared differences, normalized

SCP	Sum of cross products
SCP-N	Sum of cross products, normalized
RMS	Root mean square
RMSE	Root mean square error
<Lxx, Lyy>	Code lines from a line *Lxx* to *Lyy*
HVS	Human Visual System
SDK	Software Development Kit
$\wedge$	logical 'and'
$\vee$	logical 'or'
LRC	Left-right checking (cross-checking)
OCC	Occlusion constraint
ORD	Point ordering constraint
BMD	Bimodality rule
MGJ	Match goodness jumps
NM	Null method
GT RMS	Ground-truth RMS
WTA	Winner-takes-all
*	Convolution operator

Part I

1

Introduction

The purpose of this text on stereo-based imaging is twofold: it is to give students of computer vision a thorough grounding in the image analysis and projective geometry techniques relevant to the task of recovering three-dimensional (3D) surfaces from stereo-pair images; and to provide a complete reference text for professional researchers in the field of computer vision that encompasses the fundamental mathematics and algorithms that have been applied and developed to allow 3D vision systems to be constructed.

Prior to reviewing the contents of this text, we shall set the context of this book in terms of the underlying objectives and the explanation and design of 3D vision systems. We shall also consider briefly the historical context of optics and vision research that has led to our contemporary understanding of 3D vision.

Here we are specifically considering 3D vision systems that base their operation on acquiring stereo-pair images of a scene and then decoding the depth information *implicitly* captured within the stereo-pair as parallaxes, i.e. relative displacements of the contents of one of the images of the stereo-pair with respect to the other image. This process is termed *stereo-photogrammetry*, i.e. measurement from stereo-pair images. For readers with normal functional binocular vision, the everyday experience of observing the world with both of our eyes results in the perception of the relative distance (depth) to points on the surfaces of objects that enter our field of view. For over a hundred years it has been possible to configure a stereo-pair of cameras to capture stereo-pair images, in a manner analogous to mammalian binocular vision, and thereafter view the developed photographs to observe a miniature 3D scene by means of a stereoscope device (used to present the left and right images of the captured stereo-pair of photographs to the appropriate eye). However, in this scenario it is the brain of the observer that must decode the depth information locked within the stereo-pair and thereby experience the perception of depth. In contrast, in this book we shall present underlying mechanisms by which a computer program can be devised to analyse digitally formatted images captured by a stereo-pair of cameras and thereby recover an *explicit* measurement of distances to points *sampling* surfaces in the imaged field of view. Only by explicitly recovering depth estimates does it become possible to undertake useful tasks such as 3D measurement or reverse engineering of object surfaces as elaborated below. While the science of stereo-photogrammetry is a well-established field and it has indeed been possible to undertake 3D

An Introduction to 3D Computer Vision Techniques and Algorithms Bogusław Cyganek and J. Paul Siebert

measurement by means of stereo-pair images using a manually operated measurement device (the stereo-comparator) since the beginning of the twentieth century, we present fully automatic approaches for 3D imaging and measurement in this text.

1.1 Stereo-pair Images and Depth Perception

To appreciate the structure of 3D vision systems based on processing stereo-pair images, it is first necessary to grasp, at least in outline, the most basic principles involved in the formation of stereo-pair images and their subsequent analysis. As outlined above, when we observe a scene with both eyes, an image of the scene is formed on the retina of each eye. However, since our eyes are horizontally displaced with respect to each other, the images thus formed are not identical. In fact this stereo-pair of retinal images contains slight displacements between the relative locations of local parts of the image of the scene with respect to each image of the pair, depending upon how close these local scene components are to the point of *fixation* of the observer's eyes. Accordingly, it is possible to reverse this process and deduce how far away scene components were from the observer according to the magnitude and direction of the parallaxes within the stereo-pairs when they were captured. In order to accomplish this task two things must be determined: firstly, those local parts of one image of the stereo-pair that match the corresponding parts in the other image of the stereo-pair, in order to find the local parallaxes; secondly, the precise geometric properties and configuration of the eyes, or cameras. Accordingly, a process of *calibration* is required to discover the requisite geometric information to allow the imaging process to be inverted and relative distances to surfaces observed in the stereo-pair to be recovered.

1.2 3D Vision Systems

By definition, a stereo-photogrammetry-based 3D vision system will require stereo-pair image acquisition hardware, usually connected to a computer hosting software that automates acquisition control. Multiple stereo-pairs of cameras might be employed to allow all-round coverage of an object or person, e.g. in the context of whole-body scanners. Alternatively, the object to be imaged could be mounted on a computer-controlled turntable and overlapping stereo-pairs captured from a fixed viewpoint for different turntable positions. Accordingly, sequencing capture and image download from multiple cameras can be a complex process, and hence the need for a computer to automate this process.

The stereo-pair acquisition process falls into two categories, active illumination and passive illumination. Active illumination implies that some form of pattern is projected on to the scene to facilitate finding and disambiguating parallaxes (also termed *correspondences* or *disparities*) between the stereo-pair images. Projected patterns often comprise grids or stripes and sometimes these are even colour coded. In an alternative approach, a random speckle texture pattern is projected on to the scene in order to augment the texture already present on imaged surfaces. Speckle projection can also guarantee that that imaged surfaces appear to be randomly textured and are therefore locally uniquely distinguishable and hence able to be matched successfully using certain classes of image matching algorithm. With the advent of 'high-resolution' digital cameras the need for pattern projection has been reduced, since the surface texture naturally present on materials, having even a matte finish, can serve to facilitate matching stereo-pairs. For example, stereo-pair images of the human face and body can be matched successfully using ordinary studio flash illumination when the pixel sampling density is sufficient to resolve the natural texture of the skin, e.g. skin-pores. A camera resolution of approximately 8–13M pixels is adequate for stereo-pair capture of an area corresponding to the adult face or half-torso.

The acquisition computer may also host the principal 3D vision software components:

- An image matching algorithm to find correspondences between the stereo-pairs.
- Photogrammetry software that will perform system calibration to recover the geometric configuration of the acquisition cameras and perform 3D point reconstruction in world coordinates.
- 3D surface reconstruction software that builds complete manifolds from 3D *point-clouds* captured by each imaging stereo-pair.

3D visualisation facilities are usually also provided to allow the reconstructed surfaces to be displayed, often *draped* with an image to provide a *photorealistic* surface model. At this stage the 3D shape and surface appearance of the imaged object or scene has been captured in explicit digital metric form, ready to feed some subsequent application as described below.

1.3 3D Vision Applications

This book has been motivated in part by the need to provide a manual of techniques to serve the needs of the computer vision practitioner who wishes to construct 3D imaging systems configured to meet the needs of practical applications. A wide variety of applications are now emerging which rely on the fast, efficient and low-cost capture of 3D surface information. The traditional role for image-based 3D surface measurement has been the reserve of *close-range* photogrammetry systems, capable of recovering surface measurements from objects in the range of a few tens of millimetres to a few metres in size. A typical example of a classical close-range photogrammetry task might comprise surface measurement for manufacturing quality control, applied to high-precision engineered products such as aircraft wings.

Close-range video-based photogrammetry, having a lower spatial resolution than traditional plate-camera film-based systems, initially found a niche in imaging the human face and body for clinical and creative media applications. 3D clinical photographs have the potential to provide quantitative measurements that reduce subjectivity in assessing the surface anatomy of a patient (or animal) before and after surgical intervention by providing numeric, possibly automated, scores for the shape, symmetry and longitudinal change of anatomic structures. Creative media applications include whole-body 3D imaging to support creation of human avatars of specific individuals, for 3D gaming and cine special effects requiring virtual actors. Clothing applications include body or foot scanning for the production of custom clothing and shoes or as a means of sizing customers accurately. An innovative commercial application comprises a 'virtual catwalk' to allow customers to visualize themselves in clothing prior to purchasing such goods on-line via the Internet.

There are very many more emerging uses for 3D imaging beyond the above and commercial 'reverse engineering' of premanufactured goods. 3D vision systems have the potential to revolutionize autonomous vehicles and the capabilities of robot vision systems. Stereo-pair cameras could be mounted on a vehicle to facilitate autonomous navigation or configured within a robot workcell to endow a 'blind' pick-and-place robot, both object recognition capabilities based on 3D cues and simultaneously 3D spatial quantification of object locations in the workspace.

1.4 Contents Overview: The 3D Vision Task in Stages

The organization of this book reflects the underlying principles, structural components and uses of 3D vision systems as outlined above, starting with a brief historical view of vision research in Chapter 2. We deal with the basic existence proof that binocular 3D vision is possible, in an overview of the human visual system in Chapter 3. The basic projective geometry techniques that underpin 3D vision systems are also covered here, including the geometry of monocular and binocular image formation which relates how binocular parallaxes are produced in stereo-pair images as a result of imaging scenes containing variation in depth. Camera calibration techniques are also presented in Chapter 3, completing the introduction of the role of image formation and geometry in the context of 3D vision systems.

We deal with fundamental 2D image analysis techniques required to undertake image filtering and feature detection and localization in Chapter 4. These topics serve as a precursor to perform image matching, the process of detecting and quantifying parallaxes between stereo-pair images, a prerequisite to recovering depth information. In Chapter 5 the issue of spatial scale in images is explored, namely how to structure algorithms capable of efficiently processing images containing structures of varying scales which are unknown in advance. Here the concept of an image *scale-space* and the *multi-resolution image pyramid* data structure is presented, analysed and explored as a precursor to developing matching algorithms capable of operating over a wide range of visual scales. The core algorithmic issues associated with stereo-pair image matching are contained in Chapter 6 dealing with distance measures for comparing image patches, the associated parametric issues for matching and an in-depth analysis of area-based matching over scale-space within a practical matching algorithm. Feature-based approaches to matching are also considered and their combination with area-based approaches. Then two solutions to the stereo problem are discussed: the first, based on the *dynamic programming*, and the second one based on the *graph cuts* method. The chapter ends with discussion of the *optical flow* methods which allow estimation of local displacements in a sequence of images.

Having dealt with the recovery of disparities between stereo-pairs, we progress logically to the recovery of 3D surface information in Chapter 7. We consider the process of *triangulation* whereby 3D points in world coordinates are computed from the disparities recovered in the previous chapter. These 3D points can then be organized into surfaces represented by *polygonal meshes* and the *3D point-clouds* recovered from *multi-view* systems acquiring more than one stereo-pair of the scene can be fused into a coherent surface model either directly or via volumetric techniques such as *marching cubes*. In Chapter 8 we conclude the progression from theory to practice, with a number of case examples of 3D vision applications covering areas such as face and body imaging for clinical, veterinary and creative media applications and also 3D vision as a visual prosthetic. An application based only on image matching is also presented that utilizes motion-induced inter-frame disparities within a cine sequence to synthesize missing or damaged frames, or sets of frames, in digitized historic archive footage.

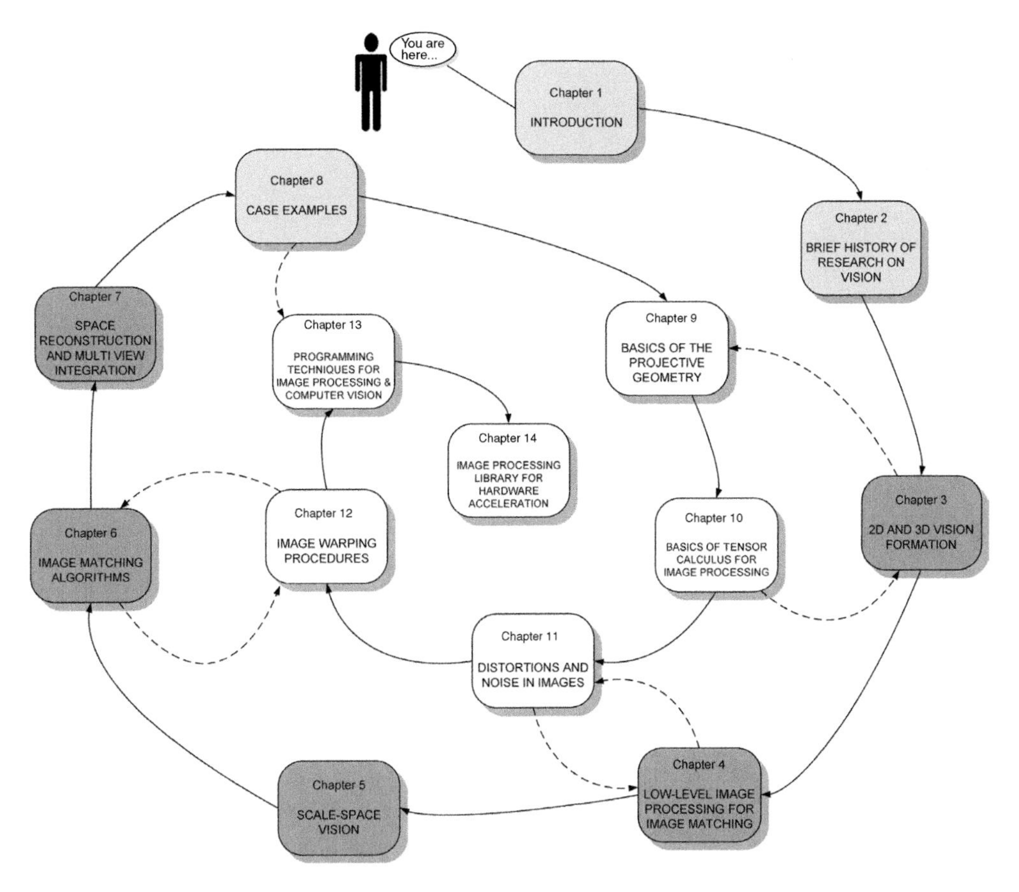

Figure 1.1 Organization of the book

The remaining chapters provide a series of detailed technical tutorials on projective geometry, tensor calculus, image warping procedures and image noise. A chapter on programming techniques for image processing provides practical hints and advice for persons who wish to develop their own computer vision applications. Methods of object oriented programming, such as design patterns, but also proper organization and verification of the code are discussed. Chapter 14 outlines the software presented in the book and provides the link to the recent version of the code.

Figure 1.1 depicts possible order of reading the book. All chapters can be read in number order or selectively as references to specific topics. There are five main chapters (Chapters 3–7), three auxiliary chapters (Chapters 1, 2 and 8) as well as five technical tutorials (Chapters 9–13). The latter are intended to aid understanding of specific topics and can be read in conjunction with the related main chapters, as indicated by the dashed lines in Figure 1.1.

2

Brief History of Research on Vision

2.1 Abstract

This chapter is a brief retrospective on vision in art and science. 3D vision and perspective phenomena were first studied by the architects and artists of Ancient Greece. From this region and time comes *The Elements* by Euclid, a treatise that paved the way for geometry and mathematics. Perspective techniques were later applied by many painters to produce the illusion of depth in flat paintings. However, called an 'evil trick', it was denounced by the Inquisition in medieval times. The blooming of art and science came in the Renaissance, an era of Leonardo da Vinci, perhaps the most ingenious artist, scientist and engineer of all times. He is attributed with the invention of the camera obscura, a prototype of modern cameras, which helped to acquire images of a 3D scene on a flat plane. Then, on the 'shoulders of giants' came another 'giant', Sir Isaac Newton, whose *Opticks* laid the foundation for modern physics and also the science of vision. These and other events from the history of research on vision are briefly discussed in this chapter.

2.2 Retrospective of Vision Research

The first people known to have investigated the phenomenon of depth perception were the Ancient Greeks [201]. Probably the first writing on the subject of disparity comes from Aristotle (380 BC) who observed that, if during a prolonged observation of an object one of the eyeballs is pressed with a finger, the object is experienced in double vision.

The earliest known book on optics is a work by Euclid entitled *The Thirteen Books of the Elements* written in Alexandria in about 300 BC [116]. Most of the definitions and postulates of his work constitute the foundations of mathematics since his time. Euclid's works paved the way for further progress in optics and physiology, as well as inspiring many researchers over the following centuries. At about the same time as Euclid was writing, the anatomical structure of human organs, including the eyes, was examined by Herofilus from Alexandria. Subsequently Ptolemy, who lived four centuries after Euclid, continued to work on optics.

Many centuries later Galen (AD 180) who had been influenced by Herofilus' works, published his own work on human sight. For the first time he formulated the notion of the *cyclopean* eye, which 'sees' or visualizes the world from a common point of intersection within the optical nervous pathway that originates from each of the eyeballs and is located perceptually at an intermediate position between the eyes. He also introduced the notion of parallax and described the process of creating a single view of an object constructed from the binocular views originating from the eyes.

The works of Euclid and Galen contributed significantly to progress in the area of optics and human sight. Their research was continued by the Arabic scientist Alhazen, who lived around AD 1000 in the lands of contemporary Egypt. He investigated the phenomena of light reflection and refraction, now fundamental concepts in modern geometrical optics.

Based on Galen's investigations into anatomy, Alhazen compared an eye to a dark chamber into which light enters via a tiny hole, thereby creating an inverted image on an opposite wall. This is the first reported description of the *camera obscura*, or the pin-hole camera model, an invention usually attributed to Roger Bacon or Leonardo da Vinci. A device called the camera obscura found application in painting, starting from Giovanni Battista della Porta in the sixteenth century, and was used by many masters such as Antonio Canal (known as Canaletto) or Bernaldo Bellotto. A painting by Canaletto, entitled *Perspective*, is shown in Figure 2.1. Indeed, his great knowledge of basic physical properties of light and projective

Figure 2.1 *Perspective* by Antonio Canal (Plate 1). (1765, oil on canvas, Gallerie dell'Accademia, Venice)

Figure 2.2 Painting by Bernardo Bellotto entitled *View of Warsaw from the Royal Palace* (Plate 2). (1773, oil on canvas, National Museum, Warsaw)

geometry allowed him to reach mastery in paintings. His paintings are very realistic which was a very desirable skill of a painter, since we have to remember that these were times when people did not yet know of photography.

Figure 2.2 shows a view of eighteenth-century Warsaw, the capital of Poland, painted by Bernaldo Bellotto in 1773. Just after, due to invasion of the three neighbouring countries, Poland disappeared from maps for over a century.

Albrecht Dürer was one of the first non-Italian artists who used principles of geometrical perspective in his art. His famous drawing *Draughtsman Drawing a Recumbent Woman* is shown in Figure 2.3.

However, the contribution of Leonardo da Vinci cannot be overestimated. One of his famous observations is that a light passing through a small hole in the camera obscura allows the

Figure 2.3 A drawing by Albrecht Dürer entitled *Draughtsman Drawing a Recumbent Woman.* (1525, woodcut, Graphische Sammlung Albertina, Vienna)

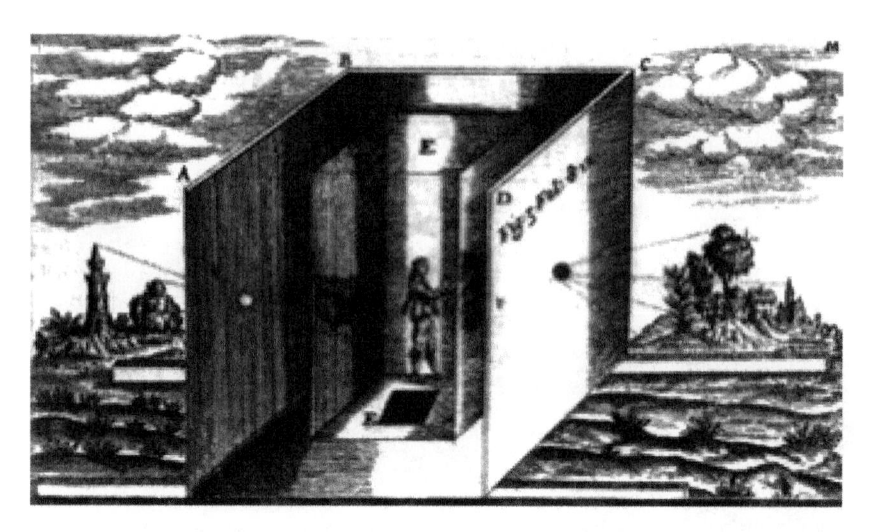

Figure 2.4 Drawing of the camera obscura from the work of the Jesuit Athanasius Kircher, around 1646

observation of all surrounding objects. From this he concluded that light rays passing through different objects cross each other in any point from which they are visible. This observation suggests also the wave nature of light, rather than light comprising a flow of separate particles as was believed by the Ancient Greeks. Da Vinci's unquestionable accomplishment in the area of stereoscopic vision is his analysis of partial and total occlusions, presented in his treatise entitled *Trattato della Pittura*. Today we know that these phenomena play an important role in the human visual system (HVS), facilitating correct perception of depth [7] (section 3.2).

Other accomplishments were made in Europe by da Vinci's contemporaries. For instance in 1270 Vitello, who lived in Poland, published a treatise on optics entitled *Perspectiva*, which was the first of its kind. Interestingly, from almost the same time comes a note on the first binoculars, manufactured probably in the glassworks of Pisa.

Figure 2.4 depicts a drawing of a camera obscura by the Jesuit Athanasius Kircher, who lived in the seventeenth century.

In the seventeenth century, based on the work of Euclid and Alhazen, Kepler and Descartes made further discoveries during their research on the HVS. In particular, they made great contributions towards understanding of the role of the retina and the optic nerve in the HVS.

More or less at the same time, i.e. the end of the sixteenth and beginning of the seventeenth centuries, the Jesuit Francois D'Aguillon made a remarkable synthesis of contemporary knowledge on optics and the works of Euclid, Alhazen, Vitello and Bacon. In the published treatise *Opticorum Libri Sex*, consisting of six books, D'Aguillon analysed visual phenomena and in particular the role of the two eyes in this process. After defining the locale of visual convergence of the two eyeballs, which he called the horopter, D'Aguillon came close to formulating the principles of stereovision which we still use today.

A real breakthrough in science can be attributed to Sir Isaac Newton who, at the beginning of the eighteenth century, published his work entitled *Opticks* [329]. As first, he correctly described a way of information passing from the eyes to the brain. He discovered that visual sensations from the "inner" hemifields of the retina (the mammalian visual field is split along the vertical meridian in each retina), closest to the nose, are sent through the optic nerves directly to the corresponding cerebral hemispheres (cortical lobes), whereas sensations coming from the "outer" hemifields, closest to the temples, are crossed and sent to the opposite hemispheres. (The right eye, right hemifield and left eye, left hemifield cross, while the left eye, right hemifield and the right eye, left hemifield do not cross.) Further discoveries in this area were made in the nineteenth century not only thanks to researchers such as Heinrich Müller and Bernhard von Gudden, but also thanks to the invention of the microscope and developments in the field of medicine, especially physiology.

In 1818 Vieth made a precise explanation of the horopter, being a spherical placement of objects which cause a focused image on the retina, a concept that was already familiar to D'Aguillon. At the same time this observation was reported by Johannes Müller, and therefore the horopter is termed the Vieth–Müller circle.

In 1828 a professor of physics of the Royal Academy in London, Sir Charles Wheatstone, formulated the principles underlying stereoscopic vision. He also presented a device called a *stereoscope* for depth perception from two images. This launched further observations and discoveries; for instance, if the observed images are reversed, then the perception of depth is also reversed. Inspired by Wheatstone's stereoscope, in 1849 Sir David Brewster built his version of the stereoscope based on a prism (Figure 2.5), and in 1856 he published his work on the principles of stereoscopy [56].

The inventions of Wheatstone and Brewster sparked an increased interest in three-dimensional display methods, which continues with even greater intensity today due to the invention of the random dot autostereograms, as well as the rapid development of personal computers. Random dot stereograms were analysed by Bela Julesz who in 1960 showed that

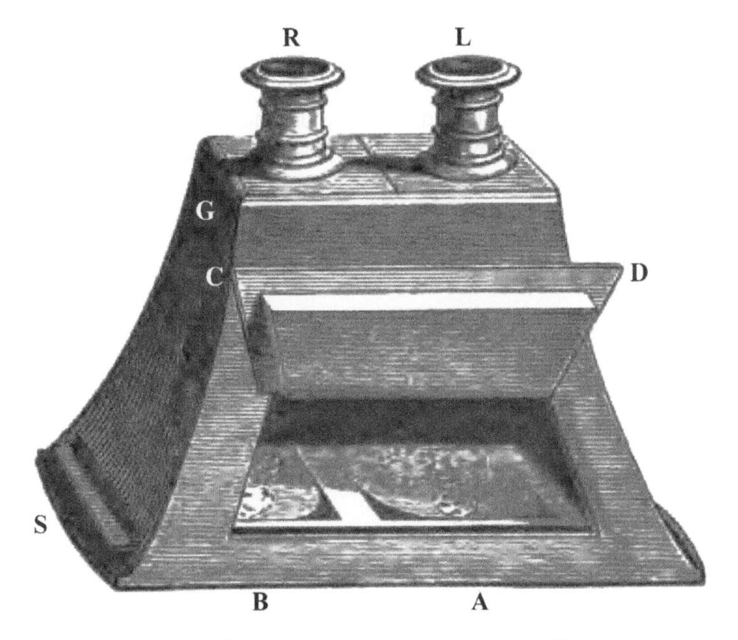

Figure 2.5 Brewster's stereoscope (from [56])

depth can be perceived by humans from stereo-pairs of images comprising only random dots (the dots being located with relative shifts between the images forming the stereo-pair) and no other visible features such as corners or edges.

Recent work reported by the neurophysiologists Bishop and Pettigrew showed that in primates special cells, which react to disparity signals built from images formed on two retinas of the eyes, are already present in the input layer (visual area 1, V1) of the visual cortex. This indicates that depth information is processed even earlier in the visual pathway than had been thought.

2.3 Closure

In this chapter we have presented a very short overview of the history of studies on vision in art and science. It is a very wide subject which could have merited a separate book by itself. Nevertheless, we have tried to point out those, in our opinion, important events that paved the way for contemporary knowledge on vision research, which also inspired us to write this book. Throughout the centuries, art and science were interspersed and influenced each other. An example of this is the camera obscura which, first devised by artists, after centuries became a prototype of modern cameras. These are used to acquire digital images, then processed with vision algorithms to infer knowledge on the surrounding environment, for instance. Further information on these fascinating issues can be found in many publications, some of which we mention in the next section.

2.3.1 Further Reading

There are many sources of information on the history of vision research and photography. For instance the Bright Bytes Studio web page [204] provides much information on camera obscuras, stereo photography and history. The Web Gallery of Art [214] provides an enormous number of paintings by masters from past centuries. The book by Brewster mentioned earlier in the chapter can also be obtained from the Internet [56]. Finally, Wikipedia [215] offers a wealth of information in many different languages on most of the subjects, including paintings, computer vision and photography.

Part II

3

2D and 3D Vision Formation

3.1 Abstract

This chapter is devoted mainly to answering the question: "What is the difference between having one image of a scene, compared to having two images of the same scene taken from different viewpoints?" It appears that in the second case the difference is a fundamental one: with two (or more) views of the same scene, taken however at different camera positions, we can infer depth information by means of geometry: three-dimensional (3D) information can be recovered through a process known as *triangulation*. This is why having two eyes makes a difference.

We start with a brief overview of what we know about the human visual system which is an excellent example of precision and versatility. Then we discuss the image acquisition process using a single camera. The main concept here is the simple pin-hole camera model which is used to explain the transformation from 3D world-space to the 2D imaging-plane as performed by a camera. The so-called extrinsic and intrinsic parameters of a camera are introduced next. When images of a scene are captured using two cameras simultaneously, these cameras are termed a *stereo-pair* and produce stereo-pairs of images. The properties of cameras so configured are determined by their *epipolar geometry*, which tells us the relationship between world points observed in their fields of view and the images impinging on their respective sensing planes. The image-plane locations of each world point, as sensed by the camera pair, are called corresponding or matched points. Corresponding points within stereo-pair images are connected by the fundamental matrix. If known, it provides fundamental information on the epipolar geometry of the stereo-pair setup. However, finding corresponding points between images is not a trivial task. There are many factors which can confound this process, such as occlusions, limited image resolution and quantization, distortions, noise and many others. Technically, matching is said to be *under constrained*: there is not sufficient information available within the compared images to guarantee finding a unique match. However, matching *can* be made easier by applying a set of rules known as *stereo constraints*, of which the most important is the *epipolar constraint*, and this implies that corresponding points always lie on corresponding epipolar lines. The epipolar constraint limits the search for corresponding points from the entire 2D space to a 1D space of epipolar lines. Although the positions of the epipolar lines are not known in advance, in the special case when stereo-pair cameras are

An Introduction to 3D Computer Vision Techniques and Algorithms Bogusław Cyganek and J. Paul Siebert

configured with parallel optical axes – called the canonical, fronto-parallel, or standard stereo system – the epipolar lines follow the image (horizontal) scan-lines. The problem of finding corresponding points is therefore one of the essential tasks of computer vision.

It appears that by means of point correspondences the extrinsic and intrinsic parameters of a camera can be determined. This is called camera calibration and is also discussed in this chapter. We conclude with a discussion of a practical implementation of the presented concepts, with data structures to represent images and some C++ code examples which come from the image library provided with this book.

3.2 Human Visual System

Millions of years of evolution have formed the human visual system (HVS) and within it the most exquisite, unattainable and mysterious stereoscopic depth perception engine on planet Earth. The vision process starts in the eye, a diagram of which is depicted in Figure 3.1.

Incident light at first passes through the pupil which controls the amount of light passing to the lens of the eye. The size of the pupil aperture is controlled by the iris pupilliary sphincter muscles. The larger this aperture becomes, the larger the spherical aberration and smaller the depth of focus of the eye. The visual axis joins a point of fixation and the fovea. Although an eye is not rotationally symmetric, an approximate optical axis can be defined as a line joining the centre of curvature of the cornea and centre of the lens. The angle between the two axes is about 5°. It should be noted that the eye itself is not a separate organ but a 150 mm extension of the brain. In the context of computer vision, the most important part of the eye is the retina which is the place of exchange that converts an incoming stream of photons into corresponding neural excitations.

In the context of binocular vision and stereoscopic perception of depth, it is important that the eyes are brought into convergence such that the same scene region is projected onto the respective foveae of each eye. Figure 3.2 presents a model of binocular vision: an image of a certain point H is created in the two eyes, exactly in the centres of their foveae.

On each retina images of the surrounding 3D points are also created. We mark the distance of those images in respect to the corresponding fovea. Under this assumption, the two image points on each of the retinas are *corresponding* when their *distances* to their corresponding

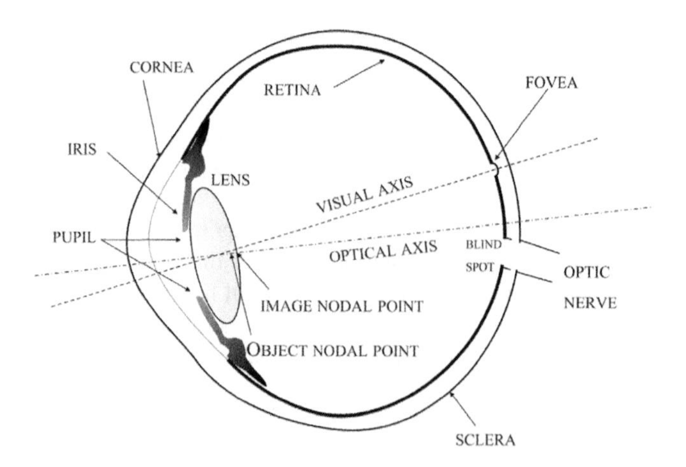

Figure 3.1 Schematic of a human eye

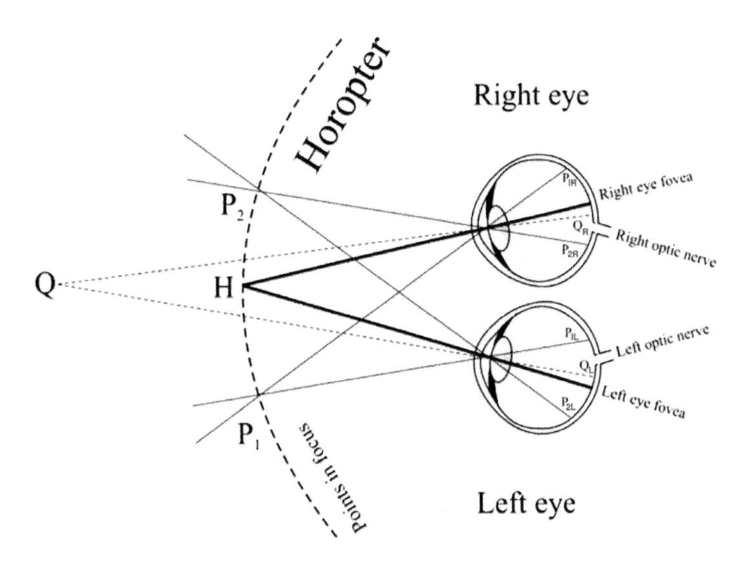

Figure 3.2 Disparity on the retina of an eye. The horopter is denoted by a broken line. H is a point of fixation

foveae are the *same*. In Figure 3.2 this condition is fulfilled for the points P_1 and P_2, but not for Q. That is, the distances P_{1R} and P_{1L} are the same. This holds also for P_{2R} and P_{2L} but not for the Q_R and Q_L which are in opposite directions from the foveae. However, the latter property allows the HVS to conclude that Q is further from the horopter. Conducting now the reverse reasoning, i.e. looking for 3D points such that their retinal images are the same distance from the two foveae, we find the 3D region known as the *horopter*. Retinal images of all points other than those belonging to the horopter are said to be non-corresponding. The relative difference in distance from the fovea for of each these non-corresponding points is termed *retinal disparity* [201, 442]. It is evident now that the horopter points have zero retinal disparity. The retinal disparity is used by the HVS to assess distance to 3D locations in the world.

The signals induced on the fovea are transferred to the input of the primary visual cortex of the brain, labelled by neuro-anatomists as Visual Area 1 (V1). This area of the visual cortex is the first location in the entire structure where individual neurons receive binocular input. It was also discovered that some neurons in V1 respond exclusively to mutual excitations from the two eyes. Those neurons, called disparity detectors, are sensitive to stereoscopic stimuli [442].

In addition, the relationship between the *firing rates* of these disparity detecting neurons, measured in units of *impulses per second*, and input retinal disparity is called the disparity-tuning function. It has an evident maximum for zero retinal disparity (i.e. it is "tuned" to respond best to zero disparity), that is for 3D points lying on the horopter [201].

Many experiments have been conducted to achieve a better understanding of the stereoscopic processes in the HVS. A phenomenon first noticed during such research was the influence of luminance variation on the process of associating corresponding visual stimuli from each eye, i.e. disparity detection. In the simplest case this concerns the detection, i.e. correlation, of corresponding image edges in each retina, while correlation of corresponding textured areas is more complex. In 1979 Marr and Poggio [299] put forward a theory that stereoscopic matching relies on the correlation of retinal image locations in which the second derivative of the luminance signal is crossing a zero value; these are the so-called zero-crossings.

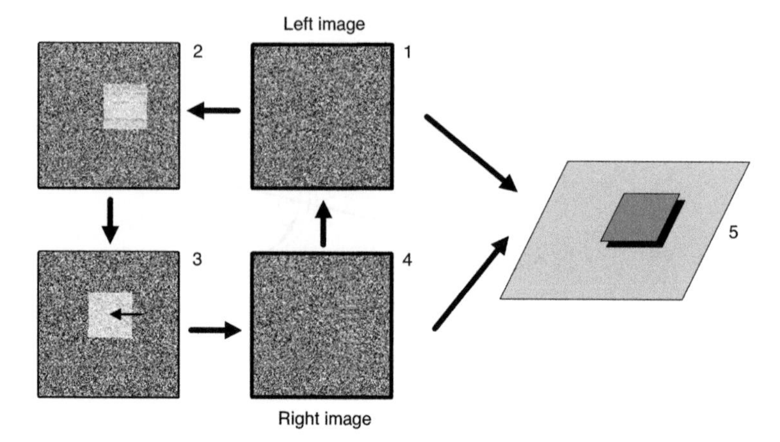

Figure 3.3 Construction of a random dot stereogram: (1) left image; (2) extracted region in the left image; (3) shift of this region; (4) right image; (5) depth effect when observed by two eyes

Zero-crossings corresponds to those regions in an image that exhibit the greatest change in the signal instead of the greatest absolute value of the signal itself. Further research undertaken by Mayhew and Frisby [302] showed that stereoscopic correlation in the HVS does not depend exclusively on the zero-crossings but on a more generalised matching mechanism applied to the spectral components of the two-dimensional luminance signal. Mallot *et al.* [291] revealed the possibility of a secondary correlation mechanism being invoked when the luminance signal is changing very slowly. Based on these results it can be stated that the HVS prefers to correlate more general features, if available, in the image. This relates correlation based on zero-crossings and also correlation based on signal value maxima. However, correlation based on matching spectral components of the luminance signal dominates when these are the most distinctive features found in the images. When there are neither significant zero-crossings nor other signal differences, the HVS is capable of estimating disparity values based on correlating the maximal values of the low-pass components of the luminance signal.

A qualitatively new development was reported by Julesz in 1960 [235] when he demonstrated the so-called random dot stereogram.[1] A random dot stereogram comprises a stereo-pair of images in which the first image of the pair is created by generating a field of random points. The second image of the stereo-pair is generated by copying the first image and then selecting and displacing by a small amount a specific region within the copy. Figure 3.3 outlines steps of this construction. Table 3.4 (page 62) contains another example of a random dot stereogram. When constructing random stereograms the random dots can be substituted by random lines [201].

When observed by two eyes, the random dot stereogram allows perception of depth, as seen in Figure 3.3 in a form of a rectangle closer to the observer. Further research on this subject has shown that the stereoscopic effect is attained even if one of the random images

[1] This type of stereogram was already known, however, among artists.

is disturbed, e.g. by adding some spurious dots or by low-pass filtering. On the other hand, a change of luminance polarity (i.e. light and dark regions are exchanged in one image of the stereogram) leads to a loss of the stereo effect.

Research on depth perception based exclusively on a perception of colours has shown that colour information also affects this process to a limited degree [201].

It has been discovered that the stereo correlation process depends also on other factors, leading to a theory that predicts that those compared locations which conform in size, shape, colour, and motion are more privileged during stereo matching. It would also explain why it takes more time for the HVS to match random dot stereograms which do not possess such features. This theory can also be interpreted in the domain of computational stereo matching methods: if a certain local operator can gather enough information in a given neighbourhood of pixels, such as local frequency, orientation or phase, then subsequent matching can be performed more reliably and possibly faster than would otherwise be possible when such information is missing. This rather heuristic rule can be justified by experiment. An example of a tensor operator that quantifies local image structure is presented in section 4.6.

Another known stereo matching constraint adopted by the HVS is so-called most related image matching. It implies that if there is a choice, an image or an image sub-region is considered to be 'matched' if it gives the highest number of meaningful matches. Otherwise the preferred image is one which contains the highest number of space point projections. Due to this strategy, the HVS favours those images, or their sub-regions, that are potentially the most interesting to an observer, since they are closest to him or her.

Yet another constraint discovered by Julesz [235], is the disparity gradient limit. This concept, explained in more detail in section 3.5, is very often used in computer image matching.

Other constraints are based on experience acquired from daily observations of the surrounding space. One of which is that the daily environment usually is moderately 'dense', since we have to move in it somehow. A similar observation indicates that surrounding objects are not transparent either. From these observations we can draw other matching constraints based on: surface continuity, figural smoothness, matching point ordering and matching point uniqueness (section 3.5). Their function in and influence on the HVS, although indicated by many experiments, have not yet been completely explained.

Yet another phenomenon plays an important role in both human and machine stereovision, namely that of occlusions which are explained in Figure 3.4.

How partial occlusions of observed objects influence their binocular perception was investigated by Leonardo Da Vinci [93]. Recent work by Anderson indicates that the occlusion phenomenon has a major influence on the stereovision perception process [7]. The area visible exclusively to the left eye is called the left visible area. Similarly for the right eye we get the right visible area. In Figure 3.4 these areas are marked in light grey. The area observable to both eyes simultaneously can be perceived in full stereo vision. In contrast, the dark area to the left of the object in Figure 3.4 presents a totally occluded location to both eyes. Far beyond the object there is again an area visible to both eyes, so effectively an object does not occlude the whole space behind it, only a part. It is also known and easily verified that the half-occluded regions seen by the right eye falls close to the right edge of the occluding object. Similarly, the half-occluded regions seen by the left eye fall near the left edge of such an occluding object. This situation is portrayed in Figure 3.4.

The effect of partial occlusions is inevitably connected with a break in the smoothness (continuity) of a perceived surface in depth. Thus, due to the presence of partial occlusions,

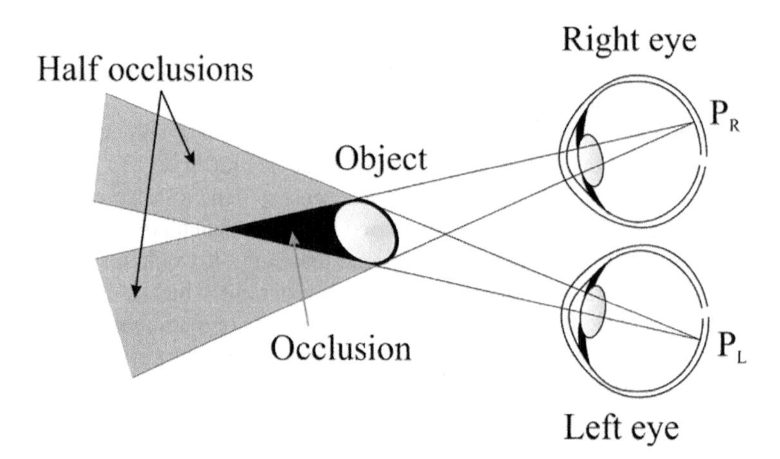

Figure 3.4 Phenomenon of occlusions. Partial occlusions are lighter. The dark area remains totally occluded by an object

it is possible to distinguish depth discontinuities from gradually changing surfaces which, in turn, are limited by the maximum allowable disparity gradient. These and other facts show that the HVS actively decomposes vertical and horizontal image parallaxes into disparities and half-occlusions [7]. They form two complementary sources of visual information. Retinal disparities provide information about the relative depth of observed surfaces visible to both eyes simultaneously. On the other hand, partial-occlusions which are visible to each eye separately, give sufficient data for segmentation of the observable scene into coherent objects at object boundaries.

It is interesting to mention that also the gradient of the *vertical* disparity can be used to infer distance from observed objects, as has been shown by Mayhew and Longuet-Higgins [303] and discussed also by Brenner *et al.* [55]. However, recent psychophysical experiments have indicated that such information is not used by the HVS. Indeed, vertical image differences are not always vertical parallaxes. Sometimes they are caused by half-occlusions. Based on these observations and psychophysical experiments, Anderson [7] suggests that interocular differences in vertical position can influence stereoscopic depth perceived by the HVS by signalling the presence of occluding contours.

Depth perception by the HVS is not only induced purely by stereovision mechanisms, it is also supported by the phenomena of head and eye movements, as well as by motion parallax.

Many psychophysical experiments lead to the observation that there is continuous rivalry between the different vision cues that impinge on the HVS. Then the HVS detects such objects that arise from maxima in the density of goodmatches, when simultaneously in agreement with daily experience.

Depth information acquired by the HVS, as well as other visual cues such as information on colour, edges, shadows and occlusions are only ingredients gathered by the brain to generate inferences about the world. How these visual inferences are then integrated and interpreted into a unified percept is still not known, although hypotheses and models have been proposed by researchers. Knowledge of the function of the visual system has been garnered indirectly by means of observations of two different sets of phenomena known from medicine

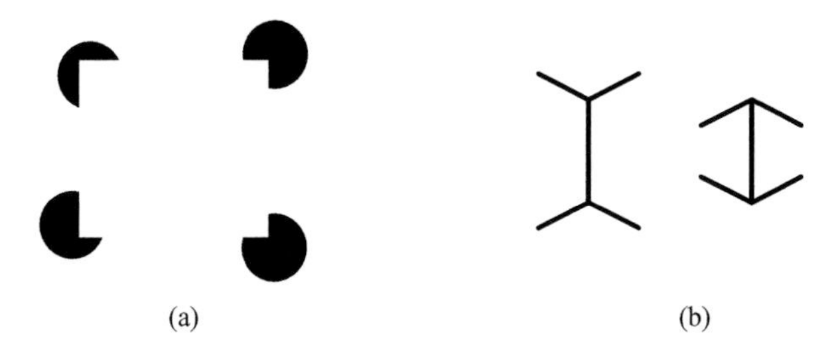

Figure 3.5 Visual illusions. (a) An artificial rectangle is clearly visible although not drawn directly. (b) The two vertical lines are exactly the same length (which can be verified with a ruler), although the left one is perceived to be longer

and psychophysiology. The first set of phenomena are described in case studies that record sight diseases and their subsequent cures. It was clinically observed that those patients who were visually impaired from birth and then had their ability to perceive visual sensations restored, had difficulties learning how to perceive objects and how to interpret scenes, although they can easily detect basic features [201, 442]. Indirectly this provides us with some insight into the conceptual stages and complexity of the seeing mechanisms of our brains.

Visual illusions comprise a second set of phenomena that help us understand how the visual pathways translate retinal images into the perception of objects. There are many illusions that trick our visual system by providing visual cues that do not agree with the physics of the 3D world learned by daily experience [125, 161, 360]. Two simple illusions apparently related to the human perception of depth are presented in Figure 3.5. The first example (Figure 3.5(a)) illustrates the role of occlusions in visual perception. Our acquired experience on transparency of objects makes us perceive an illusory figure whose existence is only cued (i.e. made apparent) by the presence of occluding contours overlaid on other visible objects in the image.

The second example (Figure 3.5(b)) shows two lines of *exactly the same* length, which terminate with an arrow-head at each line end. However, the arrow head pairs for correspondng line ends point in opposite directions. None the less, the first line gives an impression of being longer. This phenomenon can be explained by daily experience. The left configuration in Figure 3.5(b) suggests that the central line is further from the observer compared to the right hand line configuration. This makes us believe that the left line has to be longer in the 3D world.

What seems a common observation about such illusions in 2D images is that we experience some false interpretation of the ‘flat’ patterns because our visual system always tries to interpret image data as if it were views of real 3D objects [442].

In other words the heuristics we have evolved for visual perception are grounded in the assumption that we observe scenes embedded in 3D space. An understanding of these heuristics may provide the potential means by which we can craft binocular depth recovery algorithms that perform as robustly as those depth perception mechanisms of the HVS.

3.3 Geometry and Acquisition of a Single Image

In this section we provide an introduction to the geometry and image acquisition of a single camera. More specifically, we start with an explanation of the projective transformation with basic mathematics describing this process. Then, the so-called pin-hole model of a camera is presented. Finally, we discuss the extrinsic and intrinsic parameters of acquisition with a single camera.

3.3.1 Projective Transformation

Every image acquisition system, either the human or machine visual system, by its nature performs some kind of transformation of real 3D space into 2D local space. Finding the parameters of such a transformation is fundamental to describing the acquisition system.

For most cameras a model that describes the space transformation they perform is based either on the parallel or central perspective projections. The linear parallel projection is the simplest approach. However, it only roughly approximates what we observe in real cameras [185]. Therefore the parallel projection, although linear, can be justified only if the observed objects are very close to the camera.

A better approach to describing the behaviour of real optical systems can be obtained using the perspective projective transformation which can be described by a linear equation, in a higher dimensional space of so-called homogeneous coordinates [95, 119, 122, 180]. Additionally, when describing real optical elements a simple projective transformation has to be augmented with nonlinear terms to take into account physical parameters of these [113, 185].

3.3.2 Simple Camera System: the Pin-hole Model

The simplest form of real camera comprises a pinhole and an imaging screen (or plane). Because the pinhole lies between the imaging screen and the observed 3D world scene, any ray of light that is emitted or reflected from a surface patch in the scene is constrained to travel through the pinhole before reaching the imaging screen. Therefore there is correspondence between each 2D area on the imaging screen and the area in the 3D world, as observed "through the pinhole" from the imaging screen. It is the solid angle of rays that is subtended by the pinhole that relates the field of view of each region on the imaging screen to the corresponding region imaged in the world. By this mechanism an image is built up, or projected (derived from the Latin projicere from *pro* "forward" and *jacere* "to throw") from world space to imaging space. A mathematical model of the simple pin-hole camera is illustrated in Figure 3.6. Notice that the imaging screen is now in front of the pin-hole. This formulation simplifies the concept of projection to that of magnification. In order to understand how points in the real world are related mathematically to points on the imaging screen two coordinate systems are of particular interest:

1. The external coordinate system (denoted here with a subscript 'W' for 'world') which is independent of placement and parameters of the camera.
2. The camera coordinate system (denoted by 'C', for 'camera').

The two coordinate systems are related by a translation, expressed by matrix **T**, and rotation, represented by matrix **R**.

The point $\mathbf{O_c}$, called *a central* or *a focal point*, together with the axes X_c, Y_c and Z_c determine the coordinate system of the camera. An important part of the camera model is the image plane $\mathbf{\Pi}$. We can observe in Figure 3.6 that this plane $\mathbf{\Pi}$ has been tessellated into rectangular elements, i.e. tiled, and that within an electronic camera implementation these tiles will form discrete photosensing locations that sample any image projected onto the plane. Each tile is

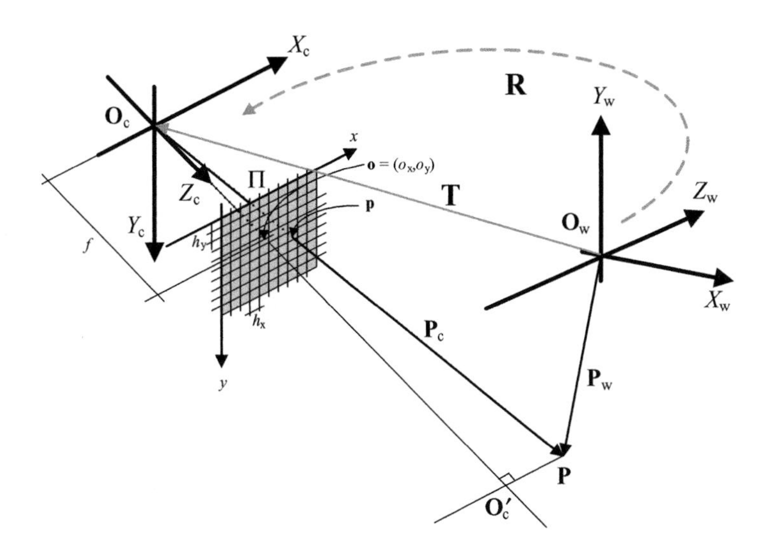

Figure 3.6 Pin-hole model of the perspective camera with two coordinate systems: external W and internal C

called a *pixel*, i.e. picture element, and is indexed by a pair of coordinates expressed by natural numbers. Figure 3.6 depicts the plane $\mathbf{\Pi}$ with a discrete grid of pixels. The projection of the point $\mathbf{O_c}$ on the plane $\mathbf{\Pi}$ in the direction of Z_c determines *the principal point* of local coordinates (o_x, o_y). The principal axis is a line between points $\mathbf{O_c}$ and $\mathbf{O'_c}$. The distance from the image plane to the principal point is known as *the focal length*. Lastly, the values h_x and h_y determine physical dimensions of a single pixel.

Placement of a given point $\mathbf{P}$ from the 3D space depends on the chosen coordinate system: in the camera coordinate system it is a column vector $\mathbf{P}_c$; in the external coordinate system it is a column vector $\mathbf{P}_w$.

Point $\mathbf{p}$ is an image of point $\mathbf{P}$ under the projection with a centre in point $\mathbf{O_c}$ on to the plane $\mathbf{\Pi}$. Coordinates of the points $\mathbf{p}$ and $\mathbf{P}$ in the camera coordinate system are denoted as follows:[2]

$$\begin{aligned} \mathbf{P} &= [X, Y, Z]^T \\ \mathbf{p} &= [x, y, z]^T . \end{aligned} \tag{3.1}$$

Since the optical axis is perpendicular to the image plane, then taking into account that the triangles $\Delta\mathbf{O}_c\mathbf{po}$ and $\Delta\mathbf{O}_c\mathbf{PO'}_c$ are similar and placing $z = f$, we obtain immediately

$$x = f\frac{X}{Z}, \quad y = f\frac{Y}{Z}, \quad z = f. \tag{3.2}$$

Equation (3.2) constitutes a foundation of the pin-hole camera model.

[2] Points are denoted by letters in bold, such as $\mathbf{p}$. Their coordinates are represented either by the same letter in italic and indexed starting from 1, such as $\mathbf{p} = (p_1, p_2, p_3, p_4)$, or as $\mathbf{p} = (x, y)$ and $\mathbf{p} = (x, y, z)$ for 2D or 3D points, respectively. When necessary, points are assumed to be column vectors, such as $\mathbf{p} = [x, y, z]^T$.

The pin-hole camera model can be defined by providing two sets of parameters.

1. The extrinsic parameters.
2. The intrinsic parameters.

In the next sections we discuss these two sets in more detail.

3.3.2.1 Extrinsic Parameters

The mathematical description of a given scene depends on the chosen coordinate system. With respect to the chosen coordinate system and based solely on placement of the image plane we determine an exact placement of the camera. Thereafter, it is often practical to select just the camera coordinate system as a reference. The situation becomes yet more complicated, however, if we have more than one camera since the exact (relative) position of each camera must be determined.

A change from the camera coordinate system 'C' to the external world coordinate system 'W' can be accomplished providing a translation $\mathbf{T}$ and a rotation $\mathbf{R}$ (Figure 3.6). The translation vector $\mathbf{T}$ describes a change in position of the coordinate centres $\mathbf{O_c}$ and $\mathbf{O_w}$. The rotation, in turn, changes the corresponding axes of each system. This change is described by the orthogonal[3] matrix $\mathbf{R}$ of dimensions 3×3 [132, 430].

For a given point $\mathbf{P}$, its coordinates related to the camera 'C' and external coordinates related to the external world 'W' are connected by the following formula:

$$\mathbf{P}_{\mathrm{c}} = \mathbf{R}(\mathbf{P}_{\mathrm{w}} - \mathbf{T}), \tag{3.3}$$

where $\mathbf{P}_{\mathrm{c}}$ expresses placement of a point $\mathbf{P}$ in the camera coordinate system, $\mathbf{P}_{\mathrm{w}}$ is its placement in the external coordinate system, $\mathbf{R}$ stands for the rotation matrix and $\mathbf{T}$ is the translation matrix between origins of those two coordinate systems. The matrices $\mathbf{R}$ and $\mathbf{T}$ can be specified as follows:

$$\mathbf{R} = \begin{bmatrix} \mathbf{R}_1 \\ \mathbf{R}_2 \\ \mathbf{R}_3 \end{bmatrix}_{3\times 3} = \begin{bmatrix} R_{11} & R_{12} & R_{13} \\ R_{21} & R_{22} & R_{23} \\ R_{31} & R_{32} & R_{33} \end{bmatrix}_{3\times 3}, \quad \mathbf{T} = \mathbf{O_w} - \mathbf{O_c} = \begin{bmatrix} T_1 \\ T_2 \\ T_3 \end{bmatrix}_{3\times 1}, \tag{3.4}$$

where $\mathbf{R}_i$ denotes an i-th row of the rotation matrix $\mathbf{R}$, i.e. $\mathbf{R} = [R_{i1}, R_{i2}, R_{i3}]_{1\times 3}$.

Summarizing, we say that *the extrinsic parameters of the perspective camera* are all the necessary geometric parameters that allow a change from the camera coordinate system to the external coordinate system and vice versa. Thus, the extrinsic parameters of a camera are just introduced matrices $\mathbf{R}$ and $\mathbf{T}$.

[3] That is, $\mathbf{RR}^{\mathrm{T}} = 1$.

3.3.2.2 Intrinsic Parameters

The intrinsic camera parameters can be summarized as follows.

1. The parameters of the projective transformation itself: For the pin-hole camera model, this is given by the focal length f.
2. The parameters that map the camera coordinate system into the image coordinate system: Assuming that the origin of the image constitutes a point $\boldsymbol{o} = (o_x, o_y)$ (i.e. a central point) and that the physical dimensions of pixels (usually expressed in μm) on a camera plane in the two directions are constant and given by h_x and h_y, a relation between image coordinates x_u and y_u and camera coordinates x and y can be stated as follows (see Figure 3.6):

$$\begin{aligned} x &= (x_u - o_x)h_x \\ y &= (y_u - o_y)h_y, \end{aligned} \tag{3.5}$$

   where a point (x, y) is related to the camera coordinate system 'C', whereas (x_u, y_u) and (o_x, o_y) to the system of a local camera plane. It is customary to assume that $x_u \geq 0$ and $y_u \geq 0$. For instance, the point of origin of the camera plane $(x_u, y_u) = (0, 0)$ transforms to the point $(-o_x h_x, -o_y h_y)$ of the system 'C'. More often than not it is assumed also that $h_x = h_y = 1$. A value of h_y/h_x is called an aspect ratio. Under this assumption a point from our example is simply $(-o_x, -o_y)$ in the 'C' coordinates, which can be easily verified analysing Figure 3.6.
3. Geometric distortions that arise due to the physical parameters of the optical elements of the camera: Distortions encountered in real optical systems arise mostly from the nonlinearity of these elements, as well as from the dependence of the optical parameters on the wavelength of the incident light [185, 343, 382]. In the first case we talk about spherical aberration, coma, astigmatism, curvature of the view field and distortions. The second case is related to the chromatic aberration [50, 185, 382]. In the majority of practical situations, we can model these phenomena as radial distortions, the values of which increase for points more distant from the image centre. The radial distortions can be modelled by providing a nonlinear correction (offset) to the real coordinates of a given image point. This can be accomplished by adding even-order polynomial terms, as follows:

$$x_v = \frac{x_u}{1 + k_1 r^2 + k_2 r^4}, \quad y_v = \frac{y_u}{1 + k_1 r^2 + k_2 r^4}, \tag{3.6}$$

   where $r^2 = x_v^2 + y_v^2$, k_1 and k_2 are the new intrinsic parameters of the perspective camera that model the influence of the radial distortions of the optical system; x_u and y_u are the ideal (i.e. as if there were no distortions) coordinates of a given image point; and x_v and y_v are modified coordinates reflecting the radial distortions.

An iterative algorithm for finding x_v and y_v is provided by Klette *et al.* [246]. Trucco and Verri suggest that for most real optical systems with a CCD sensor of around 500×500 image elements, setting k_2 to 0 does not introduce any significant change to the quality of the camera model [430].

3.3.3 Projective Transformation of the Pin-hole Camera

Substituting (3.3) and (3.5) into (3.2) and disregarding distortions (3.6) we obtain the linear equation of the pin-hole camera:[4]

$$\mathbf{p} = \mathbf{MP}, \tag{3.7}$$

where $\mathbf{p}$ is an image of the point $\mathbf{P}$ under transformation $\mathbf{M}$ performed by the pin-hole camera. Linearity in (3.7) is due to the homogeneous[5] transformation of the point coordinates.

The matrix $\mathbf{M}$ in (3.7), called a projection matrix, can be partitioned into the following product of two matrices:

$$\mathbf{M} = \mathbf{M}_\mathrm{i}\mathbf{M}_\mathrm{e}, \tag{3.8}$$

where

$$\mathbf{M_i} = \begin{bmatrix} \frac{f}{h_x} & 0 & o_x \\ 0 & \frac{f}{h_y} & o_y \\ 0 & 0 & 1 \end{bmatrix}_{3\times 3}, \quad \mathbf{M_e} = \begin{bmatrix} \mathbf{R}_1 & -\mathbf{R}_1\mathbf{T} \\ \mathbf{R}_2 & -\mathbf{R}_2\mathbf{T} \\ \mathbf{R}_3 & -\mathbf{R}_3\mathbf{T} \end{bmatrix}_{3\times 4}. \tag{3.9}$$

The matrices $\mathbf{R}$ and $\mathbf{T}$ are given in (3.4). $\mathbf{M}_\mathrm{i}$ defines the intrinsic parameters of the pin-hole camera, that is, the distance of the camera plane to the centre of the camera's coordinate system, as well as placement of the central point o and physical dimensions of the pixels on the camera plane – these are discussed in section 3.3.2.2. The matrix $\mathbf{M}_\mathrm{e}$ contains the extrinsic parameters of the pin-hole camera and relates the camera and the external 'world' coordinate systems (section 3.3.2.1).

The three equations above can be joined together as follows:

$$\mathbf{p} = \begin{bmatrix} x_{uh} \\ y_{uh} \\ z_{uh} \end{bmatrix} = \underbrace{\begin{bmatrix} \frac{f}{h_x} & 0 & o_x \\ 0 & \frac{f}{h_y} & o_y \\ 0 & 0 & 1 \end{bmatrix}}_{\mathbf{M}_\mathrm{i}} \underbrace{\begin{bmatrix} \mathbf{R}_1 & -\mathbf{R}_1\mathbf{T} \\ \mathbf{R}_2 & -\mathbf{R}_2\mathbf{T} \\ \mathbf{R}_3 & -\mathbf{R}_3\mathbf{T} \end{bmatrix}}_{\mathbf{M}_\mathrm{e}} \mathbf{P}, \tag{3.10}$$

where $\mathbf{P} = [\mathbf{P}_\mathrm{w}\ 1]^\mathrm{T}$ is a point $\mathbf{P}_\mathrm{w}$ expressed in the homogeneous coordinates.

Let us observe that

$$x_u = \frac{x_{uh}}{z_{uh}}, \quad y_u = \frac{y_{uh}}{z_{uh}}. \tag{3.11}$$

[4] Derivation of the equations for the projective transformation of a camera can be found in section 3.8.

[5] Before further study, readers not familiar with the concept of homogeneous coordinates are asked to read section 10.1.

As already alluded to, it is often assumed that $(o_x, o_y) = (0,0)$, and also $h_x = h_y = 1$. With these assumptions (3.10) takes on a simpler form

$$\mathbf{p} = \begin{bmatrix} f & 0 & 0 \\ 0 & f & 0 \\ 0 & 0 & 1 \end{bmatrix} \begin{bmatrix} \mathbf{R}_1 & -\mathbf{R}_1\mathbf{T} \\ \mathbf{R}_2 & -\mathbf{R}_2\mathbf{T} \\ \mathbf{R}_3 & -\mathbf{R}_3\mathbf{T} \end{bmatrix} \mathbf{P}. \tag{3.12}$$

Equation (3.7) defines a transformation of the projective space $\wp^3$ into the projective plane $\wp^2$. However, note that this transformation changes each point of a line into exactly one and the same image point of the image plane. This line is given by the central point O_c and any other point from the projective space. Therefore the projective transformation (3.7) assigns exactly the same image point to *all* the points belonging to the mentioned line. This fact can be embedded into (3.7) by introduction of an additional scaling parameter, as follows:

$$\gamma \mathbf{p} = \mathbf{MP}, \tag{3.13}$$

where γ is a scalar. Equations (3.7) and (3.13) differ only by the scalar γ. It can also be said that (3.7) is a version of (3.13) after dividing both sides by a nonzero scalar γ. Thus, without loss of generality we will assume henceforth that (3.7) holds, where the matrix $\mathbf{M}$ is defined only up to a certain multiplicative parameter γ.

3.3.4 Special Camera Setups

For some camera setups it is possible to assume that distances among observed objects are significantly smaller than the average distance $\bar{z}$ from those objects to the centre of projection. Under this assumption we obtain a simplified camera model; termed *weak perspective* [314, 322, 430]. In this model the perspective projection simplifies to the parallel projection by the scaled magnification factor $f/\bar{z}$. Equations (3.2) transform then to the following set of equations:

$$x = \frac{f}{\overline{Z}}X, \quad y = \frac{f}{\overline{Z}}Y, \quad z = f, \tag{3.14}$$

where $\overline{Z}$ is assumed to be much larger than f and constant for the particular setup of a camera and a scene. This simplification makes (3.14) independent of the current depth of an observed point $\mathbf{P}_w$. Thus, in the case of a camera with a simplified perspective the element at indices 3×1 of the matrix $\mathbf{M}_e$ in (3.9) changes to $\mathbf{0}$, and the element 3×2 of this matrix changes to $\overline{Z}$ (section 3.8). The latter, in turn, can be defined selecting an arbitrary point $\mathbf{A}_w$, which is the same for acquisition of the whole scene

$$\overline{Z} = \mathbf{R}_3(\mathbf{A}_w - \mathbf{T}). \tag{3.15}$$

The mathematical extension to this simplification is a model of an *affine camera* in which proportions of distances measured alongside parallel directions are invariant [122, 314, 322, 430]. There are also other camera models that take into consideration parameters of real lenses, e.g. see Kolb *et al.* [251]. Finally, more information on design of real lenses can be found in [113, 382].

3.3.5 Parameters of Real Camera Systems

The quality of the images obtained by real acquisition systems depends also on many other factors beyond those already discussed. These are related to the physical and technological phenomena which influence the acquisition process. In this section we briefly discuss such factors.

1. *Limited dynamics of the system.* The basic photo-transducer element within a modern digital camera converts the number of photons collected over a specific time interval (the integration interval of the device, analogous to the exposure time in a film camera) within each pixel within the sensor array into a voltage. While this voltage is linearly proportional to the intensity of the input photon flux arriving at a given pixel, the following analog-to-digital converter circuitry is limited to a finite number of bits of precision with which to represent the incoming voltage. Therefore, in order to extend the allowable input signal range nonlinear limiting circuits are introduced prior to digitisation. One such limiter is the pre-knee circuit [246] whose circuit characteristic causes a small degree of saturation for higher values of the input signal. As a result, the input range of the system is increased but at a cost of a slight nonlinearity.
2. *Resolution of the CCD element and aliasing.* In agreement with the sampling theory, to avoid aliasing, a device converting continuous signals into a discrete representation must fulfil the Nyquist sampling criterion (i.e. the sampling frequency has to be at least twice the value of the highest frequency component of the sampled signal). In the rest of this book we assume that this is the case and that aliasing is not present [312, 336]. In real imaging systems there are two factors that can help to alleviate the problem of aliasing. The first consists of the application of low-pass filters at the input circuitry. The second is the natural low-pass filter effect due to the lens itself, manifest as a point spread function (PSF) or modulation transfer function (MTF) which naturally limits the high spatial frequencies present prior to digitisation [66, 430].
3. *Noise.* Each image acquisition channel contains many sources of noise. In the CCD device there is a source of noise in the form of cross-talk. This is the phenomenon of charge leakage between neighbouring photoreceptors in each row of the CCD. Another source of noise comes from the filters and the analogue-to-digital converter. The latter adds so-called quantization noise which is a result of the finite length of bit streams representing analogue signals. The most frequently encountered types of image noise can be represented by Poison and Gaussian distributions. Schott Noise is by far the most significant source of noise in a modern imaging sensor. This noise source results from the statistical variation of the photon arrival rate from any illumination source. In any fixed time interval the standard deviation of the photon flux rate is proportional to the square of the illumination intensity. Other sources of noise are now becoming less significant than Schott noise, hence this fundamental limit of physics now tends to dominate image capture performance. The interested reader is referred to the ample literature [95, 158, 172, 183, 224, 226, 247, 346, 430]. Different types of noise are also discussed in Chapter 11.
4. *Signal saturation.* The phenomenon of signal saturation results from an excessive signal level being applied to the input of the acquisition channel. Such a signal is nonlinearly attenuated and cannot be accurately converted by A/C converters due to their limited dynamic range. Where there is insufficient scene illumination, as can be caused by shadows, the

image signal "bottoms-out" providing no visual information by which to compute stereo matches.

5. *Blooming effect.* Blooming is caused by an excessive charge appearing in certain areas of a CCD device. In effect, some parts of this charge spread out to the neighbouring CCD cells causing visible distortions [54]. This phenomenon is usually caused by light reflections entering the lens of the camera.
6. *Scene lighting conditions.* There are many different sources of light that can be used to illuminate an observed scene and different photometric models apply [95, 173, 224, 226, 343]. In the rest of this book we assume the Lambertian photometric model, i.e. each point of the illuminated surface is perceived from each direction as being equally light. This model is a reasonable approximation for many real situations, especially if we consider only the scenes comprising matt and opaque objects. In the case of the stereoscopic systems non-uniformity in scene illumination can potentially lead to an increase in false matches (6.4). This happens if the stereo method does not attempt to compensate for local inequalities in the average illumination of the stereo-pair images.

The above mentioned parasitic phenomena arising in real image acquisition systems are even more severe when capturing colour images, since they can be present in each colour component independently.

The last question concerns accuracy of the pin-hole model when applied to real camera systems. Many experimental results with simple camera systems help to answer this question [314, 408]. For example, for a camera system with a sensor resolution of 512×512 pixels, the difference in accuracy obtained between the real camera and the pin-hole model is about 1/20 of a pixel. Such results justify the application of the pin-hole model in many image processing methods, including those presented in this book.

More information on different technologies of CCD devices, their manufacturing processes and application in real machine vision tasks can be found in many publications, some available also from the Internet. For instance the 'CCD Primer' by Eastman Kodak Company gives a nice introduction to CCD technology [111]. A discussion of CCD versus CMOS devices for image acquisition can be found in Janesick [228] or in one of the technical reports by Dalsa Corporation [94]. Information on special imagers using amorphous silicon can be found in Böhm [54]. Finally, Baldock and Graham [23] discuss CCDs and image acquisition systems for microscopic imaging systems.

3.4 Stereoscopic Acquisition Systems

In this section we discuss the basic properties of stereoscopic acquisition systems. When two (static) cameras observe the same scene from different viewpoints, a qualitatively new kind of observation can be made that is not possible using a single (static) camera alone – this is the perception of depth by triangulation.

3.4.1 Epipolar Geometry

Figure 3.7 depicts an imaging configuration comprising two projective systems. They create a stereoscopic image acquisition system. It is based on two pin-hole cameras, each composed of

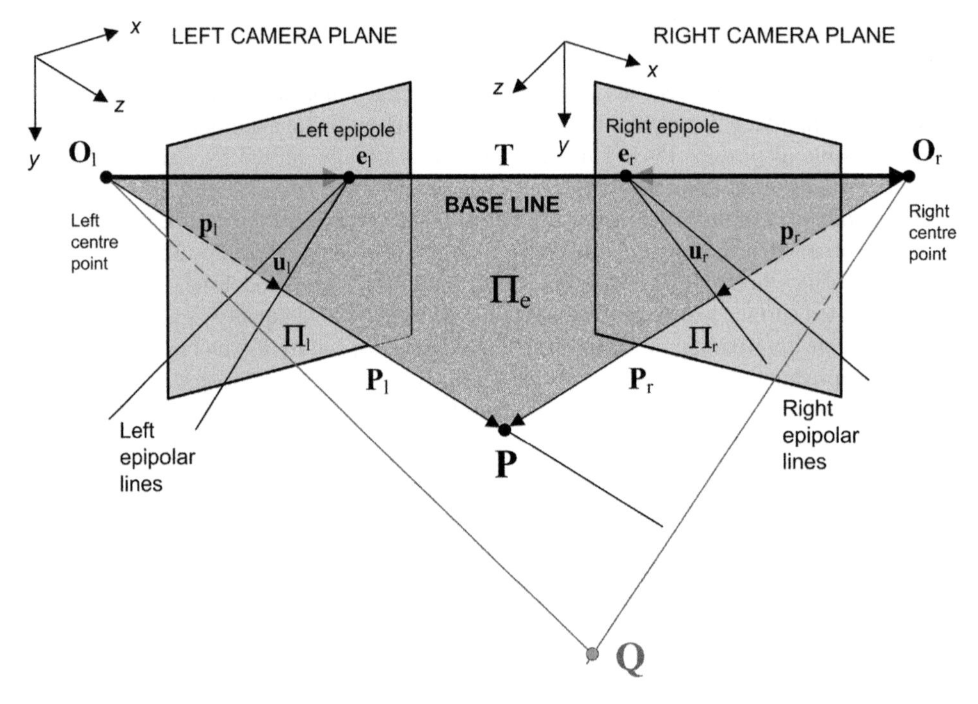

Figure 3.7 Epipolar geometry

the projective plane $\mathbf{\Pi}_i$ (where subscript "i" is changed to l for the left and to "r" for the right camera respectively) with respective projective centre point $\mathbf{O}_i$. The line coming through the point $\mathbf{O}_i$ and perpendicular to the plane $\mathbf{\Pi}_i$ crosses this plane in a point called *the principal point*. The distance from this point to the centre point $\mathbf{O}_i$ is called *the focal length f*.

The line $\overline{\mathbf{O}_l\mathbf{O}_r}$ connecting the centres $\mathbf{O}_l$ and $\mathbf{O}_r$ is called the *base line*. Points of its crossing with the image planes $\mathbf{\Pi}_i$ determine the *epipolar points*. In the special case, when the line $\overline{\mathbf{O}_l\mathbf{O}_r}$ does not cross the image planes $\mathbf{\Pi}_i$, the corresponding epipolar points lie in infinity (section 9.4).

A plane determined by a given 3D point **P** and the projective centres $\mathbf{O}_l$ and $\mathbf{O}_r$ is called the *epipolar plane* $\mathbf{\Pi}_e$. The epipolar plane $\mathbf{\Pi}_e$ intersects the image planes $\mathbf{\Pi}_l$ and $\mathbf{\Pi}_r$ – their intersections are the *epipolar lines*. The role of the epipolar lines can be understood, for example, by analysing the left image $\mathbf{p}_l$ (a point is represented as a vector) of the 3D point **P**. In this case, the central point $\mathbf{O}_l$ and the left point $\mathbf{p}_l$ define a certain ray $\overline{\mathbf{O}_l\mathbf{p}_l}$. It can be seen that the point $\mathbf{p}_l$ is an image of the point **P** but also of all the other points on the ray $\overline{\mathbf{O}_l\mathbf{p}_l}$. This means that the point **P** can lie anywhere on this ray, still having the same image. Therefore determination of its exact space position is not possible having only one image. To clarify space position we need a second image point, viewed from another position. This is, for example, an image point $\mathbf{p}_r$ on the plane $\mathbf{\Pi}_r$. The point p_r and the second central point $\mathbf{O}_r$ determine the second ray $\overline{\mathbf{O}_r\mathbf{p}_r}$. This ray is fixed at $\mathbf{O}_r$ and simultaneously it can slide through the ray $\overline{\mathbf{O}_l\mathbf{p}_l}$, crossing it in the space point **P**. Moreover, the crossing point of each ray $\overline{\mathbf{O}_l\mathbf{p}_l}$ or $\overline{\mathbf{O}_r\mathbf{p}_r}$ with their respective image planes $\mathbf{\Pi}_l$ or $\mathbf{\Pi}_r$ lies on the epipolar lines. Similarly, projections of these rays on the opposite image planes constitute epipolar lines as well. Hence the very important conclusion, called *an epipolar constraint*:

> Each image point $\mathbf{p}_i$ of a space point $\mathbf{P}$ lies in the image plane *only* on the corresponding epipolar line.

The consequence of this constraint plays an important role when searching for the image points (not knowing their space points) limiting this process from the whole image plane to the search along the epipolar line. In the general case positions of the epipolar lines are not known beforehand. However, for special camera configurations, such as a canonical one, positions of the epipolar lines are known and this fact can greatly facilitate the search for corresponding points.

With each of the cameras of the stereo system we associate a separate coordinate system with its centre coinciding with the central point of the camera. The Z axis of such a coordinate system is collinear with the optical axis of the camera. In both coordinate systems, the vectors $\mathbf{P}_l = [X_l, Y_l, Z_l]^T$ and $\mathbf{P}_r = [X_r, Y_r, Z_r]^T$ represent the same 3D point $\mathbf{P}$. On the other hand, on the respective image planes, the vectors $\mathbf{p}_l = [x_l, y_l, z_l]^T$ and $\mathbf{p}_r = [x_r, y_r, z_r]^T$ determine two different images of the space point $\mathbf{P}$. Additionally we notice that $z_l = f_l$ and $z_r = f_r$, where f_l and f_r are the focal lengths of the left and right cameras, respectively.

As was already pointed out in section 3.3, each camera is described by a set of extrinsic parameters. They determine placement of a camera in respect to the external coordinate system. On the other hand, in the stereo camera setup each camera has its associated local coordinate system. Similarly to (3.3), it is possible to change from one coordinate system to the other by a translation $\mathbf{T} = \mathbf{O}_r - \mathbf{O}_l$ and rotation determined by an orthogonal matrix $\mathbf{R}$. Thus, for the two vectors $\mathbf{P}_l$ and $\mathbf{P}_r$ pointing at the same point $\mathbf{P}$ from 3D space the following holds [430]:

$$\mathbf{P}_r = \mathbf{R}(\mathbf{P}_l - \mathbf{T}). \tag{3.16}$$

The epipolar plane $\mathbf{\Pi}_e$ in the coordinate system associated with the left camera is spanned by the two vectors $\mathbf{T}$ and $\mathbf{P}_l$. Therefore, also the vector $\mathbf{P}_l - \mathbf{T}$ belongs to this plane. This means that their mixed product must vanish, that is

$$(\mathbf{P}_l - \mathbf{T}) \cdot (\mathbf{T} \times \mathbf{P}_l) = 0. \tag{3.17}$$

The product (3.17) can be written in matrix form as a product of a certain matrix $\mathbf{A}$ and the vector $\mathbf{P}_l$, which is presented by the following equation:

$$\begin{aligned}
\mathbf{T} \times \mathbf{P}_l &= \begin{vmatrix} T_1 & T_2 & T_3 \\ P_{l1} & P_{l2} & P_{l3} \\ \mathbf{i} & \mathbf{j} & \mathbf{k} \end{vmatrix} \\
&= \mathbf{i}(T_2 P_{l3} - T_3 P_{l2}) - \mathbf{j}(T_1 P_{l3} - T_3 P_{l1}) + \mathbf{k}(T_1 P_{l2} - T_2 P_{l1}) \qquad (3.18) \\
&= \begin{bmatrix} -T_3 P_{l2} + T_2 P_{l3} \\ T_3 P_{l1} - T_1 P_{l3} \\ -T_2 P_{l1} + T_1 P_{l2} \end{bmatrix} = \begin{bmatrix} 0 & -T_3 & T_2 \\ T_3 & 0 & -T_1 \\ -T_2 & T_1 & 0 \end{bmatrix} \begin{bmatrix} P_{l1} \\ P_{l2} \\ P_{l3} \end{bmatrix} = \mathbf{A}\mathbf{P}_l
\end{aligned}$$

where $\mathbf{i} = [1, 0, 0]^T$, $\mathbf{j} = [0, 1, 0]^T$, $\mathbf{k} = [0, 0, 1]^T$ are the unit vectors and $\mathbf{A}$ is a skew symmetric matrix (section 9.3). Now, substituting (3.16) and (3.18) into (3.17), we obtain

$$(\mathbf{R}^{-1}\mathbf{P}_r)^T \mathbf{A}\mathbf{P}_l = 0. \tag{3.19}$$

Taking into account that $\mathbf{R}$ is orthogonal and after simple rearrangements we have

$$\mathbf{P}_r^T \mathbf{R}\mathbf{A}\mathbf{P}_l = 0, \tag{3.20}$$

$$\mathbf{P}_r^T \mathbf{E}\mathbf{P}_l = 0, \tag{3.21}$$

where the matrix

$$\mathbf{E} = \mathbf{R}\mathbf{A} \tag{3.22}$$

is called the *essential matrix* which due to the rank of the matrix $\mathbf{A}$ in (3.18) is also of rank two.

3.4.1.1 Fundamental Matrix

The points $\mathbf{p}_l$ and $\mathbf{P}_l$, as well as $\mathbf{p}_r$ and $\mathbf{P}_r$, are connected by relation (3.2). Thus (3.21) can be written as

$$\mathbf{p}_r^T \mathbf{E}\mathbf{p}_l = 0, \tag{3.23}$$

where $\mathbf{p}_r$ and $\mathbf{p}_l$ are image points on the image planes. Since the corresponding points can lie only on the corresponding epipolar lines, $\mathbf{E}\mathbf{p}_l$ in (3.23) is an equation of the epipolar line on the right image plane that goes through the point $\mathbf{p}_r$, and, as all the epipolar lines, through the epipole. Therefore both epipolar lines can be expressed as

$$\mathbf{u}_r = \mathbf{E}\mathbf{p}_l, \tag{3.24}$$

$$\mathbf{u}_l = \mathbf{E}^T\mathbf{p}_r. \tag{3.25}$$

For a given point $\mathbf{p}_k$ we find its pixel coordinates from (3.117) as follows:

$$\mathbf{M}_{ik}\mathbf{p}_k = \overline{\mathbf{p}_k}, \tag{3.26}$$

where $\mathbf{M}_{ik}$ is an intrinsic matrix for the k-th image, $\mathbf{p}_k$ a point in the camera coordinate system and $\overline{\mathbf{p}_k}$ homogeneous pixel coordinates.

Equation (3.23) can be written as

$$\left(\mathbf{M}_{ir}^{-1}\overline{\mathbf{p}_r}\right)^T \mathbf{E}\mathbf{M}_{il}^{-1}\overline{\mathbf{p}_l} = 0, \tag{3.27}$$

$$\overline{\mathbf{p}_r^T}\mathbf{M}_{ir}^{-T}\mathbf{E}\mathbf{M}_{il}^{-1}\overline{\mathbf{p}_l} = 0. \tag{3.28}$$

Finally we obtain

$$\overline{\mathbf{p}_r^T}\mathbf{F}\overline{\mathbf{p}_l} = 0, \tag{3.29}$$

where the matrix

$$\mathbf{F} = \mathbf{M}_{i\mathrm{r}}^{-\mathrm{T}} \mathbf{E} \mathbf{M}_{i\mathrm{l}}^{-1} \tag{3.30}$$

is called the *fundamental matrix*. It describes the epipolar geometry in terms of the pixel coordinates in contrast to the essential matrix in (3.21) and (3.23) where the homogeneous camera coordinates are used.

The two matrices **E** and **F** are related by (3.30). Substituting (3.22) into (3.30) we obtain also that

$$\mathbf{F} = \mathbf{M}_{i\mathrm{r}}^{-\mathrm{T}} \mathbf{R} \mathbf{A} \mathbf{M}_{i\mathrm{l}}^{-1}. \tag{3.31}$$

Taking into account that the rank of the matrix **E** is two, the rank of the matrix **F** is also two. Further analysis of the matrices **E** and **F** can be found in the ample literature on this subject, e.g. in Luong and Faugeras [288] and Hartley and Zisserman [180]. The method of representation of a scene by means of the images and their fundamental matrices is discussed by Laveau and Faugeras [267].

3.4.1.2 Epipolar Lines and Epipoles

Because the point $\overline{\mathbf{p}_\mathrm{r}}$ lies on the epipolar line $\overline{\mathbf{u}_\mathrm{r}}$ in the right image plane, then the following equation must hold (section 9.3):

$$\overline{\mathbf{p}_\mathrm{r}^\mathrm{T}} \overline{\mathbf{u}_\mathrm{r}} = 0. \tag{3.32}$$

Now based on (3.29), similarly to (3.24), it is possible to write the equation of the right epipolar line as

$$\overline{\mathbf{u}_\mathrm{r}} = \mathbf{F} \overline{\mathbf{p}_\mathrm{l}}. \tag{3.33}$$

Analogously, the equation of the left epipolar line can be expressed as

$$\overline{\mathbf{u}_\mathrm{l}} = \mathbf{F}^\mathrm{T} \overline{\mathbf{p}_\mathrm{r}}. \tag{3.34}$$

Let us take a look at Figure 3.7 and notice that all epipolar planes have one common line – the base line; similarly, all epipolar lines from a given image plane have one common point – the epipole. Since (3.29) holds for all points from the image plane then we can consider the case when the left point $\overline{\mathbf{p}_\mathrm{l}}$ in (3.29) is at the same time the left epipole $\overline{\mathbf{e}_\mathrm{l}}$. Then (3.29) takes the form

$$\overline{\mathbf{p}_\mathrm{r}^\mathrm{T}} \mathbf{F} \overline{\mathbf{e}_\mathrm{l}} = 0. \tag{3.35}$$

However, the above is obviously fulfilled for all points from the right image plane that lie on the base line. Therefore, and taking into account that the matrix **F** is of rank two, we conclude that the following must hold:

$$\mathbf{F}\overline{\mathbf{e}_l} = 0. \tag{3.36}$$

Based on (3.36) this means that $\overline{\mathbf{e}_l}$ must be the kernel transformation defined by the matrix $\mathbf{F}$. Similarly, $\overline{\mathbf{e}_r}$ is the kernel of $\mathbf{F}^T$:

$$\mathbf{F}^T\overline{\mathbf{e}_r} = \overline{\mathbf{e}_r^T}\mathbf{F} = 0. \tag{3.37}$$

Thus, the left and right epipoles can be computed by finding kernels of the transformations described by $\mathbf{F}$ and $\mathbf{F}^T$, respectively. This, in turn, can be accomplished by the singular value decomposition (SVD) of the respective matrix [154, 317, 352, 355, 425]. Changing the matrix $\mathbf{F}$ into the form

$$\mathbf{F} = \mathrm{SVD}^T, \tag{3.38}$$

we notice that $\overline{\mathbf{e}_l}$ is a column of the matrix $\mathbf{D}$ that corresponds to the zero-valued element of the diagonal matrix $\mathbf{V}$. By the same token, $\overline{\mathbf{e}_r}$ is a column of the matrix $\mathbf{S}$ that corresponds to the zero-valued element of the diagonal matrix $\mathbf{V}$. This can be seen quite easily when substituting (3.38) into (3.36) and (3.37), respectively, to obtain

$$\begin{aligned} \mathrm{SVD}^T\overline{\mathbf{e}_l} &= 0, \\ \overline{\mathbf{e}_r^T}\mathrm{SVD}^T &= 0. \end{aligned}$$

Let us recall that all columns of $\mathbf{D}$ (i.e. rows of $\mathbf{D}^T$) are orthogonal, so if $\overline{\mathbf{e}_l}$ is set to one of them then all multiplications of this column with other columns of $\mathbf{D}$ with different indices will also be 0. However, the multiplication of $\overline{\mathbf{e}_l}$ with itself corresponds to the lowest eigenvalue of $\mathbf{V}$ (possibly close to 0). The same analysis can be applied to the computation of the second epipole $\overline{\mathbf{e}_r}$. Nevertheless, in practice computation of the epipoles is sometimes burdened with numerical instabilities.

3.4.2 Canonical Stereoscopic System

We introduce a notion of disparity in respect of the canonical stereo setup (Figure 3.8).

Considering the similar triangles $\Delta p_L o_L O_L$ and ΔPXO_L, as well as $\Delta p_R o_R O_R$ and ΔPXO_R, we obtain the formula for the horizontal disparity $D_x(\mathbf{p}_l, \mathbf{p}_r)$ between two points $\mathbf{p}_l$ and $\mathbf{p}_r$ as

$$D_x(\mathbf{p}_l, \mathbf{p}_r) = p_{r1} - p_{l1} = x_l - x_r = \frac{bf}{Z}, \tag{3.39}$$

where the points $\mathbf{p}_l = [p_{l1}, p_{l2}]^T$, $\mathbf{p}_r = [p_{r1}, p_{r2}]^T$ are images of a certain 3D point $\mathbf{P}$, b is the base distance between the cameras, f is the camera focus length and Z is the distance from point $\mathbf{P}$ to the base line (depth of a scene). Let us observe that because bf/Z is positive, then from (3.39) it follows that $x_l \geq x_r$. This limits search range on the epipolar lines and simplifies the matching algorithms.

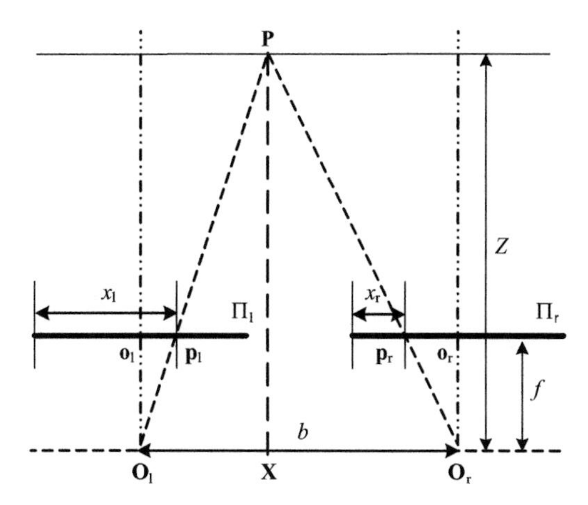

Figure 3.8 Standard (canonical) system of two cameras with focal lengths f, displaced by a base distance b. The difference between coordinates x_l and x_r is called a (horizontal) disparity between points $\mathbf{p}_l$ and $\mathbf{p}_r$

In the same way we can define the vertical disparity D_y in a direction perpendicular to D_x:

$$D_y(\mathbf{p}_l, \mathbf{p}_r) = p_{r2} - p_{l2}. \tag{3.40}$$

Certainly D_y in the canonical stereo setup is zero. Finally, in the canonical stereo setup the fundamental matrix takes on the following form [122]:

$$\mathbf{F}_C = \begin{bmatrix} 0 & 0 & 0 \\ 0 & 0 & c \\ 0 & -c & 0 \end{bmatrix}, \tag{3.41}$$

where c is a constant value different from 0. Substituting $\mathbf{F}_C$ into (3.29) we obtain

$$\begin{bmatrix} p_{r1} & p_{r2} & 1 \end{bmatrix} \begin{bmatrix} 0 & 0 & 0 \\ 0 & 0 & c \\ 0 & -c & 0 \end{bmatrix} \begin{bmatrix} p_{l1} \\ p_{l2} \\ 1 \end{bmatrix} = 0,$$

which is equivalent to

$$p_{l2} = p_{r2}. \tag{3.42}$$

The equation above states simply that the second coordinates of the matched points are the same whereas their first coordinates can change, as in (3.39), which gives us information on parallax. However, we have to remember that the coordinates are expressed in the local coordinate systems of the two (or more) cameras observing a scene. We need also to take into account their orientation. In Figure 3.7 these were chosen to comply with common practice of computer graphics, i.e. the coordinate systems' axes start from the top left corner of an image on the screen. Algorithmic aspects of representation of images are discussed further in section 3.7.1.2.

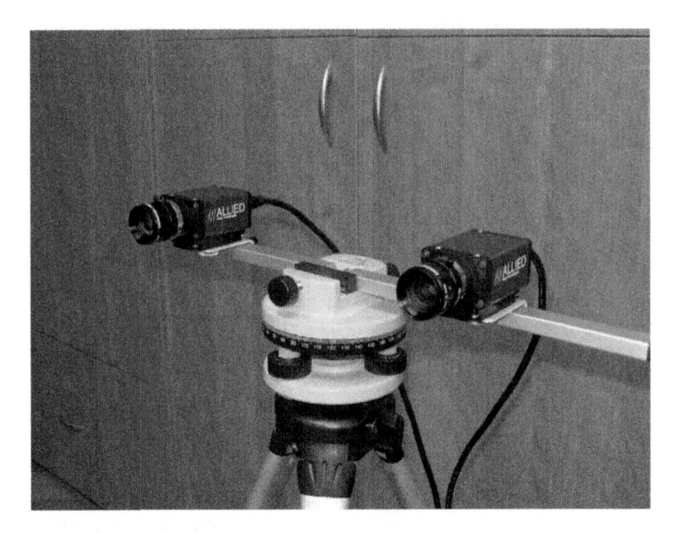

Figure 3.9 Stereo setup with cameras in the canonical position

Similarly, starting from (3.33) we obtain a formula of an epipolar line in the right image of the canonical stereo setup for a certain point in the left image:

$$\overline{\mathbf{u}_{\mathrm{r}}} = \mathbf{F}_{\mathrm{C}}\overline{\mathbf{p}_{\mathrm{l}}} = \begin{bmatrix} 0 & 0 & 0 \\ 0 & 0 & c \\ 0 & -c & 0 \end{bmatrix} \begin{bmatrix} p_{l1} \\ p_{l2} \\ 1 \end{bmatrix} = \begin{bmatrix} 0 \\ c \\ -cp_{l2} \end{bmatrix}. \tag{3.43}$$

We see that in this case $\overline{\mathbf{u}_{\mathrm{r}}}$ denotes simply a horizontal line.

Figure 3.9 depicts a canonical stereo setup, i.e. one in which the camera axes are parallel. The two Marlin® F033C cameras are mounted on to a geodesic tripod which allows precise control of the positions due to the three levels installed on it. The cameras are connected through IEEE 1394 connections to a PC with installed SDK for image acquisition (from Allied Vision®). In practice it is not so easy to set up all the cameras to their canonical positions, however. The best results are obtained when the calibration is done after presenting the grid calibration pattern to the two cameras and then setting the tripod positions manually to align corresponding grids. Nevertheless, each movement of the system results in the necessity of a new calibration; hence the importance of stereo methods that do not require precise canonical setups. This is especially true for vision systems operating in difficult conditions, such as the ones assembled on moving vehicles, for instance.

3.4.3 Disparity in the General Case

Having the horizontal and vertical disparities it is possible to define the common disparity $D(\mathbf{p}_{\mathrm{l}}, \mathbf{p}_{\mathrm{r}})$ as [246]

$$D(\mathbf{p}_{\mathrm{l}}, \mathbf{p}_{\mathrm{r}}) = \sqrt{D_x^2(\mathbf{p}_{\mathrm{l}}, \mathbf{p}_{\mathrm{r}}) + D_y^2(\mathbf{p}_{\mathrm{l}}, \mathbf{p}_{\mathrm{r}})}. \tag{3.44}$$

For the definition of disparity in the general case, see [122].

The term 'disparity' used henceforth denotes exclusively the horizontal disparity D_x as defined by (3.39) unless stated otherwise.

3.4.4 Bifocal, Trifocal and Multifocal Tensors

Dependences among multiple views can be analysed with the help of tensor calculus (see Chapter 10). The perspective transformation performed by a single pin-hole camera is given by (3.13). Let us assume that simultaneously we have for instance four such cameras observing *the same* 3D point **P**. It can be easily shown that in this case we can transform (3.13) to the following compact representation [189]:

$$\underbrace{\begin{bmatrix} \mathbf{M}_1 & \mathbf{p}_1 & 0 & 0 & 0 \\ \mathbf{M}_2 & 0 & \mathbf{p}_2 & 0 & 0 \\ \mathbf{M}_3 & 0 & 0 & \mathbf{p}_3 & 0 \\ \mathbf{M}_4 & 0 & 0 & 0 & \mathbf{p}_4 \end{bmatrix}}_{\mathbf{H}_4} \begin{bmatrix} \mathbf{P} \\ -s_1 \\ -s_2 \\ -s_3 \\ -s_4 \end{bmatrix} = 0, \tag{3.45}$$

where $\mathbf{M}_i$ is the 3×4 matrix of the projective transformation (3.13) performed by the i-th camera, **P** is a 3D point, $\mathbf{p}_i$ is its image created on the i-th camera plane and s_i is a scaling factor. In the case of four cameras $\mathbf{H}_4$ is a 12×8 matrix created from the matrices $\mathbf{M}_i$ and $\mathbf{p}_i$. Its rank has to be up to seven to have nontrivial null space. In other words, (3.45) denotes a set of homogeneous equations of eight unknowns, and to have a solution different from $[0, 0, \ldots, 0]^T$, $\det(\mathbf{H}_4)$ has to be 0 [259]. In the general case of m cameras, $\mathbf{H}_m$ is of rank at most $m + 3$. Thus, all minors of size equal to or greater than $(m + 4) \times (m + 4)$ are zero. This feature will be used soon to derive further conditions.

Let us now consider a special case of two images. From (3.45) we build a matrix $\mathbf{H}_2$, taking only two images indexed by 1 and 2. To shorten the notation, from $\mathbf{M}_1$ and $\mathbf{M}_2$ their first three rows are taken separately, with upper index denoting the number of a row:

$$\mathbf{H}_2 = \begin{bmatrix} \mathbf{M}_1^1 & p_1^1 & 0 \\ \mathbf{M}_1^2 & p_1^2 & 0 \\ \mathbf{M}_1^3 & 1 & 0 \\ \mathbf{M}_2^1 & 0 & p_2^1 \\ \mathbf{M}_2^2 & 0 & p_2^2 \\ \mathbf{M}_2^3 & 0 & 1 \end{bmatrix}_{6\times 6}, \tag{3.46}$$

where $\mathbf{M}_j^i$ is the i-th row of the j-th camera matrix and p_j^i is the i-th component of the j-th image point. From the previous discussion we know that the rank of $\mathbf{H}_2$ is at most $2 + 3 = 5$, that is

$$\det(\mathbf{H}_2) = 0, \tag{3.47}$$

since $\mathbf{H}_2$ is of size 6×6. Equation (3.47) can be computed from the Laplace development. Starting from an expansion by a row containing p_j^i, we obtain

$$\det(\mathbf{H}_2) = p_1^1 \underbrace{\det \begin{bmatrix} \mathbf{M}_1^2 & 0 \\ \mathbf{M}_1^3 & 0 \\ \mathbf{M}_2^1 & p_2^1 \\ \mathbf{M}_2^2 & p_2^2 \\ \mathbf{M}_2^3 & 1 \end{bmatrix}} - p_1^2 \det \begin{bmatrix} \mathbf{M}_1^1 & 0 \\ \mathbf{M}_1^3 & 0 \\ \mathbf{M}_2^1 & p_2^1 \\ \mathbf{M}_2^2 & p_2^2 \\ \mathbf{M}_2^3 & 1 \end{bmatrix} + 1 \det \begin{bmatrix} \mathbf{M}_1^1 & 0 \\ \mathbf{M}_1^2 & 0 \\ \mathbf{M}_2^1 & p_2^1 \\ \mathbf{M}_2^2 & p_2^2 \\ \mathbf{M}_2^3 & 1 \end{bmatrix} = 0. \quad (3.48)$$

Then we expand each of the three minors. For simplicity we write only the first one

$$\underbrace{\det \begin{bmatrix} \mathbf{M}_1^2 & 0 \\ \mathbf{M}_1^3 & 0 \\ \mathbf{M}_2^1 & p_2^1 \\ \mathbf{M}_2^2 & p_2^2 \\ \mathbf{M}_2^3 & 1 \end{bmatrix}} = p_2^1 \det \begin{bmatrix} \mathbf{M}_1^2 \\ \mathbf{M}_1^3 \\ \mathbf{M}_2^2 \\ \mathbf{M}_2^3 \end{bmatrix} - p_2^2 \det \begin{bmatrix} \mathbf{M}_1^2 \\ \mathbf{M}_1^3 \\ \mathbf{M}_2^1 \\ \mathbf{M}_2^3 \end{bmatrix} + 1 \det \begin{bmatrix} \mathbf{M}_1^2 \\ \mathbf{M}_1^3 \\ \mathbf{M}_2^1 \\ \mathbf{M}_2^2 \end{bmatrix}. \quad (3.49)$$

Inserting (3.49) and the remaining two expansions into (3.48) it becomes obvious that (3.47) can be written in the form

$$\det(\mathbf{H}_2) = \sum_{i,j=1}^{3} F_{ij} p_1^i p_2^j = F_{ij} p_1^i p_2^j = 0, \quad (3.50)$$

where F_{ij} are elements of the already introduced fundamental matrix (3.31), which in the tensor notation are elements of the *bifocal tensor* in accordance with the following.

Definition 3.1. Let ε_{ijk} be a permutation[6] symbol and let $\mathbf{M}^i{}_j$ denote the i-th row of the camera matrix $\mathbf{M}_j$ for the j-th image. The elements of the bifocal tensor, corresponding to views 1 and 2, are given as

$$F_{ij} = \varepsilon_{ii'i''} \varepsilon_{jj'j''} \det \begin{bmatrix} \mathbf{M}_1^{i'} \\ \mathbf{M}_1^{i''} \\ \mathbf{M}_2^{j'} \\ \mathbf{M}_2^{j''} \end{bmatrix}. \quad (3.51)$$

In the above the summation symbol was dropped in accordance with Einstein's summation rule (see Chapter 10). The numbers F_{ij} constitute a covariant tensor of second degree. This means that a change of coordinates systems associated with images 1 and 2 induces a concordant change of F_{ij}.

[6]The permutation symbol ε_{ijk} denotes 0, if any pair of its indices is equal, and $(-1)^p$, where p is a minimal number of index changes, leading to their normal order, i.e. 1, 2, 3...

Further extensions of the presented analysis of minors of the matrix (3.45) lead to higher degree tensors [189, 421]. For instance the *trifocal tensor* is obtained in an analogous way, based on the matrix $\mathbf{H}_3$. Then from the rank condition we obtain $\det(\mathbf{H}_3) = 0$, which leads to the following trifocal constraint:

$$T_i^{jk} \mathbf{p}_1^i \varepsilon_{jj'j''} \mathbf{p}_2^{j'} \varepsilon_{kk'k''} \mathbf{p}_3^{k'} = 0, \tag{3.52}$$

where

$$T_i^{jk} = \varepsilon_{ii'i''} \det \begin{bmatrix} \mathbf{M}_1^{i'} \\ \mathbf{M}_1^{i''} \\ \mathbf{M}_2^{j} \\ \mathbf{M}_3^{k} \end{bmatrix}. \tag{3.53}$$

The trifocal tensor is an example of a third-order mixed tensor (section 10.4) in which the order of images is also important since the first image is treated differently.

3.4.5 Finding the Essential and Fundamental Matrices

The 3×3 matrices $\mathbf{E}$ and $\mathbf{F}$ can be determined based on (3.21) and (3.29), respectively. There are nine elements to be computed. However, these formulas employ the homogeneous coordinates, and therefore any solution is determined up to a certain scaling factor (see the properties of the homogeneous coordinate transformation in section 9.2). Because of this, only eight different pairs of matched points are necessary to find $\mathbf{E}$ or $\mathbf{F}$ [278]. Thus the name of the simplest linear method is the *eight-point algorithm* [118, 121, 122, 177, 278, 430]. If more matched pairs are known, then a solution can be found by means of the least-squares method [70, 154, 352]. Remember that if coordinates of the matched points are expressed in respect to the external coordinate system then we compute the essential matrix $\mathbf{E}$, otherwise – if the coordinates are local to the image planes – we deal with the fundamental matrix $\mathbf{F}$.

The rest of this chapter is focused on determination of the fundamental matrix $\mathbf{F}$. Nevertheless, computation of the essential matrix $\mathbf{E}$ can be accomplished in the same way. Let us now observe that (3.29) can be rewritten as follows:

$$\sum_{i=1}^{3} \sum_{j=1}^{3} p_{ri} F_{ij} p_{lj} = 0, \tag{3.54}$$

where p_{ri} and p_{lj} are coordinates of the corresponding points from a matched pair, from the right and left image, respectively, and F_{ij} denotes elements of the fundamental matrix. The above equation can be rewritten again to involve only one summation:

$$\mathbf{q}^{\mathrm{T}} \mathbf{f} = r = \sum_{i=1}^{9} q_i f_i = 0, \tag{3.55}$$

where q_i denotes a component built from the point coordinates, r is called a residual and f_i denotes a coordinate of a vector of nine elements arising from the stacked elements of the matrix **F**, as follows:

$$\mathbf{q} = \left[p_{l1}p_{r1},\, p_{l2}p_{r1},\, p_{r1},\, p_{l1}p_{r2},\, p_{l2}p_{r2},\, p_{r2},\, p_{l1},\, p_{l2},\, 1\right]^{\mathrm{T}}, \tag{3.56}$$

and

$$\mathbf{f} = \left[F_{11}, F_{12}, F_{13}, F_{21}, F_{22}, F_{23}, F_{31}, F_{32}, F_{33}\right]^{\mathrm{T}}. \tag{3.57}$$

To find **F**, being now in the form of a nine-element vector f_i, we have to solve (3.55). As alluded to previously this is possible after gathering at least eight pairs of matched points. Unfortunately, although simple, the eight-point algorithm shows significant instabilities due to noise, numerical roundoff errors and mismatched points. A partial remedy to some of these problems was proposed by Hartley [177]. He suggested normalizing the point coordinates before one tries to solve (3.55). Finally, after the solution is found, the matrix has to be denormalized.

Since the matrix **F** is defined up to a certain scaling factor, it is necessary to place an additional constraint to fix the solution. It is most common here to set the norm of **f** to 1. However, other options exist which will be discussed later on.

In practice, instead of solving (3.55) exactly for eight points, solution to **f** is found for a higher number of matched pairs with a simultaneously imposed constraint on the norm of **f**. Each pair of corresponding points gives one equation of the type (3.55). Then, a $K \geq 8$ number of the corresponding points is gathered into a compound matrix $\mathbf{Q}_{K\times 9}$. Therefore the solution is obtained in the least-squares fashion, as follows:

$$\min_{\|f\|=1} \|\mathbf{Qf}\|^2, \tag{3.58}$$

where **Q** is a matrix with each row built from a pair of matched points and **f** is, as before, the sought vector of stacked elements of the matrix **F**. In accordance with definition, the norm in (3.58) can be represented as

$$\|\mathbf{Qf}\|^2 = (\mathbf{Qf})^{\mathrm{T}}(\mathbf{Qf}) = \mathbf{f}^{\mathrm{T}}\left(\mathbf{Q}^{\mathrm{T}}\mathbf{Q}\right)\mathbf{f}. \tag{3.59}$$

From **Q**, the so-called moment matrix $\mathbf{M} = \mathbf{Q}^{\mathrm{T}}\mathbf{Q}$ is created, which is of size 9×9. It can be shown using the optimization theorem of Lagrange–Euler multipliers [352] that a solution to (3.58) constitutes a minimal eigenvalue of the positive-definite matrix **M**. This can be done again by the SVD decomposition algorithm [154, 308, 352]. In this case, the matrix **F** is given by the column of the matrix **S** which corresponds to the position of the lowest singular value in the matrix **V**. However, due to discrete positions of the matched points, as well as due to noise and mismatches, when found this way matrix **F** does not have rank two.

We can take yet another approach which offers some advantages. The matrix **F** is found as an eigenvector $\boldsymbol{w}$ of **M** which corresponds to the lowest eigenvalue of **M**. Such **F** minimizes the sum of squares of algebraic residuals $E = \sum_{k=1}^{K} \rho_k$. Therefore, finding the matrix **F** can be written as the following optimization task:

$$\min\{E\}, \tag{3.60}$$

where the functional E, which derives from (3.58), is given as

$$E = \sum_{k=1}^{K} \rho_k = \sum_{k=1}^{K} \frac{r_k^2}{\mathbf{f}^\mathrm{T}\mathbf{J}\mathbf{f}} = \sum_{k=1}^{K} \frac{\mathbf{a}_k^\mathrm{T}\mathbf{F}\mathbf{b}_k}{\mathbf{f}^\mathrm{T}\mathbf{J}\mathbf{f}} = \frac{\mathbf{f}^\mathrm{T}\mathbf{M}\mathbf{f}}{\mathbf{f}^\mathrm{T}\mathbf{J}\mathbf{f}}, \tag{3.61}$$

where $\mathbf{J} = \mathbf{J}_1 = \mathrm{diag}[1, 1, \ldots, 1]$ is a normalization matrix, which in this form is equivalent to the optimization constraint $\|f\| = \Sigma_i f^2{}_i = 1$.

The denominator in (3.61) plays a role of an optimization constraint which allows a solution from the equivalence class of solutions, excluding the trivial zero results at the same time. Solution to (3.58) and (3.60) is obtained as an eigenvector $\mathbf{f}_s = \mathbf{w}$ that corresponds to the lowest eigenvalue λ_k of the moment matrix $\mathbf{M}$. To impose the rank two of the computed matrix $\mathbf{F}$ we set the smallest singular value found to 0 and then recalculate the fundamental matrix. This method was first proposed by Tsai and Huang [431]. It proceeds as follows:

$$\mathbf{F} = \mathrm{SVD} = \mathbf{S}\begin{bmatrix} v_1 & 0 & 0 \\ 0 & v_2 & 0 \\ 0 & 0 & v_3 \end{bmatrix}\mathbf{D}, \quad \text{where } v_1 \geq v_2 \geq v_3 \geq 0. \tag{3.62}$$

Then the smallest eigenvalue v_3 is set to 0 to obtain $\mathbf{F}_1$ with rank two, as follows:

$$\mathbf{F}_1 = \mathbf{S}\begin{bmatrix} v_1 & 0 & 0 \\ 0 & v_2 & 0 \\ 0 & 0 & 0 \end{bmatrix}\mathbf{D}. \tag{3.63}$$

The way to estimate the given point configuration is just to measure how close to 0 is the smallest singular value of $\mathbf{M}$. Thus optimizing for the smallest singular value leads to the quality assessment of point matching.

In the simplest case of $\mathbf{J} = \mathbf{J}_1$ (3.58) and (3.60) are equivalent. However Torr and Fitzgibbon [423] go a step further and, instead of setting $\mathbf{J} = \mathbf{J}_1$, propose to apply a constraint which is invariant to the Euclidean transformations of coordinates in the image planes. They found that the Frobenius norm of the form $f^2{}_1 + f^2{}_2 + f^2{}_4 + f^2{}_5$ fulfils the invariance requirement. This corresponds to $\mathbf{J}_2 = \mathrm{diag}[1, 1, 0, 1, 1, 0, 0, 0, 0]$. Finding $\mathbf{f}_s$ in this case is more complicated and is equivalent to solving the generalized eigenvector problem $\mathbf{f}^\mathrm{T}\mathbf{J}\mathbf{f} - \mathbf{f}^\mathrm{T}\mathbf{M}\mathbf{f} = 0$. However, a faster and more stable solution can be obtained by the procedure originally proposed by Bookstein and also cited by Torr and Fitzgibbon [423]. The methodology consists of partitioning $\mathbf{f}$ into $\mathbf{f}_1 = [f_1, f_2, f_4, f_5]$ and $\mathbf{f}_2 = [f_3, f_6, f_7, f_8, f_9]$. Then $\mathbf{f}_1$ is obtained as an eigenvector solution to the equation

$$\mathbf{D}\mathbf{f}_1 = \lambda\mathbf{f}_1, \tag{3.64}$$

where

$$\mathbf{D} = \mathbf{M}_{11} - \mathbf{M}_{12}\mathbf{M}_{22}^{-1}\mathbf{M}_{12}^{\mathrm{T}} \quad \text{and} \quad \mathbf{M} = \begin{bmatrix} \mathbf{M}_{11} & \mathbf{M}_{12} \\ \mathbf{M}_{12}^{\mathrm{T}} & \mathbf{M}_{22} \end{bmatrix}. \tag{3.65}$$

$\mathbf{M}$ is divided into $\mathbf{M}_{\mathrm{ij}}$ in such a way that

$$\mathbf{f}^{\mathrm{T}}\mathbf{M}\mathbf{f} = \mathbf{f}_1^{\mathrm{T}}\mathbf{M}_{11}\mathbf{f}_1 + 2\mathbf{f}_1^{\mathrm{T}}\mathbf{M}_{12}\mathbf{f}_2 + \mathbf{f}_2^{\mathrm{T}}\mathbf{M}_{22}\mathbf{f}_2. \tag{3.66}$$

Then $\mathbf{f}_2$ is obtained from $\mathbf{M}_{12}$, $\mathbf{M}_{22}$ and $\mathbf{f}_1$, as follows:

$$\mathbf{f}_2 = -\mathbf{M}_{22}^{-1}\mathbf{M}_{12}^{\mathrm{T}}\mathbf{f}_1. \tag{3.67}$$

Further details of this method and experimental results are presented in [423].

It is also important to notice that the linear methods cannot automatically ensure the rank constraint. It can be enforced by a proper parameterization of the fundamental matrix. This, however, leads to a nonlinear problem of finding matrices $\mathbf{E}$ and $\mathbf{F}$ [119, 180, 289, 458]. They offer much more accurate solutions and are more resistant to the false matches at a cost, however, of iterative performance and usually more complicated implementation. Nevertheless, the linear methods outlined in this section usually can be used as a first estimate which is then refined by one of the nonlinear methods.

Especially the distance minimization (nonlinear) methods, operating with the parameterized fundamental matrix, were shown to be robust and stable [288]. They start from the linear estimation of the fundamental matrix with two-rank enforcement, say $\mathbf{F}_0$. Then, a nonlinear error function is iteratively minimized with respect to the chosen parameterization and a distance measure. More details on these methods can be found in [119], for instance.

3.4.5.1 Point Normalization for the Linear Method

The first problem that arises during computation of the essential and fundamental matrices is that point coordinates are discrete (quantized) values rather than continuous ones. Thus, we have to deal with a *discrete epipolar geometry* [171]. This raises a problem of point correspondences which, even if correct on a discrete grid, do not reflect the real ground truth correspondences. This has a direct influence on the accuracy of the matrix components. In practice, sufficient resolution of images helps to lessen the influence of this phenomenon.

The second problem comes from the magnitude of the point coordinates. As already pointed out, the matrix $\mathbf{M} = \mathbf{Q}^{\mathrm{T}}\mathbf{Q}$ usually does not lead to a stable solution. Indeed, observing how its elements are composed we notice that they span quite a significant range. For an exemplary image of size 512×512 some elements of this matrix have value 1 while others can be as much as $512^2 = 262\,144$. A simple normalization of point coordinates to the range $[-1, 1]$ helps to alleviate this problem [177, 290, 460].

The normalization is done by an affine transformation, given by a matrix $\mathbf{N}$, consisting of translation and scaling. It takes the centroid of the reference points at the origin of the coordinate system and ensures that the root-mean-square distance of the points from the origin is $\sqrt{2}$. This normalization mapping is given as

$$\mathbf{p}' = \mathbf{N}\mathbf{p} = \begin{bmatrix} s_1 & 0 & -m_1 s_1 \\ 0 & s_2 & -m_2 s_2 \\ 0 & 0 & 1 \end{bmatrix} \begin{bmatrix} p_1 \\ p_2 \\ 1 \end{bmatrix} = \begin{bmatrix} s_1 (p_1 - m_1) \\ s_2 (p_2 - m_2) \\ 1 \end{bmatrix}, \tag{3.68}$$

where $\mathbf{p}$ is a point of an image, expressed with the homogeneous coordinates, $\mathbf{p}_m = [m_1, m_2, 1]$ is a mean point with the following coordinates:

$$m_1 = \frac{1}{K} \sum_{i=1}^{K} p_{1i}, \quad m_2 = \frac{1}{K} \sum_{i=1}^{K} p_{2i}; \tag{3.69}$$

K denotes a total number of points, $s_{1,2}$ are scaling factors

$$s_1 = \left[\frac{1}{K} \sum_{i=1}^{K} (p_{1i} - m_1) \right]^{-1/2}, \quad s_2 = \left[\frac{1}{K} \sum_{i=1}^{K} (p_{2i} - m_2) \right]^{-1/2}, \tag{3.70}$$

which ensure that after normalization an average point distance from the origin point [0, 0, 1] equals $\sqrt{2}$.

The aforementioned normalization is done independently in each image. Thus we have two point normalization matrices $\mathbf{N}_{\mathrm{pl}}$ and $\mathbf{N}_{\mathrm{pr}}$ in the form (3.68). Substituting (3.68) into (3.29) we obtain

$$\overline{\mathbf{p}_{\mathrm{r}}^{\mathrm{T}}} \mathbf{F} \overline{\mathbf{p}_{\mathrm{l}}} = \left(\mathbf{N}_{\mathrm{r}}^{-1} \overline{\mathbf{p}}_{\mathrm{r}}'\right)^{\mathrm{T}} \mathbf{F} \left(\mathbf{N}_{\mathrm{l}}^{-1} \overline{\mathbf{p}}_{\mathrm{l}}'\right) = \overline{\mathbf{p}}_{\mathrm{r}}'^{\mathrm{T}} \underbrace{\mathbf{N}_{\mathrm{r}}^{-\mathrm{T}} \mathbf{F} \mathbf{N}_{\mathrm{l}}^{-1}}_{\mathbf{F}'} \overline{\mathbf{p}}_{\mathrm{l}}' = 0 \tag{3.71}$$

Thus, in the *domain of transformed coordinates* we actually compute $\mathbf{F}'$ in the form

$$\mathbf{F}' = \mathbf{N}_{\mathrm{r}}^{-\mathrm{T}} \mathbf{F} \mathbf{N}_{\mathrm{l}}^{-1}. \tag{3.72}$$

Finally to recover $\mathbf{F}$ we compute

$$\mathbf{F} = \mathbf{N}_{\mathrm{r}}^{\mathrm{T}} \mathbf{F}' \mathbf{N}_{\mathrm{l}}. \tag{3.73}$$

The denormalization (3.73) is done once. Experimental results show a significant improvement in accuracy of the linear methods when preceded by the point normalization procedure [177]. Thus, in every method which estimates some parameters from point indices one should be always concerned with the influence of their magnitudes on accuracy of computations. This is especially important if coordinates are multiplied or raised to some power, as for instance in the discussed estimation of the fundamental matrix or in other computations, such as statistical moments, etc. [157, 351].

3.4.5.2 Computing F in Practice

As alluded to previously, the linear methods of computation of the fundamental or essential matrices give in practice good results if special precautions are undertaken. The two main problems one should be aware of are as follows.

1. Excessive dynamics of the products of point coordinates in (3.56) which can be alleviated by the point normalization procedure (section 3.4.5.1).
2. Incorrectly matched points (outliers) which result in gross errors if not disposed of. This problem can be mitigated by robust estimation methods, such as RANSAC (section 3.4.6) or LMedS [420].

Computation of the epipolar geometry can be arranged in a series of steps, such as those depicted in Figure 3.10. The process starts with acquisition of two (or more) images. Then the images are processed, optionally to filter out noise, change image resolution or convert from colour to monochrome representations. In Figure 3.10 this is denoted as filtering. Then the feature detectors come into action. Their role is to select salient points which are characterized by sufficient signal variations in their local neighbourhoods. This, in turn, ensures good discriminative abilities for the correlation process. The most common features are corners which are discussed in section 4.7.1. In our experiments a corner detector is used which is based on the structural tensor. It is discussed in section 4.7.2. However, other features such as SIFT can be also used for this purpose [283], at a cost of a more complicated implementation, however.

Point matching is the next stage in the block diagram in Figure 3.10. There are a variety of methods for this purpose which commonly are known as feature matching, as discussed in section 6.8. For the purpose of estimation of the epipolar geometry we test two methods here. The first one requires manual matching. This has an advantage of high reliability in avoiding large outliers. However, small inaccuracies are still possible. The second method is a version of the log-polar matching around the corner points, discussed in section 6.3.8. From a user it requires only setting of a size of a log-polar region around salient points which are then correlated by the D_{CV} covariance–variance measure (section 6.3.1). If not set explicitly, the size of a matching region is set to 17×17 pixels as a default, which was found to be a fair compromise between accuracy and speed. Additionally, a user can select a number of tiles into which the input image is divided for corner detection, as well as an allowable amount of the most prominent corners in each of them. The 'strength' of a corner is measured by the value of the lowest eigenvalue of a structural tensor in a neighbourhood of a point

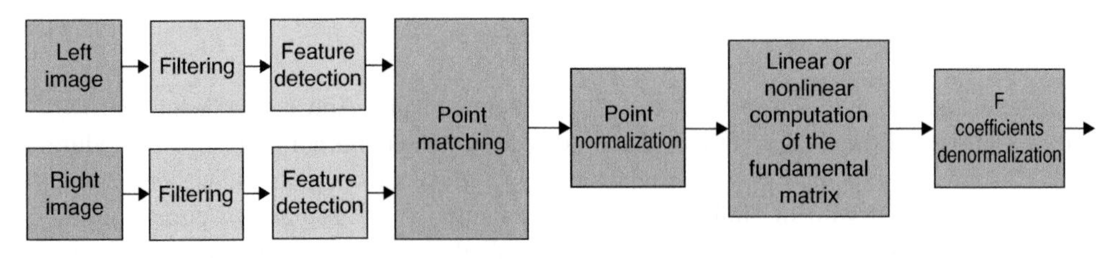

Figure 3.10 Architecture of the system for computation of the fundamental matrix

(section 4.7.2). By default an image is divided into 4×4 tiles, each containing up to two corners. The better the separation of the salient points, the more accurate the estimation of the epipolar geometry. Thus the method used of independent detection of candidate points in separate image tiles (section 4.7.2) leads to a quality improvement of the matching and estimation tasks. The additional advantage of this method is its natural ability to select the potential outliers. If there is an excessive difference in a local log-polar scale or rotation of a matched region then such a match is marked as an outlier. It has been verified experimentally that for stereo-pairs and video sequences the local scale and rotation of the log-polar representations of the corresponding points should be almost the same. The technique is safe, since even rejecting a pair of true correspondences does not influence the estimation results due to an overdetermined system of equations (i.e. usually there are many more matched pairs of points than the required minimum of eight). Contrarily, allowing one or more outliers can influence the accuracy of the estimation.

The areas found around corners in the reference (left) image are then log-polar transformed. Afterwards, the area-based correlation is applied to these transformed areas and all potential areas from the other (right) image(s) (see section 6.3.8 for details). In practice, the search for matches can be reduced to neighbouring tiles. This is true for small baseline stereo or local motion matching.

The pairs of matched points are then normalized in accordance with (3.68), after which the fundamental matrix is computed with the linear methods presented in the previous section. The original matrix is obtained by the denormalization process, given by (3.73). Finally, the epipoles are computed from (3.36) and (3.37). To check for potential numerical problems the residuals (3.61) are computed at each stage of estimation of the fundamental matrix.

Figure 3.11 depicts real test images in which eight pairs of corresponding points were manually selected. Then the epipolar geometry was computed in the system depicted in Figure 3.10. Numerical values of the computations are provided in Table 3.1 (only four decimal places are shown).

The process of automatic feature detection and matching was applied to the Pentagon stereo-pair. It is depicted in Figure 3.12 with found corner points. Only the left image was

Figure 3.11 Real test images with manually matched points

Table 3.1 Computation of the fundamental matrix **F** for the images in Figure 3.11.

Left	Right
Set of left points	Set of right points
$\mathbf{p}_{l1} = [77, 87, 1]^T$	$\mathbf{p}_{r1} = [81, 83, 1]^T$
$\mathbf{p}_{l2} = [75, 142, 1]^T$	$\mathbf{p}_{r2} = [80, 142, 1]^T$
$\mathbf{p}_{l3} = [46, 55, 1]^T$	$\mathbf{p}_{r3} = [47, 55, 1]^T$
$\mathbf{p}_{l4} = [204, 190, 1]^T$	$\mathbf{p}_{r4} = [213, 191, 1]^T$
$\mathbf{p}_{l5} = [154, 194, 1]^T$	$\mathbf{p}_{r5} = [162, 194, 1]^T$
$\mathbf{p}_{l6} = [182, 120, 1]^T$	$\mathbf{p}_{r6} = [185, 121, 1]^T$
$\mathbf{p}_{l7} = [217, 171, 1]^T$	$\mathbf{p}_{r7} = [224, 172, 1]^T$
$\mathbf{p}_{l8} = [270, 166, 1]^T$	$\mathbf{p}_{r8} = [276, 169, 1]^T$
Left normalization matrix (3.68)	Right normalization matrix (3.68)
$\mathbf{N}_l = \begin{bmatrix} 0.0134 & 0 & -2.0523 \\ 0 & 0.0214 & -3.0077 \\ 0 & 0 & 1 \end{bmatrix}$	$\mathbf{N}_r = \begin{bmatrix} 0.0131 & 0 & -2.0816 \\ 0 & 0.0210 & -2.9523 \\ 0 & 0 & 1 \end{bmatrix}$
$\mathbf{p}_m = [153.2, 140.5, 1]$	$\mathbf{p}_m = [158.9, 140.6, 1]$
Set of left points after normalization	Set of right points after normalization
$\mathbf{p}'_{l1} = [-1.0203, -1.1470, 1]^T$	$\mathbf{p}'_{r1} = [-1.0178, -1.2129, 1]^T$
$\mathbf{p}'_{l2} = [-1.0471, 0.0294, 1]^T$	$\mathbf{p}'_{r2} = [-1.0310, 0.0236, 1]^T$
$\mathbf{p}'_{l3} = [-1.4357, -1.8314, 1]^T$	$\mathbf{p}'_{r3} = [-1.4644, -1.7997, 1]^T$
$\mathbf{p}'_{l4} = [0.6819, 1.0561, 1]^T$	$\mathbf{p}'_{r4} = [0.7158, 1.0505, 1]^T$
$\mathbf{p}'_{l5} = [0.0117, 1.1416, 1]^T$	$\mathbf{p}'_{r5} = [0.0460, 1.1134, 1]^T$
$\mathbf{p}'_{l6} = [0.3870, -0.4411, 1]^T$	$\mathbf{p}'_{r6} = [0.3480, -0.4165, 1]^T$
$\mathbf{p}'_{l7} = [0.8561, 0.6497, 1]^T$	$\mathbf{p}'_{r7} = [0.8602, 0.6523, 1]^T$
$\mathbf{p}'_{l8} = [1.5664, 0.5427, 1]^T$	$\mathbf{p}'_{r8} = [1.5432, 0.5894, 1]^T$
F computed with the linear method (3.58)	**F** after rank two enforcement (3.63)
$\mathbf{F} = \begin{bmatrix} 0 & -0.0001 & 0.0111 \\ 0.0001 & 0 & -0.0057 \\ -0.0111 & 0.0048 & 0.0673 \end{bmatrix}$	$\mathbf{F} = \begin{bmatrix} 0 & -0.0001 & 0.0111 \\ 0.0001 & 0 & -0.0057 \\ -0.0111 & 0.0048 & 0.0673 \end{bmatrix}$
Left epipole (3.36)	Right epipole (3.37)
$\mathbf{e}_l = [0.4420, 0.8970, 0.0084]^T$	$\mathbf{e}_r = [-0.4125, -0.9109, -0.0088]^T$
$\Rightarrow (52.7578, 107.07)$	$\Rightarrow (46.66, 103.041)$

partitioned into 4 × 4 tiles of equal size and subjected to corner detection. A single, strongest corner response was allowed to be found in each tile. In effect 16 salient points were detected (Figure 3.12, left). Then each of the square 17 × 17 pixel neighbourhoods around each of the corners in the left image was transformed into the log-polar representation, which were then matched with the same sized log-polar versions around points in the right image. The method is very robust and reliable; therefore cross checking was not applied in this case (section 6.6.6). The matched points are denoted by the same labels in Figure 3.12, right.

Table 3.2 contains results of the computation of the epipolar parameters for the stereo-pair in Figure 3.12. The interesting observation is that a slight misalignment of the second coordinate in pair 4, as well as 12, results in the epipolar geometry not following ideally the

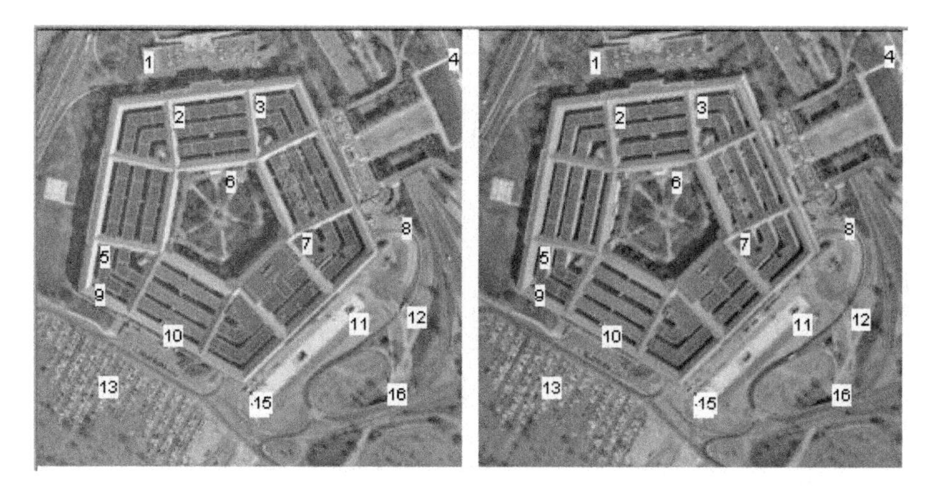

Figure 3.12 Pentagon test images with automatically matched points

canonical setup of the Pentagon pair. Thus $\mathbf{F} \neq \mathbf{F}_C$ and epipoles are out of image space but not at infinity.

Then the RANSAC procedure (discussed in the next section), controlled by the value of (3.61), was applied to the same set of points of Figure 3.12. As a best fit the following matrix was found:

$$\mathbf{F} = \begin{bmatrix} 0 & 0 & 0 \\ 0 & 0 & 0.7071 \\ 0 & -0.7071 & 0 \end{bmatrix},$$

which is exactly what we expect, i.e. now it is a fundamental matrix of a canonical stereo setup in the form (3.41).

The presented techniques were tested for monochrome images. The structural tensor and as a consequence the corner detector which is based on it, can be easily extended into a domain of multichannel images [89]. However, the tripled size of colour images does not necessarily transform into quality of the correlation process, as was verified experimentally [91]. Therefore the best way seems to be transformation of colour images into monochrome versions, as already suggested.

3.4.6 Dealing with Outliers

Finding parameters of a line based on point measurements, or components of the fundamental matrix based on point correspondences, belongs to the broader group of model estimation problems. Here we have a model, a line, a fundamental matrix, etc., the parameters of which are unknown. The only data available are measurements of point positions. In practice no measurement is free of errors, however. When determining point positions and/or their correspondences, two main types of error can be expected.

Table 3.2 Computation of the fundamental matrix **F** for the image in Figure 3.12.

Set of left points	Set of right points
$\mathbf{p}_{l1} = [63\ 12\ 1]^T$	$\mathbf{p}_{r1} = [66\ 12\ 1]^T$
$\mathbf{p}_{l2} = [82\ 44\ 1]^T$	$\mathbf{p}_{r2} = [81\ 44\ 1]^T$
$\mathbf{p}_{l3} = [131\ 38\ 1]^T$	$\mathbf{p}_{r3} = [130\ 38\ 1]^T$
$\mathbf{p}_{l4} = [247\ 10\ 1]^T$	$\mathbf{p}_{r4} = [244\ 8\ 1]^T$
$\mathbf{p}_{l5} = [37\ 126\ 1]^T$	$\mathbf{p}_{r5} = [36\ 126\ 1]^T$
$\mathbf{p}_{l6} = [113\ 82\ 1]^T$	$\mathbf{p}_{r6} = [115\ 82\ 1]^T$
$\mathbf{p}_{l7} = [159\ 118\ 1]^T$	$\mathbf{p}_{r7} = [156\ 118\ 1]^T$
$\mathbf{p}_{l7} = [219\ 109\ 1]^T$	$\mathbf{p}_{r8} = [220\ 109\ 1]^T$
$\mathbf{p}_{l9} = [34\ 148\ 1]^T$	$\mathbf{p}_{r9} = [33\ 148\ 1]^T$
$\mathbf{p}_{l10} = [75\ 172\ 1]^T$	$\mathbf{p}_{r10} = [73\ 172\ 1]^T$
$\mathbf{p}_{l11} = [187\ 165\ 1]^T$	$\mathbf{p}_{r11} = [188\ 165\ 1]^T$
$\mathbf{p}_{l12} = [222\ 161\ 1]^T$	$\mathbf{p}_{r12} = [224\ 162\ 1]^T$
$\mathbf{p}_{l13} = [36\ 202\ 1]^T$	$\mathbf{p}_{r13} = [37\ 202\ 1]^T$
$\mathbf{p}_{l14} = [127\ 213\ 1]^T$	$\mathbf{p}_{r14} = [128\ 213\ 1]^T$
$\mathbf{p}_{l15} = [129\ 211\ 1]^T$	$\mathbf{p}_{r15} = [130\ 211\ 1]^T$
$\mathbf{p}_{l16} = [210\ 206\ 1]^T$	$\mathbf{p}_{r16} = [212\ 206\ 1]^T$
Left normalization matrix (3.68)	Right normalization matrix
$\mathbf{N}_l = \begin{bmatrix} 0.0144 & 0 & -1.8613 \\ 0 & 0.0146 & -1.8349 \\ 0 & 0 & 1 \end{bmatrix}$	$\mathbf{N}_r = \begin{bmatrix} 0.0144 & 0 & -1.8609 \\ 0 & 0.0145 & -1.8274 \\ 0 & 0 & 1 \end{bmatrix}$
$\mathbf{p}_m = [129.26,\ 125.68,\ 1]$	$\mathbf{p}_m = [129.23,\ 126.03,\ 1]$
Set of left points after normalization	Set of right points after normalization
$\mathbf{p}'_{l1} = [-0.9554, -1.6602, 1]^T$	$\mathbf{p}'_{r1} = [-0.9129, -1.6534, 1]^T$
$\mathbf{p}'_{l2} = [-0.6821, -1.1944, 1]^T$	$\mathbf{p}'_{r2} = [-0.6975, -1.1893, 1]^T$
$\mathbf{p}'_{l3} = [0.0225, -1.2818, 1]^T$	$\mathbf{p}'_{r3} = [0.0063, -1.2763, 1]^T$
$\mathbf{p}'_{l4} = [1.6905, -1.6893, 1]^T$	$\mathbf{p}'_{r4} = [1.6436, -1.7114, 1]^T$
$\mathbf{p}'_{l5} = [-1.3292, -0.0009, 1]^T$	$\mathbf{p}'_{r5} = [-1.3438, 0, 1]^T$
$\mathbf{p}'_{l6} = [-0.2364, -0.6413, 1]^T$	$\mathbf{p}'_{r6} = [-0.2092, -0.6382, 1]^T$
$\mathbf{p}'_{l7} = [0.4251, -0.1174, 1]^T$	$\mathbf{p}'_{r7} = [0.3797, -0.1160, 1]^T$
$\mathbf{p}'_{l8} = [1.2879, -0.2483, 1]^T$	$\mathbf{p}'_{r8} = [1.2989, -0.2466, 1]^T$
$\mathbf{p}'_{l9} = [-1.3724, 0.3193, 1]^T$	$\mathbf{p}'_{r9} = [-1.3869, 0.3191, 1]^T$
$\mathbf{p}'_{l10} = [-0.7828, 0.6686, 1]^T$	$\mathbf{p}'_{r10} = [-0.8124, 0.6672, 1]^T$
$\mathbf{p}'_{l11} = [0.8277, 0.5667, 1]^T$	$\mathbf{p}'_{r11} = [0.8393, 0.5656, 1]^T$
$\mathbf{p}'_{l12} = [1.3310, 0.5085, 1]^T$	$\mathbf{p}'_{r12} = [1.3564, 0.5221, 1]^T$
$\mathbf{p}'_{l13} = [-1.3436, 1.1053, 1]^T$	$\mathbf{p}'_{r13} = [-1.3295, 1.1023, 1]^T$
$\mathbf{p}'_{l14} = [-0.0351, 1.2654, 1]^T$	$\mathbf{p}'_{r14} = [-0.0224, 1.2618, 1]^T$
$\mathbf{p}'_{l15} = [-0.0063, 1.2363, 1]^T$	$\mathbf{p}'_{r15} = [0.0063, 1.2328, 1]^T$
$\mathbf{p}'_{l16} = [1.1585, 1.1635, 1]^T$	$\mathbf{p}'_{r16} = [1.1840, 1.1603, 1]^T$
F computed with the linear method (3.58)	**F** after rank two enforcement (3.63)
$\mathbf{F} = \begin{bmatrix} 0 & 0 & -0.0015 \\ 0 & 0 & 0.0114 \\ 0.0016 & -0.0114 & -0.0103 \end{bmatrix}$	$\mathbf{F} = \begin{bmatrix} 0 & 0 & -0.0015 \\ 0 & 0 & 0.0114 \\ 0.0016 & -0.0117 & -0.0103 \end{bmatrix}$
Left epipole (3.36)	Right epipole
$\mathbf{e}_l = [-0.9909, -0.1350, -0.00078]^T$ $\Rightarrow (1284.86,\ 174.9)$	$\mathbf{e}_r = [-0.9912, -0.1326, -0.0007]^T$ $\Rightarrow (1336.29,\ 178.7)$

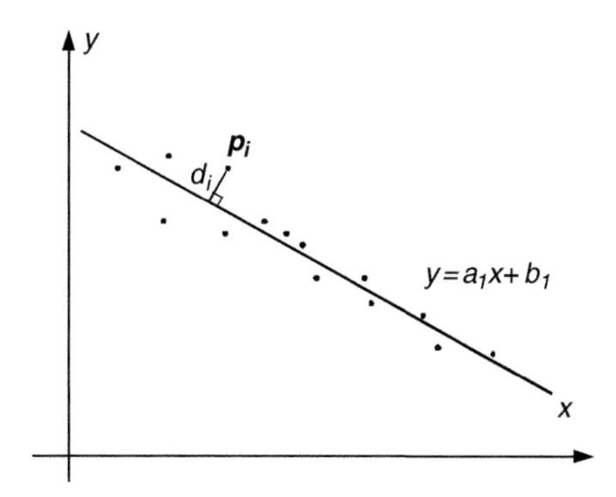

Figure 3.13 Fitting a line to points. The task is to estimate parameters a_1 and b_1 for which the line $y = a_1x + b_1$ is the closest to all measured points, computed as a minimal cumulative distance $\Sigma_i {d^2}_i$

1. The systematic error associated with the imprecise measurement of point positions. This follows the Gaussian distribution.
2. The large errors associated with erroneous matches. These mismatched points are called outliers and usually do not follow the Gaussian distribution.

Especially errors of the second type are severe since even a single outlier can greatly divert computed estimates from the real parameters. An estimate of a model from data containing some outliers can be obtained by the smoothing technique which finds an initial estimate from all the points and then tries to eliminate the invalid ones. However, in practice this method does not lead to precise estimates, due to a small group of points which differ significantly from the model. Therefore it is essential to find a method of sieving 'good' points, i.e. the *inliers*, from the erroneous ones, i.e. from the *outliers*.

A very successful method called random sample consensus (RANSAC) was proposed by Fischler and Bolles [126]. Their idea was simple: randomly choose a number of samples from the set of all measurements, try to fit a model to them, and check how many other points are in consensus with this model estimate. The process is repeated and the best fit, i.e. an estimate supported by the maximal number of measurements, is left as a solution. All other points are treated as outliers.

The best way to illustrate the RANSAC method is to consider a problem of a line fit to some points on a plane. This is depicted in Figure 3.13. The problem of a line estimate is to find such parameters a_1 and b_1 of a line model given by $y = a_1x + b_1$, for which a cumulative distance to the data is minimized. This can be written as[7]

$$\min \sum_i d_i^2. \tag{3.74}$$

[7] It is easy to show that $d_i = |a_1x_i - y_i + b_1| / \sqrt{a_1^2 + 1}$.

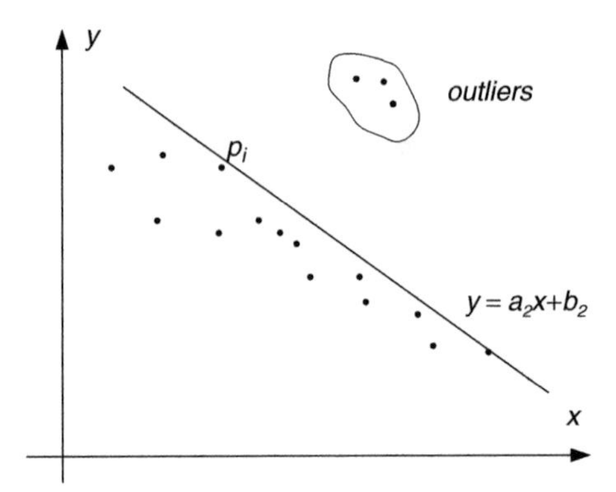

Figure 3.14 Fitting a line to points in the presence of outliers. New line estimate $y = a_2x + b_2$ is biased towards the group of outliers if not removed from the dataset

The situation is changed if the dataset contains some outliers, i.e. points which are erroneous for some reason. This is depicted in Figure 3.14.

In most real datasets the problem is that we can expect some outliers; however, we do not know their values (coordinates) nor their number. Thus, Figure 3.14 shows a rather convenient situation in which outliers are marked *a priori*. The true power of RANSAC is that it is able to partition data into inliers and outliers when no other information on data is given.

Figures 3.15 and 3.16 depict two different attempts of estimation of line parameters with the RANSAC method. In each step two points are randomly selected and used to place a hypothesis on the line parameters. These are a_3, b_3 in Figure 3.15 and a_4, b_4 in Figure 3.16.

Then, for each hypothesis about a line, the number of points which are in consensus with that line is counted. Separation of points is achieved simply by placing a constant threshold τ on a maximal allowable distance d_i of a point from the estimated line (grey regions in

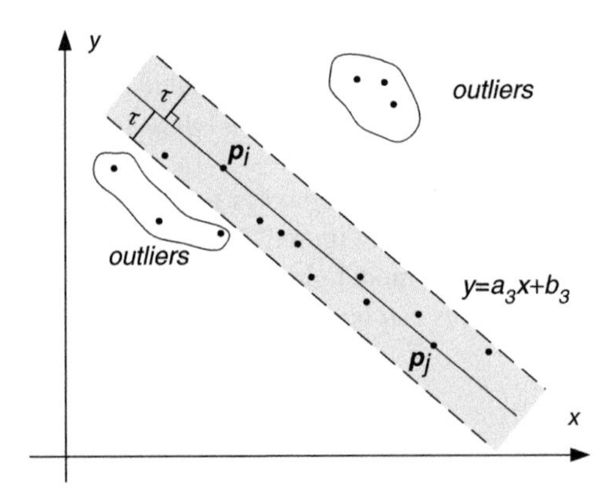

Figure 3.15 Fitting a line to points: a step of the RANSAC algorithm. Randomly selected pair of points x_i and x_j serves an initial estimate $y = a_3x + b_3$. Distances of all other points to this estimate are checked and only those within a predefined threshold (grey area) are considered as inliers

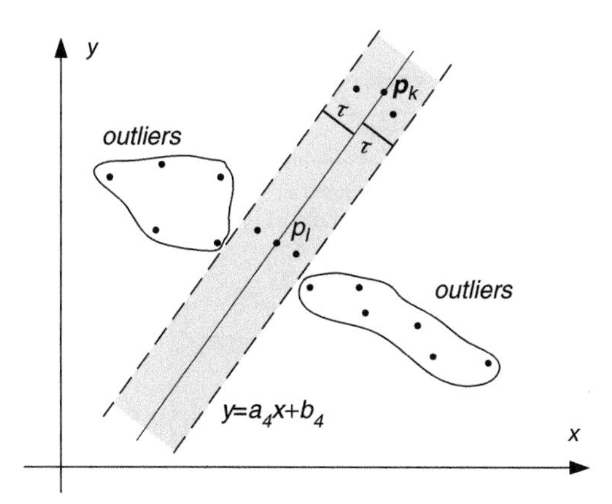

Figure 3.16 Fitting a line to points by the RANSAC algorithm. Another randomly selected pair of points x_k and x_l serves another estimate $y = a_4x + b_4$. New region of inliers is denoted in grey

Figures 3.15 and 3.16). In the first example (Figure 3.15) there are 11 inliers and 6 outliers for a given value of τ.

If another pair of points $\mathbf{p}_k$ and $\mathbf{p}_l$ is randomly selected then a new estimate a_4, b_4 is computed (Figure 3.16). Now, the number of inliers is 6 and the number of outliers 11. Thus, the first estimate in Figure 3.15 provides a more 'consistent' estimate.

This process is repeated a number of times until the most consistent estimate is found, or until a preset number of steps, or until a sufficiently large set of inliers is found. The flow chart of the RANSAC method is presented in Algorithm 3.1.

As usual, the immediate question is how to measure the 'consensus' of the points. This is achieved by choosing a suitable threshold value τ. Then a point is classified as an inlier if its distance is less from τ, and an outlier otherwise. A procedure for finding τ as a function of a probability distribution of inliers from the model is discussed in [180]. If we assume that the point measurements follow a Gaussian distribution with zero mean and a variance σ, then τ can be related to σ as well as to the co-dimension of the estimated model, which is 1 for a line or fundamental matrix, 2 for homography, 3 for trifocal tensor, and so on. For instance, in the case of estimates of a line or a fundamental matrix, Hartley and Zisserman provide estimation of $\tau = 1.96\sigma$ which was computed with an assumption of 95% probability that a point drawn from a dataset is an inlier [180].

The other parameters to set are the number of samples drawn at each step of the algorithm, as well as the stopping criteria, i.e. maximal number of steps and/or count of the acceptable consensus set. As a rule of thumb the number of samples should be rather small: i.e. two points for a line instead of three or four, for instance. For the stopping criteria the maximal number of iterations can be set empirically as a tradeoff between accuracy of an estimate and computation time. Additionally, *a priori* knowledge of the proportion of inliers to outliers in data can be of help. Settings of these parameters and further properties of the RANSAC method are presented in more details in [126, 180].

The RANSAC method has been shown to be very robust in many practical applications and therefore it should be considered in all cases when determining parameters of a model from the empirical data for which a number of outliers is also expected.

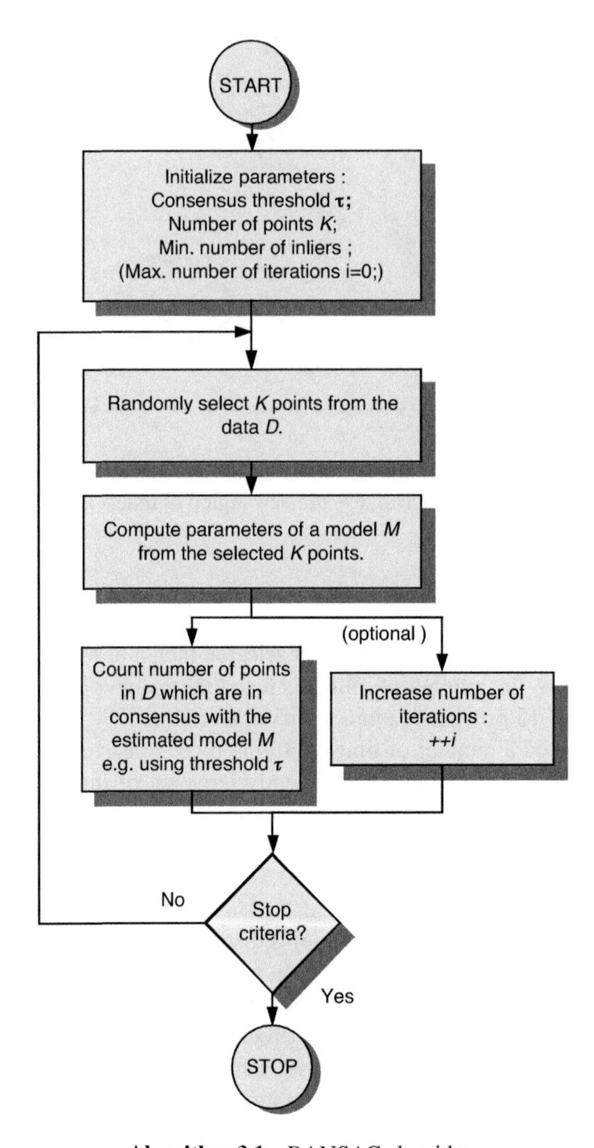

Algorithm 3.1 RANSAC algorithm

3.4.7 Catadioptric Stereo Systems

By using optical devices[8] that bend and reflect the direction of light rays it is possible to construct cameras with much broader fields of view and also stereo systems which employ only *single* cameras [148, 149, 151, 328, 408]. Figure 3.17 depicts such a stereo system that utilizes two flat mirrors and a single camera.

[8]Called also catadioptric elements.

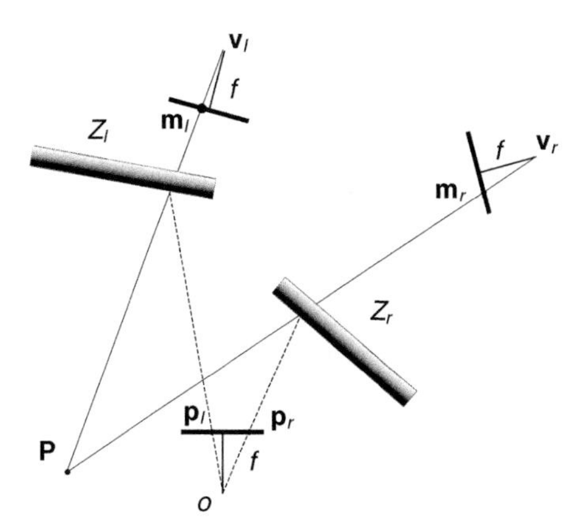

Figure 3.17 Catadioptric stereo system with two mirrors and a single camera

A 3D point **P** in Figure 3.17 is projected on to the camera plane as two image points $\mathbf{p}_l$ and $\mathbf{p}_r$. This is possible due to reflection of the light rays performed by two flat mirrors Z_l and Z_r. The points $\mathbf{v}_l$ and $\mathbf{v}_r$ are centres of the virtual cameras while the points $\mathbf{m}_l$ and $\mathbf{m}_r$ are virtual images of the real point **P**.

Catadioptric systems exhibit some differences in respect to classic stereo systems.

- Identical system parameters. In a catadioptric system there is only one analogue or digital channel transferring the acquired image. Therefore automatically such parameters as bandwidth characteristics, distortions, parameters of the CCD or CMOS photodetectors, etc., are identical (section 3.3.5).
- Calibration. There is only one set of camera parameters to be found. However, when using nonlinear optical elements, such as parabolic or hyperbolic mirrors, we have to use special computation methods. Also the epipolar geometry of such catadioptric systems is different [150, 408].
- Broader field of view.
- Usually simpler and cheaper construction. This is especially valuable for systems working with multiple (two or more) images.

A practical realization of a catadioptric stereo system working in real time was presented by Gluckmann *et al.* [148, 149, 151]. Their systems perform image acquisition followed by a matching stage. Due to the features and parameters of the systems, matching was possible with a simple SAD measure (section 6.3.1) since the two acquisition channels have the same characteristics. In effect they obtained high matching quality and fast computation. Gluckmann *et al.* reported about 20 frames per second with resolution 320×240. The hardware they used consisted of a Pentium® II 300 MHz, Sony XC-77 single camera with a 12.5 mm Computar® lens, and 5-inch Melles Griot® mirrors.

3.4.8 Image Rectification

Stereo image rectification is a process of image transformations in such a way that the corresponding epipolar lines in all images become collinear with each other and with the image scanning lines [17, 122, 139, 340, 430]. In rectified images all optical axes are parallel as well. The stereo setups that comply with these conditions are called standard or canonical stereo setups. A very important feature of the stereo setups is an inherent constraint of the search space to one dimension only (the so-called epipolar constraint; see also section 3.5), but in rectified systems we know beforehand the positions of the epipolar lines which is the direction of the scanning lines. This is a very desirable feature from the computational point of view [142, 369].

The other interesting feature inherent to the rectified stereo system is a shift of the epipoles to infinity. Thus, rectification of images can be thought of as a process of changing positions of epipoles to infinity. The rectification process is limited to the search for the transformation of the planes $\mathbf{\Pi}_{l0}$ and $\mathbf{\Pi}_{r0}$ to the planes $\mathbf{\Pi}_{l1}$ and $\mathbf{\Pi}_{r1}$, respectively (Figure 3.18). The transformation sought can be described as a composition of the following transformations [430].

1. Rotation of the left and right camera planes in such a way that the epipoles go to infinity (and thus the epipolar lines become parallel). This rotation is described by a rotation matrix **Q**.
2. Rotation of the right camera according to the transformation described by a matrix **R** from (3.16).

Additionally, without lost of generality, we assume the following.

1. The focal length f of the two cameras is the same.
2. The origin of the local camera coordinate system is the camera principal point (i.e. the cross point of the optical axis with the image plane).

The matrix **Q** can be found by considering three mutually orthogonal unit vectors: $\mathbf{q}_1$, $\mathbf{q}_2$ and $\mathbf{q}_3$. The vector $\mathbf{q}_1$ is collinear with the translation vector **T** between the focus points of the two cameras (Figure 3.7) and is given as

$$\mathbf{q}_1 = \frac{\mathbf{T}}{\|\mathbf{T}\|}. \tag{3.75}$$

The vector $\mathbf{q}_2$ is orthogonal to the vector $\mathbf{q}_1$. Because

$$[-T_2, T_1, 0] \cdot [T_1, T_2, T_3]^{\mathrm{T}} = 0 \tag{3.76}$$

then $\mathbf{q}_2$ takes the form

$$\mathbf{q}_2 = \frac{[-T_2, T_1, 0]^{\mathrm{T}}}{\sqrt{T_2^2 + T_1^2}}. \tag{3.77}$$

The third vector $\mathbf{q}_3$ has to be simultaneously orthogonal to the vectors $\mathbf{q}_1$ and $\mathbf{q}_2$. Therefore it can be set to the vector product

$$\mathbf{q}_3 = \mathbf{q}_1 \times \mathbf{q}_2. \tag{3.78}$$

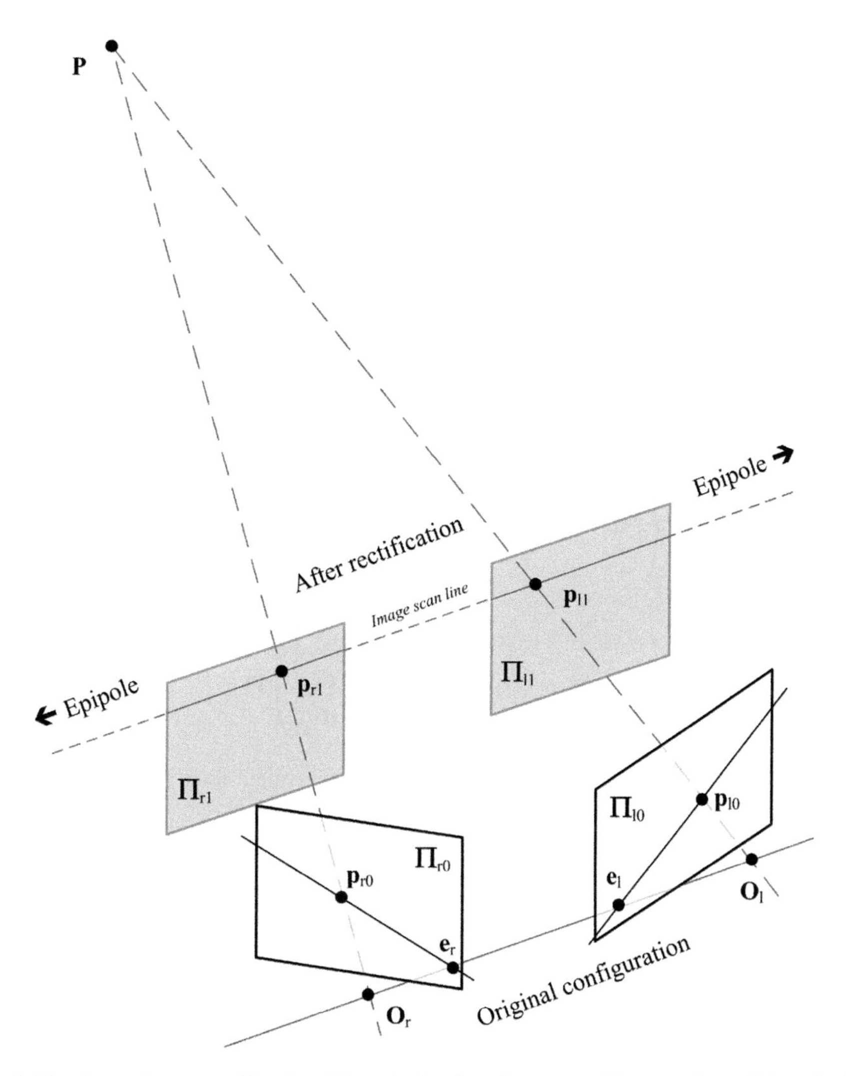

Figure 3.18 Stereo image rectification. The epipolar lines become collinear and parallel to the image scanning lines

The vectors $\mathbf{q}_1$, $\mathbf{q}_2$ and $\mathbf{q}_3$ determine the following rotation matrix $\mathbf{Q}$:

$$\mathbf{Q} = \begin{bmatrix} \mathbf{q}_1^{\mathrm{T}} \\ \mathbf{q}_2^{\mathrm{T}} \\ \mathbf{q}_3^{\mathrm{T}} \end{bmatrix}_{3\times 3} . \tag{3.79}$$

In practice, to obtain integer values of coordinates in the new (i.e. rectified) camera setup, the rectification process should be performed backwards, i.e. starting from the new coordinates and applying the inverse transformation $\boldsymbol{Q}^{-1}$. This way, the new intensity values in the 'new' system can be determined, for example, by the bilinear interpolation of the original values from the 'old' setup (see Chapter 12).

The stereo rectification problem can be approached in another way, taking as a starting point computation of the fundamental matrix (section 3.4.1.1). This can be done with one of the methods discussed in section 3.4.5. As alluded to previously, image rectification is a process which takes epipoles of an original stereo setup into infinity; thus the system becomes a 'canonical' one, i.e. its fundamental matrix $\mathbf{F}$ becomes $\mathbf{F}_C$. However, let us start from the basic epipolar equation (3.29) with $\mathbf{F}$ decomposed into singular values (3.38), as follows:

$$\overline{\mathbf{p}_r^T}\left(\mathbf{SVD}^T\right)\overline{\mathbf{p}_l} = 0,$$

which can be written

$$\overline{\mathbf{p}_r^T}\left(\begin{bmatrix}\mathbf{e}_r & \mathbf{s}_1 & \mathbf{s}_2\end{bmatrix}\begin{bmatrix}0 & 0 & 0\\ 0 & v_1 & 0\\ 0 & 0 & v_2\end{bmatrix}\begin{bmatrix}\mathbf{e}_l\\ \mathbf{d}_1\\ \mathbf{d}_2\end{bmatrix}\right)\overline{\mathbf{p}_l} = 0,$$

$$\overline{\mathbf{p}_r^T}\left(\begin{bmatrix}\mathbf{e}_r & \mathbf{s}_1 & \mathbf{s}_2\end{bmatrix}\begin{bmatrix}0 & 0 & 0\\ 0 & 1 & 0\\ 0 & 0 & v\end{bmatrix}\begin{bmatrix}\mathbf{e}_l\\ \mathbf{d}_1\\ \mathbf{d}_2\end{bmatrix}\right)\overline{\mathbf{p}_l} = 0, \tag{3.80}$$

where we put $v = v_2/v_1$. The above can be written in an equivalent form [119]

$$\overline{\mathbf{p}_r^T}\left(\underbrace{\begin{bmatrix}\mathbf{e}_r & \mathbf{s}_1 & \sqrt{v}\mathbf{s}_2\end{bmatrix}}_{\mathbf{H}_r}\underbrace{\begin{bmatrix}0 & 0 & 0\\ 0 & 0 & -1\\ 0 & 1 & 0\end{bmatrix}}_{\mathbf{F}_C}\underbrace{\begin{bmatrix}\mathbf{e}_l\\ \sqrt{v}\mathbf{d}_2\\ -\mathbf{d}_1\end{bmatrix}}_{\mathbf{H}_l}\right)\overline{\mathbf{p}_l} = 0, \tag{3.81}$$

where we notice the canonical fundamental matrix $\mathbf{F}_C$ as well as two matrices $\mathbf{H}_r$ and $\mathbf{H}_l$, each of dimensions 3×3, which denote the two homographies (section 9.5.3). From the above we obtain finally

$$\left(\mathbf{H}_r^T\overline{\mathbf{p}_r}\right)^T \mathbf{F}_C\left(\mathbf{H}_l\overline{\mathbf{p}_l}\right) = 0. \tag{3.82}$$

Thus, when points $\overline{\mathbf{p}_r}$ and $\overline{\mathbf{p}_l}$ from the original stereo images are transformed by the homographies $\mathbf{H}_r$ and $\mathbf{H}_l$, then the obtained system is described by the canonical fundamental matrix and its epipoles are at infinity.

In practice, however, rectification can lead to excessive and unwanted image distortions. Therefore care must be taken to alleviate this problem. For instance one can try to design a transformation that acts as a rigid transformation in the neighbourhood of a certain (e.g. central) image point [179] or find such transformation that minimizes the effect of resampling [152].

A Matlab procedure for a linear rectification of a general unconstrained stereo setup is provided in the paper by Fusiello *et al.* [143]. It is assumed that the system is calibrated, that is, the intrinsic parameters of the cameras are known, as well as parameters of the mutual positions of the cameras (section 3.6.4). Thus, the procedure takes the two perspective projection matrices **M** (3.7) of the cameras and outputs a pair of rectifying projection matrices.

3.4.9 Depth Resolution in Stereo Setups

Figure 3.19 explains the phenomenon of diminishing accuracy of depth measurement with increasing distance from the camera planes. This is a geometrical limitation since it depends exclusively on geometrical parameters of a stereo system.

The dependence of the depth accuracy versus camera resolution and distance to the observed scene can be found analysing Figure 3.20. Observing the similarity of triangle △ABC to △ADF, as well as △AEF to △AHG, we obtain the following relations:

$$\frac{\overline{DF}}{\overline{AF}} = \frac{\overline{BC}}{\overline{AC}} \qquad \frac{\overline{EF}}{\overline{AF}} = \frac{\overline{GH}}{\overline{AG}}.$$

Let us now introduce new shorter symbols:

$$\overline{AF} = b, \quad \overline{BC} = \overline{GH} = f, \quad \overline{EF} = Z, \quad \overline{DE} = R.$$

We obtain

$$r = \overline{AG} - \overline{AC}, \quad r = \frac{fb}{Z} - \frac{bf}{Z+R}.$$

Thus, after a simple change

$$Rfb = rZ(Z+R).$$

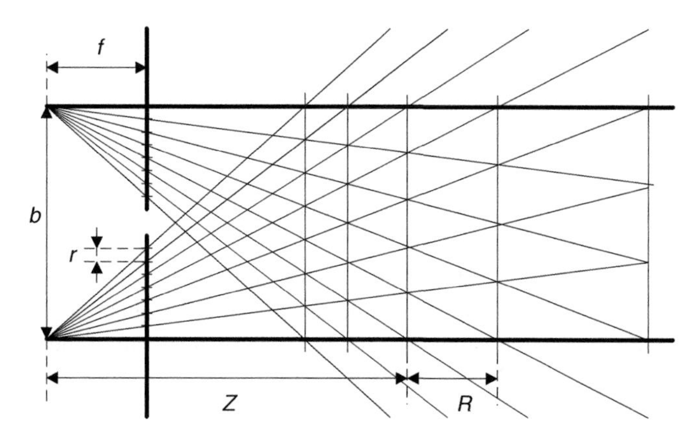

Figure 3.19 Phenomenon of a limited accuracy of depth measurement with increasing distance from the camera

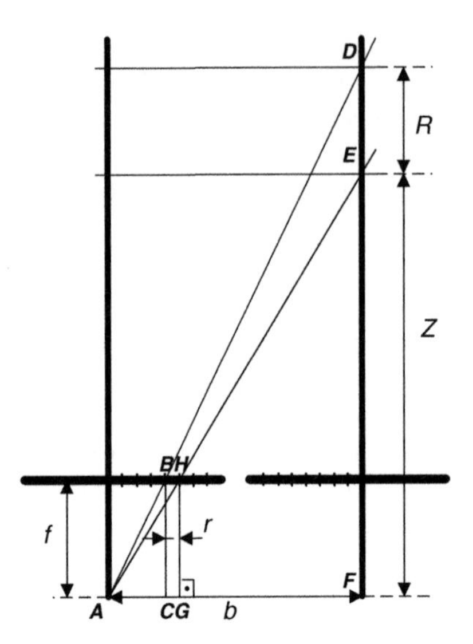

Figure 3.20 Relation of depth measurement accuracy in respect to camera resolution

Finally we obtain the following formula:

$$R = \frac{rZ^2}{fb - rZ}. \tag{3.83}$$

Assuming now that fb/Z is much larger than the pixel resolution r, we obtain the following approximation:[9]

$$R \approx \frac{rZ^2}{fb}. \tag{3.84}$$

Analysing (3.83) and (3.84), the following conclusions can be drawn. Equation (3.83) is true under the following condition:

$$fb \neq rZ.$$

Gradually as Z approaches the limit value

$$Z = \frac{bf}{r},$$

the depth measurement resolution value R approaches infinity. For most image acquisition systems, the values of r, b and f are constant, at least for a single acquisition. This means that there is such a value Z for which it is not possible to measure the depth of the observed scene due to geometrical limitations of the stereo camera setup.

[9] This assumption is justified for relatively small values of Z (Figure 3.19). The focal length f, as well as base distance b can also change but here they are assumed to be constant at least for a single exposition.

Table 3.3 Exemplary values of the depth resolution R [m] for a stereo setup with constant parameters (horizontal pixel resolution $R_h = 1024$ pixels, view angle $\alpha = 60°$, base line $b = 5$ and 30 cm)

b[m] \ Z[m]	0.1	0.5	1.0	5	10
0.05	0.000226	0.0057	0.023	**0.635**	**2.91**
0.3	0.000038	0.00094	0.0038	0.096	**0.39**

For reasonably small ranges of Z and fixed values of r, b and f, the approximate relation (3.84) exhibits a quadratic relation R of Z. This means that if it is necessary to measure absolute position of real objects with an *a priori* assumed accuracy, then the parameters of the stereo setup must be chosen in such a way that R would be at least an order of magnitude less than the assumed measurement accuracy.

From the following diagram

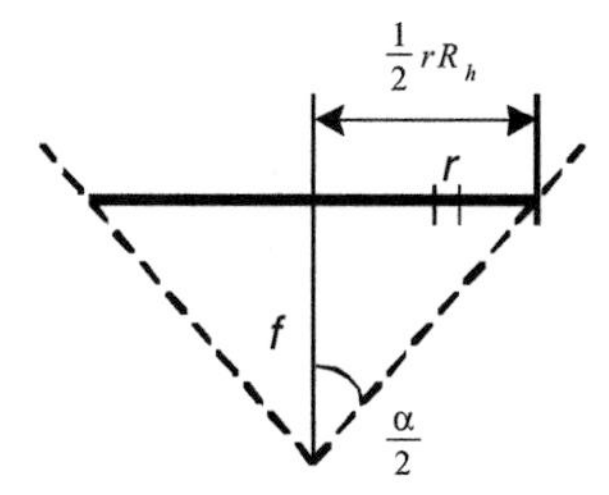

we easily notice that

$$\frac{f}{r} = \frac{R_h}{2\tan(\alpha/2)},$$

where R_h is the horizontal resolution of the camera (in pixels) and α the horizontal view angle of the camera. Table 3.3 presents depth resolution values of R for an exemplary stereo setup. The values were computed based on (3.83) converted to the following formula:

$$R = \frac{Z^2}{[R_h b/2\tan(\alpha/2)] - Z}. \tag{3.85}$$

Examining Table 3.3 it becomes evident that for a distance Z of only 10 m and distance between cameras of 5 cm, the depth measurement resolution is as much as 3 m! Moving the cameras apart, for example to 30 cm, allows for an improvement of R to be in this example about 40 cm.

3.4.10 Stereo Images and Reference Data

Table 3.4 contains pairs of stereo images used for testing of stereo matching algorithms. All of them are artificial images supplied with the ground-truth data. Apart from the true depth

Table 3.4 Artificial test stereo-pairs with ground-truth information. One can observe depth by placing a blank sheet between the two images and observing the left image with the left and the right image with the right eye

Name	Left image	Right image
Random dots stereogram (AGH University)		
Corridor (courtesy Bonn University [206])		
Tsukuba (courtesy Tsukuba University)		
Venus (courtesy Middlebury [370])		

Table 3.4 (*Continued*)

Name	Left image	Right image
Sawtooth (courtesy Middlebury [371])		
Map (courtesy Middlebury [371])		
Teddy (courtesy Middlebury [371])		
Cones (courtesy Middlebury [371])		

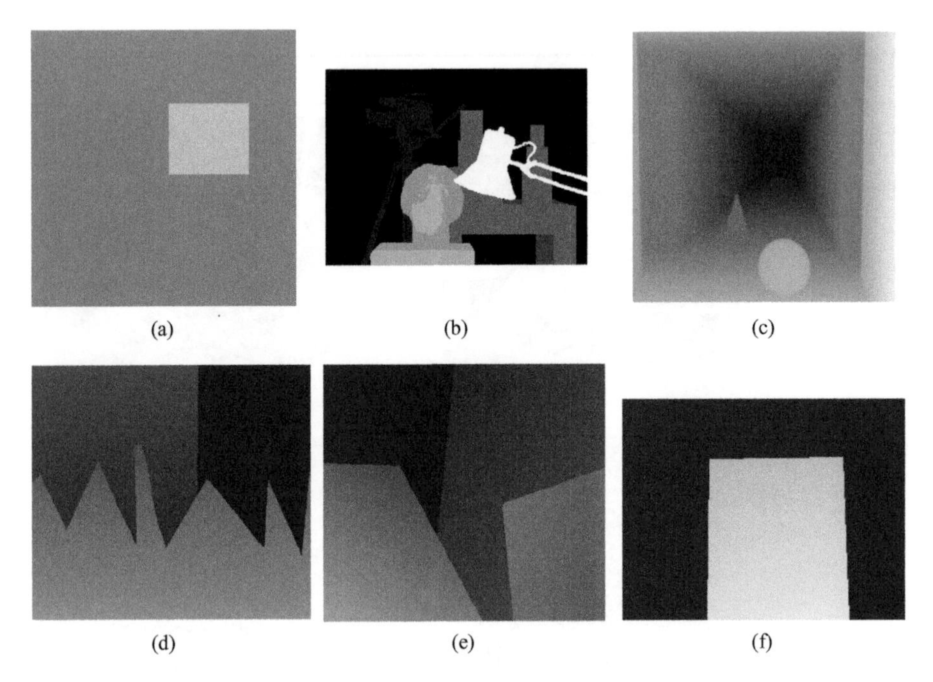

(a) (b) (c)

(d) (e) (f)

Figure 3.21 Ground-truth data of (a) 'Random dots', (b) 'Tsukuba', (c) 'Corridor', (d) 'Sawtooth', (e) 'Venus' and (f) 'Map'. (Images (b–f) courtesy of Middlebury University [209])

values, information on half-occluding areas is also provided. Because of such a common platform it is possible to compare qualitatively results of many different stereo matching methods.[10] For a given method, the closer its output is to the ground-truth data, the better the quality of the method.

The reader can easily experience the stereo effect observing the presented stereo-pairs. This can be done by ensuring that the left eye is watching exclusively the left image of the stereo-pair, and the right eye the right image, for example by placing a separating sheet between the two images. For beginners it can take some time to see the results of this experiment, i.e. the depth of a scene due to stereovision. Less than 2% of the human population has some problems with perception of depth [201].

In the case of the images presented in Table 3.4 the ground-truth maps are also available. They are presented in Figure 3.21. However, most real images do not have ground-truth data available. Some examples of such stereo-pairs are presented in Table 3.5.

Lack of ground-truth data poses a problem when measuring quality of stereo algorithms. Discussion of some evaluation methods, other than comparing with ground-truth data, is contained in section 6.4.3.

[10]The image 'Tsukuba' courtesy of Prof. Yuichi Ohta from Tskukuba University, Japan. The image 'Corridor' belongs to Bonn University. The images 'Venus', 'Sawtooth' and 'Map' are from the Middlebury Stereo Vision Page [209], courtesy of Prof. Richard Szeliski. The Middlebury web page provides invaluable source of information on all aspect of stereo vision algorithms, their comparison, test data as well as reference papers, such as [194, 370–372].

Table 3.5 Examples of real stereo-pairs. No ground-truth data available. ('Trees' courtesy of SRI, 'Pentagon' courtesy of CMU/VASC [212])

Trees (courtesy SRI)		
Pentagon (courtesy CIL CMU [212])		
Park (AGH University)		
Street (AGH University)		

3.5 Stereo Matching Constraints

Table 3.6 lists the most common assumptions, constraints and simplifications for the point matching process in stereo images [47, 68, 122, 246, 413, 454]. These very important relations can greatly facilitate the matching task or help in clarification of point matches. For instance, the already discussed epipolar constraint limits the search space from the general 2D to 1D alongside the epipolar lines (section 3.4.1). However, is was already mentioned that the position of these is not given beforehand, except for the canonical stereo setup. The epipolar lines can be determined from the fundamental matrix which, in turn, can be computed by one of the techniques presented in section 3.4.5.

Other frequently used assumptions are the uniqueness and the ordering constraints. They can be applied if some photometric and geometric characteristics of objects are fulfilled. These usually hold for a diverse group of real images; therefore the uniqueness and ordering constraints can greatly simplify the matching algorithms.

The third group of constraints concerns assumptions on disparity values. These are disparity continuity, absolute disparity value and disparity gradient limits.

Table 3.6 Stereo matching constraints and assumptions

Name of constraint	Description
Epipolar constraint	A plane of a 3D point and its image points in the two camera planes contains the base line, i.e. the line joining two camera centres and the two epipoles. The plane created this way is called the epipolar plane. The crossings of the epipolar plane and image planes of the cameras give epipolar lines (section 3.4.1). As a consequence the corresponding image points lie always on the corresponding epipolar lines. If the latter are known *a priori* then the matching search reduces to a 1D search, i.e. along the epipolar lines. In the canonical stereo system the epipolar lines are collinear with image scanlines (section 3.4.2).
Uniqueness constraint	A given pair of the matched points, one lying on the left and second on the right camera planes, respectively, corresponds *at most to the one* 3D point. This constraint is fulfilled for opaque objects. This assumption can greatly simplify a matching process. For transparent objects it is possible that many different 3D points have the same image on one or more camera planes. In other words this constraint means that a 3D point, which belongs to an opaque object in a scene, is allowed to have only zero or one image point on each camera plane. The case with zero image points happens if for some reason the 3D point is not visible for a camera, e.g. due to occlusions.
Photometric compatibility constraint	Two regions U_{l1} and U_{l2}, belonging to the left image, and regions U_{r1} and U_{r2}, belonging to the right image, are corresponding if the following conditions hold: $$\underset{i}{\forall} \underset{\tau}{\exists} \left| \sum_{(x,y)\in U_{li}} I_{li}(x,y) - \sum_{(x,y)\in U_{ri}} I_{ri}(x,y) \right| < \tau \quad (3.86)$$

Table 3.6 (*Continued*)

Name of constraint	Description
	and $$\exists_{\tau} \left\| \left| \sum_{(x,y)\in U_{l1}} I_{l1}(x,y) - \sum_{(x,y)\in U_{l2}} I_{l2}(x,y) \right| - \left| \sum_{(x,y)\in U_{r1}} I_{r1}(x,y) - \sum_{(x,y)\in U_{r2}} I_{r2}(x,y) \right| \right\| < \tau \quad (3.87)$$ where $I_{ki}(x,y)$ is an intensity value of the k-th image in the i-th region, τ is a threshold value. It is not assumed that the regions of consideration are compact or not.
Geometric similarity constraint	The geometric constraint is usually defined in respect to the similarity of angles as well as edge segments: 1. A segment S_l with spatial orientation W_l, belonging to the left image, corresponds to a segment S_r with orientation W_r in the right image, if the following holds: $$\|W_l - W_r\| < \tau, \quad (3.88)$$ where τ is a threshold value. 2. A segment S_l of length L_l, belonging to the left image, corresponds with a segment S_r of length L_r in the right image, if $$\|L_l - L_r\| < \tau \quad (3.89)$$ where τ is a threshold value.
Ordering constraint (local gradient constraint)	The ordering constraint concerns the order of the corresponding image points. That is, the corresponding points from the left and right images have the same order. This constraint is fulfilled only if the specific conditions are met for the 3D objects of that scene, for instance if all visible objects are located at about the same distance from the cameras or a continuous surface is observed. It was shown by Faugeras that by eliminating a *forbidden zone* from considerations the ordering constraint holds for all other points of that scene [122]. The forbidden zone contains cones given by lines connecting a certain 3D point **M** with its image points m_l, m_r on the camera planes. However, determination of the forbidden zone is not a trivial task since it requires *a priori* knowledge of the scene geometry.

(*continued*)

Table 3.6 Stereo matching constraints and assumptions (*Continued*)

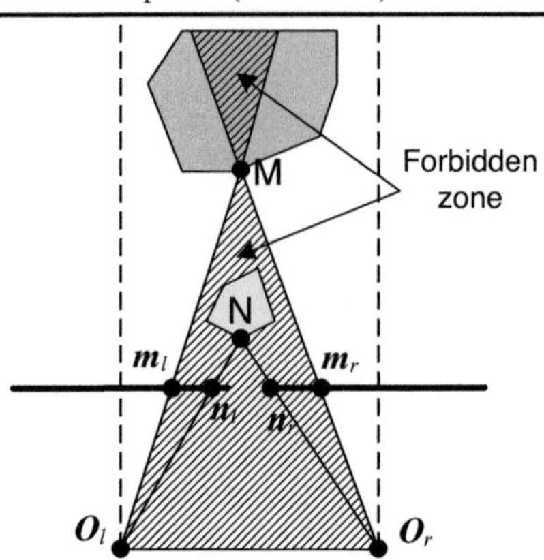

Figure 3.22 Exemplary scene for which the ordering constraint is not fulfilled

Figure 3.22 presents two points **M** and **N**, each belonging to different 3D objects, for which the ordering constraint does not hold. It is evident that for the left camera the order of image points is m_l, n_l, whereas for the right camera it is n_r, m_r.

The ordering constraint can be expressed also as a constraint on the *local gradient of disparity*. Assuming that for a certain point $\mathbf{p}_l(x, y)$ there is a corresponding point $\mathbf{p}_r(x + D(x), y)$, then after advancing the x coordinate by a positive and nonzero value δ we obtain the new correspondences: $\mathbf{p'}_l(x + \delta, y)$ and $\mathbf{p'}_r(x + \delta + D(x + \delta), y)$.

Assuming further that the point ordering constraint is fulfilled – the point order is $\mathbf{p}_l$, $\mathbf{p'}_l$ then also it is $\mathbf{p}_r$, $\mathbf{p'}_r$ – from the second relation we obtain that the following holds:

$$(x + \delta) + D(x + \delta) > x + d(x). \tag{3.90}$$

This, after dividing by δ and taking the limit $\delta \to 0$ leads to

$$\frac{\partial D(x)}{\partial x} > -1 \tag{3.91}$$

where $D(x)$ is a disparity value in the standard stereo setup (section 3.4.2). The last equation places a constraint on a horizontal gradient of disparity if point ordering is to be fulfilled.

Disparity continuity constraint

Assume that $\mathbf{p}_{l1} = [x_{l1}, y_{l1}]^T$ and $\mathbf{p}_{r1} = [x_{r1}, y_{r1}]^T$ is a pair of corresponding points, from the left and right images respectively. Let us assume that a point $\mathbf{p}_{l2} = [x_{l2}, y_{l2}]^T$, from a certain local neighbourhood of the point $\mathbf{p}_{l1}$, corresponds to a point $\mathbf{p}_{r2} = [x_{r2}, y_{r2}]^T$ in the right image. Then the disparity continuity constraint states that the following inequality should be preserved:

$$\underset{\tau}{\exists} |D(\mathbf{p}_{l1}, \mathbf{p}_{r1}) - D(\mathbf{p}_{l2}, \mathbf{p}_{r2})| < \tau \tag{3.92}$$

where $D(\mathbf{p}_i, \mathbf{p}_j)$ is disparity between points $\mathbf{p}_i$ and $\mathbf{p}_j$ and τ is a certain threshold value.

This constraint should be applied with great care since it can break on image boundaries.

Table 3.6 *(Continued)*

Name of constraint	Description
Figural continuity constraint	This is a version of the disparity continuity constraint but applied only to the edge points. This formulated figural continuity constraint assumes that edges found in images correspond to the continuous boundaries of real objects.
Feature compatibility constraint	The feature compatibility constraint states that the two points correspond to each other if certain image features around these points arise from the same source in the two images. This constraint is frequently used for edge points, i.e. an edge point in the one image can correspond only to an edge point in the second image. In the case of the feature compatibility constraint applied to the edge points, information on edges is augmented by a type of signal change (i.e. a sign of the local image gradient) [162, 172, 247]. Figure 3.23 presents an example of the feature compatibility constraint applied to ensure compatibility of curves in the two images. 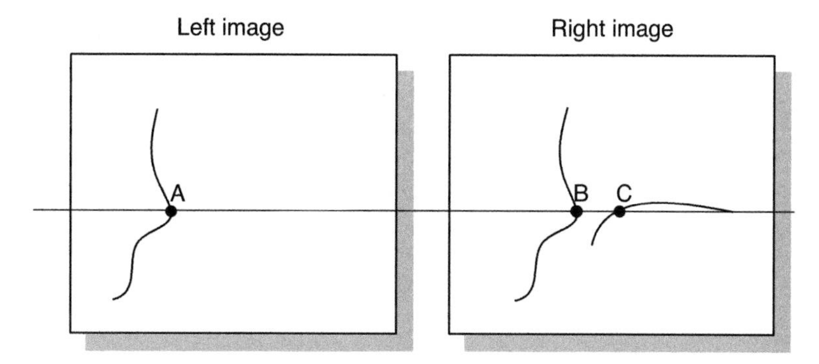 **Figure 3.23** Example of the figural compatibility constraint for proper contour matching Around the points **A** and **B** the contours are very similar. This cannot be observed for the points **A** and **C**. Therefore the first pair will be preferred in this case.
Disparity limit	The disparity limit constraint imposes a global limit on the allowable disparity between images. This can be written as follows: $$\underset{i}{\forall}\underset{\tau}{\exists}\lvert D(\mathbf{p}_{li}, \mathbf{p}_{ri})\rvert < \tau \quad (3.93)$$ where $D(\mathbf{p}_i, \mathbf{p}_j)$ is disparity between points $\mathbf{p}_i$ and $\mathbf{p}_j$ and τ is a threshold value. This constraint is always present in the matching algorithms; however, precise determination of the threshold value is usually not possible.
Disparity gradient limit	The *disparity gradient* concept for the two pairs of point correspondences is given by the following formula [122,162,302]: $$\Gamma(\mathbf{A}, \mathbf{B}) = \frac{D(\mathbf{A}) - D(\mathbf{B})}{G(\mathbf{A}, \mathbf{B})} \quad (3.94)$$

(continued)

Table 3.6 Stereo matching constraints and assumptions (*Continued*)

where $\mathbf{A} = (\mathbf{p}_{lA}, \mathbf{p}_{rA})$ and $\mathbf{B} = (\mathbf{p}_{lB}, \mathbf{p}_{rB})$ represent two pairs of corresponding points, $D(\mathbf{A})$ is a disparity value between points from the pair $\mathbf{A}$, $G(\mathbf{A}, \mathbf{B})$ is a cyclopean distance between pairs of points $\mathbf{A}$ and $\mathbf{B}$. The latter is defined as [246]

$$G(\mathbf{A}, \mathbf{B}) = \left| \frac{\mathbf{p}_{lA} + \mathbf{p}_{rA}}{2}, \quad \frac{\mathbf{p}_{lB} + \mathbf{p}_{rB}}{2} \right|. \tag{3.95}$$

$G(\mathbf{A}, \mathbf{B})$ is a length of the distance between middle points of the segments $\mathbf{p}_{lA}$ and $\mathbf{p}_{rA}$ as well as $\mathbf{p}_{lB}$ and $\mathbf{p}_{rB}$ respectively. This concept is illustrated in Figure 3.24.

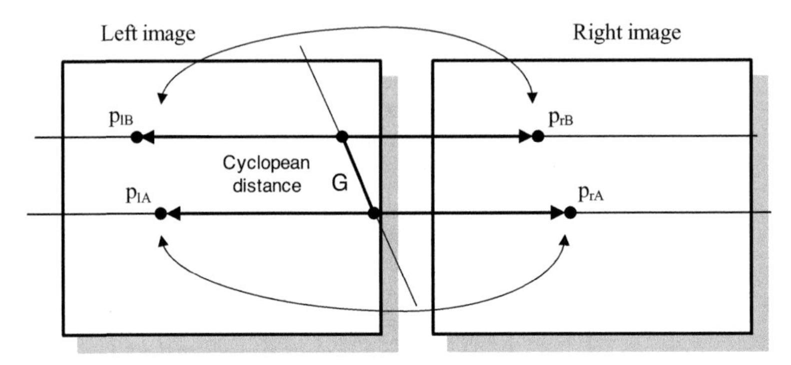

Figure 3.24 Cyclopean distance G

With the help of the aforementioned definitions it is possible to formulate the disparity gradient constraint as follows:

$$\underset{\tau}{\exists} |\Gamma(A, B)| < \tau \tag{3.96}$$

where $\Gamma(\mathbf{A}, \mathbf{B})$ denotes disparity gradient between two pairs $\mathbf{A}$ and $\mathbf{B}$ of matched points and τ is a threshold value (in practice it is in the range 0.5 to 2).

Psychophysical experiments verified that the HVS is limited more by the disparity gradient than by the absolute value of disparity [164, 201, 302, 442].

3.6 Calibration of Cameras

Camera calibration is a process of finding the intrinsic (section 3.3.2.2) and extrinsic (section 3.3.2.1) parameters of a camera or a subset of these. Because camera calibration usually precedes depth reconstruction this subject has attracted great attention among researchers resulting in ample literature, for instance [120, 122, 166, 186, 257, 282, 286, 287, 353, 364, 426, 427, 457].

Special interest has been devoted to the development of fast calibration methods for simple cameras. For instance, one method proposed by Zhang allows camera calibration using a very simple pattern which can be obtained from a laser printer [459].

The influence of the measurement accuracy of the calibration patterns on the accuracy of the computed intrinsic and extrinsic camera parameters was analysed by Lavest *et al.* [268].

The theory and implementation of the iterative algorithm to the precise camera calibration by means of the control circular patterns was presented by Heikkilä [186]. An evaluation of the three common calibration methods of Tsai, Heikkilä and Zhang can be found in the paper by Sun and Cooperstock [407]. It provides practical details as well as serving as an introduction to the field of camera calibration with a brief review of the recent literature on the subject. Also the book by Gruen and Huang provides an overview of the methods for camera calibration [166].

The calibration methods for cameras with long focal length (i.e. telelenses), as well as the methods of creation of the calibration patterns for such optical systems, are discussed by Li and Lavest [275]. Calibration of such cameras is more difficult mostly due to the change of the system parameters in time as well as because the simple pin-hole camera model cannot be applied in this case. Finally, calibration of cameras with wide view angle is analysed by Swaminathan and Nayar [409].

3.6.1 Standard Calibration Methods

The classic calibration methods are based on specially prepared calibration patterns, i.e. objects with known dimensions and position in a certain coordinate system. Then features, such as corners, lines, etc., are extracted from an image of the calibration pattern. Usually the calibration objects are chosen to have prominent features, which are easy for unambiguous localization and measurement of their positions. A simple chessboard can serve this purpose (Figure 3.25).

There is a large number of methods of computation of the internal and external camera parameters. Most of them rely on the already presented formulas (3.3)–(3.13). One such classic method was proposed by Tsai [122, 162, 246, 430]. This method uses (3.7) directly to find out the matrix **M**, denoting the projective transformation performed by a camera. However,

Figure 3.25 Chessboard as a camera calibration pattern

the elements of $\mathbf{M}$ contain linear combinations of the intrinsic $\mathbf{M}_i$ and extrinsic $\mathbf{M}_e$ parameters, according to the formula (3.8). Therefore it is necessary to partition matrix $\mathbf{M}$ into a product $\mathbf{M}_i\mathbf{M}_e$, which can be done analytically as discussed, for instance, in [119, 180]. Considering (3.7) and (3.11) we obtain the conditions joining coordinates of the image points with the coordinates of the observed real point:

$$\begin{aligned} x_u &= \frac{x_{uh}}{z_{uh}} = \frac{p_1}{p_3} = \frac{m_{11}X_1 + m_{12}X_2 + m_{13}X_3 + m_{14}}{m_{31}X_1 + m_{32}X_2 + m_{33}X_3 + m_{34}} \\ y_u &= \frac{y_{uh}}{z_{uh}} = \frac{p_2}{p_3} = \frac{m_{21}X_1 + m_{22}X_2 + m_{23}X_3 + m_{24}}{m_{31}X_1 + m_{32}X_2 + m_{33}X_3 + m_{34}} \end{aligned} \tag{3.97}$$

where (recall (3.5) and (3.11)) x_u and y_u are coordinates of image points expressed in the local coordinate system associated with the camera plane, whereas $\mathbf{P}_{wh} = [X_1, X_2, X_3, 1]^T = [\mathbf{P}_w, 1]^T = [X_w, Y_w, Z_w, 1]^T$, so in effect X_1, X_2, X_3 are 'world' coordinates of an observed 3D point.

Because the matrix $\mathbf{M}$ is given up to a certain scaling factor (3.13), in the general case there are 11 free parameters that have to be determined. They are connected by the formula (3.97). Taking at least six points, the coordinates of which are already known in the external as well as camera coordinate system, we are able to solve (3.97) in respect to the unknown m_{ij}. In practice, using a certain calibration pattern, such as the one presented in Figure 3.25, we obtain more well matched image points. In such a case we have the following set of equations:

$$\mathbf{Qm} = 0, \tag{3.98}$$

where the matrix $\mathbf{Q}$ is composed as follows:

$$\mathbf{Q} = \begin{bmatrix} X_{11} & X_{12} & X_{13} & 1 & 0 & 0 & 0 & 0 & -x_{u1}X_{11} & -x_{u1}X_{12} & -x_{u1}X_{13} & -x_{u1} \\ 0 & 0 & 0 & 0 & X_{11} & X_{12} & X_{13} & 1 & -y_{u1}X_{11} & -y_{u1}X_{12} & -y_{u1}X_{13} & -y_{u1} \\ \cdot & \cdot & \cdot & \cdot & \cdot & \cdot & \cdot & \cdot & \cdot & \cdot & \cdot & \cdot \\ 0 & 0 & 0 & 0 & X_{N1} & X_{N2} & X_{N3} & 1 & -y_{uN}X_{N1} & -y_{uN}X_{N2} & -y_{uN}X_{N3} & -y_{uN} \end{bmatrix}, \tag{3.99}$$

where the coordinates of the k point are denoted as $P_k = [X_{k1}, X_{k2}, X_{k3}]^T$, $p_{uk} = [x_{uk}, y_{uk}]^T$. The vector $\mathbf{m}$ is built as a linear composition of elements of the matrix $\mathbf{M}$:

$$\mathbf{m} = [m_{11}, \ldots, m_{14}, m_{21}, \ldots, m_{24}, m_{31}, \ldots, m_{34}]^T. \tag{3.100}$$

Solution of (3.98) can be done by means of the singular value decomposition $\mathbf{Q} = \mathrm{SVD}^T$ [154, 352]. It is simply a column of $\mathbf{D}$ which corresponds to an entry of $\mathbf{V}$ with a smallest value. SVD has been discussed already for computation of the fundamental matrix (section 3.4.5) and is discussed also in section 4.2.2. However, let us stress that also in this case a proper data normalization is very important to avoid excessive numerical errors. Similarly we have to be sure that there are no outliers in the calibration data.

3.6.2 Photometric Calibration

Most of the image processing methods assume the existence of a photosensor with a linear characteristic of the output signal in respect to the light intensity gathered by the sensor [272]. The following two phenomena concerning real image acquisition systems require proper calibration.

1. *The gamma correction.* Real cameras, although endowed with fairly linear CCD devices, usually contain the so-called gamma correction circuit. It is employed for a proper signal representation for an output display (such as a monitor screen). In such cameras, to obtain an undistorted image, it is necessary to perform an inverse process to the gamma correction. However, this is possible if the parameters of this correction are known beforehand.
2. *Polarization in zero light conditions.* The CCD device generates electrons even if the whole device is situated in a totally dark chamber. This is a thermal generation that causes nonzero output even without any incident light. The level of this signal is called the *black initial level.* Although in most applications this does not cause nonlinearities, it should be taken into consideration when designing a method of image processing.

In stereovision systems an additional photometric calibration is required which consists of equalization of the average amplification level of the two cameras. Such a calibration should eliminate any differences of the intensities in the output stereo images. Otherwise an increase of so-called false matches can be expected, especially if simple matching methods are used (section 6.6).

3.6.3 Self-calibration

Much research has been devoted to answer the question whether it is possible to calibrate a camera solely from image sequences taken by the camera. Solution to this problem, known as camera self-calibration, allows determination of the camera's intrinsic parameters (section 3.3.2.2). Although this is not a full camera calibration, it allows scene reconstruction up to a certain scaling factor (section 7.2.2) which is sufficient in many computer vision applications. Thus, camera self-calibration methods allow computation of the intrinsic camera parameters based on the matches among series of images of the same scene, taken by a single camera, but with changed view parameters, such as camera position (translation, rotation, or both), its focal length or a combination of these [10, 120, 124, 180, 190, 282, 286, 314, 353, 426]. The main advantage of this approach is that a special calibration pattern is not used (section 3.6.1). However, not all methods and camera motions used to take a given sequence can be used for self-calibration [180].

Knowledge of images of the *absolute conics* as well as the *dual absolute conics* (sections 9.4.2.1 and 9.4.2.2), allows determination of the matrix with intrinsic camera parameters with the help of so-called Kruppa equations [180, 380]. Hartley showed a direct relation of the Kruppa equations and elements of the fundamental matrix $\mathbf{F}$ (section 3.4.1.1) [176]. He showed also that knowledge of at least three images, which allows determination of the three fundamental matrices joining mutual pairs of images, is sufficient to find a solution to the Kruppa equations, and finally to determine matrix $\mathbf{M}_i$ of the intrinsic camera parameters (section 3.3.2.2).

Let us assume that the fundamental matrix **F** can be factored as

$$\mathbf{F} = \mathbf{U} \begin{bmatrix} r & 0 & 0 \\ 0 & s & 0 \\ 0 & 0 & 1 \end{bmatrix} \begin{bmatrix} 0 & -1 & 0 \\ 1 & 0 & 0 \\ 0 & 0 & 0 \end{bmatrix} \begin{bmatrix} 0 & 1 & 0 \\ -1 & 0 & 0 \\ 0 & 0 & 1 \end{bmatrix} \mathbf{V}^{\mathbf{T}}. \tag{3.101}$$

Then the Kruppa equations can be expressed by the equations [176]

$$\frac{\mathbf{v}_2^{\mathbf{T}}\mathbf{C}\mathbf{v}_2}{r^2\mathbf{u}_1^{\mathbf{T}}\mathbf{C}\mathbf{u}_1} = \frac{-\mathbf{v}_2^{\mathbf{T}}\mathbf{C}\mathbf{v}_1}{rs\mathbf{u}_1^{\mathbf{T}}\mathbf{C}\mathbf{u}_2} = \frac{\mathbf{v}_1^{\mathbf{T}}\mathbf{C}\mathbf{v}_1}{s^2\mathbf{u}_2^{\mathbf{T}}\mathbf{C}\mathbf{u}_2}, \tag{3.102}$$

where r and s are elements of the **F** factorization (3.101), $\mathbf{u}_i$ stands for an i-th column of the matrix **U**, $\mathbf{v}_i$ is an i-th column of **V** whereas the matrix **C** depends *exclusively* on the matrix $\mathbf{M}_\mathrm{i}$, i.e. the matrix of intrinsic camera parameters. The latter relation is given as

$$\mathbf{C} = \mathbf{M}_\mathrm{i}\mathbf{M}_\mathrm{i}^{\mathbf{T}}. \tag{3.103}$$

As was shown by Hartley, to determine the matrix **C** from (3.102) the three fundamental matrices (each joining a different pair of images) have to be computed. This can be done with one of the methods already discussed in section 3.4.5. What is left is a final factorization of the matrix **C** in accordance with (3.103). This can be done with help of the methods of numerical linear algebra [154, 352].

As alluded to previously the self-calibration methods have their limitations related to the camera positions. Triggs [426], as well as Sturm [402], showed the existence of degenerate camera positions such that Kruppa equations lead to false solutions. One such position is when the optical centres of consecutive camera positions move on the sphere while their optical exes go through a centre of this sphere [402].

The work by Lourakis and Deriche [282] presents an alternative approach to using the Kruppa equations for determination of the intrinsic parameters. There are also other methods [426] for camera self-calibration that do not have internal limitations associated with (3.102). However, they are nonlinear and computationally more complicated [180].

The last but not least issue of camera self-calibration is computation of the fundamental matrices of consecutive views. This process involves matching of consecutive pairs of images, which is a very common step for the majority of methods studied as so far. Therefore this problem is addressed separately in Chapter 6.

3.6.4 Calibration of the Stereo Setup

The problem of a stereo setup calibration consists of determination of the parameters of the two cameras and the two matrices from (3.16). The former can be computed based on the already presented methods in the previous section. The latter concerns computations of the rotation matrix **R**, describing a relative rotation between coordinate systems of the two cameras, and the vector **T** that describes translation of the two camera centres.

Let us assume now that the extrinsic parameters are already known for the two cameras of the stereo system. These are given by four matrices: $\mathbf{R}_\mathrm{l}$ and $\mathbf{T}_\mathrm{l}$ for the left camera, and $\mathbf{R}_\mathrm{r}$ and

$\mathbf{T}_r$ for the right one. Using (3.3), which connects coordinates of a certain 3D point $\mathbf{P}_w$ from an external coordinate system with the camera-related coordinate system, we obtain

$$\begin{aligned} \mathbf{P}_l &= \mathbf{R}_l\left(\mathbf{P}_w - \mathbf{T}_l\right) \\ \mathbf{P}_r &= \mathbf{R}_r\left(\mathbf{P}_w - \mathbf{T}_r\right), \end{aligned} \tag{3.104}$$

where $\mathbf{P}_w$ determines coordinates of a certain 3D point in respect of an external coordinate system, and $\mathbf{P}_l$ and $\mathbf{P}_r$ are coordinates of this point but in the left and right camera coordinate systems, respectively. Finally $\mathbf{R}_l$, $\mathbf{T}_l$, $\mathbf{R}_r$, $\mathbf{T}_r$ are the rotation and translation matrices between an external coordinate system and left and right camera coordinate systems, respectively.

On the other hand, the matrices $\mathbf{P}_l$ and $\mathbf{P}_r$ are related in the stereo system by (3.16). After factoring out $\mathbf{P}_w$ from (3.104) we obtain

$$\mathbf{P}_r = \mathbf{R}_r\mathbf{R}_l^T\left[\mathbf{P}_l - \mathbf{R}_l\left(\mathbf{T}_r - \mathbf{T}_l\right)\right], \tag{3.105}$$

which compared with (3.16) leads to the following relations:

$$\begin{aligned} \mathbf{R} &= \mathbf{R}_r\mathbf{R}_l^T \\ \mathbf{T} &= \mathbf{R}_l\left(\mathbf{T}_r - \mathbf{T}_l\right), \end{aligned} \tag{3.106}$$

where $\mathbf{R}$ and $\mathbf{T}$ are the sought calibration matrices of the stereo system.

3.7 Practical Examples

In this section we discuss some propositions of practical realizations of the concepts already discussed in this chapter. For this purpose two software platforms are used. The first is a software layer of the Hardware Image Library provided with this book [216]. It is written in C++, taking advantage of its features such as objects, components, templates, etc. (Chapter 13). The second platform is Matlab software, commonly used in many scientific and engineering developments.

3.7.1 Image Representation and Basic Structures

Images are many-dimensional arrays of discrete signals. For image processing with computers special models have to be created that represent images as data structures that fit into computer resources. The models should be able to represent different types of images and best if in a uniform way. Fortunately these requirements can be fulfilled using one of the modern programming languages, such as C++ [401]. Thus, the C++ template class mechanism has been used as the main design tool for image models. Such an approach has many benefits which will be clarified in the following sections. The most obvious is the possibility of automatic generation of new types of images based only on a template parameter which is a type of pixel, even for pixel types which will be defined in the future. This makes design very flexible and easily extendable. The other important feature that follows from the philosophy of C++ is that designed this way images will become *strongly typed objects*. This is a very advantageous feature that allows avoidance of many programming errors already at the compilation time.

3.7.1.1 Computer Representation of Pixels

A pixel denotes an atomic element of an image. It is characteristic of a value and an addressable location within the image [157, 351]. Thus a pixel is an entity with attributes of:

- scalar or nonscalar value or values;
- a position in an image.

A value of a pixel can be a single integer, set of integers, positive integers, real or complex numbers, etc., but also a vector of integers or real values or even another image, for instance. In practice, however, the pixel value must be modelled by the best representation available on a given computer platform. The value attribute of a pixel defines a set of allowable operations on an image.

The second attribute, the addressable position of a pixel within the image, is a vector whose dimension reflects the dimension of an image to which the pixel belongs. Due to the nature of display systems, still images are 2D and video signals are 3D. However, there is no obstacle in defining other (higher) dimensional images. Because images are digitized it is a common practice to locate pixels on an integer grid – thus coordinate values of position vectors belong to the set of integers [356]. However, there are image processing techniques that operate on fractional displacements or pixel positions (e.g. disparity maps, optical flow, image warping). In this case pixel positions (or their displacements) need not be integer values any more and should belong to the real domain. Based on the aforementioned analysis we notice that pixels form a discrete or continuous *vector field* which must then be modelled in hardware or software resources based on applications.

In practical computer realization there are many different types of pixel values, such as monochrome pixels (represented usually with 8–10 bits), colour pixels (e.g. $3 \times$ 8–10 bits), but also fixed and floating point values, etc. On the other hand, choice of a data structure representing pixels has a crucial effect on robustness of the computations since for instance access and processing of the floating point data is much more complex than it is for simple bytes. Therefore the chosen computer representation should fit as closely as possible the physical values of pixels.

In the presented software platform the following data models are used to represent pixels.

1. For scalar pixels (i.e. one value) the C++ built-in types (e.g. *unsigned char*, *int*, *long*, etc.).
2. For static length vector valued pixels (i.e. more than one value for a pixel but addressed linearly) the template class parameterized by a type of a single coefficient and number of such coefficients for a pixel (i.e. pixel depth).
3. For multidimensional valued pixels custom data structures.

Algorithm 3.2[11] presents the basic implementation of the *MMultiPixelFor<>* class for representation of static length vector valued pixels. Pixel values are stored in the *fData* array. This is shown to be faster or equal in run-time performance than a simple structure with a separate data member for each pixel coefficient (e.g. *struct {char a; char b; char c;};*). At the same time, an array in C and C++ (and other languages) allows uniform algorithmic access when extending the class to higher dimensions, or manipulating pixels with different

[11] A discussion of programming techniques is provided in Chapter 13.

```
// Multi-pixel models a pixel with "Channels" values of a type "P".
template< typename P, int Channels = kDefaultChannels >
class MMultiPixelFor
{
	public:

		enum { kChannels = Channels };
		typedef P	PixelClusterArray[ Channels ];
		PixelClusterArray fData;

	public:
		// A default constructor (all to 0)
		MMultiPixelFor( void );

		// Constructor
		explicit MMultiPixelFor( const P * data );

		MMultiPixelFor( const P singleVal );

		// Copy constructors
		template< class R >
		MMultiPixelFor( const MMultiPixelFor< R, kChannels > & r );

		template<>		// let's make a specialization for its own type
		MMultiPixelFor( const MMultiPixelFor & r );

		// Assignement
		template< class R >
		MMultiPixelFor & operator
				= (const MMultiPixelFor< R, kChannels > & r );

		template<>		// let's make a specialization for its own type
		MMultiPixelFor & operator = ( const MMultiPixelFor & r );

	public:
		bool operator == ( const MMultiPixelFor & r );

		//////////////////////////////////////////////////////
		MMultiPixelFor & operator += ( const MMultiPixelFor & r );
		MMultiPixelFor & operator -= ( const MMultiPixelFor & r );
		MMultiPixelFor & operator *= ( const MMultiPixelFor & r );
		MMultiPixelFor operator + ( const MMultiPixelFor & r );
		MMultiPixelFor operator - ( const MMultiPixelFor & r );
		// Point-wise multiplication
		MMultiPixelFor operator * ( const MMultiPixelFor & r );
		//////////////////////////////////////////////////////

		MMultiPixelFor & operator += ( const P val );
		MMultiPixelFor & operator -= ( const P val );
		// Point-wise multiplication
		MMultiPixelFor & operator *= ( const P val );
		MMultiPixelFor operator + ( const P val );
		MMultiPixelFor operator - ( const P val );
		// Point-wise multiplication
		MMultiPixelFor operator * ( const P val );
};
```

Algorithm 3.2 The template class *MMultiPixelFor<>* which models all types of nonscalar valued pixels. The class is parameterized by pixel coefficient type and number of coefficients. (Reproduced by permission of Pandora Int. Inc., London)

dimensions. There is no easy way of enumerating member fields of *struct*. There are three main groups of function-members in the *MMultiPixelFor<>*.

1. Construction and assignment.
2. Arithmetic operations among pixel objects.
3. Arithmetic operations among pixel object and a scalar value.

The above define basic functionality of the pixel objects. This can be interpreted also in terms of vector operations.

There are also different specializations for most common pixel types: e.g. three-channel monochrome and single monochrome representation. The purpose of specialization is to provide an implementation trimmed to a particular data type. In the case of three channels it appears to be faster to address all channels directly than in a software loop, for instance.

3.7.1.2 Representation of Images

A flexible data structure for image representation is crucial for efficient image processing. In this design the data structures used for image representation were implemented in the form of template classes. There is a tradeoff between different input formats of images and their internal representation. Further, considering allowable size of images and time complexity of algorithms, it was decided to represent images as square matrices, programmatically denoted by the base template class *TImageFor<T>*, where *T* stands for a given data type chosen for representation of a single pixel. Such representation was also chosen to fit requirements for the envisaged cooperation of the software layer with the hardware acceleration boards which require frequent DMA transfers of the whole structure to and from the operating memory of the computer. Figure 3.26 presents the basic class hierarchy proposed to represent digital images.

The heart of the hierarchy presented in Figure 3.26 is the base template class *TImageFor<>*. It defines the interface for all images used in the library. The class is parameterized by pixel type. For different pixels we obtain different instantiations with the same semantics. So, we can easily create images of bytes, real value representation, colour pixels, etc., by defining the following types:

```
typedef TImageFor< unsigned char >                          MonochromeImage;
typedef TImageFor< double >                                 RealImage;
typedef TImageFor< MMultiPixelFor< unsigned char, 3 > >     ColorImage;
```

Single pixels can be represented by built-in types or by system defined classes, as shown in Figure 3.26. For special pixels separate classes have to be defined. These are: *MMultiPixelFor<>*, which stands for pixels in a form of vectors, and *FixedFor<>*, which add custom fixed-point type.

From the base *TImageFor<>* the three classes have been derived. These are the following.

1. The *TDanglingImageFor<>* which models special images with pixels being vectors, possibly of different length (this is in contrast to the already discussed *MMultiPixelFor<>*).
2. The *TProxyImageFor<>* which implements the proxy pattern embedded into a base image. Objects of this class are used to model regions-of-interest in images. In other words, the

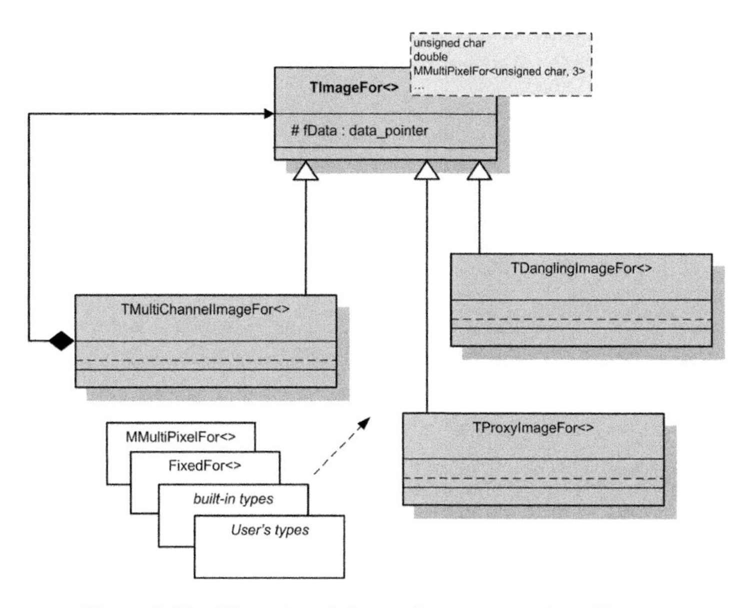

Figure 3.26 Hierarchy of classes for representation of images

proxy-image objects allow one to treat part of an image as a separate image with its own local coordinate system (kind of local discrete manifold), although there is only one set of 'real' pixels (see also section 13.2.7).

3. The *TMultiChannelImageFor<>* for multichannel images, i.e. images which are composed of a number of simpler type images. Objects of this class model for instance noninterlaced colour images.

Definition of the class *TImageFor<>*, with the most important members of its interface highlighted, is presented in Algorithm 3.3. Its template parameter defines a type of pixel for that image (extension to this is discussed in section 13.3.5). The internal representation of an image is a 2D matrix (an array). There are three distinct groups of members of the interface.

1. Constructors, used to create images based on their dimensions and initial values of pixels.
2. Pixel access routines (*GetPixel*, *SetPixel* and their reference-based counterparts).
3. Image operators in a form of a set of overloaded operators.

Interestingly, pixels can also be other images with their own pixel types. This is the concept behind template images [356], an example of which is depicted in Figure 3.27(a). In terms of the already introduced mechanisms template images can be created directly from the *TImageFor<>* base, providing its template parameter being another *TImageFor<>*, and so on. An example for single byte pixels is presented below:

```
typedef TImageFor< MonochromeImage > MonochromeImageImage;
// an image with pixels being... images
```

```
////////////////////////////////////////////////////////////////////////
// The basic structure for images.
//
template< class T >
class TImageFor
{
public:

        typedef typename T PixelType;

protected:

    Dimension fRow;            // contains number of rows of this image

    Dimension fCol;            // contains number of columns of this image

    T * fData;                 // pointer to the pixels

public:

    // Default constructor.
    TImageFor( void ) : fRow( 0 ), fCol( 0 ), fData( 0 )
    {
    }

public:

    //========================================

    ////////////////////////////////////////////////////////////
    // Class constructor
    ////////////////////////////////////////////////////////////
    //
    // INPUT:
    //        col - number of columns
    //        row - number of rows
    //
    // OUTPUT:
    //
    //
    // REMARKS:
    //        Memory for data is allocated but
    //        data is NOT initialized.
    //
    TImageFor( Dimension col, Dimension row );

    ////////////////////////////////////////////////////////////
    // Class constructor
    ////////////////////////////////////////////////////////////
    //
    // INPUT:
    //        col - number of columns
    //        row - number of rows
    //        init_val - initial value for each pixel
    //
    // OUTPUT:
    //
    //
    // REMARKS:
    //
    //
    TImageFor( Dimension col, Dimension row, const T init_val );
```

Algorithm 3.3 Template base class *TImageFor<>* for computer representation of images. The template parameter determines the type of pixels. (Reproduced with permission of Pandora Int. Inc., London)

```
///////////////////////////////////////////////////////////////
// Template copy constructor - mixed copy constructor
///////////////////////////////////////////////////////////////
template < typename U >
TImageFor( const TImageFor< U > & ref );

///////////////////////////////////////////////////////////////
// Copy constructor
///////////////////////////////////////////////////////////////
TImageFor( const TImageFor< T > & ref );

///////////////////////////////////////////////////////////////
// Template assignment operator (mixed copy)
///////////////////////////////////////////////////////////////
template < class U >
TImageFor< T > & operator = ( const TImageFor< U > & ref );

///////////////////////////////////////////////////////////////
// An assignment operator
///////////////////////////////////////////////////////////////
TImageFor< T > & operator = ( const TImageFor< T > & ref );

//========================================

///////////////////////////////////////////////////////////////
// Overloaded binary operators.
///////////////////////////////////////////////////////////////
//
// INPUT:
//        image - a constant reference to the second image
//
// OUTPUT:
//        result image (by reference or a local copy,
//           what should be avoided)
//
// REMARKS:
//
//

TImageFor< T > operator + ( const TImageFor< T > & image ) const;
TImageFor< T > & operator += ( const TImageFor< T > & image );
TImageFor< T > operator - ( const TImageFor< T > & image ) const;
TImageFor< T > & operator -= ( const TImageFor< T > & image );
TImageFor< T > operator * ( const TImageFor< T > & image ) const;
TImageFor< T > & operator *= ( const TImageFor< T > & image );
TImageFor< T > operator / ( const TImageFor< T > & image ) const;
TImageFor< T > & operator /= ( const TImageFor< T > & image );

// Returns true if the two pictures are the same
bool       operator == ( const TImageFor< T > & image ) const;

//========================================

///////////////////////////////////////////////////////////////
// This function sets a pixel at position (x,y) or (col,row)
// of this image.
///////////////////////////////////////////////////////////////
//
// INPUT:
//      xPixPosition - the horizontal (or column) position of a pixel
//      yPixPosition - the vertical (or row) position of a pixel
//      value - a value to be set at pixel position
//
// OUTPUT:
//      none
//
// REMARKS:
//      From the OOP point of view this function should be virtual.
//      However, to avoid run-time panalty it is not virtual.
//
void SetPixel( Dimension xPixPosition, Dimension yPixPosition,
               const Tvalue ) const;
```

Algorithm 3.3 *(Continued)*

```
//////////////////////////////////////////////////////////////
// This function sets a pixel at position (x,y) or (col,row)
// of this image.
//////////////////////////////////////////////////////////////
//
// INPUT:
//     xPixPosition - the horizontal (or column) position of a pixel
//     yPixPosition - the vertical (or row) position of a pixel
//     value - a value to be set at pixel position,
//        passed by REFERENCE!
//
// OUTPUT:
//        none
//
// REMARKS:
//     From the OOP point of view this function should be virtual.
//     However, to avoid run-time panalty it is not virtual.
//
void SetRefPixel( Dimension xPixPosition, Dimension yPixPosition,
        const T & value ) const;

//////////////////////////////////////////////////////////////
// This function gets a VALUE of a pixel at position
// (x,y) or (col,row) of this image.
//////////////////////////////////////////////////////////////
//
// INPUT:
//     xPixPosition - the horizontal (or column) position of a pixel
//     yPixPosition - the vertical (or row) position of a pixel
//
// OUTPUT:
//     a copy of a pixel, of type T, from the given position
//
// REMARKS:
//     The xPixPosition should span from 0 to max_columns-1, while
//     the yPixPosition from 0 to max_rows-1.
//
//     From the OOP point of view this function should be virtual.
//     However, to avoid run-time penalty it is not virtual.
//
T GetPixel( Dimension xPixPosition, Dimension yPixPosition ) const ;

//////////////////////////////////////////////////////////////
// This function gets a REFERENCE to a pixel at position
// (x,y) or (col,row) of this image.
//////////////////////////////////////////////////////////////
//
// INPUT:
//     xPixPosition - the horizontal (or column) position of a pixel
//     yPixPosition - the vertical (or row) position of a pixel
//
// OUTPUT:
//     a reference to a pixel, of type T, from the given position
//
// REMARKS:
//     The xPixPosition should span from 0 to max_columns-1, while
//     the yPixPosition from 0 to max_rows-1.
//
//     From the OOP point of view this function should be virtual.
//     However, to avoid run-time penalty it is not virtual.
//
T & GetRefPixel( Dimension xPixPosition, Dimension yPixPosition );
///////////////
```

Algorithm 3.3 *(Continued)*

```
   ////////////////////////////////////////////////////////////
   // This function sets a pixel at position (x,y)or (col,row)
   // of this image.
   ////////////////////////////////////////////////////////////
   //
   // INPUT:
   //     xPixPosition - the horizontal (or column) position of a pixel
   //     yPixPosition - the vertical (or row) position of a pixel
   //     value - a value to be set at pixel position
   //
   // OUTPUT:
   //     none
   //
   // REMARKS:
   //     All positive input values are allowed for a pixel position.
   //
   void SetPixel_Modulo ( Dimension xPixPosition, Dimension yPixPosition,
              const T value ) const ;

   ////////////////////////////////////////////////////////////
   // This function gets a VALUE of a pixel at position
   // (x,y) or (col,row) of this image.
   ////////////////////////////////////////////////////////////
   //
   // INPUT:
   //     xPixPosition - the horizontal (or column) position of a pixel
   //     yPixPosition - the vertical (or row) position of a pixel
   //
   // OUTPUT:
   //     a copy of a pixel, of type T, from the given position
   //
   // REMARKS:
   //     All positive input values are allowed for a pixel position.
   //
   T GetPixel_Modulo( Dimension xPixPosition, Dimension yPixPosition );

   ////////////////////////////////////////////////////////////
   // This function gets a REFERENCE to a pixel at position
   // (x,y) or (col,row) of this image.
   ////////////////////////////////////////////////////////////
   //
   // INPUT:
   //     xPixPosition - the horizontal (or column) position of a pixel
   //     yPixPosition - the vertical (or row) position of a pixel
   //
   // OUTPUT:
   //     a reference to a pixel, of type T, from the given position
   //
   // REMARKS:
   //     All positive input values are allowed of a pixel position.
   //
   T & GetRefPixel_Modulo( Dimension xPixPosition,
                Dimension yPixPosition );

   //============================================================


   Dimension GetRow( void ) const { return fRow; }
   Dimension GetCol( void ) const { return fCol; }
};
```

Algorithm 3.3 *(Continued)*

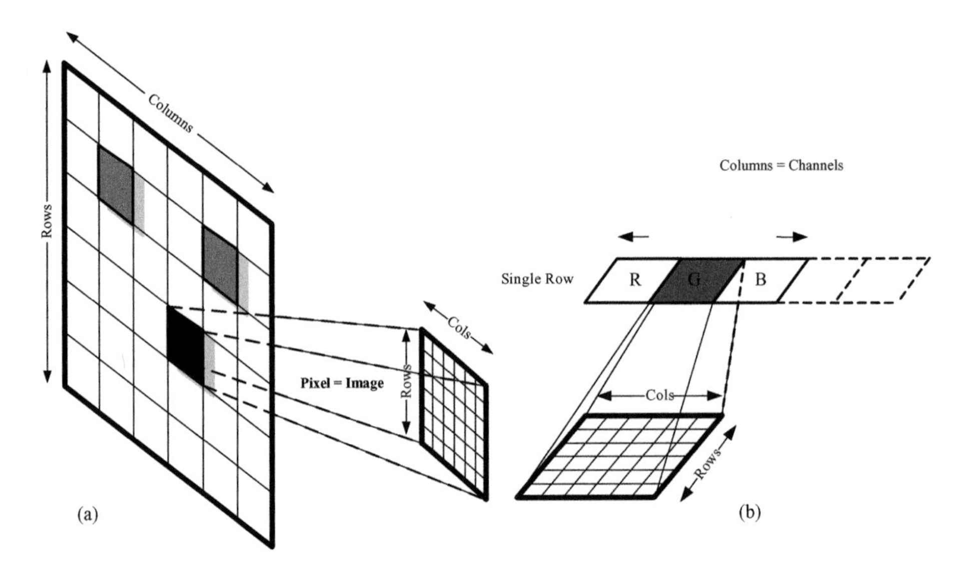

Figure 3.27 Explanation of template images, i.e. images with pixels which are also images

However, very frequently we are interested in template images with a *linear* number of channels (Figure 3.27(b)). To meet this requirements a special version of the template images was created. This is a *TMultiChannelImageFor<>* template class which is derived from the base *TImageFor<>*, for which a pixel type has been set to be another image, i.e. a type of *TImageFor<>*, and so on. Attention should be paid to the *SetPixel* and *GetPixel* members which, in contrast to the directly created template images, operate on multichannel pixels rather than on pixel images. However, the pixel images (i.e. elements of the channels) can also be accessed by calling *SetPixel* and *GetPixel* of the *base* class.

For most applications only three channels are required which model RGB or HSI colour images, for instance (thus the default template parameter, denoting number of channels, is three).

However, multichannel images are not just any collection of channels (or images). The important feature is that all images must be of the same size. Moreover, the elements (pixels) located at the same position in each channel compose a multichannel pixel which can be accessed at once. Thus, the concept of multichannel images can be visualized as presented in Figure 3.28.

Each multichannel pixel in Figure 3.28 can be accessed providing the number of co-ordinates is the same as the number of channels. Definition of the image template class *TMultiChannelImageFor<>* is presented in Algorithm 3.4 (*kDefaultChannels* is set to three).

The semantics of the *SetPixel* and *GetPixel* methods is different from the corresponding methods in the base *TImageFor<>* class. In the derived class they operate on multichannel pixels, whereas in the base class they refer to the base-class pixels which are images – i.e. the channels that constitute the derived class.

Real implementation of the aforementioned classes is endowed with a mechanism of class traits which allows definition of specific class behaviour (such as accessing objects by reference or by value) depending on a concrete type of pixel. These issues are treated in section 13.3.5.

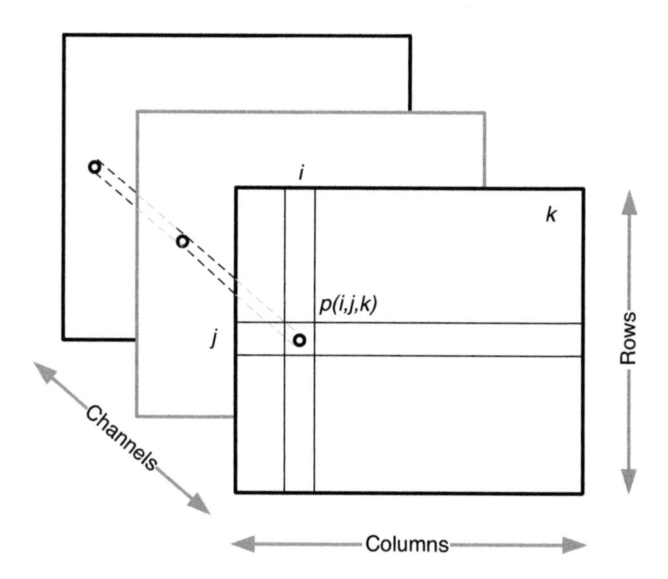

Figure 3.28 Scheme of a multichannel image

```
///////////////////////////////////////////////////////////////////////////
// This type of images can be used for multi-channel images with
// non-interlaced pixels. Implemented as a template-image pattern
// (i.e. image with pixels being images)
///////////////////////////////////////////////////////////////////////////
template< class T, int Channels = kDefaultChannels >
class TMultiChannelImageFor : public TImageFor< TImageFor< T > >
{
	public:

		// For each instantiation of the enclosing class
		// there will be its own type of MMultiPixel.

		enum { kChannels = Channels };

		typedef MMultiPixel< T, Channels >	MultiPixel;

		typedef TImageFor< T >			SingleChannelImage;

		typedef TImageFor< TImageFor< T > >	BaseClass;

	public:

		//========================================

		TMultiChannelImageFor( Dimension col, Dimension row );

		TMultiChannelImageFor( Dimension col, Dimension row,
					const MultiPixel & init_val );
```

Algorithm 3.4 Definition of the *TMultiChannelImageFor<>* class representing multichannel images. (Reproduced with permission of Pandora Int. Inc., London)

```
public:

    ////////////////////////////////////////////////////////////
    // This function sets a multi-pixel "multPixel"
    // at position (x,y) or (col,row) of this multi-channel image.
    // The multi-pixel is conveyed by reference thus there
    // is no "reference version" (i.e. SetRefPixel) of this function.
    ////////////////////////////////////////////////////////////
    //
    // INPUT:
    //      xPixPosition - the horizontal (or column) position of a pixel
    //      yPixPosition - the vertical (or row) position of a pixel
    //      multPixel - a reference to the multi-pixel value to be
    //      set at pixel position
    //
    // OUTPUT:
    //                     none
    //
    // REMARKS:
    // There is a question of passing a mutli-pixel argument either by
    // reference of by value. In the base implementation of TImageFor<>
    // the latter was chosen, and the second function SetRefPixel also
    // supplied. In this class a passing by reference was chosen
    // already for SetPixel and SetRefPixel is not supplied
    // However, this can be changed (e.g. by creating a derived
    // version)after proper PROFILING and measuring performance in
    // both cases.
    //
    void   SetPixel(   Dimension   xPixPosition,   Dimension   yPixPosition,
                   const MultiPixel & multPixel ) const ;

    ////////////////////////////////////////////////////////////
    // This function gets a multi-pixel "multPixel"
    // at position (x,y)or (col,row) of this multi-channel image.
    // The multi-pixel is conveyed by value (a copy is created).
    ////////////////////////////////////////////////////////////
    //
    // INPUT:
    //      xPixPosition - the horizontal (or column) position of a pixel
    //      yPixPosition - the vertical (or row) position of a pixel
    //
    // OUTPUT:
    //      multPixel - a multi-pixel value from pixel position
    //
    // REMARKS:
    //      The xPixPosition should span from 0 to max_columns-1, while
    //                  the yPixPosition from 0 to max_rows-1.
    //
    //
    MultiPixel GetPixel( Dimension xPixPosition, Dimension yPixPosition
                          ) const;
```

Algorithm 3.4 *(Continued)*

```
    public:

        ///////////////////////////////////////////////////////////////
        // Overloaded binary operators.
        ///////////////////////////////////////////////////////////////
        //
        // INPUT:
        //              image - a constant reference to the second
        //                      multi-channel image
        //
        // OUTPUT:
        //              result image (by reference or a local copy,
        //                      what should be avoided)
        //
        // REMARKS:
        //              Each operation is performed on each channel
        //              separately. For each channel, action is delegated
        //              to the base implementation of the corresponding
        //              operator.
        //
        //              Additional operators can be added based
        //              on the supplied ones, either changing this
        //              class, or (presumably better) by deriving a new class.
        //

        // Returns true if the two pictures are the same
        bool operator == ( const TMultiChannelImageFor< T, Channels > &
                        image );

        // .........
};
```

Algorithm 3.4 *(Continued)*

3.7.1.3 Image Operations

Having defined image representations it is possible to provide some operations on them. Figure 3.29 presents a class hierarchy for this purpose. *TImageOperation* is the base class for all operations. It defines the common function operator which is extended in derived classes (section 13.3.1). There are four major derived classes that define each type of image operations and operation compositions. The most general template solution was chosen that allows for *any* type to be supplied for arguments of an operation. Wherever possible all parameters of operators are treated as images. So, an image is a more ample notion than a classic 'visible' image. For example, an image can store in its pixels a value of a just found maximum pixel in another image, as well as x and y coordinates of that pixel, as its next pixels. This is analogous to a matrix-processing context where each value is treated as a matrix.

It is interesting to notice that the base *TImageOperation* is a pure virtual class (i.e. it can serve only to be derived from, no objects of this class allowed) but it is not a template class, whereas its derived classes are (Algorithm 3.5).

In the class hierarchy (Figure 3.29) there are two types of classes.

1. The base operations of the library (shaded shapes) which consists of the following pure virtual classes: *TImageOperation*, *TUnaryImageOperationFor*, *TBinaryImageOperationFor*, *TImageTemplateOperationFor* and *TComposedImageOperationFor*.
2. The classes for specific image library operations; for example, *FindMaxVal_OperationFor*, *FormatConvert_OperationFor*, *Add_OperationFor*, *_2D_Convolve_OperationFor* and *_Horz_Convolve_OperationFor*.

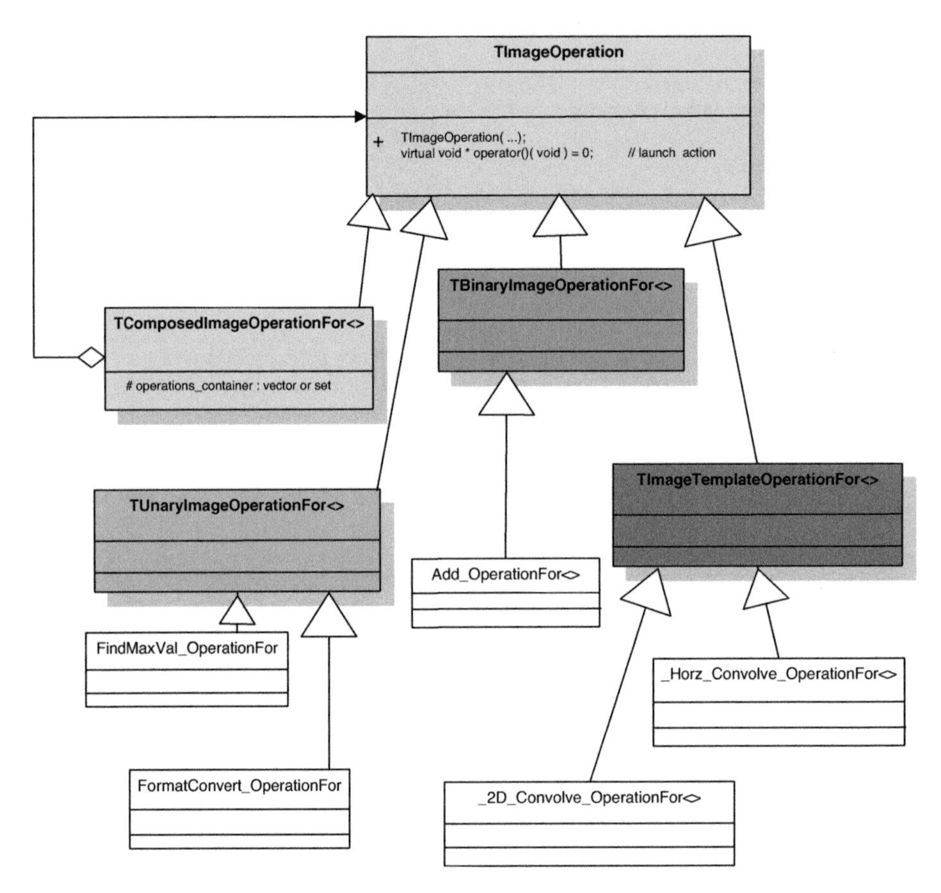

Figure 3.29 Class hierarchy for the image operators. *TImageOperation* is the base class that provides a common functionality which is then specialized in the derived classes. There are four major derived classes

The skeleton of the hierarchy of image operations is composed of the following classes.

1. The *TImageOperation* class: its responsibility is to define a common interface for all operations. The most important part of its interface consists of the pure virtual overloaded functional operator in the following form:

```
virtual void * operator()( void ) = 0;
```

   This is a common operator for all image operations. For the virtual functions it is required that its declaration is the same throughout the whole hierarchy. Therefore all the necessary parameters of specific operations are supplied to the constructors of their classes. This is a common strategy assumed in the whole library.

2. The *TUnaryImageOperationFor* class: its responsibility is to compose a branch of specific unary image operations, i.e. such operations that need only one image as input (Figure 3.29).
3. The *TBinaryImageOperationFor* class: its responsibility is to start a branch of specific binary image operations, i.e. such operations that need two images of the same size as input (Figure 3.29).

```
class TImageOperation
{
  protected:

    TThreadSecurity &                 fResourceAccessPolicy;
    TOperationCompletionCallback &    fOperationCompletionCallback;

  public:

    ///////////////////////////////////////////////////////////
    // Base class constructor
    ///////////////////////////////////////////////////////////
    //
    // INPUT:
    // resourceAccessPolicy - optional reference to
    //    the thread security object (derivative
    //    of the TThreadSecurity class); by default
    //    the static kgThreadSecurity object is supplied
    //    which does nothing
    // opCompCallback - optional reference to the callback
    //    object which is called upon completion of operation;
    //    by default the static kgOperationCompletionCallback
    //    object is supplied which does nothing
    //
    // OUTPUT:
    //
    //
    // REMARKS:
    //
    //
    TImageOperation ( TThreadSecurity & resourceAccessPolicy
                        = kgThreadSecurity,
                      TOperationCompletionCallback opCompCallback &
                        = kgOperationCompletionCallback );
  protected:

    ///////////////////////////////////////////////////////////
    // This function should be called at the beginning
    // of each operator()
    ///////////////////////////////////////////////////////////
    //
    // INPUT:
    //        none
    //
    // OUTPUT:
    //        none
    //
    // REMARKS:
    //        It calls resource access function of the
    //        supplied thread security object
    //
    virtual void operator_begin( void );
```

Algorithm 3.5 Definition of the pure virtual *TImageOperation* class. It is a root class for all other image operations. (Reproduced with permission of Pandora Int. Inc., London)

```
   ////////////////////////////////////////////////////////////
   // This function should be called at the end
   // of each operator()
   ////////////////////////////////////////////////////////////
   //
   // INPUT:
   //       none
   //
   // OUTPUT:
   //       none
   //
   // REMARKS:
   //       It calls resource release function of the
   //       supplied thread security object
   //       and the callback
   //
   virtual void operator_end( void );

  ///////////////////////////////////////////////////////////////////////
  // A helper class...
  // Just create a local object of this class and the operator_begin()
  // operator_end() will be called automatically due to auto object
  // semantics of C++
  ///////////////////////////////////////////////////////////////////////
   class MImageOperationRetinue
   {
      private:

         TImageOperation & fImageOperation;

         void * operator new ( size_t );
   // make the operator new private to disable creation on the heap

      public:

         MImageOperationRetinue( TImageOperation & imOper ) :
            fImageOperation( imOper )
         {
            fImageOperation.operator_begin();
            // resource acquisition is initialization
         }
         ~MImageOperationRetinue()
         {
            fImageOperation.operator_end();
            // destructor de-initializes automatically
         }
   };

   friend class MImageOperationRetinue;

public:

      ////////////////////////////////////////////////////////////
      // The function operator which - in a derived class -
      // defines an image operation.
      ////////////////////////////////////////////////////////////
      //
      // INPUT:
      //       none
      //
      // OUTPUT:
      //       user defined (in a derived class) void pointer
      //
      // REMARKS:
      //       The necessary input parameters should be supplied
      //       to an appropriate CONSTRUCTOR of a derived class.
      //
      virtual void * operator()( void ) = 0;
};
```

Algorithm 3.5 *(Continued)*

4. The *TImageTemplateOperationFor* class: it is a specialization of the *TImageOperation* class where the second image is assumed to be an image template (Figure 3.29).
5. The *TComposedImageOperationFor* class: this class creates a specific composite – it is a kind of image operation that is built up from the other image operations, i.e. those that are derived from the base *TImageOperation* class (Figure 3.29). Composite is discussed in section 13.3.3.

The multiparameter template technique used allows flexible creation of specific operations with strict type checking of its input arguments. This way, an operation defined for a given type of input image constitutes a type different from the same operation but defined for different input images.

3.8 Appendix: Derivation of the Pin-hole Camera Transformation

Let us rewrite Equations (3.3)

$$\mathbf{P}_{\mathrm{c}} = \mathbf{R}(\mathbf{P}_{\mathrm{w}} - \mathbf{T}), \tag{3.107}$$

then (3.5)

$$\begin{aligned} x_{\mathrm{c}} &= (x_u - o_{ux})h_x \\ y_{\mathrm{c}} &= (y_u - o_{uy})h_y, \end{aligned} \tag{3.108}$$

and finally (3.2)

$$x_{\mathrm{c}} = f\frac{X_{\mathrm{c}}}{Z_{\mathrm{c}}}, \quad y_{\mathrm{c}} = f\frac{Y_{\mathrm{c}}}{Z_{\mathrm{c}}}, \quad z_{\mathrm{c}} = f \tag{3.109}$$

with proper coordinate systems explicitly indicated by a subscript letter: 'w' for world coordinates, 'c' for camera coordinates and 'u' for coordinates associated with the local camera plane. An additional subscript 'h' denotes homogeneous coordinates (section 9.2) in contrast to Cartesian ones. We wish to relate world coordinates of a 3D point $\mathbf{P}_{\mathrm{w}}$ with its image point $\mathbf{p}_{\mathrm{u}}$ on a camera plane. For this purpose let us write (3.107) as

$$\mathbf{R}(\mathbf{P}_{\mathrm{w}} - \mathbf{T}) = \begin{bmatrix} \mathbf{R}_1 \\ \mathbf{R}_2 \\ \mathbf{R}_3 \end{bmatrix} [\mathbf{P}_{\mathrm{w}} - \mathbf{T}] = \begin{bmatrix} \mathbf{R}_1 (\mathbf{P}_{\mathrm{w}} - \mathbf{T}) \\ \mathbf{R}_2 (\mathbf{P}_{\mathrm{w}} - \mathbf{T}) \\ \mathbf{R}_3 (\mathbf{P}_{\mathrm{w}} - \mathbf{T}) \end{bmatrix} = \begin{bmatrix} X_{\mathrm{c}} \\ Y_{\mathrm{c}} \\ Z_{\mathrm{c}} \end{bmatrix} = \mathbf{P}_{\mathrm{c}}, \tag{3.110}$$

where $\mathbf{R}_i$ denotes the i-th row of the matrix $\mathbf{R}$ (i.e. it is a row vector of dimensions 1×3). Inserting the above into (3.109) the following set of equations is obtained:

$$\begin{cases} x_{\mathrm{c}} = f\dfrac{\mathbf{R}_1 (\mathbf{P}_{\mathrm{w}} - \mathbf{T})}{\mathbf{R}_3 (\mathbf{P}_{\mathrm{w}} - \mathbf{T})} \\ y_{\mathrm{c}} = f\dfrac{\mathbf{R}_2 (\mathbf{P}_{\mathrm{w}} - \mathbf{T})}{\mathbf{R}_3 (\mathbf{P}_{\mathrm{w}} - \mathbf{T})} \\ z_{\mathrm{c}} = f \end{cases}. \tag{3.111}$$

Observe that $\mathbf{R}_i(\mathbf{P}_w - \mathbf{T})$ is a scalar value. Now, inserting (3.111) into (3.108) we find coordinates of a point expressed in a local coordinate system of the camera's plane:

$$\begin{cases} x_u = \dfrac{f}{h_x}\dfrac{\mathbf{R}_1(\mathbf{P}_w - \mathbf{T})}{\mathbf{R}_3(\mathbf{P}_w - \mathbf{T})} + o_{ux} \\ y_u = \dfrac{f}{h_y}\dfrac{\mathbf{R}_2(\mathbf{P}_w - \mathbf{T})}{\mathbf{R}_3(\mathbf{P}_w - \mathbf{T})} + o_{uy} \end{cases}. \tag{3.112}$$

The above set of *two* equations for (x_u, y_u) can be extended into a set of *three* equations for the homogeneous coordinates (x_{uh}, y_{uh}, z_{uh}):

$$\begin{cases} x_{uh} = x_u \mathbf{R}_3(\mathbf{P}_w - \mathbf{T}) = \dfrac{f}{h_x}\mathbf{R}_1(\mathbf{P}_w - \mathbf{T}) + o_{ux}\mathbf{R}_3(\mathbf{P}_w - \mathbf{T}) \\ y_{uh} = y_u \mathbf{R}_3(\mathbf{P}_w - \mathbf{T}) = \dfrac{f}{h_y}\mathbf{R}_2(\mathbf{P}_w - \mathbf{T}) + o_{uy}\mathbf{R}_3(\mathbf{P}_w - \mathbf{T}) \\ z_{uh} = \mathbf{R}_3(\mathbf{P}_w - \mathbf{T}) \end{cases}. \tag{3.113}$$

Thanks to this transformation a nonlinearity due to division is avoided at the cost of an additional coordinate – this is the main idea behind homogeneous coordinates. From (3.113) we easily observe that

$$x_u = \frac{x_{uh}}{z_{uh}}, \quad y_u = \frac{y_{uh}}{z_{uh}}. \tag{3.114}$$

It is now easy to see that (3.113) can be rewritten as

$$\mathbf{p}_{uh} = \begin{bmatrix} x_{uh} \\ y_{uh} \\ z_{uh} \end{bmatrix} = \begin{bmatrix} \frac{f}{h_x} & 0 & o_{ux} \\ 0 & \frac{f}{h_y} & o_{uy} \\ 0 & 0 & 1 \end{bmatrix} \begin{bmatrix} \mathbf{R}_1(\mathbf{P}_w - \mathbf{T}) \\ \mathbf{R}_2(\mathbf{P}_w - \mathbf{T}) \\ \mathbf{R}_3(\mathbf{P}_w - \mathbf{T}) \end{bmatrix} = \begin{bmatrix} \frac{f}{h_x} & 0 & o_{ux} \\ 0 & \frac{f}{h_y} & o_{uy} \\ 0 & 0 & 1 \end{bmatrix} \begin{bmatrix} \mathbf{R}_1 & -\mathbf{R}_1\mathbf{T} \\ \mathbf{R}_2 & -\mathbf{R}_2\mathbf{T} \\ \mathbf{R}_3 & -\mathbf{R}_3\mathbf{T} \end{bmatrix} \underbrace{\begin{bmatrix} \mathbf{P}_w \\ 1 \end{bmatrix}}_{\mathbf{P}_{wh}},$$

and after some rearrangements

$$\mathbf{p}_{uh} = \underbrace{\begin{bmatrix} \frac{f}{h_x} & 0 & o_{ux} \\ 0 & \frac{f}{h_y} & o_{uy} \\ 0 & 0 & 1 \end{bmatrix}}_{\mathbf{M}_i} \underbrace{\begin{bmatrix} \mathbf{R}_1 & -\mathbf{R}_1\mathbf{T} \\ \mathbf{R}_2 & -\mathbf{R}_2\mathbf{T} \\ \mathbf{R}_3 & -\mathbf{R}_3\mathbf{T} \end{bmatrix}}_{\mathbf{M}_e} \mathbf{P}_{wh}. \tag{3.115}$$

Let us also notice the assumed orientations of the coordinate systems in Figure 3.6 which is left-handed. This comes from a common practice in algorithmic image processing of placing the origin of a coordinate system in the left top corner of the camera's plane. Then the coordinates are assumed to be always positive and increasing in the directions of the x and y axes of this system. Such an orientation is also assumed in all of the algorithms presented in this book.

Let us also analyse the following relations among coordinate systems. Assume that $\mathbf{P}_w = [X_w, Y_w, Z_w]^T$ is a certain point from the 3D 'world' space. Let $\mathbf{P}_c = [X_c, Y_c, Z_c]^T$ be an image of the point $\mathbf{P}_w$, expressed, however, in the camera coordinate system. Lastly, the point $\mathbf{p}_u = [x_u, x_u]^T$ is a point with coordinates related to the local image plane of the camera (pixel coordinates). Now, transforming Cartesian into homogeneous coordinates we obtain respectively the following relations (subscript h again means homogeneous coordinate): $\mathbf{P}_{wh} = [X_{wh}, Y_{wh}, Z_{wh}, 1]^T$, $\mathbf{P}_{ch} = [X_{ch}, Y_{ch}, Z_{ch}, 1]^T$ and $\mathbf{p}_{uh} = [x_{uh}, y_{uh}, z_{uh}]^T$. Therefore, apart from the point $\mathbf{p}_{uh}$, we have that: $X_w = X_{wh}, \ldots, Z_w = Z_{wh}, X_c = X_{ch},' \ldots, Z_c = Z_{ch}$. Considering now the projective transformation (3.7), expressed in homogeneous coordinates, and taking into an account (3.8) we obtain the following conditions:

$$\mathbf{P_c} = \mathbf{M_e P_{wh}}, \quad \mathbf{P}_{ch} = \begin{bmatrix} \mathbf{M_e} \\ \mathbf{O} \quad 1 \end{bmatrix}_{4\times4} \mathbf{P_{wh}}, \tag{3.116}$$

$$\mathbf{p_{uh}} = \mathbf{M_i P_c}, \tag{3.117}$$

$$\mathbf{p_u} = \begin{bmatrix} x_u \\ y_u \end{bmatrix} = \begin{bmatrix} \frac{x_{uh}}{z_{uh}} \\ \frac{y_{uh}}{z_{uh}} \end{bmatrix}. \tag{3.118}$$

3.9 Closure

An image is created either by an eye or by a camera. Both transform visual information about the surrounding world, which in its nature is 3D, into 2D images. In this chapter we discuss the basic principles of this process and also of recovery of the 3D information. This can be achieved with two images of the same scene, taken however from different locations. This process is called stereovision. Further we discuss the basics of the epipolar geometry, point correspondence, different stereo systems as well as stereo matching constraints. The subject of calibration of a single camera and stereo systems follows. Finally, theory meets practice in the proposition of a C++ library. We discuss computer representations of pixels, images and image operations and provide their C++ implementations.

3.9.1 Further Reading

Direct references to particular topics discussed in this chapter are placed in the text of the sections. Here we try to give some hints on further reading or 'where to go next' to find more information on the main topics touched upon in this chapter.

The human visual system can be discussed from many aspects. A well-balanced but thorough discussion of the subject can be found in the excellent monograph by Wandell [442]. A more psychologically oriented approach is presented in the classic text by Gregory [161]. High-level vision is discussed in the book by Ullman [432]. The book by Howard and Rogers [201] is a seminal work on the psychology of binocular vision and stereopsis. Problems of computational vision are addressed in the book by Mallot [292].

In this chapter we did not discuss the physical formation of images in the optical systems nor the aspects of light, photometry, colorimetry, colour representations, etc. These can be found in many textbooks on physics, such as the one by Halliday *et al.* [170]. There are also books devoted solely to the problems of optics, of which the book by Hecht [185] can be recommended as a first source. The work by Born and Wolf [50] contains advanced information on all aspects of optics. A nice introduction to colour imaging is provided in the book by Lee [272].

A very intuitive and clear introduction to most of the fundamental tasks of computer vision is provided in the excellent book by Trucco and Verri [430].

A must in geometry of multiple views is the excellent book by Hartley and Zisserman [180], and also that by Faugeras and Luong [119]. The book by Ma *et al.* [290] is another source of knowledge on many aspects of computer vision. More information on panoramic image formation and related topics can be found in the book edited by Benosman and Kang [34]. Problems and methods of camera calibration are addressed in the book by Gruen and Huang [166].

An excellent work on applied numerical methods is the book by Press *et al.* [352]. In a unique way it provides both a concise theory and C++ implementations. On the other hand, Matlab [208] and its toolboxes offer myriads of scientific and engineering methods packed in a single 'software laboratory'. Basic information on using Matlab can be found in many books, such as the one by Gilat [146]. Matlab in the context of image processing is stressed in the book by Gonzalez *et al.* [158].

There are few publications on software development for image processing and computer vision. Some image processing algorithms are provided in the book by Parker [342]. Image procedures written in C can be found in the somehow dated compendium by Myler and Weeks [325].

3.9.2 Problems and Exercises

1. Derive the formula on point distance d_i used in the RANSAC method in section 3.4.6.
2. Implement the RANSAC method for line fitting in accordance with Algorithm 3.1.
3. Using any graphic software, create a random dot stereogram with differently shifted hidden shapes. Observe the stereo effect when shifting hidden shapes to the left and then to the right.
4. Verify Equation (3.45).
5. Find the null space of the following fundamental matrix:

$$\mathbf{F} = \begin{bmatrix} 0 & 0 & a \\ 0 & 0 & b \\ c & d & e \end{bmatrix}$$

   which is called an affine fundamental matrix and arises from affine cameras used instead of projective ones (section 3.3.4).
6. What denotes a null space of $\mathbf{F}$ from the above example?
7. Find equations for the epipolar lines for the affine fundamental matrix.
8. What is the rank of the fundamental matrix?

4

Low-level Image Processing for Image Matching

4.1 Abstract

In this chapter we discuss the basic concepts of signal processing which aim at the detection of features in digital images. What are features? In simple words, we can assume that features are any well-distinguishable signal patterns. For example 'good' features are lines and corners, since they are conspicuous and well resistant to noise and some other distortions. However, in this chapter we focus mostly on methods of feature detection which facilitate 3D computer vision. For instance we are interested in features which allow most reliable matching of images.

We start with the basic concepts of digital signal processing that lead to feature detection: convolution, filtering, mask separability, discrete differentiation; Gaussian and binomial filters are discussed as well. Discussion of some methods of edge detection follows, presenting also the concepts of Laplacian of Gaussian and difference of Gaussians. Then we introduce the structural tensor which is a powerful technique for low-level feature detection. The chapter is augmented with examples of implementations of the basic techniques as well as with results of their application. A literature review is also provided, followed by exercises for the reader.

4.2 Basic Concepts

4.2.1 Convolution and Filtering

We can say that signal filtering is a process of changing spectral properties of the signal. In the frequency domain linear filtering can be done by simple multiplication of the signal spectrum by some filter function which cuts off specific band(s) of the input signal. The spectrum of the filter usually is a kind of window-like function in the frequency domain. It can easily be shown that this multiplication in the frequency domain translates into convolution in the time domain. Mathematical details of signal processing in time and frequency domains, as well as discussion of different types of filters can be found in the classic texts on signal processing (section 4.9.1 contains a discussion of the literature) [9, 312, 336]. In this book we will mostly

An Introduction to 3D Computer Vision Techniques and Algorithms Bogusław Cyganek and J. Paul Siebert

utilize the concept of 2D convolution in the discrete domain, an approach that is more suitable to processing of digital images. To define this we start from the definition of the 1D discrete convolution of a discrete signal $x[i]$ with a filter given by discrete series $f[j]$, as follows:

$$y\,[m] = f * x = \sum_{i=-r}^{r} f\,[i]x\,[m-i]\,, \tag{4.1}$$

where $y[m]$ is a filter response at index m, while we assume also that f is defined for all indices from $-r$ up to r, and the input signal x is defined at least for all from $m-r$ up to $m+r$. Let us note that in (4.1) indices in $f[\cdot]$ and $x[\cdot]$ go in different directions, i.e. if i reaches $-r$ then $m-i=m+r$; however, when i comes to r, then $m-i=m-r$. This can be seen in the following example. Let us compute the response of the following system

$$f = \begin{bmatrix} -\frac{1}{2}, & 0, & \frac{1}{2} \end{bmatrix}_{\substack{-1 \quad 0 \quad 1}}, \qquad x = \ldots, \underset{-1}{112}, \underset{\substack{0\\ \uparrow \\ m}}{250}, \underset{+1}{154}, \ldots,$$

then response y at m is

$$\begin{aligned} y\,[m] &= f\,[-1]\,x\,[m+1] + f\,[0]\,x\,[m] + f\,[1]\,x\,[m-1] \\ &= -\frac{1}{2}\cdot 154 + 0\cdot 250 + \frac{1}{2}\cdot 112 = -21. \end{aligned}$$

Filters with symmetrical masks play a special role, i.e. for which

$$f\,[-m] = f\,[m]\,. \tag{4.2}$$

The symmetry can be even, which is identical to (4.2), or odd if the right side of (4.2) is negated. Their importance comes from the fact of pure real or imaginary spectral representation and thus they exhibit desirable linear phase properties [312]. The advantage is even twofold since a symmetrical mask can be stored and processed more efficiently due to repetition of data. We notice also that in this case if the mask is odd we can substitute the minus sign in (4.1) with a plus, and still have the same result.

Let us now extend easily our analysis to the case of digital images. In the 2D signal space, the convolution takes the following form:

$$y\,[m,n] = \sum_{j=-q}^{q} \sum_{i=-r}^{r} f\,[i,j]x\,[m-i,n-j], \tag{4.3}$$

where again we assume that f and x are defined for all index runs in the above sums. The very important case arises when the 2D filter mask can be represented as a product of two 1D masks, as follows:

$$f\,[i,j] = f_1\,[i]\,f_2\,[j]\,. \tag{4.4}$$

In this special case – discussed also in the next section – the 2D convolution can be written as

$$y[m,n] = \sum_{j=-q}^{q} f_2[j] \underbrace{\left(\sum_{i=-r}^{r} f_1[i] x \left[m-i, \underbrace{n-j}_{n'} \right] \right)}_{=x'}. \tag{4.5}$$

In the above we can split the two sums into two separate runs: first with mask f_1, then with f_2. In the inner summation the second index $n - j = n'$ is fixed. Let us note also that the inner summation produces the intermediate signal x'. This can be done in a separate run of the 1D convolution. Then the outer summation takes place – this is the second 1D convolution. This is a very desirable property, since if the 2D mask can be split into two 1D masks, the computations can be speeded up significantly, as will be discussed in the following sections.

Computational aspects of convolution are discussed, for example, by Jähne [224]. Especially cumbersome are computations on borders of an image – there are no perfect solutions to this problem. However, we can always leave a margin of width equal to half the size of the convolution mask and consider only this created 'inner' area of an image. Our C++ implementation, which follows this idea, is discussed in section 4.8.1.1.

If in (4.3) we substitute the minus sign with plus we obtain the so-called *cross-correlation* value, defined as [172]

$$y[m,n] = \sum_{j=-q}^{q} \sum_{i=-r}^{r} f[i,j] x[m+i, n+j]. \tag{4.6}$$

We already know that cross-correlation is equal to the convolution in the case of symmetrical masks. It is often used as a simple measure of similarity between two images.[1] In this case, $f[i, j]$ can be a constant mask (e.g. an object template) or an another image. For instance, in the work by Antonini *et al.* it is used in a system for pedestrian tracking to correlate image patches in consecutive video frames [8].

4.2.2 Filter Separability

Let us assume that an impulse response of a filter is given by a matrix $\mathbf{A}$ of dimension $m \times n$. We know from linear algebra that any matrix $\mathbf{A} \in \Re^{m \times n}$ can be decomposed as follows [308]:

$$\mathbf{A} = \mathbf{SVD}^{\mathrm{T}}, \tag{4.7}$$

where $\mathbf{V}$ is a diagonal matrix containing nonnegative singular values and $\mathbf{S}$ and $\mathbf{D}$ are unitary matrices, so the following holds:

$$\mathbf{SS}^{\mathrm{T}} = \mathbf{1}, \quad \mathbf{DD}^{\mathrm{T}} = \mathbf{1}. \tag{4.8}$$

[1] A more in-depth discussion on similarity measures is provided in section 3.3.

Equivalently we can write

$$\mathbf{A} = \sum_{i=1}^{r} v_i \mathbf{s}_i \mathbf{d}_i^{\mathrm{T}}, \tag{4.9}$$

where the matrices **S** and **D**,

$$\mathbf{S} = [\mathbf{s}_1, \quad \mathbf{s}_2, \ldots, \quad \mathbf{s}_m], \quad \mathbf{D} = [\mathbf{d}_1, \quad \mathbf{d}_2, \ldots, \quad \mathbf{d}_n], \tag{4.10}$$

are composed of the vectors $\mathbf{s}_i$ and $\mathbf{d}_i$, respectively, where v_i is the i-th singular value of $\mathbf{A}$ of rank r, $\mathbf{s}_i$ is an $m \times 1$ column vector and $\mathbf{d}_{\mathrm{i}}$ is a $1 \times n$ column vector. Thus, all $\mathbf{s}_i$ are orthogonal to each other, and the same holds for $\mathbf{d}_i$. It follows also that $v_i{}^2$ are positive eigenvalues of $\mathbf{A}^{\mathbf{T}}\mathbf{A}$, for which $v_1 \geq v_2 \geq \cdots \geq v_r > 0$ and $v_{r+1} = \cdots = v_k = 0$, for $k = \min(m, n)$, $\mathbf{s}_i$ are eigenvectors of $\mathbf{A}\mathbf{A}^{\mathrm{T}}$, whereas $\mathbf{d}_i$ are eigenvectors of $\mathbf{A}^{\mathbf{T}}\mathbf{A}$ corresponding to the eigenvalues $v_i{}^2$, respectively.[2] For symmetrical matrices it holds also that $v_1 = |\lambda_1|$, where λ_1 are eigenvalues of such symmetric matrices.

Decomposition (4.9) is another form of the SVD which was already discussed in the case of the fundamental matrix **F** (section 3.4.5).

Let us now consider image filtering with discrete filter mask given by **A**. It follows that such a computation requires at least mn multiplications and almost the same additions per pixel. This can take a considerable amount of time, especially for large values of m and n. However, observing (4.9) we notice that the full 2D convolution with matrix **A** can be exchanged into r 1D convolutions with vectors: $\sqrt{v_i}\mathbf{s}_i$ and $\sqrt{v_i}\mathbf{d}_i$, respectively. This takes approximately $r(m + n)$ multiplications and a similar number of additions per pixel. If rank r of **A** is small (preferably one) this can save much time. More precisely, convolutions with **A** separated into a series (4.9) saves computations when its rank r fulfils

$$r \leq \left\lfloor \frac{mn}{m+n} \right\rfloor, \tag{4.11}$$

where $\lfloor \cdot \rfloor$ denotes a floor operation (i.e. the nearest integer that is equal to or less than its argument). For example, for equal and odd values $m = n \in \{3, 5, 7, 9\}$ we obtain $r_{\max} = 1, 2, 3, 4$, respectively. It is obvious that the lower the rank the faster performance – in this respect $r = 1$ is of special interest since it guarantees the fastest computations.

Concluding, let us analyse some practical filter masks and their decompositions:

$$\mathbf{A}_1 = \frac{1}{32}\begin{bmatrix} 3 & 0 & -3 \\ 10 & 0 & -10 \\ 3 & 0 & -3 \end{bmatrix} = \frac{1}{32}\begin{bmatrix} -3 \\ -10 \\ -3 \end{bmatrix} \cdot \begin{bmatrix} -1 & 0 & 1 \end{bmatrix}$$
$$= \begin{bmatrix} -0.09375 \\ -0.3125 \\ -0.09375 \end{bmatrix} \cdot \begin{bmatrix} -1 & 0 & 1 \end{bmatrix} = \begin{bmatrix} -0.1875 \\ -0.625 \\ -0.1875 \end{bmatrix} \cdot \begin{bmatrix} -0.5 & 0 & 0.5 \end{bmatrix}. \tag{4.12}$$

[2] We assume that **A** is a real-valued matrix, therefore $\mathbf{A}^T\mathbf{A} = \mathbf{A}^*\mathbf{A}$, where $\mathbf{A}^*$ denotes a conjugate matrix. In some texts $\mathbf{A}^*\mathbf{A}$ is called a covariance matrix of **A**.

Let us analyse what practical implications can be derived from the different representations of $\mathbf{A}_1$ in (4.12). The first representation requires 2D convolution, however with integer data. The second is equivalent to two 1D convolutions with integer masks, then followed with data scaling. The third is also a twofold 1D convolution; however the first mask requires fractional data. The last one requires fractional data in the second mask as well. What solution is the most appropriate? It depends on the platform of implementation, although separation into two 1D convolutions always looks more promising. When implementing in a programming language (like C++ or Java) we have no problem with floating point representation, so we can use any representation. However, if we implemented this filter in hardware we would probably adhere to the integer representation. It is interesting to note that $\mathbf{A}_1$ is a mask of a filter proposed by Jähne [224] as an example of the optimized and regularized vertical edge detector with a minimum angle error. $\mathbf{A}_1{}^{\mathrm{T}}$ is a detector with direction orthogonal to $\mathbf{A}_1$.

The mask $\mathbf{A}_2$ is a smoothing mask. Its decomposition into lower rank representations can be given as follows:[3]

$$\begin{aligned}\mathbf{A}_2 &= \frac{1}{11}\begin{bmatrix}1&1&1\\1&3&1\\1&1&1\end{bmatrix} = \frac{4}{11}\cdot\begin{bmatrix}\frac{1}{\sqrt{6}}\\2\frac{1}{\sqrt{6}}\\\frac{1}{\sqrt{6}}\end{bmatrix}\cdot\begin{bmatrix}\frac{1}{\sqrt{6}}\\2\frac{1}{\sqrt{6}}\\\frac{1}{\sqrt{6}}\end{bmatrix}^{\mathrm{T}} + \frac{1}{11}\cdot\begin{bmatrix}\frac{1}{\sqrt{3}}\\-\frac{1}{\sqrt{3}}\\\frac{1}{\sqrt{3}}\end{bmatrix}\cdot\begin{bmatrix}\frac{1}{\sqrt{3}}\\-\frac{1}{\sqrt{3}}\\\frac{1}{\sqrt{3}}\end{bmatrix}^{\mathrm{T}}\\ &= \frac{2}{33}\begin{bmatrix}1\\2\\1\end{bmatrix}\cdot\begin{bmatrix}1\\2\\1\end{bmatrix}^{\mathrm{T}} + \frac{1}{33}\begin{bmatrix}1\\-1\\1\end{bmatrix}\cdot\begin{bmatrix}1\\-1\\1\end{bmatrix}^{\mathrm{T}}\end{aligned} \tag{4.13}$$

Thus, $\mathbf{A}_2$ can be represented as two different twofold 1D convolutions, the results of which are then added together. However, let us notice that the condition (4.11) is not fulfilled in this case, contrary to the example (4.12).

Finally we should also be aware that such decompositions can introduce additional computational errors which are due to finite length representation of real-valued data. This should be separately analysed in concrete realizations.

4.3 Discrete Averaging

Discrete averaging refers to the process of low-pass filtering of discrete signals. Explaining in simple words, this is a process of substituting a value of a pixel with a value computed as an average of its surrounding pixels, usually multiplied by some weighting parameters. However, other algorithms can also be employed for this task. Sometimes these are nonlinear as a median filter, etc.

Such low-pass filtering is ubiquitous in all areas of digital signal processing, and also in computer vision. The most common application is removal of the unwanted component of a signal, commonly known as noise. Different types of the latter are discussed in Chapter 11. In image matching averaging around a central pixel is employed for the aggregation of a support in local regions (section 6.5.1.1).

[3] See also the Matlab example in section 4.8.3.

Image averaging is also discussed in the section on Savitzky–Golay filters (section 4.4.2). Some simple averaging masks were also presented when discussing filter separability (4.2.2). In the next two sections we discuss two types of important low-pass filters: Gaussian and binomial filters.

4.3.1 Gaussian Filter

The 2D Gaussian kernel $G(x,y,\sigma)$ is given by the following formula [259]:

$$G(x, y, \sigma) = \frac{1}{2\pi\sigma^2} \exp\left(-\frac{x^2+y^2}{2\sigma^2}\right), \tag{4.14}$$

where x, y denote two free coordinates and σ is a parameter. Figure 4.1 depicts two plots of the Gaussian kernel $G(x,y,\sigma)$ with $\sigma = 1.0$ and 5.0, respectively. It can be observed that the Gaussian kernels are isotropic, i.e. their characteristic is perfectly symmetric in all directions.

After a short scrutiny of (4.14) we notice that this formula can be expressed in the following form:

$$G(x, y, \sigma) = g(x, \sigma)\, g(y, \sigma), \tag{4.15}$$

where $g(\cdot, \sigma)$ is a *one-dimensional* Gaussian function given as follows:

$$g(t, \sigma) = \frac{1}{\sqrt{2\pi}\,\sigma} \exp\left(-\frac{t^2}{2\sigma^2}\right). \tag{4.16}$$

The formulas (4.15) and (4.16) mean that the 2D Gaussian kernel (4.14) can be *separated* into two operations of the 1D kernel. This very important feature allows much faster implementation of multidimensional Gaussian filtering.

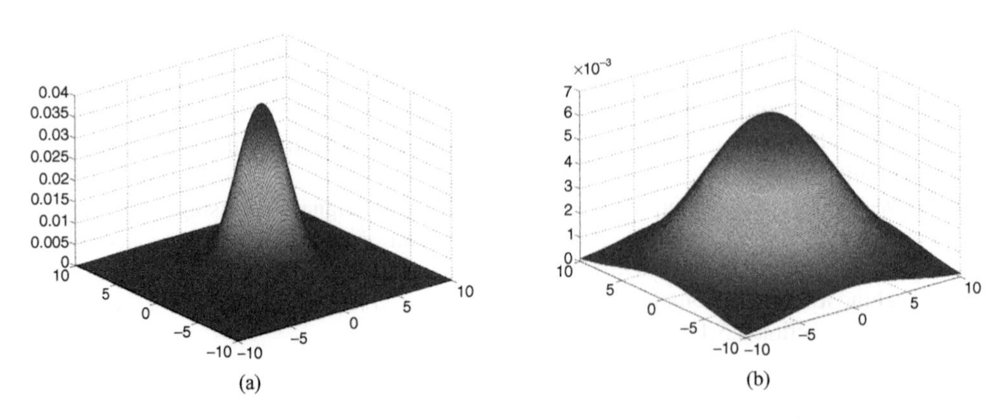

Figure 4.1 Plots of 2D Gaussian kernels: (a) $\sigma = 1$; (b) $\sigma = 5$

4.3.2 Binomial Filter

Implementation and complexity of the Gaussian filters (4.14) can be cumbersome in some applications (e.g. in a hardware realization). This is an outcome of the fractional arithmetic which requires at least fixed point representation.[4] However, there is another type of filter whose spectral response approximates the Gaussian while the implementation can be much simpler: the binomial filter [226, 351].

4.3.2.1 Specification of the Binomial Filter

The binomial filter is a low-pass filter which shows the following properties:

1. Isotropic response.
2. Separable mask for higher dimensional filters.
3. Approximation of the Gaussian response for sufficiently large masks.
4. Simple implementation.

The ideal isotropic response means that smoothing is the same in all directions. This means that all image directions are treated in the same way and the transfer function in respect to the magnitude of the wave number is uniform in all directions, as will be shown in the next section.

The kernel of the binomial filter is based on the following filtering element:

$$\mathbf{F} = \tfrac{1}{2}\begin{bmatrix} 1 & 1 \end{bmatrix}. \tag{4.17}$$

The effect of applying $\mathbf{F}$ to a digital signal is a simple averaging of each pair of its samples. If we use the same mask $\mathbf{F}$ q times we obtain the following response:

$$\mathbf{F}_i^n = \mathbf{F}_i^1 * \mathbf{F}_i^1 * \ldots * \mathbf{F}_i^1, \tag{4.18}$$

where q is a filter order, i denotes filter direction, $*$ stands for convolution (4.2.1) and by definition

$$\mathbf{F}_i^1 \equiv \mathbf{F} = \tfrac{1}{2}\begin{bmatrix} 1 & 1 \end{bmatrix}. \tag{4.19}$$

Expanding (4.18) for some integer values of q, the following filters are obtained:

$$\begin{gathered}
\mathbf{F}^2 = \tfrac{1}{4}\begin{bmatrix} 1 & 2 & 4 \end{bmatrix}, \\
\mathbf{F}^3 = \tfrac{1}{8}\begin{bmatrix} 1 & 3 & 3 & 1 \end{bmatrix}, \\
\mathbf{F}^4 = \tfrac{1}{16}\begin{bmatrix} 1 & 4 & 6 & 4 & 1 \end{bmatrix}, \\
\mathbf{F}^5 = \tfrac{1}{32}\begin{bmatrix} 1 & 5 & 10 & 10 & 5 & 1 \end{bmatrix}, \\
\mathbf{F}^6 = \tfrac{1}{64}\begin{bmatrix} 1 & 6 & 15 & 20 & 15 & 6 & 1 \end{bmatrix}, \\
\ldots
\end{gathered} \tag{4.20}$$

[4]More on numerical data representation in computers and seminumerical algorithms can be found in the books by Knuth [249] and by Koren [258].

Table 4.1 Coefficients of binomial filters

Rank q	Scaling factor 2^{-q}	Mask	Variance $\sigma^2 = q/4$
0	1	1	0
1	1/2	1 1	1/4
2	1/4	1 2 1	1/2
3	1/8	1 3 3 1	3/4
4	1/16	1 4 6 4 1	1
5	1/32	1 5 10 10 5 1	5/4
6	1/64	1 6 15 20 15 6 1	3/2
7	1/128	1 7 21 35 35 21 7 1	7/4
8	1/256	1 8 28 56 70 56 28 8 1	2
9	1/512	1 9 36 84 126 126 84 36 9 1	9/4
10	1/1024	1 10 45 120 210 252 210 120 45 10 1	5/2
11	1/2048	1 11 55 165 330 462 462 330 165 55 11 1	11/4
...	...	...	...

From (4.20) it is evident that the mask coefficients follow coefficients of the binomial distribution $(a + b)^q$ and can be easily computed from Newton's expansion formula or Pascal's triangle [259]. For an q rank filter, the number of its coefficients equals $q + 1$. Table 4.1 provides parameters of the binomial filter for different rank q.

The nice feature of the binomial filter comes from its integer mask and the scaling factor being a power of two. Thus, the necessary scaling can be easily implemented as a shift of a computer word by q bits to the right.

Given filter variance σ we can find its corresponding filter rank q from the following formula:

$$q = 4\sigma^2. \tag{4.21}$$

For the Laplacian pyramids frequently employed in image matching it is sufficient to use filters with $\sigma = 0.5$, and thus the nearest rank for a symmetrical mask is $q = 2$.

4.3.2.2 Spectral Properties of the Binomial Filter

The spectral properties of a digital system can be analysed based on the Fourier transform. The Fourier representation of a digital signal $x[n]$ can be obtained from the formula [312, 336]

$$X\left(e^{j\omega}\right) = \sum_{n=-\infty}^{+\infty} x[n]e^{-j\omega n}, \tag{4.22}$$

where $x[n]$ is a digital signal (a series of samples) and ω is a digital frequency. The digital frequency ω is commonly expressed in terms of a wave number k, as follows:

$$\omega = \pi k \quad \text{and} \quad -1 \le k \le +1. \tag{4.23}$$

The spectral representation of the basic binomial filter $\mathbf{F}^1$ can be obtained applying (4.22) to (4.19) and assuming that the last sample is at $n = 0$:

$$\tilde{\mathbf{F}}^1 = \tfrac{1}{2}\left(e^{-j\omega} + 1\right). \tag{4.24}$$

However, with the assumption that the first sample in (4.19) is at $n = 0$ we obtain the following representation:

$$\tilde{\mathbf{F}}^1 = \tfrac{1}{2}\left(1 + e^{j\omega}\right). \tag{4.25}$$

From (4.24) and (4.25) we conclude that the spectral characteristics of the basic binomial filter $\mathbf{F}^1$ (observe a nonsymmetric mask) belong to the complex domain which means it is not recommended for practical filtering of images (which by their nature belong to the real domain).

Based on the above analysis we conclude that the nearest practical mask is the symmetrical $\mathbf{F}^2$. Assuming that the central sample is at $n = 0$ the spectral response of $\mathbf{F}^2$ can be obtained directly from (4.20) and (4.22) – or by multiplication of (4.24) and (4.25):

$$\tilde{\mathbf{F}}^2 = \tfrac{1}{4}\left(e^{j\omega} + 2 + e^{-j\omega}\right) = \tfrac{1}{2}\left(1 + \cos\omega\right) = \cos^2\tfrac{\omega}{2}, \tag{4.26}$$

or alternatively, after applying (4.23):

$$\tilde{\mathbf{F}}^2 = \cos^2\frac{\pi k}{2}. \tag{4.27}$$

We see that (4.27) is characteristic of a desirable real response. It can be further composed in cascades the spectral responses of which follow the simple formula

$$\tilde{\mathbf{F}}^{2q} = \cos^{2q}\frac{\pi k}{2}, \tag{4.28}$$

where q denotes the number of cascaded $\mathbf{F}^2$ structures.

The analysis above concerned a 1D signal. For multidimensional signals, such as images, thanks to the feature of separable filters, the multidimensional binomial filter can be composed directly from (4.28), as a simple multiplication of 1D spectral responses:

$$\tilde{\mathbf{F}}^{2q} = \tilde{\mathbf{F}}_x^{2q}\tilde{\mathbf{F}}_y^{2q} = \cos^{2q}\frac{\pi k_x}{2} \cdot \cos^{2q}\frac{\pi k_y}{2}. \tag{4.29}$$

It can be shown that (4.29) approaches quickly the Gaussian response for larger values of q [224]. In practice, a reasonable approximation is chosen as $q \geq 4$. The two responses for the

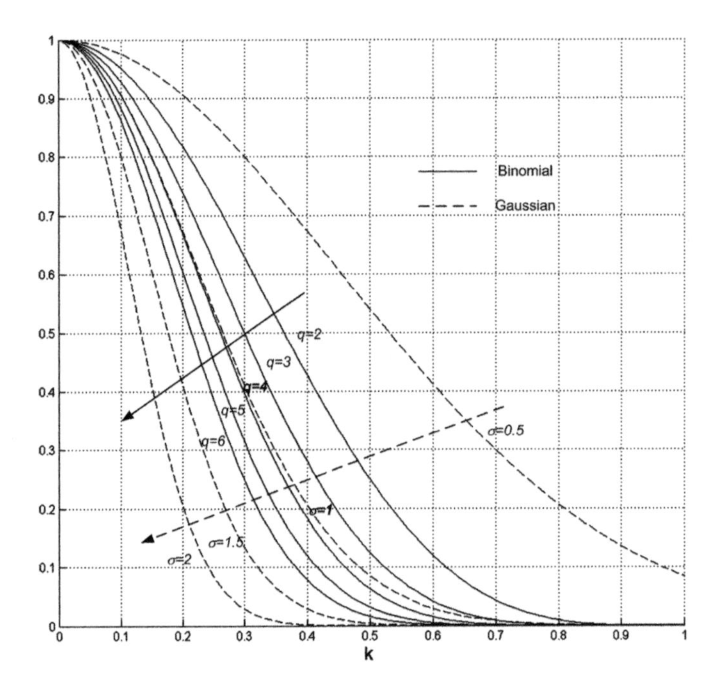

Figure 4.2 Comparison of spectral responses of the Gaussian and binomial kernels. The horizontal axis denotes the wave number k. The Gaussian plots are for σ from 0.5 to 2 with step 0.5. The binomial plots are for q from 2 to 6 with step 1. The binomial plot is almost identical with the Gaussian for $\sigma = 1$ and $q = 4$

1D case are compared in Figure 4.2 (up to a scaling value). It can be noticed that for some parameters the two plots are almost identical.

Figure 4.3 depicts spectral characteristics of the 2D binomial kernels for $q = 2$ and 4. Comparing these with the plots from Figure 4.1, some isotropic differences are noticeable, however.

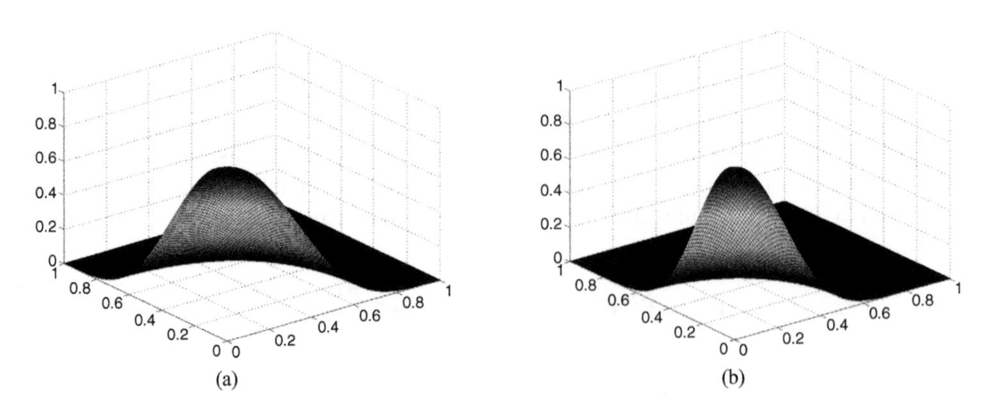

Figure 4.3 Plots of 2D binomial kernels: (a) $q = 2$; (b) $q = 4$

4.4 Discrete Differentiation

In image processing, differentiation of discrete signals (images) is often abused by the easiest approximation of differentiation by finite differences. Such an approach is a result of a simple dropping of the limit in the definition of the continuous differential. Thus, assuming that $I(x)$ represents a continuous function (i.e. not sampled), for the 1D case we start from the classic definition

$$\frac{\mathrm{d}}{\mathrm{d}x} I(x) \equiv \lim_{\delta \to 0} \frac{I(x+\delta) - I(x)}{\delta}, \tag{4.30}$$

assuming that such a limit exists. Then, after dropping the limit we obtain the following [352]:

$$\frac{\mathrm{d}}{\mathrm{d}x} I(x) \approx \frac{I(x+\Delta x) - I(x)}{\Delta x}. \tag{4.31}$$

The above approximation is commonly used in numerical analysis and is justified by a choice of an appropriately small step Δx in (4.31). However, in image processing Δx is fixed and unknown. It is only assumed that consecutive samples $I[n]$ of $I(x)$ are close enough to fulfil the sampling theorem. Moreover, it is frequently assumed that $\Delta x = 1$, which leads to the following (not symmetrical) approximation:

$$\frac{\mathrm{d}}{\mathrm{d}x} I(x) \approx I[n+1] - I[n]. \tag{4.32}$$

On the other hand, the simplest way to obtain a symmetrical formula is as follows:

$$\frac{\mathrm{d}}{\mathrm{d}x} I(x) \approx \frac{I[n+1] - I[n-1]}{2}. \tag{4.33}$$

These, in turn, lead directly to the most common discrete differentiators: $[-1 \quad +1]$ and $1/2[-1 \quad 0 \; +1]$, respectively. Unfortunately, such approximations are not sufficiently accurate for most of the image processing tasks that require precise directional computations of gradients (e.g. optical flow, structural tensor, edge and corner detection, etc.).

4.4.1 Optimized Differentiating Filters

An interesting solution to the problem of differentiation of discrete signals was proposed by Farid and Simoncelli [117] and Simoncelli [391]. They propose an alternative approach to differentiation which is based on differentiating continuous signals that are interpolated from their initial discrete versions. Figure 4.4 depicts flow charts of the two approaches.

Let us assume that the samples $I[n]$ come from sampling of a (usually unknown) continuous image signal $I(x)$ with a rate of T samples/length. Let us assume further that in the input signal there are no higher frequencies than $2\pi/T$ cycles/length, so no aliasing is introduced during sampling. Then from the sampling theorem [147, 312, 317, 336] we know that the continuous

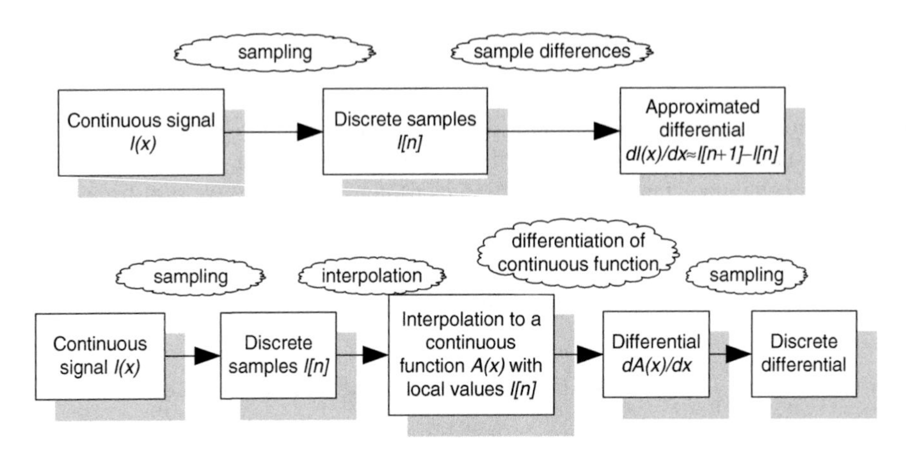

Figure 4.4 Two approaches to differentiation of discrete signals: approximation with simple differences (upper); approach of Farid and Simoncelli (lower)

signal can be precisely reconstructed from those samples:

$$I(x) = \sum_{n=-\infty}^{+\infty} I[n] \cdot A(x - nT), \tag{4.34}$$

where we assume that this series is convergent and $A(x)$ is a continuous reconstructing function.[5] Now we can differentiate the continuous function $I(x)$ given by (4.34) by means of the classic definition for differentiation of continuous functions (4.30). We obtain

$$\begin{aligned}\frac{\mathrm{d}}{\mathrm{d}x} I(x) &= \frac{\mathrm{d}}{\mathrm{d}x}\left[\sum_{n=-\infty}^{+\infty} I[n] \cdot A(x - nT)\right] = \sum_{n=-\infty}^{+\infty} I[n] \cdot \frac{\mathrm{d}}{\mathrm{d}x}[A(x - nT)] \\ &= \sum_{n=-\infty}^{+\infty} I[n] \cdot \mathrm{D}_A(x - nT).\end{aligned} \tag{4.35}$$

Finally, sampling (4.35) with the original sampling frequency, we obtain a formula for discrete differentiation [117]:

$$\begin{aligned}\frac{\mathrm{d}}{\mathrm{d}x} I[k] &= \sum_{n=-\infty}^{+\infty} I[n] \cdot \mathrm{D}_A(x - nT)\,|_{x=kT} = \sum_{n=-\infty}^{+\infty} I[n] \cdot \mathrm{D}_A[(k - n)T] \\ &= \sum_{n=-\infty}^{+\infty} I[n] \cdot \mathrm{d}_A[k - n] = I[k] * \mathrm{d}_A[k],\end{aligned} \tag{4.36}$$

[5]We know that after sampling a continuous signal with sampling frequency above the Nyquist threshold, its basic spectrum gets periodic with T. The reconstructing function should have a low-pass spectrum to select the primary band from the infinite series. The simplest solution to this is a function with a box-like spectrum – in the time domain, this is a *sinc* function. Thus in (4.34) we have $A(x) = sinc(x) = (sin(\pi x/T))/(\pi x/T)$. However, other choice of $A(\cdot)$ is also possible [312].

where $d_A[k-n]$ is a *sampled derivative* D_A *of the approximating function* $A(x)$. This is a very important result which indicates that to compute a discrete derivative of I we need to convolve I with a derivative mask.

For the 2D case, by the same token we obtain from (4.35)

$$\begin{aligned}\frac{d}{dx} I(x, y) &= \frac{d}{dx}\left[\sum_{n,m=-\infty}^{+\infty} I[n, m] \cdot A(x - nT, y - mT)\right] \\ &= \sum_{n,m=-\infty}^{+\infty} I[n, m] \cdot \frac{d}{dx}[A(x - nT, y - mT)] \end{aligned} \tag{4.37}$$

and the sampled version

$$\begin{aligned}\frac{d}{dx} I[i, j] &= \sum_{n,m=-\infty}^{+\infty} I[n, m] \cdot \frac{d}{dx}[A(x - nT, y - mT)]\bigg|_{x=iT, y=jT} \\ &= \sum_{n,m=-\infty}^{+\infty} I[n, m] \cdot d_A[i - n, j - m] = I[i, j] * d_A[i, j]. \end{aligned} \tag{4.38}$$

The above is not free from a number of practical problems, however. The first one comes from the fact that in general the discrete convolution in (4.37) spans over infinity and if a reconstructing functions $A(x)$ has a long support – which, for example, is true for the $sinc(x)$ function – then this approach gets less practical. The second problem when using (4.37) with $A(x) = sinc(x)$ comes from somewhat awkward computation of gradients in an arbitrary direction $\vec{v}$. In this case we would like to be able to use the linear property of gradients which says that the derivative in an arbitrary direction $\vec{v}$ can be found as a linear combination of derivatives in the direction of the axis:

$$\frac{d}{d\vec{v}} I(x, y) = v_x \frac{d}{dx} I(x, y) + v_y \frac{d}{dy} I(x, y), \tag{4.39}$$

where $\vec{v} = (v_x, v_y)^T$ is a unit vector. With the mentioned reconstruction function, taking for example $\vec{v} = (1, 1)^T$, we obtain quite sparse, although infinite response of such a filter, which is not easy to be applied in practical computations. Therefore Farid and Simoncelli propose to look for another type of reconstruction function $A(x)$ in (4.35) and (4.37) and their analogues in higher dimensions. The following assumptions are imposed:

1. The interpolation function A is separable.
2. The interpolation function A is symmetric about the origin.

For the 2D case this implies that

$$A(x, y) = p(x) \cdot p(y), \tag{4.40}$$

and in consequence

$$\frac{\mathrm{d}}{\mathrm{d}x} A(x, y) = \mathrm{d}_1(x) \cdot p(y), \tag{4.41}$$

where $\mathrm{d}_1(\cdot)$ is the first derivative of the 1D function $p(\cdot)$

$$\frac{\mathrm{d}}{\mathrm{d}x} I[i, j] = \sum_{n,m=-\infty}^{+\infty} I[n, m] \cdot \mathrm{d}_1[i-n] \cdot p[j-m]. \tag{4.42}$$

The last expression means that the discrete derivatives are computed with two 1D filters, $\mathrm{d}_1[\cdot]$ and $p[\cdot]$, which are discrete representations (sampled) of their continuous representations $\mathrm{d}_1(\cdot)$ and $p(\cdot)$, respectively. From (4.42) we note also that differentiation in x direction is achieved by separable convolution with the differentiation filter $\mathrm{d}_1[\cdot]$ along the x axis, and with the interpolation filter $p[\cdot]$ in the direction of the y axis (i.e. vertical to x).

The last question is a choice of the $p(\cdot)$. In this respect Farid and Simoncelli propose to look for such functions that ensure rotation-invariance property which means that (4.39) holds. Choice of $p(\cdot)$ different from the *sinc*$(\cdot)$ function means that the reconstruction filter will not be spectrally flat. In consequence, the derivative filters will compute the derivatives of the spectrally changed signal, instead of the original one. It is also interesting to notice that the directional filters computed in accordance with (4.39) in general are not rotated versions of a common filter. This means that we will not obtain steerable filters.

Starting from (4.39), transformed into the Fourier domain, Farid and Simoncelli built an error functional $E\{P, D_1\}$ which after minimization leads to the sought filters P and D_1 [117]. Tables 4.2–4.5 present numerically found values for some filters commonly used in practice.

As presented in [117] optimized differentiating filters outperform the most common differentiators in respect of the accuracy in estimating local orientation in images. This is also crucial when computing the structural tensor, optical flow, etc., which we shall discuss later in this chapter.

4.4.2 *Savitzky–Golay Filters*

Having samples of a signal, the idea behind Savitzky–Golay filtering is *to fit a polynomial* of a certain degree around each sample point. Then, a filter response is taken as the value of this polynomial computed at the point of interest. Once a polynomial is found we can get even

Table 4.2 Symmetrical differentiating Simoncelli–Farid filters of order 1, 2 samples (p, symmetric prefilter; d_i, i-th order differentiating antisymmetric filter). Differentiation in x direction is obtained by separate application of the interpolation filter p in the direction of y axis followed by the differentiation filter d_1 along the x axis

p	0.229879	0.540242	0.229879
d_1	0.425287	0	−0.425287

Table 4.3 Symmetrical differentiating Simoncelli–Farid filters of order 1, 3 samples

p	0.037659	0.249153	0.426375	0.249153	0.037659
d_1	0.109604	0.276691	0	−0.276691	−0.109604

Table 4.4 Symmetrical differentiating Simoncelli–Farid filters of order 2, 3 samples

p	0.030320	0.249724	0.439911	0.249724	0.030320
d_1	0.104550	0.292315	0	−0.292315	−0.104550
d_2	−0.232905	−0.002668	−0.471147	0.002668	0.232905

Table 4.5 Symmetrical differentiating Simoncelli–Farid filters of order 2, 4 samples

p	0.004711	0.069321	0.245410	0.361117	0.245410	0.069321	0.004711
d_1	0.018708	0.125376	0.193091	0	−0.193091	−0.125376	−0.018708
d_2	−0.055336	−0.137778	0.056554	−0.273118	−0.056554	0.137778	0.055336

more from such a representation; for example, we can compute a derivative of a certain order at a given point from the domain of this polynomial.

The polynomial is fitted around a chosen 'central' point which means that for a 1D function a number of its left and right neighbour positions have also to be evaluated. Computation of the coefficients of polynomials is done by a least-squares fitting method. Details of this procedure, as well as a computer code, are provided in an excellent book by Press *et al.* [352], for instance. However, the real beauty of this method comes from the fact that the values of the polynomial can be obtained as a *linear* combination of some constant coefficients and data samples. The coefficients depend only on the kind of chosen polynomial and number of neighbours around a point, but not signal values at these locations. Hence, they can be precomputed and stored in a look-up table. Furthermore, having found the coefficients, the already presented linear filtering scheme in the form of (4.1) can be used. To see this maybe surprising result, let us rewrite (4.1) as follows:

$$y\,[m] = \sum_{i=-r}^{r} f\,[r+i]x\,[m-i]\,, \tag{4.43}$$

where $x[i]$ denotes a series of data samples (a discrete signal), $y[m]$ is a filter response at index m and f defines a vector of the filter coefficients (sometimes called filter weights). Now, around a chosen sample at index m and in a certain window W we wish to approximate all values of $x[i]$ with a polynomial $\tilde{x}\,(i)$ of order N. This situation, with two different polynomials $\tilde{x}_1\,(i)$ and $\tilde{x}_2\,(i)$, is depicted in Figure 4.5.

If such a polynomial is known then its value at index m, i.e. at $i = 0$ in the local coordinate system of the window W, gives a noise smoothed value of a signal $x[i]$. It is also straightforward to find a derivative of a certain order in that point, since $\tilde{x}\,(i)$ is assumed to be continuous in the window W. This procedure is repeated again in new window positions, that is, in each step W is shifted by one index. The advantage of the Savitzky–Golay filters is preservation of the higher order statistical moments of the filtered signal. This, however, depends on the chosen order N of the interpolating polynomial, which in practice is two or four. Actually N determines the highest preserved moment. Moreover, the window W need not be symmetrical in both directions [352].

Without a loss of generality let us now assume that the fitted polynomial $\tilde{x}\,(i)$ is of order four, i.e. $N = 4$, and that the index iterates from -2 up to 2, i.e. there are five elements in the

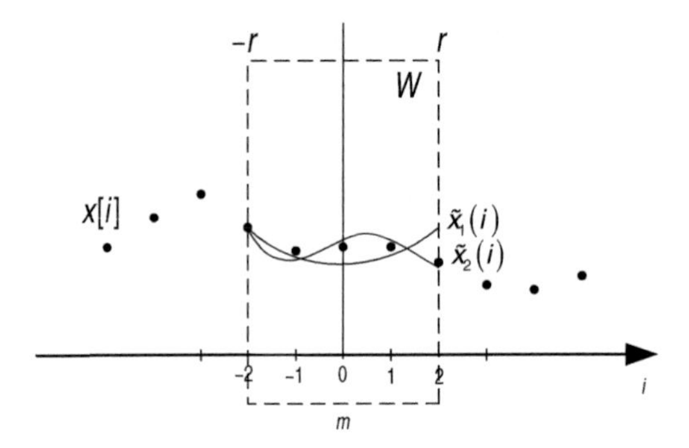

Figure 4.5 Explanation of the Savitzky–Golay filter principle. The idea is to least-squares fit a polynomial to signal data $x[i]$ in a certain window W. Then a smoothed value or a derivative at a point m is obtained based on found coefficients of the polynomial

window W, as depicted in Figure 4.5. The interpolation problem can be now written as

$$\tilde{x}(i) = f_0 + f_1 i + f_2 i^2 + f_3 i^3 + f_4 i^4, \quad \text{for } -2 \leq i \leq 2. \tag{4.44}$$

Thus, a smoothed value of the signal $x[i]$ at a point m will be *a polynomial value at index* 0 in the local coordinate system W (see Figure 4.5), that is

$$\tilde{x}(0) = f_0. \tag{4.45}$$

However, to find f_0 we need to know all values of $x[i]$ in W, as well as needing to at least partially solve a set of linear equations, as we shall see soon. In a similar fashion we easily obtain values of the derivatives:

$$\tilde{x}'(0) = f_1, \quad \tilde{x}''(0) = 2f_2, \quad \tilde{x}'''(0) = 6f_3, \ldots \tag{4.46}$$

We are interested in finding such parameters $f_0, \ldots, f_4$ for which

$$x[i] \approx \tilde{x}(i). \tag{4.47}$$

The equation above should hold for each data point in the window W. This leads to the following set of equations:

$$\begin{cases} x[-2] \approx \tilde{x}(-2) = f_0 + f_1(-2) + f_2(-2)^2 + f_3(-2)^3 + f_4(-2)^4 \\ \cdots \\ x[2] \approx \tilde{x}(2) = f_0 + f_1(2) + f_2(2)^2 + f_3(2)^3 + f_4(2)^4 \end{cases} \tag{4.48}$$

which is *linear* in respect to yet unknown values $f_0, \ldots, f_4$. This can be written in a matrix representation:

$$\mathbf{Pf} = \mathbf{X}, \tag{4.49}$$

where $\mathbf{f}$ is a column matrix of the coefficients f_k, $\mathbf{X}$ is a column matrix with the signal values $x[k]$, for $k \in W$, and $\mathbf{P}$ is a matrix of powers of indices, i.e.

$$P_{ij} = i^j, \quad \text{for } j < N. \tag{4.50}$$

The number of equations in (4.49) should be equal or greater to $N + 1$, since this is the number of the coefficients of the polynomial $\tilde{x}(i)$. Thus, in our example we have five equations. However, the number of equations can be greater if the number of samples in W exceeds the number of coefficients. Due to (4.47), (4.49) can be solved for $\mathbf{f}$ by the least-squares method, discussed also in section 12.8. Based on (12.21)[6] the solution is given as

$$\mathbf{f} = \underbrace{\left(\mathbf{P}^{\mathrm{T}}\mathbf{P}\right)^{-1} \mathbf{P}^{\mathrm{T}}}_{\mathbf{P}'} \mathbf{X}, \tag{4.51}$$

where $\mathbf{P}'$ is the pseudo-inverse matrix which depends exclusively on the local coordinate values and *not* on the data samples. In consequence each coefficient of the polynomial $\tilde{x}$ is computed as the inner product of one row of the matrix $\mathbf{P}'$ and the vector of discrete signal values $\mathbf{X}$. Thus, depending on the operation of our filter, whether it is smoothing (4.45) or differentiation (4.46), only one corresponding row of $\mathbf{P}'$ needs to be computed. As suggested in [352] this can be done by the LU decomposition. A procedure described in section 12.8 can also be used.

We can easily extend our methodology to polynomials with an arbitrary order N and higher dimensions than one. For a 2D case the polynomial (4.44) takes the form

$$\tilde{x}(i, j) = \sum_{k_1,k_2=0}^{N} f_{k_1k_2} i^{k_1} j^{k_2}, \quad \text{for } k_1 + k_2 \leq N \quad \text{and} \quad -2 \leq i, j \leq 2. \tag{4.52}$$

The parameters $f_{k_1k_2}$ are solved in exactly the same way as before after arranging in a single vector $\mathbf{f}$. For example, for $N = 4$ we have

$$\begin{aligned} \tilde{x}(i, j) = {} & f_{00} + f_{01} j + f_{02} j^2 + f_{03} j^3 + f_{04} j^4 + \\ & f_{10} i + f_{11} ij + f_{12} ij^2 + f_{13} ij^3 + \\ & f_{20} i^2 + f_{21} i^2 j + f_{22} i^2 j^2 + \\ & f_{30} i^3 + f_{31} i^3 j + \\ & f_{40} i^4. \end{aligned} \tag{4.53}$$

Thus, each k-th row P_k of the matrix $\mathbf{P}$ takes the form

$$P_k = \begin{bmatrix} 1 & j & j^2 & j^3 & j^4 & i & ij & ij^2 & ij^3 & i^2 & i^2 j & i^2 j^2 & i^3 & i^3 j & i^4 \end{bmatrix} \tag{4.54}$$

[6] See p. 424.

for all possible values $\{(i, j) : -2 \le i, j \le 2\}$, which in this case results in 25 different rows. Generally, each P_k contains

$$S = \tfrac{1}{2}(N+1)(N+2) \tag{4.55}$$

elements, i.e. 6 for $N = 2$, 10 for $N = 3$, 15 for $N = 4$, 21 for $N = 5$, and so on. Hence, in (4.49) **P** is of size 25×15, **f** of size 15×1 and **X** of size 25×1. In analogy to (4.45), a smoothed value of the 2D signal corresponds to a value of the polynomial $\tilde{x}(i, j)$ at the point (0, 0), that is

$$\tilde{x}(i, j)|_{(0,0)} = f_{00}, \tag{4.56}$$

which is given by **f**(0) in (4.51). In other words, to find the value f_{00} all we need to do is multiply a first row of the matrix **P**′ in (4.51) and the input data.

Similarly, from (4.52) we find the first partial derivatives at point of indices (0, 0) as follows:

$$\left.\frac{\partial}{\partial i}\tilde{x}(i, j)\right|_{(0,0)} = f_{10} \quad \text{and} \quad \left.\frac{\partial}{\partial j}\tilde{x}(i, j)\right|_{(0,0)} = f_{01}, \tag{4.57}$$

which correspond to entries **f**(6) and **f**(2), respectively, and also

$$\left.\frac{\partial^2}{\partial i^2}\tilde{x}(i, j)\right|_{(0,0)} = 2f_{20}, \quad \left.\frac{\partial^2}{\partial i \partial j}\tilde{x}(i, j)\right|_{(0,0)} = f_{11} \quad \text{and} \quad \left.\frac{\partial^2}{\partial j^2}\tilde{x}(i, j)\right|_{(0,0)} = 2f_{02}, \tag{4.58}$$

which correspond to entries **f**(10), **f**(7) and **f**(3), respectively.

Getting back to our exemplary settings the few initial rows of **P** are

$$\begin{aligned}
P_1 &= \begin{bmatrix} 1 & -2 & 4 & -8 & 16 & -2 & 4 & -8 & 16 & 4 & -8 & 16 & -8 & 16 & 16 \end{bmatrix} \\
P_2 &= \begin{bmatrix} 1 & -2 & 4 & -8 & 16 & -1 & 2 & -4 & 8 & 1 & -2 & 4 & -1 & 2 & 1 \end{bmatrix} \\
P_3 &= \begin{bmatrix} 1 & -2 & 4 & -8 & 16 & 0 & 0 & 0 & 0 & 0 & 0 & 0 & 0 & 0 & 0 \end{bmatrix} \\
&\vdots
\end{aligned}$$

Having found **P**, from (4.51) we easily obtain the following 5×5 filter masks.

- For smoothing f_{00}. Obtained by multiplication of the following mask with the input signal (actually $m(f_{00})$ is a formatted first row of the matrix **P**′ in (4.51)).

$M(f_{00})$

0.04163265306122	−0.08081632653061	0.07836734693878	−0.08081632653061	0.04163265306122
−0.08081632653061	−0.01959183673469	0.20081632653061	−0.01959183673469	−0.08081632653061
0.07836734693878	0.20081632653061	0.44163265306122	0.20081632653061	0.07836734693878
−0.08081632653061	−0.01959183673469	0.20081632653061	−0.01959183673469	−0.08081632653061
0.04163265306122	−0.08081632653061	0.07836734693877	−0.08081632653061	0.04163265306122

- First horizontal derivative f_{10} (4.57). Obtained by multiplication of the following mask with the input signal (actually this is a formatted sixth row of the matrix $\mathbf{P}'$ in (4.51)).

$M(f_{10})$

0.07380952380952	−0.10476190476190	0.00000000000000	0.10476190476190	−0.07380952380952
−0.01190476190476	−0.14761904761905	0.00000000000000	0.14761904761905	0.01190476190476
−0.04047619047619	−0.16190476190476	0.00000000000000	0.16190476190476	0.04047619047619
−0.01190476190476	−0.14761904761905	0.00000000000000	0.14761904761905	0.01190476190476
0.07380952380952	−0.10476190476191	0.00000000000000	0.10476190476190	−0.07380952380952

- First vertical derivative f_{01} (4.57).

$M(f_{01})$

0.07380952380952	−0.01190476190476	−0.04047619047619	−0.01190476190476	0.07380952380952
−0.10476190476190	−0.14761904761905	−0.16190476190476	−0.14761904761905	−0.10476190476190
0.00000000000000	0.00000000000000	0.00000000000000	0.00000000000000	0.00000000000000
0.10476190476190	0.14761904761905	0.16190476190476	0.14761904761905	0.10476190476191
−0.07380952380952	0.01190476190476	0.04047619047619	0.01190476190476	−0.07380952380952

- Mixed derivative f_{11} (4.58).

$M(f_{11})$

−0.07333333333333	0.10500000000000	0.00000000000000	−0.10500000000000	0.07333333333333
0.10500000000000	0.12333333333333	0.00000000000000	−0.12333333333333	−0.10500000000000
0.00000000000000	0.00000000000000	0.00000000000000	0.00000000000000	0.00000000000000
−0.10500000000000	−0.12333333333333	0.00000000000000	0.12333333333333	0.10500000000000
0.07333333333333	−0.10500000000000	0.00000000000000	0.10500000000000	−0.07333333333333

- Second horizontal derivative f_{20} (4.58).

$M(f_{20})$

−0.04914965986395	0.15374149659864	−0.20918367346939	0.15374149659864	−0.04914965986395
0.01207482993197	0.12312925170068	−0.27040816326531	0.12312925170068	0.01207482993197
0.03248299319728	0.11292517006803	−0.29081632653061	0.11292517006803	0.03248299319728
0.01207482993197	0.12312925170068	−0.27040816326531	0.12312925170068	0.01207482993197
−0.04914965986395	0.15374149659864	−0.20918367346939	0.15374149659864	−0.04914965986395

- Second vertical derivative f_{02} (4.58).

$M(f_{02})$

−0.04914965986395	0.01207482993197	0.03248299319728	0.01207482993197	−0.04914965986395
0.15374149659864	0.12312925170068	0.11292517006803	0.12312925170068	0.15374149659864
−0.20918367346939	−0.27040816326531	−0.29081632653061	−0.27040816326531	−0.20918367346939
0.15374149659864	0.12312925170068	0.11292517006803	0.12312925170068	0.15374149659864
−0.04914965986395	0.01207482993197	0.03248299319728	0.01207482993197	−0.04914965986395

To verify the above theory the Savitzky–Golay filter has been applied to the image 'Lena', commonly used for testing of image algorithms.

Results of smoothing with Savitzky–Golay filters with polynomials of order $N = 3$ and $N = 4$ are depicted in Figure 4.6(a) and (b), respectively. The two smoothed versions do not differ significantly from the original one. Indeed, computed PSNR (11.12) values between

(a)

(b)

Figure 4.6 Low-pass filtering with Savitzky–Golay filter of order (a) $N = 3$, PSNR = 32.88 and (b) $N = 4$, PSNR = 39.77. (Source: www.lenna.org/USC-SIPI Image Database)

each of the smoothed versions and the input image are very high, i.e. 32.88 and 39.77 dB, respectively.

This means that the higher the order of the polynomial (4.52) used, the stronger the adaptation to the local data in images. Thus, for low-pass smoothing lower powers of the polynomial should be chosen.

Results of the convolution with masks $M(f_{11})$, $M(f_{10})$ and $M(f_{01})$ are depicted in Figure 4.7(b)–(d), respectively. For proper visualization, values of the derivatives have been linearly transformed into a viewable range of 0 to 255.

4.4.2.1 Generation of Savitzky–Golay Filter Coefficients

The matrix **P** (4.49) can be easily computed in C++ or Matlab. Algorithm 4.1 presents implementation of the *Generate_SavGol_2D_Coordinate_Matrix*() function which accepts an order of the polynomial $\tilde{x}\,(i, j)$, as well as the number of samples in directions top-left and bottom-right from the point (0, 0). A more elaborate version would allow four different parameters for the latter values. They define the span of the window W. The function returns an image with integer pixels – a matrix **P** of size $\#W \times S$, where $\#W$ denotes the number of samples in the window W, and S is given by (4.55).

To find coefficients of the Savitzky–Golay filter we need to compute a pseudo-inverse of the matrix **P** in accordance with (4.51). This can be done with the *Orphan_PseudoInv_Matrix*() presented in section 12.8.1 or *pinv*() command in Matlab (Algorithm 4.2).

Certainly, the filter coefficients have to be computed once, which can be done offline, and stored for further usage. In applications these precomputed (constant) values are used as a mask argument in the convolution operation (section 4.8.1.1).

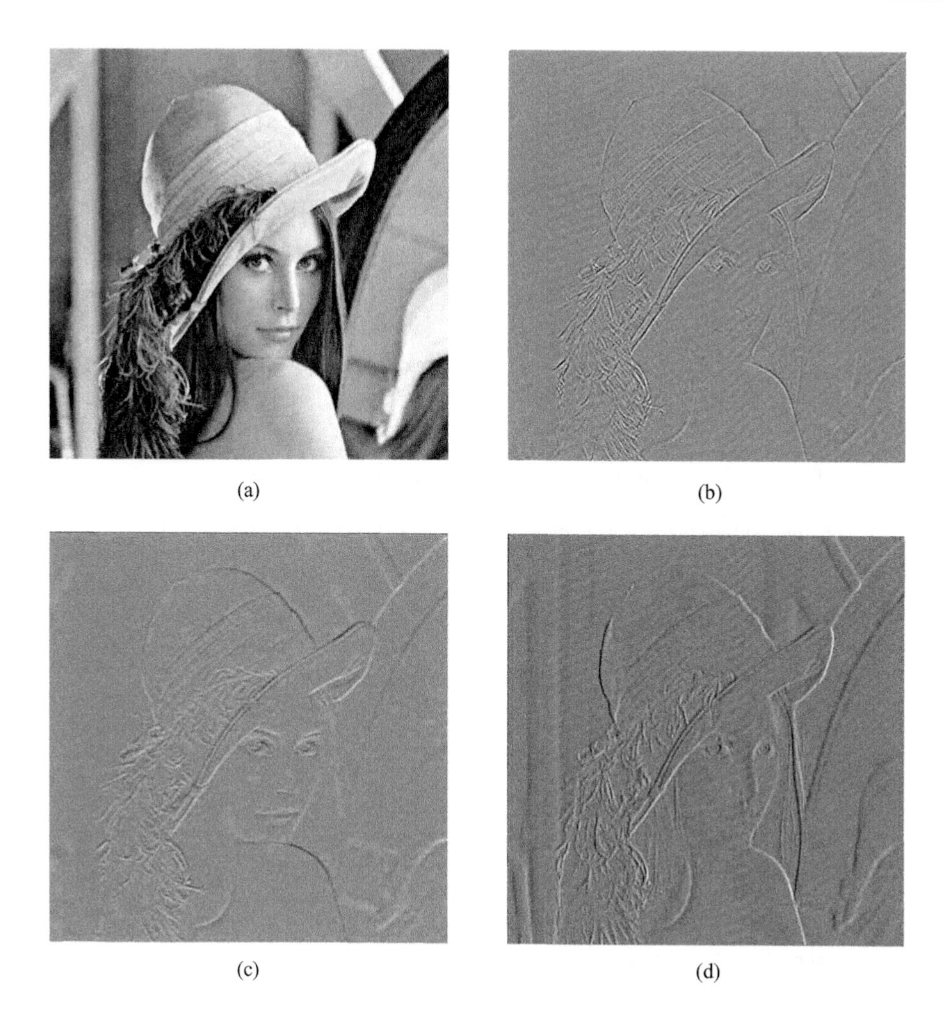

Figure 4.7 Results of Savitzky–Golay filter of order $N = 4$ applied to (a) the 'Lena' image: (b) mixed derivative f_{11}; (c) first horizontal derivative f_{10}; (d) first vertical derivative f_{01}

4.5 Edge Detection

Edges are important features of images. They reflect edges of real objects observed in a scene, or other types of edges such as the ones created by occluding objects, shades or other physical phenomena in the observed scene. Edges are characteristic of strong signal variations, a property that is used for their detection. Coordinates of edges, their length and orientation convey important information characteristics to the contents of an image. Therefore many matching methods rely only on comparison of edge pixels (sections 6.8.1 and 6.8.3). These features are also frequently used in contour matching or object detection in images [135, 157, 224, 351].

An observation of luminance values in monochrome images reveals that edges correspond to areas with significant change of the luminance signal. An example is depicted in Figure 4.8.

```
//////////////////////////////////////////////////////////////
// This function orphans a coordinate matrix for the
// Savitzky-Golay filter.
//////////////////////////////////////////////////////////////
//
// INPUT:
//       kN - degree of the interpolating polynomial
//       kIndexFrom - number of samples to the left
//          from the central one
//       kIndexTo - number of samples from to the right
//          from the central one
//
// OUTPUT:
//        Orphaned image - a matrix P
//
// REMARKS:
//         The returned object has to be deleted by
//         a calling part.
//
LongImage * Generate_SavGol_2D_Coordinate_Matrix( const long kN,
                          const long kIndexFrom, const long kIndexTo )

{
   const long kElemsInRow = ( ( kN + 1 ) * ( kN + 2 ) ) / 2;
   const long kNumOfRows =       ( kIndexTo - kIndexFrom + 1 ) *
                                 ( kIndexTo - kIndexFrom + 1 );
   register long i, j, a, b;

   //                                                    cols         rows
   LongImage * indexMatrix = new LongImage( kElemsInRow,
                                                       kNumOfRows, 0.0 );


   int row_counter = 0;

   ////////////////////////////////////////////////////////////////////
   // These loops run through all possible indexes in the window W
   // i runs horizontally
   // j runs vertically
   for ( j = kIndexFrom; j <= kIndexTo; ++ j )
   {
        for( i = kIndexFrom; i <= kIndexTo; ++ i )
        {
           //////////////////////////////////////////////
           // These two loops generate a single row Pk
           int col_counter = 0;

           for( a = 0; a <= kN; ++ a )
           {
              for( b = 0; b <= kN; ++ b )
              {
                  if( a + b <= kN )
                  {
                     long Power( long x, long a );
                     long theElement = Power( i, a ) * Power( j, b );
                     indexMatrix->SetPixel( col_counter ++,
                                                row_counter, theElement );
                  }
           }
   }
                        REQUIRE( col_counter == kElemsInRow );
```

Algorithm 4.1 Function for generation of the matrix P (4.49)

```
            ///////////////////////////////////////////////

            // changing any value of i or j moves us to a new row
            ++ row_counter;                         }

        }
        ////////////////////////////////////////////////////////////////

        REQUIRE( row_counter == kNumOfRows );

        return indexMatrix;
}

////////////////////////////////////////////////////////////////
// This function computes a power of integer values: x^a
////////////////////////////////////////////////////////////////
//
// INPUT:
//        x - argument value
//        a - power value
//
// OUTPUT:
//        x ^ a
//
// REMARKS:
//
//
long Power( long x, long a )
{
    register long retVal = 1;
    while( a -- > 0 )
        retVal *= x;
    return retVal;
}
```

Algorithm 4.1 (*Continued*)

In this section we present a signal-based approach to edge detection. A more in-depth treatment of the important problem of edge detection can be found in the literature such as the paper by Canny [60] or the books by Jähne [224], Gonzalez and Woods [157], Forsyth and Ponce [135] and Pratt [351].

From this perspective there are two basic ways of signal analysis for edge detection.

1. Computation of the modulus of the signal gradient which involves computation of the first derivatives.
2. Analysis of zero-crossings which is based on the second derivatives of a signal.

4.5.1 Edges from Signal Gradient

Let us assume that the function $I(p, q)$ takes on discrete values of luminance at the image point given by coordinates p and q. Without lost of generality we can assume also that the luminance function is a 2D continuous function $I(x, y)$.[7] With these assumptions we can use

[7] Change of a discrete representation into a continuous one is possible, e.g. by interpolation, preserving original values at the discrete points. This is discussed in section 4.4.

```
////////////////////////////////////////////////////////////////
// This function orphans coefficients of the Savitzky-Golay
// filter.
////////////////////////////////////////////////////////////////
//
// INPUT:
//        kN - degree of the interpolating polynomial
//        kIndexFrom - number of samples to the left
//           from the central one
//        kIndexTo - number of samples from to the right
//           from the central one
//
// OUTPUT:
//        Orphaned image - a matrix ~P
//
// REMARKS:
//         The returned object has to be deleted by
//         a calling part.
//
TRealImage * Compute_SavGol_Filer( const long kN,
                       const long kIndexFrom, const long kIndexTo )
{
        LongImage * SG_CoordMatrix =
        Generate_SavGol_2D_Coordinate_Matrix( kN,kIndexFrom,kIndexTo );

        REQUIRE( SG_CoordMatrix != 0 );

        TRealImage tmp_SG_CoordMatrix( * SG_CoordMatrix );
        delete SG_CoordMatrix;

        TRealImage * SG_FilterCoeffs =
        Orphan_PseudoInv_Matrix< double, double >( tmp_SG_CoordMatrix );

        return SG_FilterCoeffs;
}
```

Algorithm 4.2 Function for computation of the pseudo-inverse matrix $\mathbf{P}'$ (4.51)

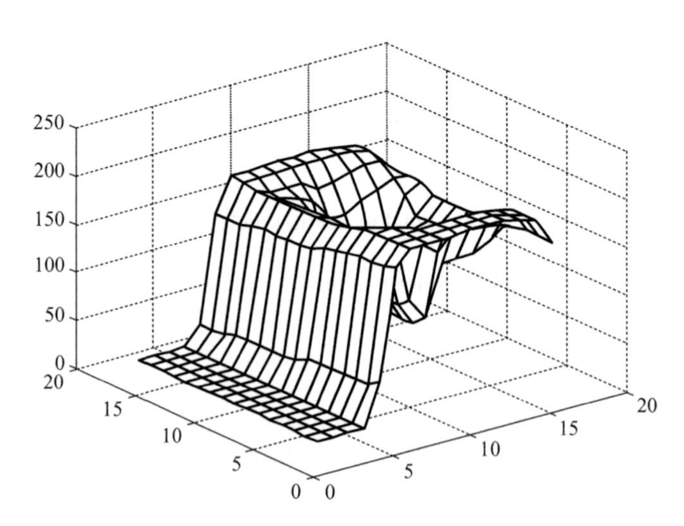

Figure 4.8 Fragment of an image containing an edge – strong variation of the luminance signal (vertical axis)

the intensity gradient vector defined as [259]

$$\nabla I(x, y) = \left[\frac{\partial I(x, y)}{\partial x}, \frac{\partial I(x, y)}{\partial y}\right]^{\mathrm{T}}. \tag{4.59}$$

For each image point the triple $(x, y, I(x, y))$ defines a plane, for which the normal vector is given as [246, 247]

$$\mathbf{n} = \left[\frac{\partial I(x, y)}{\partial x}, \frac{\partial I(x, y)}{\partial y}, 1\right]^{\mathrm{T}}. \tag{4.60}$$

For edge detection moduli of the gradient and normal to the gradient vectors,

$$\|\nabla I\| \quad \text{and} \quad \|\mathbf{n}\|, \tag{4.61}$$

are analysed. The most common here is application of the L_2 norm which leads to the following formulas:

$$\|\nabla I\|_{L_2} = \sqrt{\left(\frac{\partial I}{\partial x}\right)^2 + \left(\frac{\partial I}{\partial y}\right)^2} \quad \text{and} \quad \|\mathbf{n}\|_{L_2} = \sqrt{\left(\frac{\partial I}{\partial x}\right)^2 + \left(\frac{\partial I}{\partial y}\right)^2 + 1}. \tag{4.62}$$

However, a simplification of computations can be achieved employing the L_1 norm in (4.61). In this case the following approximation of $\|\nabla I\|$ is obtained:

$$\|\nabla I\|_{L_1} = \left|\left(\frac{\partial I}{\partial x}\right)\right| + \left|\left(\frac{\partial I}{\partial y}\right)\right|. \tag{4.63}$$

The modulus of the gradient vector takes on its minimal value for areas with constant luminance values, for which all gradients are zero. It grows in areas with much variation of the luminance signal. The latter happens just in the case of edges. Therefore a value $||\nabla I||$ is commonly used for edge detection. In the case of binary images components of the gradient are computed by means of one of the methods of discrete differentiation (section 4.4).

4.5.2 Edges from the Savitzky–Golay Filter

As alluded to previously, finding edges can be accomplished with computation of a norm of the signal gradient vector [157, 224]. In the case of the Savitzky–Golay filter (section 4.4.2) this can be stated in the following form:

$$\|\nabla \tilde{x}(i, j)\| \quad \text{at a point} \quad (i, j) = (0, 0), \tag{4.64}$$

where $\tilde{x}(i, j)$ is the signal interpolating polynomial.

Depending on the chosen norm, we obtain the formula

$$\|\nabla \tilde{x}(i, j)\|_{L_1} = \left|\frac{\partial}{\partial i}\tilde{x}(i, j)\right| + \left|\frac{\partial}{\partial j}\tilde{x}(i, j)\right| \tag{4.65}$$

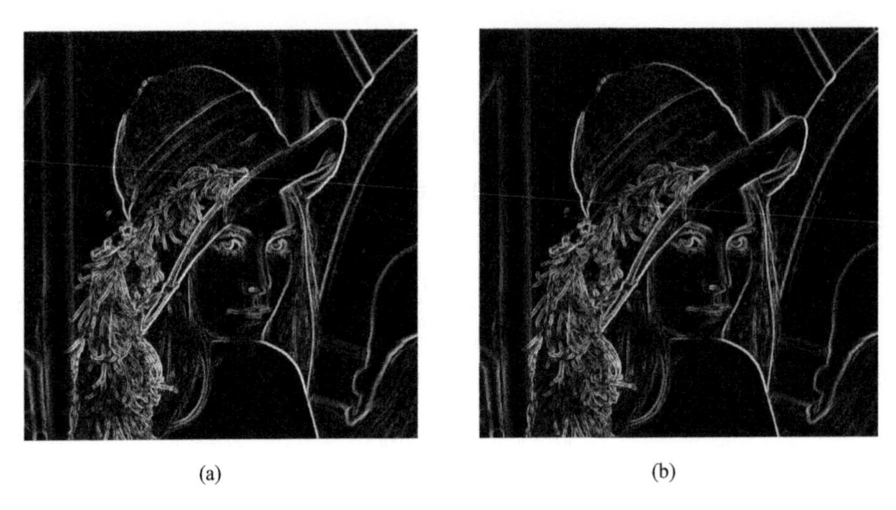

(a) (b)

Figure 4.9 Edge detection with Savitzky–Golay filter of order $N = 4$: (a) using the L_1 norm; (b) using the L_2 norm

for the norm L_1 and

$$\|\nabla\tilde{x}(i,j)\|_{L_2} = \sqrt{\left(\frac{\partial}{\partial i}\tilde{x}(i,j)\right)^2 + \left(\frac{\partial}{\partial j}\tilde{x}(i,j)\right)^2} \tag{4.66}$$

for L_2. Exemplary results of the two applied to the 'Lena' test image are presented in Figure 4.9.

Obviously computations with L_1 require less effort since we avoid multiplications and the square root.

4.5.3 Laplacian of Gaussian

A detector with much better isotropic[8] characteristics can be obtained from the *Laplace operator*,[9] which is defined as [224]

$$\nabla^2 I(x,y) = \frac{\partial^2 I}{\partial x^2} + \frac{\partial^2 I}{\partial y^2}. \tag{4.67}$$

In the case of a signal change the operators with the first derivative, e.g. the gradient modulus operator (4.61), exhibit one extreme while the operators employing the second derivate, e.g.

[8] We call a feature detector isotropic if its response does not depend on the local direction of a detected feature.

[9] Let us recall the rules of repeated applications of the operator ∇ (pronounced 'del' or 'delta'):

$$\nabla^2(\bullet) \equiv \nabla(\bullet)\cdot\nabla(\bullet) \equiv (\nabla\cdot\nabla)(\bullet) \equiv \Delta(\bullet) \equiv \left(\frac{\partial^2(\bullet)}{\partial x_1^2} + \cdots + \frac{\partial^2(\bullet)}{\partial x_N^2}\right) \equiv div \quad grad(\bullet)$$

where $(\bullet)$ denotes a differentiable N-dimensional function.

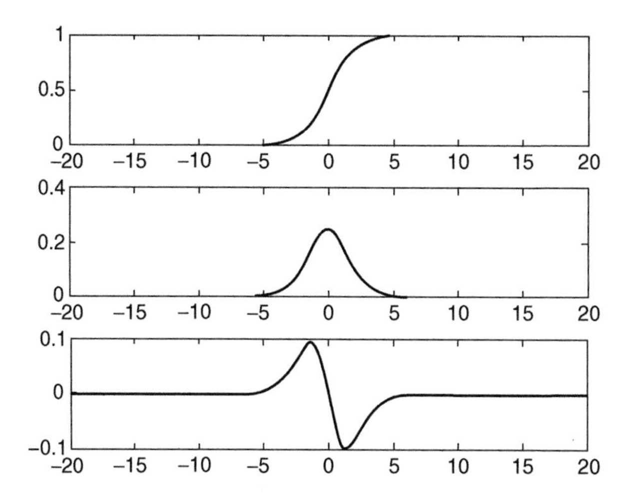

Figure 4.10 Exemplary edge detector for the 1D continuous case. From top to bottom: input signal, its first derivative, its second derivative. The latter crosses the zero value at the inflection point

like the Laplace operator (4.67), show two extrema with different signs which implies a zero value between these two.

Hence, a method of edge detection is based on finding zero crossings of the second derivative of a signal (so-called *zero crossing* operator). This is a very useful feature when it comes to computations since the search for zero crossing is easier than the search for an extreme value. Figure 4.10 presents an example of an edge signal (top curve) and its first (middle curve) and second derivatives (bottom curve). We notice that at a central point 0, where we spot an edge, the first derivative reaches its maximum, whereas the second derivative crosses the zero value.

In the case of real images, the flaw of the Laplace operator comes from its susceptibility to noise, which is ubiquitous in images (Chapter 11). This is caused by transmittance function of this operator in the frequency domain, which is proportional to the square of the frequency argument. Therefore the response of the Laplace operator is significant also for high frequencies which are characteristic of the noise spectrum. Thus, it is necessary to limit somehow the noise level in the input signal. This can be done by a prefiltering of the input signal with a low-pass filter, such as the Gaussian filter (section 4.3.1) or its similar binomial filter (section 4.3.2). The connection of these two modules, i.e. the Gaussian filter with the Laplace operator, is called *the Laplacian of Gaussian* (*LoG*) [351]. Mathematically, this idea can be expressed as

$$\nabla^2 (G * I), \tag{4.68}$$

where $G(x, y, \sigma)$ is a 2D Gaussian function given by (4.14). In the case of continuous function, (4.68) and (4.14) can be put in the following form:

$$\nabla^2 (G * I) = \left(\nabla^2 G\right) * I, \tag{4.69}$$

which means connection of the smoothing action, done by the Gaussian filter, with the second order differentiation, accomplished by the Laplace operator, into a single operator. Then the compound operator is applied to the input image for edge extraction.

The connection of the Gaussian (4.14) with Laplacian (4.67) leads to the following expression:

$$\nabla^2 G(x, y, \sigma) = -\frac{1}{2\pi\sigma^4}\left(2 - \frac{x^2 + y^2}{\sigma^2}\right)\exp\left(-\frac{x^2 + y^2}{2\sigma^2}\right). \tag{4.70}$$

Computation of (4.70) can be greatly simplified after noticing [173, 247] that this 2D operator can be decomposed into an equivalent combination of two 1D operators (section 4.2.2):

$$g(x, y) = g_1(x)g_2(y) + g_2(x)g_1(y), \tag{4.71}$$

where

$$g_1(t) = -\frac{1}{2\pi\sigma^4}\left(1 - \frac{t^2}{\sigma^2}\right)\exp\left(-\frac{t^2}{2\sigma^2}\right), \quad g_2(t) = \exp\left(-\frac{t^2}{2\sigma^2}\right). \tag{4.72}$$

Plots of the functions $g_1(t)$ and $g_2(t)$ are shown in Figure 4.11.

A transition of the kernel (4.70) to the discrete domain requires proper choice of parameters in (4.70). The most important is the size of the chosen discrete mask. We obtain width w of the central negative part of the *LoG* convolution kernel by comparison of (4.70) with zero and doubling module of that result (why?):

$$w = 2\sqrt{2}\sigma. \tag{4.73}$$

The analysis we have presented so far relates to the case of continuous signals. Now it is time to scrutinize the discrete case. The application of the *LoG* operator to the discrete signals (images) requires determination of the size of its discrete mask. Such a mask is obtained from uniform sampling with a proper sampling frequency of the continuous function *LoG* in a certain finite interval. Analysing Figure 4.11 we conclude that a reasonable size of the finite interval can be set to 3w $\times$ 3w, since outside this range the value of the *LoG* function

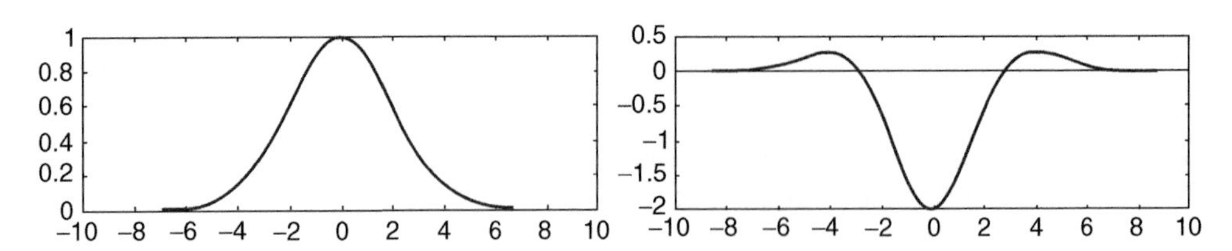

Figure 4.11 Plots of the functions $g_2(t)$ and $g_1(t)$ for $\sigma = 2.0$

practically reaches zero.[10] The unknown value left is the mentioned sampling frequency. It can be determined by means of spectral analysis with the Fourier transform of the *LoG* function. The latter can be found, at first, by determining the Fourier transform of the Gaussian function, then with the help of the theorem of the Fourier transform of the function derivative [312]. The Fourier transform from the Gaussian function, and up to the multiplicative constant, is given as

$$\begin{aligned}
\hat{G}(\omega) &= \int_{-\infty}^{+\infty} \exp\left(-\frac{t^2}{2\sigma^2}\right)\exp(-j\omega t)\,\mathrm{d}t = \int_{-\infty}^{+\infty} \exp\left(-\frac{t^2}{2\sigma^2} - j\omega t\right)\mathrm{d}t \\
&= \int_{-\infty}^{+\infty} \exp\left(-\frac{1}{2\sigma^2}(t^2 + 2\sigma^2 j\omega t)\right)\mathrm{d}t = \int_{-\infty}^{+\infty} \exp\left(-\frac{1}{2\sigma^2}\left[(t+\sigma^2 j\omega)^2 + \sigma^4\omega^2\right]\right)\mathrm{d}t \\
&= \int_{-\infty}^{+\infty} \exp\left(-\frac{\sigma^2\omega^2}{2}\right)\exp\left(-\frac{1}{2\sigma^2}(t+\sigma^2 j\omega)^2\right)\mathrm{d}t \\
&= \exp\left(-\frac{\sigma^2\omega^2}{2}\right)\int_{-\infty}^{+\infty} \exp\left(-\frac{1}{2\sigma^2}(t+\sigma^2 j\omega)^2\right)\mathrm{d}t \\
&= \left\{\begin{matrix} t+\sigma^2 j\omega = x \\ \mathrm{d}t = \mathrm{d}x \end{matrix}\right\} = \exp\left(-\frac{\sigma^2\omega^2}{2}\right)\int_{-\infty}^{+\infty} \exp\left(-\frac{1}{2\sigma^2}x^2\right)\mathrm{d}x.
\end{aligned} \tag{4.74}$$

The last integral in the above formula can be found based on [259]

$$\int_{-\infty}^{+\infty} e^{-a^2x^2}\,\mathrm{d}x = \frac{\sqrt{\pi}}{a}, \quad a > 0. \tag{4.75}$$

From the integral (4.75), and taking the multiplicative scalar from the Gaussian (4.14), we obtain the transform

$$G(\omega) = \frac{1}{\sqrt{2\pi}\sigma}\exp\left(-\frac{\sigma^2\omega^2}{2}\right). \tag{4.76}$$

Taking (4.76) and by virtue of the Fourier transform, we obtain

$$L(\omega) = \frac{-\omega^2}{\sqrt{2\pi}\sigma}\exp\left(-\frac{\sigma^2\omega^2}{2}\right). \tag{4.77}$$

Figure 4.12 depicts a plot of the function $L(\omega)$, for $\sigma = 2$.

[10] Remember that w is a width of the central negative part of the one dimensional *LoG* convolution kernel (see Figure 4.11). In this case we consider the two dimensional discrete mask.

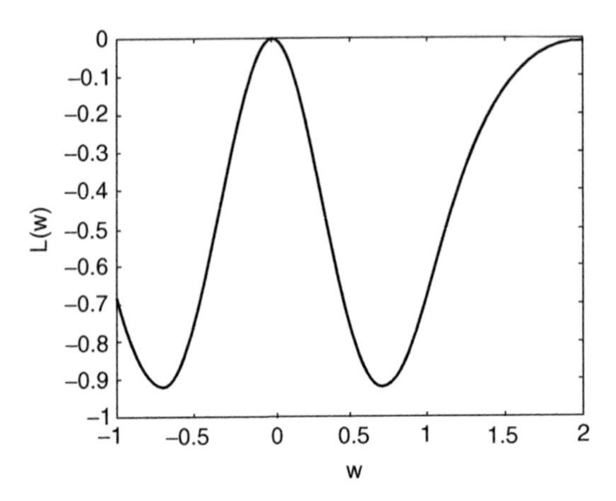

Figure 4.12 Plot of the function $L(\omega)$ (for $\sigma = 2.0$)

From Figure 4.12 it is evident that $L(\omega)$ is a band-pass filter whose parameters can be found after computing extremes of $L(\omega)$. For this purpose we compute $L'(\omega)$:

$$L'(\omega) = \frac{-2\omega}{\sqrt{2\pi}\sigma} \exp\left(-\frac{\sigma^2\omega^2}{2}\right) - \frac{\omega^2\left(-\omega\sigma^2\right)}{\sqrt{2\pi}\sigma} \exp\left(-\frac{\sigma^2\omega^2}{2}\right). \tag{4.78}$$

Equating the above to 0, we obtain the extreme points of the function $L(\omega)$ as

$$\begin{aligned} &\omega\left(2 - \sigma^2\omega^2\right) = 0 \\ &\omega_0 = 0, \quad \omega_{1,2} = \pm\frac{\sqrt{2}}{\sigma}. \end{aligned} \tag{4.79}$$

Analysing the plot of $L(\omega)$, shown in Figure 4.12, and considering the angular frequency

$$\omega_1 = \frac{\sqrt{2}}{\sigma}, \tag{4.80}$$

we can assume that the limit angular frequency of the LoG is

$$\omega_g = 3\omega_1 = \frac{3\sqrt{2}}{\sigma}. \tag{4.81}$$

From this, and based on the sampling theorem [312, 336], we obtain an expression connecting a distance Δx between consecutive samples in the following form:

$$\Delta x = \frac{1}{2f_g} = \frac{\pi}{\omega_g} = \frac{\pi\sigma}{3\sqrt{2}}. \tag{4.82}$$

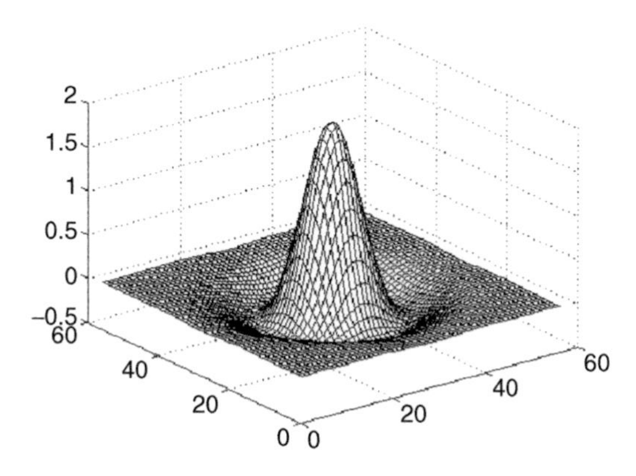

Figure 4.13 Normalized *LoG* filter mask for $n = 57$, $\sigma = 6.717$

Because the minimal mask size was set to $3w \times 3w$ then, based on (4.73) and (4.82), we obtain the *minimal* size of the discrete mask of the *LoG* filter:

$$\frac{3w}{\Delta x} \times \frac{3w}{\Delta x} \Leftrightarrow \frac{3\left(2\sqrt{2}\sigma\right)}{\Delta x} \times \frac{3\left(2\sqrt{2}\sigma\right)}{\Delta x} \Leftrightarrow 12\frac{3}{\pi} \times 12\frac{3}{\pi}. \tag{4.83}$$

Taking the minimal size of the discrete mask to be the nearest odd integer value, greater in value than in (4.83), we obtain finally that the minimal size of the discrete *LoG* mask is 13 × 13 pixels. Figure 4.13 shows an exemplary mask of the *LoG* filter for $n = 57$.

Tanaka and Kak [415] proposed an additional notation of edges found by this method to convey information on the type of local neighbourhood centred at the *LoG* zero crossing. In this notation p denotes crossing of discrete signal from large to small luminance values, whereas n denotes the opposite direction; o is set if the classification is not possible (e.g. the

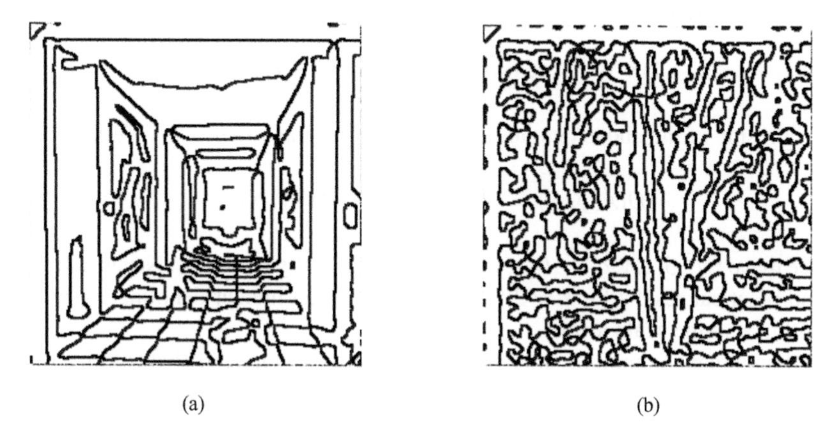

Figure 4.14 (a) 'Corridor' and (b) 'Trees' processed by the *LoG* operator with mask 27 × 27 (σ = 3.18). *Plus-minus* lines in grey, *minus-plus* in black

two neighbouring values are almost identical or at noise level). Figure 4.14 depicts 'Corridor' and 'Trees' test images processed by the *LoG* operator with 27×27 mask ($\sigma = 3.18$); *plus-minus* lines are in grey, *minus-plus* in black.

4.5.4 Difference of Gaussians

The Laplacian of Gaussian (*LoG*) $\nabla^2 G(x, y, \sigma)$, given by (4.70), plays a key role in the scale-space theory, as shown by Lindeberg [277]. He proved that the *LoG* normalized by σ^2 (*ssLoG*), given as

$$\sigma^2 \nabla^2 G(x, y, \sigma) = -\frac{1}{2\pi\sigma^2}\left(2 - \frac{x^2 + y^2}{\sigma^2}\right)\exp\left(-\frac{x^2 + y^2}{2\sigma^2}\right), \tag{4.84}$$

is required for the true scale-space invariance. It can be shown that *ssLoG* is strictly related to the difference of Gaussians (*DoG*). As presented by Lowe [283], taking the (heat) diffusion equation and exchanging the time parameter t by σ, gives

$$\frac{\partial G(x, y, \sigma)}{\partial \sigma} = \sigma \nabla^2 G(x, y, \sigma), \tag{4.85}$$

which can be approximated by expanding its left side by the finite difference, since

$$\frac{\partial G(x, y, \sigma)}{\partial \sigma} = \lim_{\Delta\sigma \to 0} \frac{G(x, y, \sigma + \Delta\sigma) - G(x, y, \sigma)}{\Delta\sigma}. \tag{4.86}$$

Placing $\nabla\sigma = \sigma(k - 1)$ in the above we obtain

$$\frac{\partial G(x, y, \sigma)}{\partial \sigma} = \lim_{k \to 1} \frac{G(x, y, k\sigma) - G(x, y, \sigma)}{\sigma\,(k - 1)}. \tag{4.87}$$

Thus, (4.85) can be approximated as

$$(k - 1)\,\sigma^2 \nabla^2 G(x, y, \sigma) \approx G(x, y, k\sigma) - G(x, y, \sigma), \tag{4.88}$$

where the right side of the above denotes the difference of Gaussians, *DoG*, defined as

$$D(x, y, \sigma) = G(x, y, k\sigma) - G(x, y, \sigma). \tag{4.89}$$

Thus, we see that $\sigma^2 \nabla^2 G(x, y, \sigma) \sim D(x, y, \sigma)$, for k sufficiently close to 1. Nevertheless, Lowe reports good practical results for $k = 2^{1/s}$, where $s > 1$ is an integer denoting the number of intervals within a single scale level.

Compared with many commonly known feature detectors, it appears that extrema of *ssLoG* give one of the most stable image features under the group of image distortions [310]. This property was used in [283] to design a detector of distinctive image features (called SIFT – Scale Invariant Feature Transform), used with great success for object detection in natural scenes.

4.5.5 Morphological Edge Detector

Image morphological operators have attracted great interest in many areas of image processing, such as filtering, segmentation, classification, contour detection, edge cleaning, texture analysis, etc. They follow mathematical operations defined on a group of sets. However, they are also equivalent to the group of rank order filters.

In the case of scalar-valued images, denoted as $f(x)$, the operations of dilation $d(x)$ and erosion $e(x)$ with the structural element denoted by $s(y)$ are defined as [396]

$$d(\mathbf{x}) = \max_{\mathbf{y} \in S} \left[f(\mathbf{x} + \mathbf{y}) + s(\mathbf{y}) \right], \tag{4.90}$$

$$e(\mathbf{x}) = \min_{\mathbf{y} \in S^*} \left[f(\mathbf{x} + \mathbf{y}) - s^*(\mathbf{y}) \right]. \tag{4.91}$$

It is assumed that S and S^* denote the support for the structural elements $s(y)$ and $s^*(y)$, respectively, where $s^*(y) = s(-y)$ for all $y \in S^*$.

The morphological gradient (so called Beucher gradient) is defined as the arithmetical difference between results of dilation and erosion, applied to the same image and with the same structural element [39, 357]. In terms of the already introduced symbols, the morphological gradient can be expressed as follows:

$$g = d - e. \tag{4.92}$$

It can be shown that the morphological gradient g is equivalent to the norm of the 'classic' gradient vector of an image, i.e. it holds that [396]

$$g(f) \equiv \|\nabla f\|, \tag{4.93}$$

where f denotes a differentiable signal of an image.

Some examples of the morphological gradient computed from different signal representations of the same image are presented in Figure 4.15. The colour version of the input image and its gradient computed separately in each colour channel depict Plate 3. The monochrome version and its Beucher gradient depict Figure 4.15(a, b)). The last pair depicts a binary version of the image from Figure 4.15(a) obtained after thresholding around its median value. The morphological gradient for that image is shown in Figure 4.15(d). In all examples the 3×3 square structural elements were used.

4.6 Structural Tensor

In this section we present a very useful technique of detecting local structures in images and their parameters, such as strength of a signal, its coherence as well as local orientation. These can be used in a variety of computer vision tasks, such as object detection, texture analysis and image matching.

(a) (b)

(c) (d)

Figure 4.15 Examples of the morphological gradient computed from (a, b) grey valued image and from (c, d) binary version of the image (gradient from a colour version of this image depicted on Plate 3)

4.6.1 Locally Oriented Neighbourhoods in Images

People easily perceive patterns in images. This is achieved easily even if the only change in an image is caused by small variation of intensity, change of scale or local orientations. The latter has been shown by many psychophysical experiments to play a very important role in perception by humans and other mammals [161, 442].

Observing each of the images depicted in Figure 4.16 we easily perceive an object contained there – the capital letter 'E' – although in each instance our knowledge comes from different phenomena. Change of the luminance signal is one of them (Figure 4.16(a)). Particularly, its nonzero gradient in a certain direction conveys sufficient information on edges. At the same time such a gradient can be used to measure local orientation in images [224]. In Figure 4.16(b) we do not spot such edges. Instead we perceive local change of texture, caused by a difference of scale in the area of the visible object and the background [398]. Nevertheless, change of scale allows precise placement of an object in respect to the background. Finally,

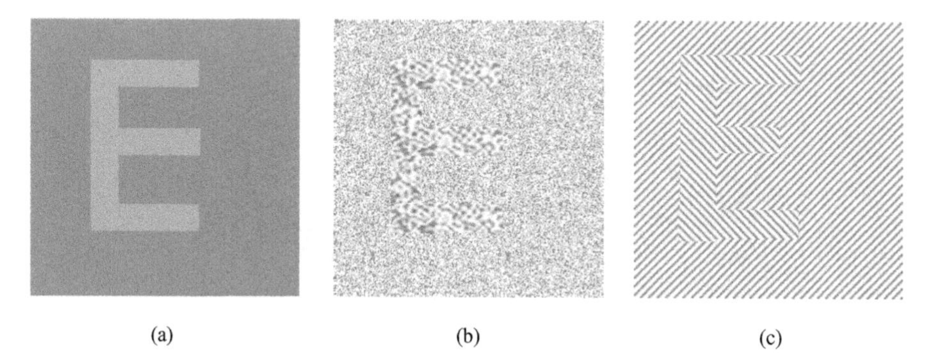

(a) (b) (c)

Figure 4.16 Recognition of the letter 'E' in images based on difference in (a) intensities, (b) scale and (c) orientation of local patterns

observing Figure 4.16(c) we come to the conclusion that even a bare change of orientation of local patterns is sufficient for us to precisely tell the letter 'E' from the rest of an image.

Apparently, our visual system is endowed with mechanisms allowing us to react to local change in intensity, scale and orientation in local neighbourhoods of pixels. Then, based on information acquired this way, we are able to draw conclusions about observed 3D space.

Let us now analyse the image in Figure 4.17 and ask what can be thought of as a local structure. Taking different areas of an image and at different scales we notice that in many of them we can spot some regular patterns. Moreover, if there is such a regularity then the whole area can be analysed after being substituted by a simple representation.

Figure 4.17 What is a local neighbourhood in an image?

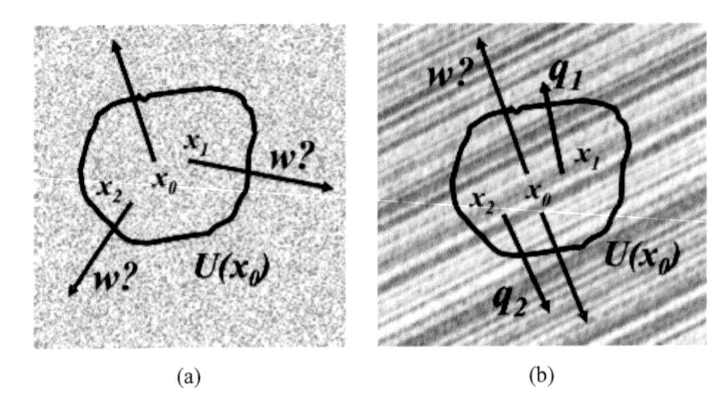

Figure 4.18 Representation of a dominating direction **w** in local neighbourhoods $U(x_0)$. (a) In a randomly changing signal it is not possible to reliably choose a dominating direction. (b) In a regular structure a dominating directional vector can be determined

4.6.1.1 Local Neighbourhood with Orientation

Let us choose a certain compact neighbourhood U of pixels around a point x_0 (Figure 4.18). Then, for each pixel $i \in U$ let us compute a gradient vector $\mathbf{q}_i$. Can we now find such a vector **w** that fits best all the other vectors $\mathbf{q}_i$? If so, then let the vector **w** represent orientation of the whole neighbourhood U. What we require is that **w** is invariant to a rotation of 180°. We need also a measure of 'how reliable' is such a representation. This can be assessed by measuring a cumulative deviation of $\mathbf{q}_i$ to **w** for all $i \in U$.

In Figure 4.18(a) a neighbourhood U contains points with randomly changing intensity signal. No uniformly oriented structure can be observed. Therefore the gradients $\mathbf{q}_i$ in U of will point more or less in random directions. As a consequence selection of its uniform representation in a form of a unique directional vector **w** is not possible. A different situation is depicted in Figure 4.18(b) in which a regular intensity pattern can be observed. As a result, the gradient vectors are highly regular reflecting common orientation of the intensity signal. The only uncertainty is their directions which can differ by 180°. The two images show us two opposite examples of local structures. The first one shows no regular orientation. The latter shows a structure with a perfect orientation. However, a unique orientation cannot be determined in a pattern with regular rings, depicted in Figure 4.19. We would like to distinguish such a case from Figure 4.18(a) as well.

For the pattern in Figure 4.19 we notice that the gradient vectors indeed exist; however, it is not possible to choose such a local orientation **w** that would represent them all.

4.6.1.2 Definition of a Local Neighbourhood of Pixels

Based on the discussion above we can conclude that the ideal *locally oriented neighbourhood* can be distinguished in an image if signal changes therein reflect a common direction. Such a case is often called linear symmetry, because luminance changes along the symmetry axis are constant, whereas those across it show quite significant variations.

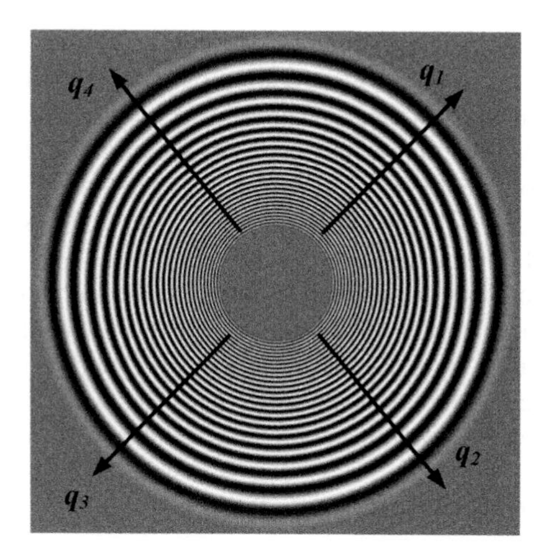

Figure 4.19 Regular ring pattern

Figure 4.18(b) presents an image fragment with evident direction of luminance change in a certain local neighbourhood of pixels. Figure 4.20 shows the same image with directional vectors **a**, **b** and **c** at a certain local neighbourhood of pixels around a point $\mathbf{x_0}$. The gradient vectors $\mathbf{q}_i$ are perpendicular to the areas (lines) of constant intensity and they point in the direction of a maximal signal change. So, in this simple case, all we need to describe this locally oriented neighbourhood of pixels is to provide a gradient vector and particular values of intensity alongside this gradient. In other words, all image points can be uniquely determined from gradient direction and one intensity value at a given point along that gradient.

Conducting a further analysis we come to a conclusion that a good approximation of a locally oriented neighbourhood of pixels could be established by introduction of an averaged gradient vector for that neighbourhood. In such a manner we would be able to substitute local neighbourhood of pixel intensities with more prominent information on their dominating direction. However, we have to be careful when thinking of gradient averaging, since a simple averaging throughout the whole neighbourhood can result in opposing gradients cancelling out each other.

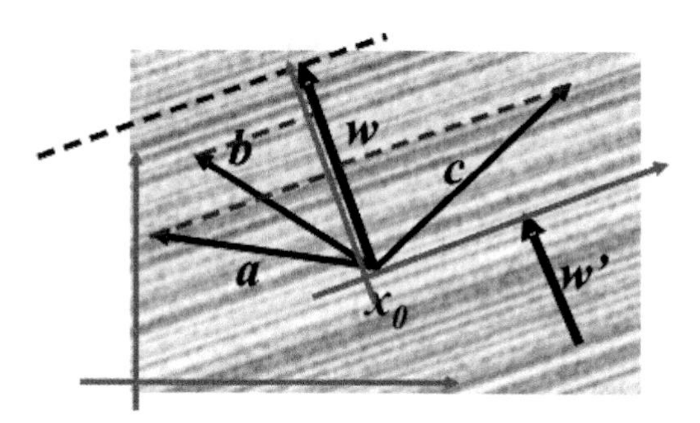

Figure 4.20 Dimensionality reduction in the case of an ideal local structure

The above discussion suggests that a good measure of a local orientation in a certain neighbourhood of pixels should be invariant to a rotation of 180°. This means for example that a local orientation of 45° is indistinguishable from a 225° orientation. At the same time it should be periodic with 180°, i.e. after reaching 179° it should reach 0° again. Thus, having a directional vector of local orientation that phase changes from 0 to 360°, then the corresponding vector of local orientation will have doubled phase. This conclusion will be used when deriving the vector of local orientation in pixel neighbourhoods.

The next constraint imposed on such a directional vector is a requirement of additional information on the type of local structure. This is a new concept that allows for differentiation between pure anisotropic areas in images from those with clear local structure. Such an indicator we call a coherence measure or just a coherence. This way we are able to selectively interpret information conveyed by an image only in places that manifest sufficient coherence coefficient.

The aforementioned postulates are grouped as follows.

1. The local structure in a certain neighbourhood of pixels should be represented by a vector whose modulus and phase correspond to local signal changes in this neighbourhood.
2. The measure of local orientation should be invariant to a rotation of 180°.
3. The measure of local orientation should be augmented by additional information on its type.

Analysing Figure 4.20 we find out that if one of the coordinate axes is oriented in the same direction as the direction of change in its local neighbourhood, then *the whole neighbourhood can be described by a 1D function.* Thus, selection of a certain direction of signal change leads to reduction of the space dimension. If there is a perfect local orientation in a local neighbourhood U of pixels around $\mathbf{x_0}$, then we can associate a local coordinate system with U, oriented with $\mathbf{w}$ and anchored at $\mathbf{x_0}$ (Figure 4.20). Now, a pixel value at different locations $\mathbf{a}$, $\mathbf{b}$ and $\mathbf{c}$ in this local system can be found from the inner product of the directional vector $\mathbf{x}$ (such as $\mathbf{a}$, $\mathbf{b}$ and $\mathbf{c}$) and one of the axes of the new coordinate system. In other words, the 1D value (i.e. the inner product) plus the directional vector is what is necessary to uniquely represent such a local neighbourhood of pixels.

These observations can be formalized by introduction of the following definition [160, 224].

Definition 4.1. A local neighbourhood of any dimension is given by a tensor[11] $\mathbf{S}$

$$\mathbf{S}(\mathbf{x}) = \mathbf{I}(\mathbf{x} \cdot \mathbf{w}), \tag{4.94}$$

where $\mathbf{I}$ denotes a tensor whose dimension depends on the luminance signal of an image, $\mathbf{x}$ is the spatial coordinate, $\mathbf{w}$ is a constant vector of direction of the maximal signal change and $(\cdot)$ denotes a scalar product, i.e. $\mathbf{x} \cdot \mathbf{w} = \mathbf{x}^T\mathbf{w}$ is a scalar value. This definition states simply that a local neighbourhood is a tensor whose dimensionality depends on the dimensionality of the input signal (through the spatial coordinate $\mathbf{x}$); however, its value is a scalar function of the tensor $\mathbf{I}$ in the direction $\mathbf{w}$.

[11] Chapter 10 gives a brief introduction to tensor analysis.

4.6.2 *Tensor Representation of Local Neighbourhoods*

Let us choose a certain local neighbourhood U around a point $\mathbf{x}_0$ (see Figure 4.18). Further, with each point $\mathbf{x} \in U$ let us associate a vector $\mathbf{q(x)}$ representing local signal direction at a point $\mathbf{x}$. We are looking for such a vector $\mathbf{w(x_0)}$ that is as close as possible to all directional vectors $\mathbf{q}$ from U. Let us notice also that the vector $\mathbf{w(x_0)}$ denotes a different thing from $\mathbf{q(x_0)}$, although it can happen that the two coincide. Very important is the choice of a measure to compare how close two vectors are to each other. For this purpose let us compare vectors with their inner product. Thus, having vectors $\mathbf{w}$ and $\mathbf{q}$ we compute their inner product ρ as follows:

$$\rho = \mathbf{q(x)} \cdot \mathbf{w(x_0)} = \mathbf{q}^{\mathrm{T}}\mathbf{(x)w(x_0)}. \tag{4.95}$$

We assume also that the direction of $\mathbf{q}$ should be irrelevant for this comparison, so vectors $\mathbf{q}$ and $-\mathbf{q}$ are treated the same. We can simply accommodate this request by taking either an absolute value $|\rho|$ or its square ρ^2, which we prefer for reasons explained later on. To find $\mathbf{w(x_0)}$ the value of ρ has to be computed for *all* directional vectors $\mathbf{q}$ and their squares summed over $U\mathbf{(x_0)}$. This way we obtain a functional Q which takes on a maximum for the sought $\mathbf{w(x_0)}$:

$$Q = \int_{U(\mathbf{x}_0)} \rho(\mathbf{x})^2 \mathbf{dx}. \tag{4.96}$$

In these terms, the task of finding $\mathbf{w(x_0)}$ can be stated as the following maximization problem:

$$\max_{\mathbf{w}}(Q) = \max_{\mathbf{w}} \left(\int_{U(\mathbf{x}_0)} \rho(\mathbf{x})^2 \mathbf{dx} \right). \tag{4.97}$$

Before we try to solve the optimization task (4.97) let us expand the functional Q given in (4.96):

$$\begin{aligned}
Q &= \int_{U(\mathbf{x}_0)} \rho(\mathbf{x})^2 \mathbf{dx} = \int_{U(\mathbf{x}_0)} \left(\mathbf{q}^{\mathrm{T}}(\mathbf{x})\mathbf{w}(\mathbf{x}_0)\right)^2 \mathbf{dx} \\
&= \int_{U(\mathbf{x}_0)} \left(\mathbf{q}^{\mathrm{T}}(\mathbf{x})\mathbf{w}(\mathbf{x}_0)\right)\left(\mathbf{q}^{\mathrm{T}}(\mathbf{x})\mathbf{w}(\mathbf{x}_0)\right)\mathbf{dx} \\
&= \int_{U(\mathbf{x}_0)} \left(\mathbf{w}^{\mathrm{T}}(\mathbf{x}_0)\mathbf{q}(\mathbf{x})\right)\left(\mathbf{q}^{\mathrm{T}}(\mathbf{x})\mathbf{w}(\mathbf{x}_0)\right)\mathbf{dx} \\
&= \mathbf{w}^{\mathrm{T}}(\mathbf{x}_0)\left[\int_{U(\mathbf{x}_0)} \mathbf{q}(\mathbf{x})\mathbf{q}^{\mathrm{T}}(\mathbf{x})\mathbf{dx}\right]\mathbf{w}(\mathbf{x}_0) = \mathbf{w}^{\mathrm{T}}(\mathbf{x}_0)\mathbf{T}(\mathbf{x}_0)\mathbf{w}(\mathbf{x}_0).
\end{aligned} \tag{4.98}$$

In the last expression we introduced the so-called structural tensor $\mathbf{T}(\mathbf{x}_0)$ at a point $\mathbf{x}_0$, defined as [160, 181, 182, 224, 226]

$$\mathbf{T}(\mathbf{x}_0) = \int_{U(\mathbf{x}_0)} \mathbf{q}(\mathbf{x})\mathbf{q}^{\mathrm{T}}(\mathbf{x})\mathrm{d}\mathbf{x}, \tag{4.99}$$

assuming that such an integral exists. Let us note that $\mathbf{q}(\mathbf{x})\mathbf{q}^{\mathrm{T}}(\mathbf{x})$ expresses the *outer* product of the directional vector with itself. This is a result of a π rotation invariance. By this token $\mathbf{T}$ is a *symmetrical* tensor whose dimension directly follows the dimension of a space of directional vectors $\mathbf{q}$. As a consequence the two important properties of $\mathbf{T}$ are obtained.

1. All elements T_{ij} of $\mathbf{T}$, given as

$$T_{ij} = \int_{U(x_0)} q_i(x)q_j(x)\mathrm{d}x, \tag{4.100}$$

   are real.
2. Eigenvectors of $\mathbf{T}$ create an orthogonal basis [308, 317].

Let us now rewrite (4.97) taking into consideration the structural tensor (4.99):

$$\max_{\mathbf{w}}(Q) = \max_{\mathbf{w}}(\mathbf{w}^{\mathrm{T}}\mathbf{T}\mathbf{w}). \tag{4.101}$$

The above can be solved for $\mathbf{w}(\mathbf{x}_0)$ by constrained optimization method, i.e. using the theorem of Lagrange multipliers [36, 127, 317, 331]. This theorem provides sufficient conditions for a function $a(\mathbf{w})$ to reach its extreme, given certain constraint $b(\mathbf{w}) = 0$. At first, the following functional is constructed:

$$L(\mathbf{w}) = a(\mathbf{w}) - \lambda b(\mathbf{w}), \tag{4.102}$$

where λ is a Lagrange multiplier. Then, sufficient conditions for a minimal or maximal value require partial derivatives to vanish, that is:

$$\frac{\partial L(\mathbf{w}, \lambda)}{\partial \mathbf{w}} = 0 \quad \text{and} \quad \frac{\partial L(\mathbf{w}, \lambda)}{\partial \lambda} = 0. \tag{4.103}$$

Let us adapt our optimization problem (4.101) to the conditions of the above theorem. At the beginning let us state a constraint $b(w)$. Without loss of generality we can start from an inner product

$$\mathbf{w}^{\mathrm{T}}\mathbf{w} = c, \tag{4.104}$$

where $c = |\mathbf{w}|^2$ is a constant. This leads directly to

$$b(\mathbf{w}) = \mathbf{w}^{\mathrm{T}}\mathbf{w} - c = 0. \tag{4.105}$$

Now, considering (4.101) and (4.105), the Lagrange multiplier (4.102) takes on the form

$$L(w) = \mathbf{w}^{\mathrm{T}}\mathbf{T}\mathbf{w} - \lambda(\mathbf{w}^{\mathrm{T}}\mathbf{w} - c), \tag{4.106}$$

where λ is a free variable, called a Lagrange multiplier.

A solution comes directly from (4.106) and (4.103) after computing partial derivatives and equating them to zero:[12]

$$\begin{aligned}\frac{\partial L(\mathbf{w},\lambda)}{\partial \mathbf{w}} &= \frac{\partial}{\partial \mathbf{w}}(\mathbf{w}^{\mathrm{T}}\mathbf{T}\mathbf{w}) - \lambda\frac{\partial}{\partial \mathbf{w}}(\mathbf{w}^{\mathrm{T}}\mathbf{w}) \\ &= \left[\mathbf{T} + \mathbf{T}^{\mathrm{T}}\right]\mathbf{w} - 2\lambda\mathbf{w} = 2\mathbf{T}\mathbf{w} - 2\lambda\mathbf{w} = 0.\end{aligned} \tag{4.107}$$

This after rewriting leads to the formula

$$\mathbf{T}\mathbf{w} = \lambda\mathbf{w}. \tag{4.108}$$

The last equation holds if $\mathbf{w} = \mathbf{w}'$ is an eigenvector of $\mathbf{T}$ and λ is a corresponding eigenvalue. The second condition in (4.103) is fulfilled immediately from (4.106) and the condition (4.104).

Up to this point we found conditions for an extreme, but to finish we have to specify conditions for a maximum. This can be done by substituting (4.108) and $\mathbf{w}'$ back into (4.101):

$$Q_{\max} = \mathbf{w}'^{T}\underbrace{\mathbf{T}\mathbf{w}'}_{=\lambda w'} = \mathbf{w}'^{T}\lambda\mathbf{w}' = \lambda\underbrace{\mathbf{w}'^{T}\mathbf{w}'}_{=c} = \lambda c, \tag{4.109}$$

which is maximized if λ is *the largest eigenvalue of* $\mathbf{T}$ and $\mathbf{w}'$ is *an eigenvector* corresponding to this eigenvalue.

A possible choice of $|\rho|$ instead of ρ^2 in (4.97) would lead to a more cumbersome functional Q' in which integration spans over space that is directly dependent on the sought directional vector $\mathbf{w}(\mathbf{x}_0)$, and which does not lead to a closed form solution like (4.99). Moreover $|\rho|$, although being a smooth function, is not differentiable at the origin which can pose some problems if used in optimization functionals.

[12]The derivatives can be verified as follows:

$$f(\mathbf{x}) = \mathbf{x}^{\mathrm{T}}\mathbf{A}\mathbf{x} = \begin{bmatrix} x_1 & \cdots & x_n \end{bmatrix}\begin{bmatrix} a_{11} & \ldots & a_{1n} \\ \vdots & \ddots & \vdots \\ a_{n1} & \cdots & a_{nn}\end{bmatrix}\begin{bmatrix} x_1 \\ \vdots \\ x_n\end{bmatrix} = \begin{bmatrix} \sum_{i=1}^{n} a_{i1}x_i & \cdots & \sum_{i=1}^{n} a_{in}x_i \end{bmatrix}\begin{bmatrix} x_1 \\ \vdots \\ x_n\end{bmatrix}$$

$$\frac{\partial f}{\partial x_1} = \left(\sum_{i=1}^{n} a_{i1}x_i + x_1 a_{11}\right) + x_2 a_{12} + \cdots + x_n a_{1n} = \sum_{i=1}^{n}(a_{i1} + a_{1i})x_i = \left(\mathbf{A}^{\mathbf{T}}_{Col\_1} + \mathbf{A}_{Row\_1}\right)\mathbf{x}$$

$$\cdots$$

$$\frac{\partial f}{\partial x_n} = x_n a_{n1} + x_n a_{n2} + \cdots + \left(x_n a_{in} + \sum_{i=1}^{n} a_{in}x_i\right) = \sum_{i=1}^{n}(a_{in} + a_{ni})x_i = \left(\mathbf{A}^{\mathbf{T}}_{Col\_n} + \mathbf{A}_{Row\_n}\right)\mathbf{x}$$

$$\nabla f(\mathbf{x}) = \begin{bmatrix} \frac{\partial f}{\partial x_1} & \cdots & \frac{\partial f}{\partial x_n}\end{bmatrix}^{\mathbf{T}} = \left[\mathbf{A}^{\mathbf{T}} + \mathbf{A}\right]\mathbf{x}.$$

Instead of building the functional Q as it is in (4.97), the problem of finding $\mathbf{w}(\mathbf{x}_0)$ can be approached in a more statistical fashion. For this purpose a histogram of local orientations (i.e. phases of gradient vectors $\mathbf{q}_i$) in a given neighbourhood can be built, from which the most frequent one is taken to represent that neighbourhood. Such an approximation was undertaken, for example, by Lowe in his SIFT detector [283].

4.6.2.1 2D Structural Tensor

Let us now focus on the 2D structural tensor $\mathbf{T}$ the components of which – given by (4.100) – can be denoted in a matrix-like fashion:

$$\mathbf{T} = \begin{bmatrix} T_{11} & T_{12} \\ T_{21} & T_{22} \end{bmatrix}, \tag{4.110}$$

where all T_{ij}, given by (4.100), are real and symmetrical, i.e.

$$T_{12} = T_{21}. \tag{4.111}$$

This special case is important for at least two reasons. The first is an obvious application to 2D images. The second comes from the close form of formulas for the eigenvalues and eigenvectors of $\mathbf{T}$ which, as we have already seen in the previous paragraphs, constitute a solution to our problem of dominating direction in small local neighbourhoods of pixels (4.97).

To find a spectrum of $\mathbf{T}$ (i.e. all its eigenvalues) we have to check the singularity of the resolving matrix: $\mathbf{T} - \lambda\mathbf{I}$, where λ stands for eigenvalues and $\mathbf{I}$ is a unit matrix (2D in this case) [259, 317]. The resolving matrix is singular if its determinant is zero:

$$\begin{aligned} \varphi(\lambda) = \det[\mathbf{T} - \lambda\mathbf{1}] &= \begin{vmatrix} T_{11} - \lambda & T_{12} \\ T_{21} & T_{22} - \lambda \end{vmatrix} \\ &= \lambda^2 - \lambda(T_{11} + T_{22}) + \left(T_{11}T_{22} - T_{12}^2\right) = 0. \end{aligned} \tag{4.112}$$

Since the above is a simple quadratic equation with respect to λ, we found its solution in a plausible closed form:

$$\begin{aligned} \lambda_1 &= \frac{1}{2}\left[(T_{11} + T_{22}) + \sqrt{(T_{11} - T_{22})^2 + 4T_{12}^2}\right], \\ \lambda_2 &= \frac{1}{2}\left[(T_{11} + T_{22}) - \sqrt{(T_{11} - T_{22})^2 + 4T_{12}^2}\right]. \end{aligned} \tag{4.113}$$

From this solution it follows easily that

$$\lambda_1 + \lambda_2 = T_{11} + T_{22} = \mathrm{Tr}(\mathbf{T}), \tag{4.114}$$

$$\lambda_1 - \lambda_2 = \sqrt{(T_{11} - T_{22})^2 + 4T_{12}^2}, \tag{4.115}$$

$$\lambda_1\lambda_2 = \det(\mathbf{T}) = T_{11}T_{22} - T_{12}^2, \tag{4.116}$$

where Tr(**T**) denotes the trace of **T**. The trace is invariant to the similarity transformation [259], i.e. given a matrix **U** for which there exists such a nonsingular matrix **R** that $\mathbf{U} = \mathbf{RTR}^{-1}$. The similar matrices **T** and **U** have the same eigenvalues. This indicates that Tr(**T**) is *invariant to the rotation*. From (4.115) we see also that $\lambda_1 \geq \lambda_2$.

Having found two eigenvalues let us consider the following special cases which are important from a practical point of view:

1. Equal eigenvalues:

$$\lambda_1 = \lambda_2 \Leftrightarrow (T_{11} = T_{22} \quad \wedge \quad T_{12} = T_{21} = 0). \tag{4.117}$$

If in addition the two are zero, then

$$\lambda_1 = \lambda_2 = 0 \Leftrightarrow T_{11} = T_{22} = 0. \tag{4.118}$$

2. Different eigenvalues with the smallest one equal to zero:

$$\lambda_2 = 0 \Leftrightarrow \{\det(\mathbf{T}) = 0 \wedge \lambda_1 = \mathrm{Tr}(\mathbf{T})\}. \tag{4.119}$$

All that we have achieved so far allows us to employ the 2D structural tensor to the analysis of characteristic patterns in images. This can be done by checking the eigenvalues of **T** or, computationally less expensive, by checking rank of **T**. Table 4.6 contains four characteristic cases of local neighbourhoods of pixels which can be deduced directly from eigenvalues or the rank of **T** [224].

Table 4.6 Types of local structures from the structural tensor

Rank of **T**	Eigenvalues	Type of local structure
0	$\lambda_1 = \lambda_2 = 0$	Constant intensity in an image – no signal change.
1	$\lambda_2 = 0, \lambda_1 > 0$	Intensity signal does not change in the direction associated with the smallest eigenvalue λ_2. However, there is a change in the direction associated with λ_1. This means that there is an ideal local structure. For vectors **q** in (4.99) which are intensity gradients, the eigenvector corresponding to λ_1 indicates a direction of maximal signal change in this neighbourhood of pixels. A special case of $\lambda_1 \gg \lambda_2 \approx 0$ can indicate lines in images.
2	$\lambda_1 > 0, \lambda_2 > 0$	The two eigenvalues are greater than zero, this means that there are changes in all directions. If one of the eigenvalues is dominating, then there is a dominating direction of signal changes. A special case of $\lambda_1 \geq \lambda_2 \gg 0$ can indicate corners in images (section 4.7.2).
2	$\lambda_1 = \lambda_2 > 0$	This is a special case of the above. Signal changes are equal in all directions. This corresponds to an *ideal isotropic* structure in the local neighbourhood of interest

We can get more quantitative information on local structures after determining eigenvectors of $\mathbf{T}$. As alluded to previously, the eigenvector associated with the greater eigenvalue λ_1 constitutes a solution to (4.97). It allows us to determine the local phase and local magnitude of a neighbourhood, as will be shown later in this chapter. We know also that for symmetrical real-valued matrices, as is the case of $\mathbf{T}$, the eigenvectors are orthogonal.

To find eigenvectors of $\mathbf{T}$, having already found its eigenvalues (4.113), one can directly solve (4.108). These eigenvectors can be found as nonzero columns of an adjoint matrix $[\mathbf{T} - \lambda\mathbf{1}]_{\text{ad}}$. In consequence there can be two eigenvectors, say $\mathbf{y}_1$ and $\mathbf{y}_2$, given as follows:

$$\begin{bmatrix} \mathbf{y}_1 & \mathbf{y}_2 \end{bmatrix} = \begin{bmatrix} y_{11} & y_{21} \\ y_{12} & y_{22} \end{bmatrix} = [\mathbf{T} - \lambda_i \mathbf{1}_2]_{\text{ad}} = \begin{bmatrix} T_{22} - \lambda_i & -T_{12} \\ -T_{12} & T_{11} - \lambda_i \end{bmatrix}, \tag{4.120}$$

where $\mathbf{y}_1$ and $\mathbf{y}_2$ are eigenvectors for a *single* eigenvalue λ_i. Substituting (4.113) into (4.120) we obtain

$$\begin{bmatrix} y_{11} & y_{21} \\ y_{12} & y_{22} \end{bmatrix} = \begin{bmatrix} \dfrac{(T_{22} - T_{11}) \pm \sqrt{(T_{11} - T_{22})^2 + 4T_{12}^2}}{2} & -T_{12} \\ -T_{12} & \dfrac{-(T_{22} - T_{11}) \pm \sqrt{(T_{11} - T_{22})^2 + 4T_{12}^2}}{2} \end{bmatrix}. \tag{4.121}$$

The sign in the above is chosen based on the eigenvalue (it is minus for λ_1 and plus for λ_2). Although for each eigenvalue we have up to two eigenvectors, we shall see that the choice of either one leads to the same result in respect to the vector $\mathbf{w}(\mathbf{x}_0)$. Moreover, each linear combination of eigenvectors for a single eigenvalue is also an eigenvector by itself.

Finally, we have to note that although we found a formula for an eigenvector of a dominating eigenvalue λ_1, it does not comply with an assumption of π rotation invariance, since we could have two different vectors with the same modulus but different signs. Therefore we have to find a vector that rotates twice the phase of the directional eigenvector – this vector will represent the local structure. Thus, it would perform a full rotation while the eigenvector traverses $0 - \pi$. Let us assume that ξ denotes the phase of the eigenvector for λ_1. Then $\mathbf{s}$ rotates with 2ξ, which can be found starting from the formula of doubled tangent:

$$\tan(2\xi) = \frac{2\tan(\xi)}{1 - \tan^2(\xi)} \quad \text{assuming that} \quad \tan(\xi) \neq 1. \tag{4.122}$$

From the definition of the tangent for any angle [259] applied to either eigenvector y_i, $i = 1, 2$

$$\tan(\xi_i) = \frac{y_{i2}}{y_{i1}}, \quad \text{for} \quad y_{i1} \neq 0,$$

and after substitution into (4.122) we obtain

$$\tan(2\xi_i) = \frac{2y_{i1}y_{i2}}{y_{i1}^2 - y_{i2}^2} \quad \text{for} \quad y_{i1} \neq y_{i2}. \tag{4.123}$$

Then, taking (4.121) into the above we obtain finally

$$\begin{aligned}\tan(2\xi) &= \frac{-4T_{12}\left[(T_{22}-T_{11}) \pm \sqrt{(T_{11}-T_{22})^2 + 4T_{12}^2}\right]}{(T_{22}-T_{11})^2 \pm 2(T_{22}-T_{11})\sqrt{(T_{11}-T_{22})^2+4T_{12}^2} + (T_{22}-T_{11})^2 + 4T_{12}^2 - 4T_{12}^2} \\ &= \frac{-4T_{12}\left[(T_{22}-T_{11}) \pm \sqrt{(T_{11}-T_{22})^2 + 4T_{12}^2}\right]}{2(T_{22}-T_{11})\left[(T_{22}-T_{11}) \pm \sqrt{(T_{11}-T_{22})^2 + 4T_{12}^2}\right]} = \frac{2T_{12}}{T_{11}-T_{22}}, \quad T_{11} \neq T_{22}.\end{aligned} \tag{4.124}$$

The last equation remains the same regardless of the chosen eigenvector since they are orthogonal, and the following holds:

$$\eta = \xi + 90^\circ; \quad \tan(2\eta) = \tan(2\xi + 180^\circ) = \tan(2\xi). \tag{4.125}$$

Immediately (4.124) gives us components of **w**:

$$\tan(\theta) = \tan(2\xi) = \frac{2T_{12}}{T_{11}-T_{22}} = \frac{w_2}{w_1}, \quad T_{11} \neq T_{22}. \tag{4.126}$$

Thus, the sought vector **w**, representing a local structure in an image, is given as

$$\mathbf{w} = \begin{bmatrix} w_1 \\ w_2 \end{bmatrix} = \begin{bmatrix} T_{11} - T_{22} \\ 2T_{12} \end{bmatrix}. \tag{4.127}$$

There are many ways to get to (4.127). For example, one can rotate the structural tensor **T** into the coordinate system of its principal axes [181, 224] or perform the SVD decomposition.

The final step consists of augmenting the structural vector **w** by a component that allows us to distinguish between two important cases: $\lambda_1 = \lambda_2 = 0$ and $\lambda_1 = \lambda_2 > 0$. This can be achieved by analysing the trace of **T** (4.114), which is 0 in the first case and $2\lambda > 0$ in the second. Thus, we obtain $\mathbf{w}^*$:

$$\mathbf{w}^* = \begin{bmatrix} \mathrm{Tr}(\mathbf{T}) \\ \mathbf{w} \end{bmatrix} = \begin{bmatrix} T_{11} + T_{22} \\ T_{11} - T_{22} \\ 2T_{12} \end{bmatrix}. \tag{4.128}$$

To find the postulated measure of coherency, the following scalar value can be used [224]:

$$c = \begin{cases} \left(\dfrac{\lambda_1 - \lambda_2}{\lambda_1 + \lambda_2}\right)^2 = \dfrac{\|\mathbf{w}\|^2}{[\mathrm{Tr}(\mathbf{T})]^2}, & \mathrm{Tr}(\mathbf{T}) \neq 0 \\ 0, & \mathrm{Tr}(\mathbf{T}) = 0 \end{cases}. \tag{4.129}$$

In the formula for c we profit from the fact that $\mathrm{Tr}(\mathbf{T})$ is invariant to the rotation. The coherence coefficient c takes on a value of 0 for ideal isotropic structures and 1 for structures with ideal linear direction. The case when $\mathrm{Tr}(\mathbf{T}) = 0$ happens for structures with constant intensity. In the case when $\|w\| = 0$ and $\mathrm{Tr}(\mathbf{T}) \neq 0$ we have the same changes of intensity in all directions instead.

Finally let us provide an equivalent representation of the directional vector $\mathbf{w}^*$, given in (4.128), which in some applications is more useful. It is also composed of three components, as follows:

$$\mathbf{s} = \begin{bmatrix} T_{11} + T_{22} \\ \angle\mathbf{w} \\ c \end{bmatrix}, \tag{4.130}$$

where c is a coherence factor given in (4.129) and $\angle w$ denotes a phase of the vector $\mathbf{w}$. It is given as

$$\angle w = \begin{cases} \arctan\left(\dfrac{2T_{12}}{T_{11} - T_{22}}\right) \; \textit{if } T_{11} \neq T_{22} \\ \dfrac{\pi}{2}, \textit{if } T_{11} = T_{22}, \textit{and } T_{12} \geq 0 \\ -\dfrac{\pi}{2}, \textit{if } T_{11} = T_{22}, \textit{and } T_{12} < 0 \end{cases} \tag{4.131}$$

If in (4.95) the vector $\mathbf{q} = \Delta I$ denotes a local gradient in an image, then it easily follows that

$$\begin{aligned} \|\nabla I\|^2 &= \nabla_x^2 I + \nabla_y^2 I, \\ \int_U \|\nabla I\|^2 \,\mathrm{d}x &= \int_U \nabla_x^2 I \mathrm{d}x + \int_U \nabla_y^2 I \mathrm{d}x = T_{11} + T_{22} = \mathrm{Tr}(\mathbf{T}). \end{aligned} \tag{4.132}$$

This means that the first component (i.e. the trace) in (4.128) can be interpreted as an averaged squared modulus of a local gradient. This property will be used later when using $\mathbf{T}$ for feature detection. It is also worth noting that coherence c (4.129) depends on trace of $\mathbf{T}$.

4.6.2.2 Computation of the Structural Tensor

Let us now assume that the directional vector $\mathbf{q}$ used to express the structural tensor (4.99) is the intensity gradient vector $\Delta I(x)$ computed at each pixel in an image. Assuming continuous signals, (4.99) can be rewritten as [224]

$$\mathbf{T}(\mathbf{x}_0) = \int_U \nabla I(\mathbf{x}) \nabla^{\mathrm{T}} I(\mathbf{x}) \mathrm{d}\mathbf{x} = \int_{-\infty}^{+\infty} h(\mathbf{x}_0 - \mathbf{x}) \nabla I(\mathbf{x}) \nabla^{\mathrm{T}} I(\mathbf{x}) \mathrm{d}\mathbf{x}, \tag{4.133}$$

where $\Delta I(\mathbf{x})$ is a gradient vector of intensity signal I at a point $\mathbf{x}$ in an image and $h(\mathbf{x})$ is a window function that models the local neighbourhood U around a pixel $\mathbf{x}_0$. With these assumptions the tensor components (4.100) take the form

$$T_{ij}(\mathbf{x}_0) = \int_{-\infty}^{+\infty} h(\mathbf{x}_0 - \mathbf{x}) \frac{\partial I(\mathbf{x})}{\partial x_i} \frac{\partial I(\mathbf{x})}{\partial x_j} \mathrm{d}\mathbf{x}, \tag{4.134}$$

where $\partial I(\mathbf{x})/\partial x_k$ denotes a set of directional derivatives of $I(\mathbf{x})$ in the direction of the k-th coordinate axis.

We easily notice that the last expression denotes a continuous convolution of a certain window function with the product of intensity gradients [312]. Because of this observation we can now switch to a domain of discrete signals, i.e. samples of the continuous signals (see also section 4.4). By this token we can transform (4.134) and achieve the following expression for tensor components in the discrete domain:

$$\hat{T}_{ij} = F(R_i R_j), \tag{4.135}$$

where $\hat{T}_{ij}$ is a discrete component of the structural tensor $\mathbf{T}$, F denotes a smoothing operator in a certain neighbourhood of pixels and R_k is a discrete differentiating operator in the k-th direction. The $R_i R_j$ operation means simple multiplication of outputs from the R_i and R_j filters, respectively.

The only one problem left in (4.135) is a practical choice of the smoothing and differentiating operators. This can be facilitated after examining some examples. The first is a binary (1/0) rectangle image, depicted in Figure 4.21(a). At the beginning, for each pixel position we compute three components of the structural tensor in accordance with (4.135). For the directional operators R_i and R_j we choose the Simoncelli 3-tap filter (section 4.4). The smoothing operator F is a 3×3 Gaussian (section 4.3). Then the three components are transformed into (4.130) for easier visualization.

Visualization of multicomponent objects, such as vectors, matrices or tensors, requires efficient transformation of the visualized values into other quantities which can be more intuitive for an observer [446]. The three components of the vector $\mathbf{s}$ (4.130) can be easily visualized in the HSI colour space, by the mapping [224]

$$\begin{aligned} \angle w &\to H \\ c &\to S \\ T_{11} + T_{22} &\to I \end{aligned} \tag{4.136}$$

By this technique we visualize the structural tensor of Figure 4.21(a) in Figure 4.21(b). We notice that smooth areas for which $\mathrm{Tr}(\mathbf{T}) = T_{11} + T_{22} = 0$ give no signal (black) areas.

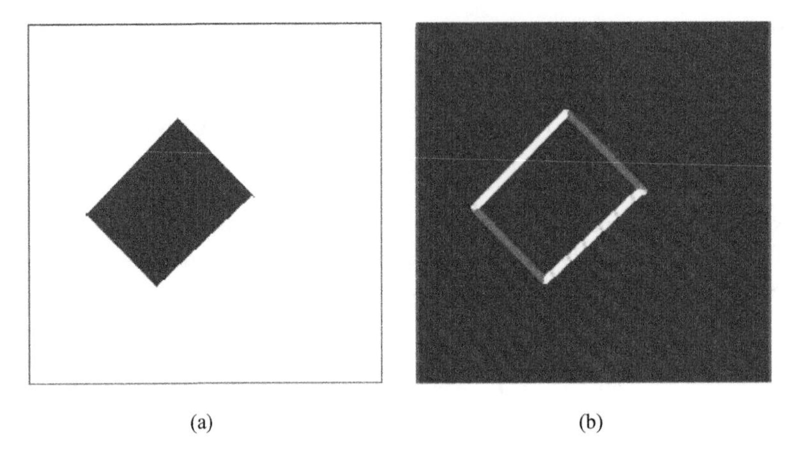

(a) (b)

Figure 4.21 (a) Binary image of a skewed rectangle and (b) colour visualization of its structural tensor. Hue H denotes a phase of local orientations, saturation S the coherence and intensity I conveys the trace of **T** (Plate 4)

The only nonzero response is at places with nonzero gradient, i.e. on the edges. The hue component corresponds to the orientation of an edge. As alluded to previously, it can be unambiguously determined up to a rotation by π. The lower edge in Figure 4.21(b) shows some irregularities which are due to irregularities in the original image. The saturation component conveys information on the coherence c.

Figure 4.22(a) depicts a monochrome grid image while Figure 4.22(b) shows the HSI colour visualization (4.136) of its structural tensor (see Plate 5). All computations in the presented examples were obtained with the C++ implementation of the structural tensor, provided in section 4.8.1.2.

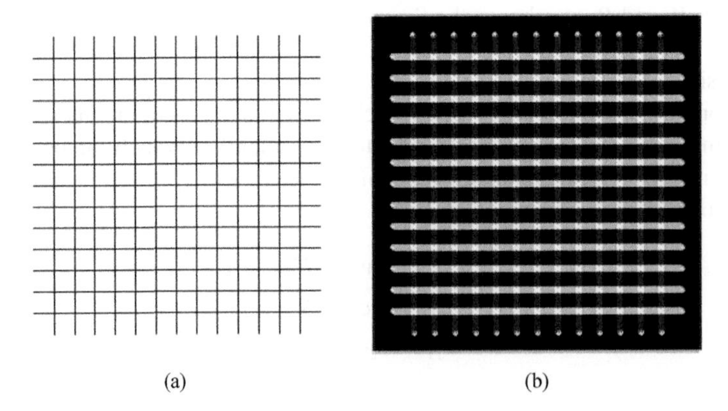

(a) (b)

Figure 4.22 (a) Monochrome image of a grid and (b) the colour visualization of its structural tensor (Plate 5)

4.6.3 Multichannel Image Processing with Structural Tensor

In the case of multichannel images, such as colour images, the question arises as to the definition of the gradient vector $\mathbf{q}(\mathbf{x})$. In this discussion we follow an approach proposed by Di Zenzo [102], which has been used also in the works by Sochen *et al.* [395] and Brox *et al.* [58] to name a few. It assumes summation of the partial gradient components, computed independently in image channels. To find the structural tensor for images with M channels we modify (4.99), as follows:

$$\mathbf{T}(\mathbf{x}_0) = \int\limits_{U(\mathbf{x}_0)} \sum_{k=1}^{M} \left(\mathbf{q}_k(\mathbf{x})\mathbf{q}_k^T(\mathbf{x})\right) d\mathbf{x}$$

$$= \sum_{k=1}^{M} \int\limits_{U(\mathbf{x}_0)} \left(\mathbf{q}_k(\mathbf{x})\mathbf{q}_k^T(\mathbf{x})\right) d\mathbf{x} = \sum_{k=1}^{M} \mathbf{T}_k(\mathbf{x}_0). \quad (4.137)$$

Thus, the summation in (4.137) spans all the gradient fields, each computed independently for every channel. This allows computation of local structures in multidimensional spaces such as multispectral (e.g. colour) images. It is also possible to employ (4.99) and (4.137) to analyse structures for physical data other than images.

A further extension of the multichannel structural tensor (4.137) is possible. This is a linear combination of the component tensors $\mathbf{T}_\mathbf{k}$:

$$\mathbf{T}(\mathbf{x}_0) = \sum_{k=1}^{M} c_k \mathbf{T}_k(\mathbf{x}_0), \quad (4.138)$$

where c_k are constants. By this it is possible to separately control the influence of each channel. A general extension to this is in the form

$$\mathbf{T}(\mathbf{x}_0) = \Gamma\left(\mathbf{T}_k\left(\mathbf{x}_0\right)\right), \quad (4.139)$$

where Γ is a function taking the component tensors $\mathbf{T}_k$.

There are also two different space dimensions involved in (4.137)–(4.139).

1. A dimension of $\mathbf{T}$ which comes directly from dimension of the gradient vector: It is 2D for single images ($\mathbf{T}$ is 2×2) or 3D for video sequences (3×3).
2. A dimension that follows the number of image channels, given by M in (4.137) and (4.138).

Similarly, there are two scale-spaces involved in (4.137)–(4.139).

1. The scale associated with the input images (i.e. in the domain of computation of tensors $\mathbf{q}_i$).
2. The scale imposed by the averaging (computation of the components T_{ij}).

Thus, we extend the discrete version of the structural tensor (4.135) to comprise the above two scale-space parameters as follows:

$$\hat{T}_{ij}(\rho, \xi) = F_\rho\big(R_i^{(\xi)} R_j^{(\xi)}\big), \tag{4.140}$$

where $R_i^{(\xi)}$ is a ξ-tap discrete directional operator (i.e. the order of the corresponding filter is $\xi - 1$) and F_ρ is a smoothing kernel of scale ρ (a second discussed type of scale).

Figure 4.23(b, c) depicts visualization of the structural tensor computed from the RGB colour image depicted in Plate 6(a). In the first case (Figure 4.23(b)) the 3-tap differentiating Simoncelli filter was used (section 4.4.1). In the second example (Figure 4.23(c)) the 5-tap Simoncelli filter was applied. We notice that different structures were detected in the two examples. As expected, a higher order of the filter results in greater smoothing and less influence of noise.

4.7 Corner Detection

Corners are very characteristic points of images. Intuitively, they are characterized by strong two- or multidirectional signal variations. Because of this feature, corner points are highly discriminative and are often used for image matching or object detection. However, many types of corners can be defined, and there are many methods for their detection in digital images.

A good overview of the corner detecting methods can be found in the paper by Zheng *et al.* [461] as well as in most classic textbooks on image analysis [157, 351]. A methodology for assessing the performance of some corner detectors is presented in the paper by Rockett [361]. It is based on the concept of the receiver operating characteristic (ROC) to check performance of the two classes – corners and noncorners – i.e. a labelling problem. A very coherent approach to the evaluation of interest point detectors is presented in the paper by Schmid *et al.* [373].

4.7.1 The Most Common Corner Detectors

Corner detection methods can be divided into three broad categories [373, 461].

1. The first group follows *parametric model fitting*. They are based on the *a priori* model of a corner which is then tried to fit to the intensity signal in an image. However, this limits potential interest points only to that model. They can be seen as a kind of template matching in which intensity values are matched to the model template. An example of a method in which a corner model is built is the method by Rohr [362]. This model depends on seven different parameters: position, angle of the symmetry axis, angle of the corner, grey level values and the blur. The optimization procedure is used then to fit a model to the local data template. Deriche and Blaszka [99] have extended this approach. For a corner model they employed an exponential function instead of a Gaussian smoothing kernel used by Rohr. A very original method from this category was proposed by Baker *et al.* [21]. It allows an automatic construction of a detector based on an arbitrary set of parameters. Each is represented as a densely sampled parametric manifold. Features are those image

Figure 4.23 (a) Examples of the structural tensor operating on an RGB colour image. (b) Visualization of the structural tensor computed with the 3-tap Simoncelli filter. (c) Version with the 5-tap Simoncelli filter (Plate 6)

points which, when projected into the lower dimensional subspace, are sufficiently close to the parametric manifold of a model.

2. Methods of the second group rely on *contour* of objects present in images. In the first step contours are detected. Then points are found where two or more contour lines meet or points are found with a maximal curvature. Sometimes inflection points in the contour chain are considered as well. An example is the method by Asada and Brady [12]. They define five groups of contour points: corners, cranks, ends, smooth joins and bumps. For each of these groups responses to convolution with Gaussian derivatives at different scales are found. Then each point of a contour is compared with the characteristic response of each group.
3. The third group constitute the *signal-based methods* (direct methods). Corners are obtained directly from the intensity (or colour) signal, analysing signal variations (i.e. its first or second derivatives). One of the first and very influential works in this area is the method of Kitchen and Rosenfeld [245] which relies on a product of grey level curvature and the magnitude of the gradient of intensity. Therefore this method is known also as a curvature-times-gradient method. Other methods, such as the one by Paler *et al.* [338] or by Harris and Stephens [174], although different, fall into this category. The former consists in subtracting a median filtered version of an image from the original one. Then corner measure is computed by multiplying the grey-level differences with the contrast over an area of interest. The latter method has gained much attention due to its properties and is described later in this section. However, many modifications to it have been proposed which show some improvements over the original proposition. The tensor-based corner detector, presented in the next section, is an example of such an improvement. Another interesting approach for corner detection has been proposed by Smith and Brady [394] in their SUSAN method (see Table 4.7 for an overview).

Table 4.7 summarizes the properties of the common signal-based corner detectors. These are the methods proposed by Beaudet [31], Harris and Stephens [174] and by Förstner [136], as well as the SUSAN method by Smith and Brady [394]. The first two rely on a Hessian matrix (second derivatives), whereas the latter two operate with first derivatives. However, the way in which they differ is the method of computing these parameters, as well as the way they are interpreted for corner detection.

Having so many methods to detect corners it is very important to have methods for their assessment. For this purpose some criteria need to be defined. These are as follows [394, 461].

1. Detection – a good corner detector should detect all corner points, even the ones that are not characterized by a strong signal response. At the same time it should be insensitive to noise.
2. Localization – corners should be detected and marked in the positions of their true occurrence.
3. Stability – detected points should persist at their locations even on multiple acquisitions under varying conditions or some geometric transformations of the same scene. Stability is often measured by a repeatability measure [373].
4. Speed – it is obvious that the faster the method the better. However, sometimes the speed factor is in opposition to other parameters of a detector, such as good localization for instance.

Table 4.7 Overview of the most common corner detectors based on direct signal analysis

Detector	Description, properties, computation
Beaudet	Corners are found based on the absolute value of the determinant of the Hessian **H**: $$\left\|\det(\mathbf{H})\right\| = \left\|H_{11}H_{22} - H_{12}H_{21}\right\| = \left\|I_{xx}I_{yy} - I_{xy}^2\right\|, \quad (4.141)$$ where $$\mathbf{H} = \begin{bmatrix} H_{11} & H_{12} \\ H_{21} & H_{22} \end{bmatrix} = \begin{bmatrix} I_{xx} & I_{xy} \\ I_{yx} & I_{yy} \end{bmatrix}, \quad (4.142)$$ and $$I_{ij}(\mathbf{x}) \approx \frac{\partial^2 \hat{I}(\mathbf{x})}{\partial x_i \partial x_j} \quad (4.143)$$ denotes a *discrete* approximation of the second order i-th and j-th derivatives of the continuous intensity signal $\hat{I}(\mathbf{x})$. These can be computed for instance with the Savitzky–Golay differentiating filters (section 4.4.2). A point **x** is classified as a corner point if $\left\|\det(\mathbf{H}(\mathbf{x}))\right\|$ is a local maximum in a closest neighbourhood of **x** and if it holds that $$\left\|\det(\mathbf{H})\right\| < \tau, \quad (4.144)$$ where τ is a specific threshold value, usually different for different images. It appears that the above determinant of the Hessian matrix is invariant to image rotations. It is also related to the Gaussian curvature of the image signal (if image surface is defined as the one containing all points which are distant from the reference plane by their intensity values) [167, 263]. The most troublesome aspect is the practical choice of the threshold value τ.
Harris and Stephens	This is one of the most popular corner detectors and operates on the smoothed *first derivatives* of the intensity signal [174]: $$\mathbf{T} = \begin{bmatrix} T_{11} & T_{12} \\ T_{21} & T_{22} \end{bmatrix} = \begin{bmatrix} F(I_x^2) & F(I_x I_y) \\ F(I_y I_x) & F(I_y^2) \end{bmatrix}, \quad (4.145)$$ where F is a smoothing operator. Usually it is a Gaussian kernel $G(0, \sigma)$ with a zero mean value and variance σ (section 4.3.1). However, in practice the binomial filter can be used as well (section 1.3.2). We easily notice that the above formula is equivalent to the *2D structural tensor* given by (4.100), for which q_i is set to the first derivative of the intensity signal, i.e. $q_1 = \partial I(x, y)/\partial x$ and $q_2 = \partial I(x, y)/\partial y$. In the method of Harris and Stevens to find a corner the following value has to be calculated [174]: $$R = \det(\mathbf{T}) - k\left[\mathrm{Tr}^2(\mathbf{T})\right], \quad (4.146)$$

(*continued*)

Table 4.7 Overview of the most common corner detectors (*Continued*)

Detector	Description, properties, computation
	where k is parameter in the range 0 to 0.25. Then, for a corner point, its value of R must constitute a local maximum and be greater than a given threshold. As reported by Rockett [361] the best results were obtained with $k = 0.04$. He also pointed out that the smoothing of the first derivatives with a Gaussian kernel in (4.145) plays a more fundamental role than simple noise filtering (as stated by Harris and Stephens). Its role is more essential since it isotropically changes the spectral response of the corner detector. In other words, if removed then **T** would be identically zero regardless of noise in the input signal. This property was explained when discussing the structural tensor (section 4.6.2).
Förstner	This method is also based on the matrix **T** given in (4.145). It is easy to show that this matrix is symmetric and positively defined, thus it can be decomposed as follows [308]: $$\mathbf{T} = \mathbf{U}^{\mathrm{T}} \begin{bmatrix} \lambda_1 & 0 \\ 0 & \lambda_2 \end{bmatrix} \mathbf{U},$$ where U is a certain unitary matrix for which $$\det(\mathbf{T}) = \lambda_1\lambda_2 \qquad \mathrm{Tr}(\mathbf{T}) = \lambda_1 + \lambda_2.$$ To classify a point $\mathbf{x}$ as a corner, two conditions have to be met. The first one, denoted by W and called a *weight* of a point, is given as $$W = \frac{\mathrm{Tr}(\mathbf{T})}{\det(\mathbf{T})} = \frac{1}{\lambda_1} + \frac{1}{\lambda_2}. \quad (4.147)$$ The second parameter q is called the roundness measure and is used to describe the *likelihood* of a point: $$q = \frac{4\det(\mathbf{T})}{Tr^2(\mathbf{T})} = 1 - \left(\frac{\lambda_1 - \lambda_2}{\lambda_1 + \lambda_2}\right)^2. \quad (4.148)$$ A corner is asserted when q and W are local maxima and are greater then certain thresholds. It is also interesting to notice that q is in the range 0 to 1.
Smith and Brady (SUSAN)	SUSAN (Smallest Univalue Segment Assimilating Nucleus) detector developed by Smith and Brady [394] presents an entirely different approach to the 1D and 2D feature detection in images, such as edges and corners, respectively. Circular mask M consisting of 37 pixels is used. The central pixel of the mask is called a nucleus. Then intensities of all pixels within a mask are compared with an intensity of a nucleus and an area of 'similar' pixels is marked. This area is called USAN (Univalue Segment Assimilating Nucleus) and it conveys the most important information on a local structure of an image. Analysing the size, centroid and the second moments of USAN the exact information on a type of local structure around a nucleus is inferred, such as edges or corners. For those regions, inverted USAN area shows strong peaks – thus the term SUSAN – i.e. the smallest USAN. This approach has an additional advantage of not using any derivatives which are cumbersome to use in the presence of noise.

Table 4.7 (*Continued*)

Detector	Description, properties, computation

Computing USAN for every pixel in the digital image leads to detection of edges or corners. The number of pixels of USAN $n(r_0)$ is computed as

$$n(r_0) = \sum_{r \in M} e^{-\left(\frac{I(r)-I(r_0)}{t}\right)^6}, \tag{4.149}$$

where t is a threshold for a difference of brightnesses and r and r_0 are distances to a pixel and to a nucleus of M, respectively. The value of USAN gets smaller near the points of interest which are located at local maxima of the following value (a SUSAN principle):

$$R(r_0) = \begin{cases} g - n(r_0), & \text{for} \quad n(r_0) < g \\ 0, & \text{for} \quad n(r_0) \geq g \end{cases}, \tag{4.150}$$

where g is half of $n_{\max}$ value of a mask M. The SUSAN corner detection procedure is outlined as follows.

1. Place a circular mask around a pixel (i.e. a nucleus).
2. Calculate the number $n(r_0)$ of pixels within the circular mask which have similar brightness to the nucleus in accordance with (4.149). Such pixels constitute the USAN.
3. Compute strength of a corner from (4.150).
4. Test for false positives by finding the centroid of USAN and its contiguity.
5. Use nonmaximum suppression to find corners (details in [394]).

The problem of scale and transformation invariance of interest points in images has been addressed in the works by Mikołajczyk [310], Mikołajczyk and Schmid [311] and Lowe [283], to name a few. It is also discussed in the already cited work by Rockett [361].

4.7.2 *Corner Detection with the Structural Tensor*

Knowing the concepts behind the 2D structural tensor (section 4.6.2.1), we can build a corner detector based on it with different properties compared to other detectors, such as that of Harris and Stevens, for instance.

We know that the type of local structure can be inferred solely from the eigenvalues of the structural tensor (see formulas (4.113) and (4.117)). Thus, the corner points (x_i, y_i) can be those which fulfil the following condition [14]:

$$\lambda_1(x_i, y_i) \gg \lambda_2(x_i, y_i) > \kappa, \tag{4.151}$$

where κ is a threshold for the lower eigenvalue of the structural tensor (discussed in section 4.6.1.2). It can be set to 0 if the point priority technique (described later in this section) is employed. The first inequality in the above equation can be written in a form which is more useful in practical realizations:

$$\lambda_1(x_i, y_i) - \lambda_2(x_i, y_i) > \kappa^*, \tag{4.152}$$

where κ^* is a second threshold value which in the simplest form can be set to 0. Based on (4.115) this can be written simply as

$$[T_{11}(x_i, y_i) - T_{22}(x_i, y_i)]^2 + 4T_{12}^2(x_i, y_i) > \kappa^{**}. \tag{4.153}$$

with a new threshold $\kappa^{**} = (\kappa^*)^2$. The main advantages of using the tensor approach to corner detection are as follows.

1. Tensors allow inherent integration of the multiple channel signals and image scale concepts into the corner detection (see section 4.6.3).
2. Application of the precise discrete signal differentiation methods, such as Simoncelli filters, results in better localization parameter (section 4.4).
3. Detection based on the prioritized queues of eigenvalues, set according to their strength, introduces a natural order among corner points.
4. Simple implementation and fast execution.

To avoid a cumbersome selection of a threshold parameter for eigenvalues in (4.151), a special *priority queue* is proposed. Additionally, the input image is partitioned into equal size tiles and corners are detected independently in each of them. This results in more uniform detection in the whole image area. Figure 4.24 depicts this technique.

The priority queue in Figure 4.24 is sorted by λ_2, denoted further without subscript. So, for the consecutive cells i and $i + 1$ it always holds that $\lambda_i \leq \lambda_{i+1}$. Whenever a new point is found that fulfils (4.151) it is tried to fit into the queue based on its lower eigenvalue. If this value is bigger than that already stored at index 0 then it is inserted into the queue at a position for which $\lambda_i \leq \lambda_{i+1}$ always holds, then the cell at 0 is removed. To avoid time consuming shifts the priority queue should be implemented as a linked list [74]. Finally, after checking all image points the priority queue contains at most M points with the biggest λ_2 values.

With this data structure it is also possible to impose additional constraints, e.g. on a minimal allowable distance among adjacent corner points. In this way we can search for more 'distributed' corner positions in an image. For instance, we can search for points of interest that are at least two pixels apart.

The other idea worth considering is to split an input image into a number of smaller size tiles and look for corners in each tile separately (see Figure 4.24). By this technique we can select corners which are not concentrated only in one part of an image.

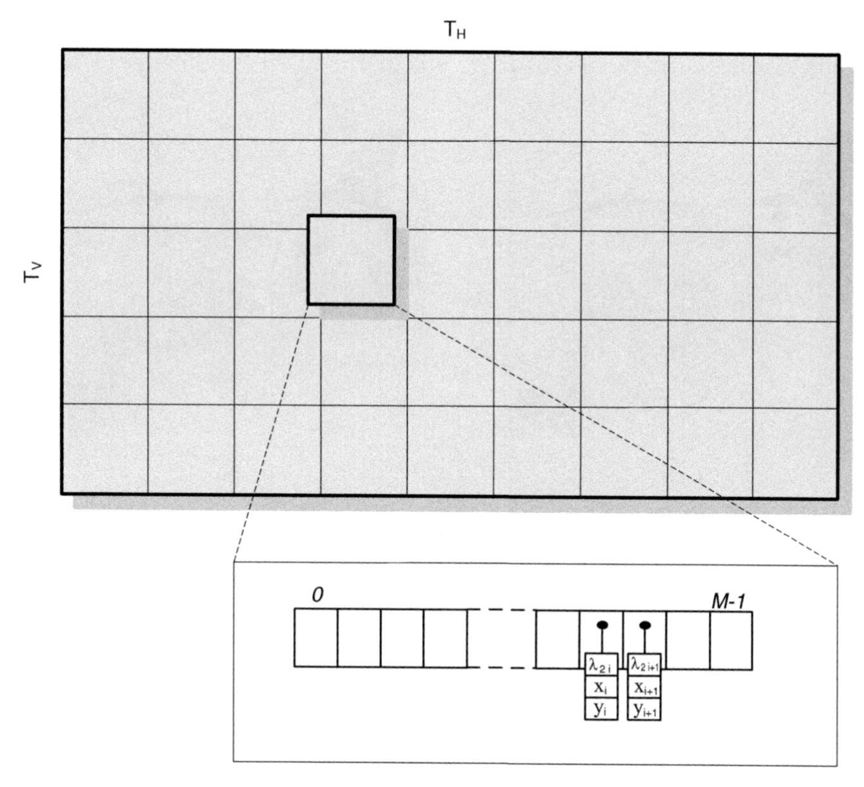

Figure 4.24 Priority queue for selection of M most prominent corner points based on the eignevalues of the structural tensor. Corners are detected independently in each image tile. An image is divided into $T_H \times T_V$ equal size tiles

Figures 4.25 and 4.26 present application of the tensor corner detection applied to the 'Airplane' test image. In the first case corners are searched in the whole image. In the second, prior to detection the image has been divided into 4×4, then into 16×16 tiles.

4.8 Practical Examples

4.8.1 C++ Implementations

4.8.1.1 Convolution

Algorithm 4.3 lists the C++ implementation of the 2D convolution. It was designed as a template class named *_Convolve*. The class contains one static member, named *Convolve*, which actually does the job. Thanks to C++ templates it is now possible to tailor the class instantiation to the pixel type *T*, which is the first type parameter of the template. The second template parameter *ATrait* helps in the choice of an appropriate type for an accumulator, i.e. a variable that stores a common sum, given pixel type. Traits are designed as a special class hierarchy *Accumulation_Trait*, also parameterized by the type of pixel. Some of the most

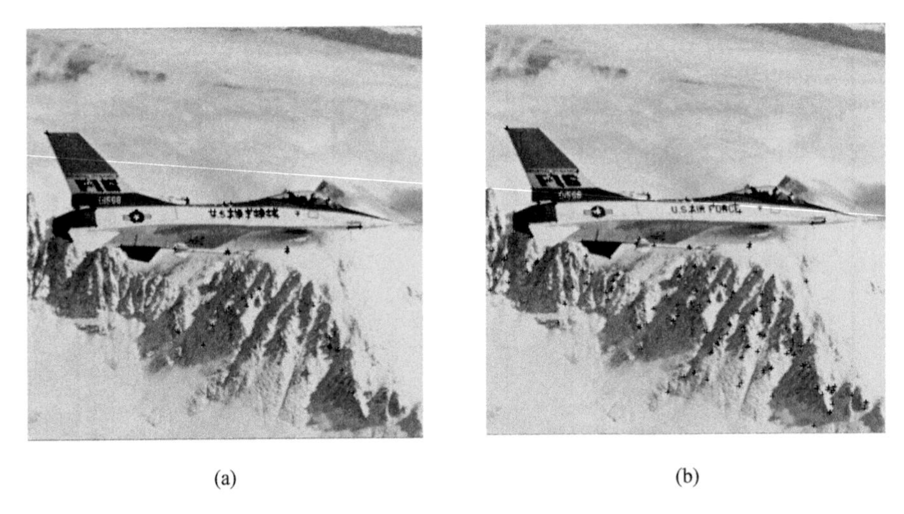

(a) (b)

Figure 4.25 Corners (black crosses) found in the whole 'Airplane' image by the tensor-based detector: (a) 200 corners, no constraints on their mutual distance; (b) 200 corners with imposed distance of one or more pixels. (Source: USC – 1 SIPI Image Database)

common traits are listed in Algorithm 4.4. We should note, however, that we need a separate trait for each different type of image (i.e. for each different type of pixel). Traits are discussed in section 13.3.5.

Implementation of the 2D convolution follows its definition given in (4.3). The algorithm is implemented in the member function *Convolve*, which takes three input parameters: a reference to the input image, to the convolution masks and to the output image. We should note that the most straightforward implementation of the 2D convolution requires two images of

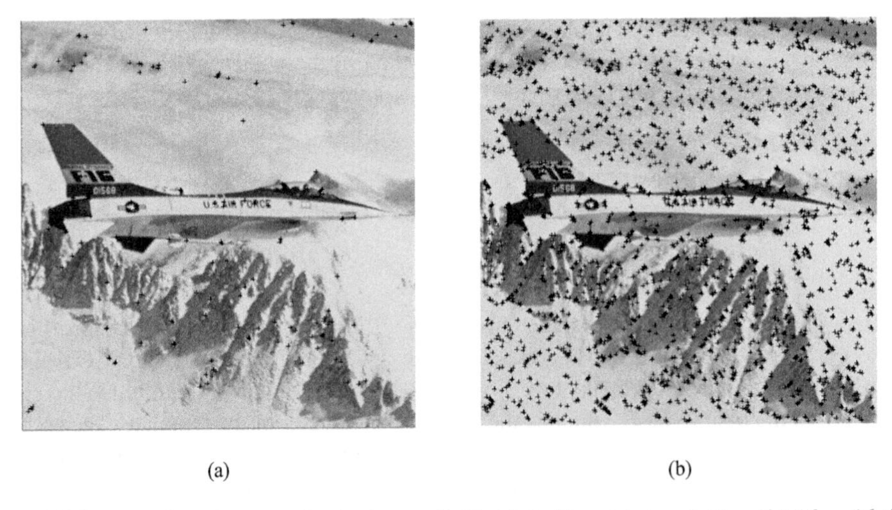

(a) (b)

Figure 4.26 Detection of corners in the image divided into tiles: (a) 4 × 4 tiles; (b) 16 × 16 tiles, maximum 10 corners in a tile. (Source: USC – 1 SIPI Image Database)

```
//////////////////////////////////////////////////////////////
// This class performs 1D horizontal convolution
//
// The type "T" defines a type of a single pixel.
// The type "ATrait" defines a type for the
// accumulator used during convolution.
//
// We create a class here instead of the template function.
// The problem is that the template functions don't allow
// default parameters. In this fashion a user would
// be forced to provied ATrait all the time. Therefore
// we have this class.
//
// For the actual convolution use the helper functions.
//
//////////////////////////////////////////////////////////////
template < typename T, typename ATrait = Accumulation_Trait< T > >
class _Convolve
{
   public:


         //////////////////////////////////////////////////////////////
         // This function does the convolution Out = In * Mask.
         // The convolution is performed symmetrically, i.e. at a
         // given point of the image convolution summation goes
         // symmetrically in all directions. If value of a image
         // or mask outside their boundaries is necessary it is
         // assumed that this value is 0.
         //////////////////////////////////////////////////////////////
         //
         // INPUT:
         //          In - a reference to the input image
         //          Mask - a reference to the convolution kernel image
         //          Out - a reference to the outcome image
         //
         // OUTPUT:
         //          Out = In * Mask
         //
         // REMARKS:
         //          Although it is not necessary, the best if mask size
         //          is odd in both directions -- in such a case the whole
         //          procedure can be simplified of the conditions.
         //
         static void Convolve( const     TImageFor< T > & In,
                               const     TImageFor< T > & Mask,
                                         TImageFor< T > & Out )

         {
            register long i,j,m,n,
            register long vertMaskIndex,horzMaskIndex,imTempCol,imTempRow;

            typename ATrait::TheAccumulatorType conv_sum;

            const long im_col = In.GetCol();
            const long im_row = In.GetRow();

            const long mask_col = Mask.GetCol();
            const long mask_row = Mask.GetRow();

            long horzHalfMask = mask_col >> 1; // == mask_col / 2;
            long vertHalfMask = mask_row >> 1; // == mask_row / 2;

            // Go through all the points of the input picture
            for( i = 0; i < im_row; i ++ )
            {
```

Algorithm 4.3 C++ implementation of the 2D convolution. (Reproduced by permission of Pandora Int. Inc., London)

```
                    for( j = 0; j < im_col; j ++ )
                    {
                        // Count the sum for the whole mask at each point
                        conv_sum = ATrait::GetZero();

                        for( m = - vertHalfMask; m <= + vertHalfMask; m ++ )
                        {
                            vertMaskIndex   = m + vertHalfMask;
                            imTempRow       = i - m;

                            for( n = - n horzHalfMask; n <= + horzHalfMask; n ++ )
                            {
                                horzMaskIndex = n + horzHalfMask;
                                imTempCol = j - n;

                                if(  imTempCol >= 0 && imTempCol < im_col &&
                                     imTempRow >= 0 && imTempRow < im_row
                                   // >>> This part of AND is necessary only
                                   // if size of the mask is even <<<
                                   &&  horzMaskIndex  <  mask_col &&
                                       vertMaskIndex  <  mask_row )
                                  conv_sum+=
                                     Mask.GetPixel( horzMaskIndex, vertMaskIndex )
                                     * In.GetPixel( imTempCol, imTempRow );

                            }

                        }

                        Out.SetPixel( j, i, (T)conv_sum );
                    }
                }
            }
};
```

Algorithm 4.3 (*Continued*)

the same type and size: the input image, which is only read and not changed, and the output image, which stores the results of convolution. Some *in situ* algorithms can be also implemented if necessary. However, we should remember not to overwrite the input pixels before they are read for all computations of the convolution. Otherwise we would end up with wrong results.

For a given type T, the sole purpose of its associated *Accumulation_Trait<T>* is to provide proper types of a zero value and a type for accumulator variable. Because of this automation we can make a versatile template for such classes as *_Convolve*. Otherwise we would need to write the number of the convolution copies, each different only by a type of one of its variables (*conv_sum* in Algorithm 4.3).

We should also remember that being universal the procedure in Algorithm 4.3 is not the most efficient. If we had a fixed size mask, or we knew it was symmetrical, then we could

```
// This trait defines a type which
// will be used when accumulating
// data of different types.
template < typename T >
class Accumulation_Trait;


// Specific traits are implemented
// as specializations.
template <>
class Accumulation_Trait< double >
{
    public:
        // For double, accumulator is: long double
        typedef long double TheAccumulatorType;
        static TheAccumulatorType GetZero( void ) { return 0.0; }
};

template <>
class Accumulation_Trait< unsigned char >
{
    public:
        // For char, accumulator is: long
        typedef unsigned long TheAccumulatorType;

};
```

Algorithm 4.4 Exemplary trait classes which can be used in *_Convolve*. (Reproduced by permission of Pandora Int. Inc., London)

write a more efficient version for that case. We should always when possible try to use the separable masks for convolution, as discussed in section 4.2.2. The attached library contains implementation of the 1D convolution which is used whenever a 2D convolution can be represented in the form of Equation (4.4), i.e. the filter mask can be separated.

4.8.1.2 Implementing the Structural Tensor

Algorithm 4.5 provides an example of using the library interface and image operators. The template function *TensorTest<T>* computes the three components of the structural tensor. The structural tensor is computed based on the simplest algorithm with separable Simoncelli and binomial filters (section 4.4).

The template function *TensorTestFor<T>* is parameterized by a type *T* of a pixel of images that will be used for computations of the tensor. However, formats of the input and output images are fixed in this example:

- monochrome images for input (8 bits per pixel);
- real images for the tensor components (float data for pixels).

Algorithm 4.6 presents an example of calling the *TensorTestFor<double>* in some other function. In the *TensorTestFor* the input image is always copied into the auxiliary image with floating point pixels. Then, computations are done in floating point precision.

```
#include "HIL_BaseDefinitions.h"
#include "HIL_ArithmeticOperators.h"
#include "HIL_ConvolveOperators.h"
#include "HIL_MultiChannelImageFor.h"

using namespace PHIL;

template< class T >
void TensorTestFor(    const MonochromeImage & inImage,
                       TRealImage & Jxx, TRealImage & Jxy, TRealImage & Jyy )
{
    const int kCols = inImage.GetCol();
    const int kRows = inImage.GetRow();

    TImageFor< T > inputImage( kCols, kRows );

    ( * FormatConverter_AP( inputImage, inImage ) )();

    // three components of the structural tensor
    TMultiChannelImageFor< T >    outputImage( kCols, kRows );


    // Simoncelli horizontal (vertical) gradient is obtained in two steps:
    // - first we do vertical (horizontal) smoothing with the prefilter
    // - then we apply the horizontal (vertical) derivative filter
    vector< T > theSimoncelliPrefilter;

    theSimoncelliPrefilter.push_back( 0.22420981526374817 );
    theSimoncelliPrefilter.push_back( 0.5515803694725037 );
    theSimoncelliPrefilter.push_back( 0.22420981526374817 );


    vector< T > theSimoncelliDerivative;

    theSimoncelliDerivative.push_back( -0.45527133345603943 );
    theSimoncelliDerivative.push_back( 0.0 );
    theSimoncelliDerivative.push_back( 0.45527133345603943 );


    vector< T > theBinomialSmoothing;

    theBinomialSmoothing.push_back( 0.25 );
    theBinomialSmoothing.push_back( 0.5 );
    theBinomialSmoothing.push_back( 0.25 );



    TImageFor< T > tmpImage_1( kCols, kRows );
    TImageFor< T > tmpImage_2( kCols, kRows );
    TImageFor< T > tmpImage_3( kCols, kRows );


    TVectorMultiImageOperation  * Jxx_Computer = new
                                        TVectorMultiImageOperation;
    TVectorMultiImageOperation  * Jxy_Computer = new
                                        TVectorMultiImageOperation;
    TVectorMultiImageOperation  * Jyy_Computer = new
    TVectorMultiImageOperation;
    // Compute the tensor components: Jij = F(Ri*Rj)

    // Precompute the horz derivative (to tmpImage_1)
    Jxx_Computer->AddAdoptNewOperation( Vert_Convolve_AP( tmpImage_3,
                    inputImage, theSimoncelliPrefilter ) );
    Jxx_Computer->AddAdoptNewOperation( Horz_Convolve_AP( tmpImage_1,
```

Algorithm 4.5 C++ implementation of the 2D structural tensor. (Reproduced by permission of Pandora Int. Inc., London)

```
                    tmpImage_3, theSimoncelliDerivative ) );

        // Precompute the vert derivative (to tmpImage_2)
        Jxx_Computer->AddAdoptNewOperation( Horz_Convolve_AP( tmpImage_3,
                        inputImage, theSimoncelliPrefilter ) );
        Jxx_Computer->AddAdoptNewOperation( Vert_Convolve_AP( tmpImage_2,
                        tmpImage_3, theSimoncelliDerivative ) );

        // Jxx
        Jxx_Computer->AddAdoptNewOperation(
                    Orphan_Mul( outputImage.GetRefPixel( 0, 0 ),
                        tmpImage_1, tmpImage_1 ) );
        Jxx_Computer->AddAdoptNewOperation( Orphan_Vert_Convolve( tmpImage_3,
                outputImage.GetRefPixel( 0, 0 ), theBinomialSmoothing ) );
        Jxx_Computer->AddAdoptNewOperation(
                    Orphan_Horz_Convolve( outputImage.GetRefPixel( 0,
                    0 ), tmpImage_3, theBinomialSmoothing ) );

        // Jxy
        Jxy_Computer->AddAdoptNewOperation(
                    Orphan_Mul( outputImage.GetRefPixel( 1, 0 ),
                    tmpImage_1, tmpImage_2 ) );
        Jxy_Computer->AddAdoptNewOperation( Orphan_Vert_Convolve( tmpImage_3,
                    outputImage.GetRefPixel( 1, 0 ), theBinomialSmoothing ) );
        Jxy_Computer->AddAdoptNewOperation(
                    Orphan_Horz_Convolve( outputImage.GetRefPixel( 1,
                    0 ), tmpImage_3, theBinomialSmoothing ) );

        // Jyy
        Jyy_Computer->AddAdoptNewOperation(
                    Orphan_Mul( outputImage.GetRefPixel( 2, 0 ),
                    tmpImage_2, tmpImage_2 ) );
        Jyy_Computer->AddAdoptNewOperation( Orphan_Vert_Convolve( tmpImage_3,
                    outputImage.GetRefPixel( 2, 0 ), theBinomialSmoothing ) );
        Jyy_Computer->AddAdoptNewOperation(
                    Orphan_Horz_Convolve( outputImage.GetRefPixel( 2,
                    0 ), tmpImage_3, theBinomialSmoothing ) );


        TVectorMultiImageOperation compoundComputer;//this object is automatic

        compoundComputer.AddAdoptNewOperation( Jxx_Computer );
        compoundComputer.AddAdoptNewOperation( Jxy_Computer );
        compoundComputer.AddAdoptNewOperation( Jyy_Computer );

        // Compute all
        compoundComputer();


        // Prepare output image
        ( * FormatConverter_AP( Jxx, outputImage.GetRefPixel( 0, 0 ) ) ) ();
        ( * FormatConverter_AP( Jxy, outputImage.GetRefPixel( 1, 0 ) ) ) ();
        ( * FormatConverter_AP( Jyy, outputImage.GetRefPixel( 2, 0 ) ) ) ();

    }
```

Algorithm 4.5 *(Continued)*

4.8.2 Implementation of the Morphological Operators

Figure 4.27 depicts a class hierarchy for the morphological image operations. All operations are derived from the *TImageTemplateOperationFor<>* template base class (section 3.7.1.3). The template image base class reflects the presence of the mandatory structural element for each operation from this group. The role of the structural element is to

```
void MyFun( void )
{
	const int kCols = 128;
	const int kRows = 100;

	MonochromeImage theMonoImage( kCols, kRows );

	// ... initialize theMonoImage

	TRealImage Jxx( kCols, kRows );
	TRealImage Jxy( kCols, kRows );
	TRealImage Jyy( kCols, kRows );

	// Compute the structural tensor
	TensorTestFor< double >( theMonoImage, Jxx, Jxy, Jyy );

	// ... the tensor components are ready

}
```

Algorithm 4.6 Example of calling the structural tensor procedure

define the structure of the local neighbourhood for morphological operation in the processed image.

Algorithm 4.7 presents the *MorphologyFor<>* template class with full implementation of the *Dilate*() member. Its organization is somewhat similar to the already discussed convolution procedures. In the innermost part of the loops the structural element is checked and, if enabled, then the maximal value in a local window is assessed. Finally, this maximal value constitutes

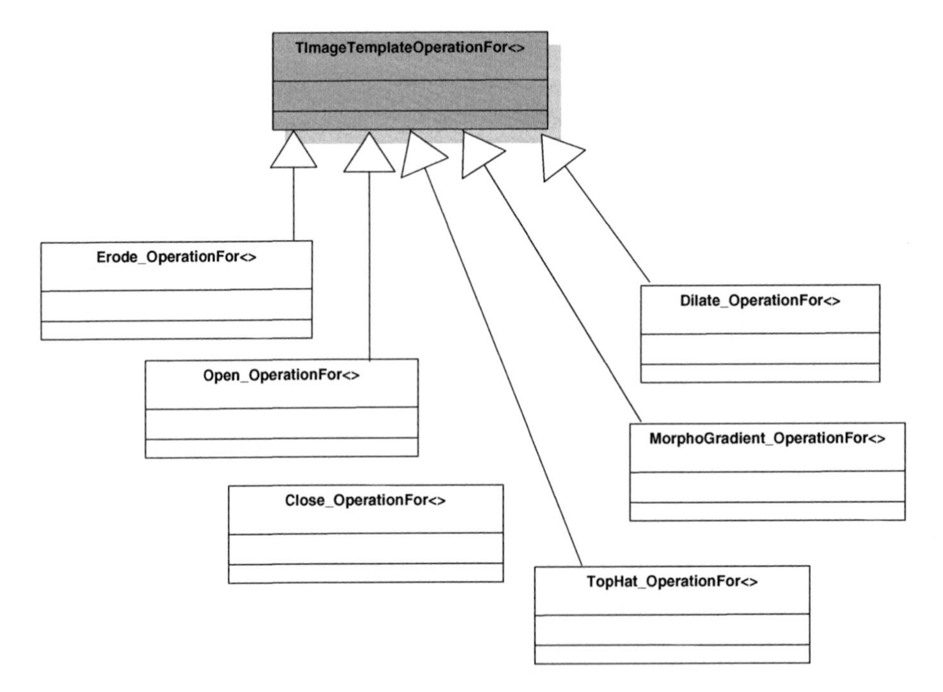

Figure 4.27 Class hierarchy of the morphological image operators

```
///////////////////////////////////////////////////////////
//
// The class for morphological operations on different
// type images.
//
// This is a general implementation that takes into account
// the grey valued definition of the morphological
// operations. This means that the presented functions
// can operate with almost any types of pixels for which
// the order relation has been defined. However, this
// means also that the implementation might not be optimal
// in many cases (e.g. for pure binary images).
//
// The structural element is always defined as the masked
// image. This means that a shape of the structural element
// is defined by a binary mask inherent to the masked image,
// whereas the pixel values of the element are taken into
// computations of the morphological functions. Certainly,
// if the pixels are set to 0, then only a shape of the
// structural element is taken into consideration.
//
// Optimization issues: the most important optimization can
// be achieved by virtue of SEPARABLE structural elements
// (the concept similar to the separable convolution masks).
// For more information see e.g. the book by Soille.
//
///////////////////////////////////////////////////////////
template < class T >
class MorphologyFor
{

	public:

		typedef std::auto_ptr< TImageFor< T > > TheImageAutoPtr;

	public:

		// ===================================================
		MorphologyFor( void ) {}			// class default constructor
		virtual ~MorphologyFor() {}			// class virtual destructor
		// ===================================================

	public:

		///////////////////////////////////////////////////////////
		// This is morphological dilation on any pixel type input
		// image. The structural element is in the form of a masked
		// image where the mask denotes which pixel belongs to the
		// that element and the values are taken to the computation.
		// The geometrical center of the sctructural element is taken
		// as the central point.
		///////////////////////////////////////////////////////////
		//
		// INPUT:
		//		theImage - the input image with pixels of type T
		//		theStructureElement - the structural element with
		//			pixels of type T and binary mask defining
		//			shape of this element
		//
		// OUTPUT:
		//		Auto ptr to the returned (orphaned) object which
		//		contains a dilated version of the input image
		//
		// REMARKS:
```

Algorithm 4.7 Definition of the class with full implementation of the dilation operator. Other members are available in the library. (Reproduced by permission of Pandora Int. Inc., London)

```
//      The returned object is orphaned!
//
TheImageAutoPtr Dilate( const TImageFor< T > & theImage,
              const TMaskedImageFor< T > & theStructureElement )

{
    register long i, j, m, n,
         vertMaskIndex, horzMaskIndex, imTempCol, imTempRow;

    const long im_col = theImage.GetCol();
    const long im_row = theImage.GetRow();

    const long mask_col = theStructureElement.GetCol();
    const long mask_row = theStructureElement.GetRow();

    // We don't assume any specific size or center of the structural
    // element here,although most often than not it is symmetrical
    // with geometrical center being a pivoting point.

    long horzHalfMask = mask_col / 2;
    long vertHalfMask = mask_row / 2;

    // This is an output image in its initial state (all pixels 0)

    TheImageAutoPtr outImage( new TImageFor< T >( im_col, im_row ));

    bool row_is_ok, mask_index_in_range;

    // Go through all the points of the input picture
    for( i = 0; i < im_row; i ++ )
    {
    for( j = 0; j < im_col; j ++ )
    {
       T maxVal( theImage.GetPixel( j, i ) );

       for( m = - vertHalfMask; m <= + vertHalfMask; m ++ )
       {
            vertMaskIndex = m + vertHalfMask;

            imTempRow = i - m;

            row_is_ok  =  imTempRow >= 0 && imTempRow < im_row;
            mask_index_in_range = vertMaskIndex < mask_row;

            for( n =- horzHalfMask; n<= + horzHalfMask; n ++ )

            {  imTempCol = j - n;
               //imTempRow = i - m;

               horzMaskIndex = n + horzHalfMask;

              if( imTempCol >= 0 && imTempCol < im_col && row_is_ok
                 && horzMaskIndex<mask_col && mask_index_in_range )

              {
                 bool is_in_structure;
                 T pixel = theStructureElement.GetPixel(
                                     horzMaskIndex,
                                     vertMaskIndex,
                                     is_in_structure );


                 if( is_in_structure == true )

                 {
```

Algorithm 4.7 (*Continued*)

```
                        pixel+=theImage.GetPixel(imTempCol,imTempRow );
                        if( pixel > maxVal )
                            maxVal = Pixel // store the maximal pixel;
                                           // value as so far
                        }
                    }

                }
            }

            outImage->SetPixel( j, i, maxVal );
        }

    }

    return outImage;
}

    // ... other members
};
```

Algorithm 4.7 (*Continued*)

a result of dilation in this local region. Such a process is repeated for local regions around each pixel in the input image.

4.8.3 Examples in Matlab: Computation of the SVD

The SVD decomposition plays a very important role in many scientific computations. Therefore, it is not a surprise that Matlab offers such a tool among its myriad others. Let us take a look at its basic call:

```
[S,V,D] = SVD(X)
```

where $\mathbf{X}$ is the matrix for which we compute SVD, $\mathbf{S}$ and $\mathbf{D}$ are unitary matrices (i.e. $\mathbf{SS}^{\mathrm{T}} = \mathbf{DD}^{\mathrm{T}} = 1$) and $\mathbf{V}$ is a diagonal matrix with singular values (which are nonnegative). It holds that $\mathbf{X} = \mathbf{SVD}^{\mathrm{T}}$ (section 4.2.2). The latter can be checked easily, writing in Matlab:

```
X1=S*V*D';
```

To see a more detailed description of this function, write in Matlab:

```
help SVD
```

To practise this technique let us decompose matrix $\mathbf{A}_2$ from (4.13) in Matlab. We can proceed as follows (here >> denotes the Matlab's prompt sign):

```
>> X= [ 1, 1, 1; 1, 3, 1; 1, 1, 1 ];
>> [S,V,D]=svd(X);
```

Then let us take a look at the matrices computed for us:

```
>> S

S =

   -0.4082    0.5774   -0.7071
   -0.8165   -0.5774   -0.0000
   -0.4082    0.5774    0.7071

>> V

V =

    4.0000         0         0
         0    1.0000         0
         0         0    0.0000

>> D

D =

   -0.4082    0.5774    0.7071
   -0.8165   -0.5774   -0.0000
   -0.4082    0.5774   -0.7071
```

Let us start by analysing the matrix **V**. We see that for **X** we have two singular values different from 0. Therefore its rank is two. Thus **X** can be decomposed in accordance with (4.9); for this purpose we take only the two first columns of **U** and **V**. Notice, however, that in numerical computations there is always a limited number of bits for number representation. Therefore, more often than not we have to deal with numerical errors. Indeed, under closer scrutiny we see that the third singular value is a very small number which, nevertheless, is different from 'pure' zero.

Let us check now the matrices **S** and **D**. We can easily find out that they are unitary matrices, so $\mathbf{S}\mathbf{S}^{\mathrm{T}} = \mathbf{1}$. Moreover, we see that the columns of **S** are orthogonal. This feature holds also for **D**.

If necessary the Matlab procedures can be also linked to users' software [208]. Alternatively, the full C++ implementation of the SVD procedure is provided in [352].

4.9 Closure

This chapter is devoted mainly to computer methods for detection of low-level features in digital images. These are intended to be used for image matching, although their applications are much broader. The signal processing approach is assumed. At first we discuss the basics of correlation and convolution with applications to image averaging and differentiation. These are examples of a much wider group of digital image filters. Differentiation of discrete signals is presented in the approach proposed by Farid and Simoncelli and also in terms of the Savitzky–Golay filters.

Then different edge detection techniques for image matching are discussed. The Laplacian of Gaussian and morphological operators are presented as well. The former belongs to the group of linear methods, the latter to the nonlinear ones.

Based on the basic operators the structural tensor is introduced. It allows compact representation of local neighbourhoods of pixels in terms of their orientation and coherence. Finally, an overview of corner detection methods is provided. These have found broad applications in image matching as well.

4.9.1 Further Reading

Good comprehensive texts on signal processing are the books by Oppenheim and Schafer [336] and by Mitra [312], for example.

Very popular is the Canny edge detector [60] which has not been discussed in this chapter. However, its description, as well as information on other feature detectors, can be found in the majority of textbooks on image processing, e.g. in the books by Forsyth and Ponce [135], by Pratt [351] or by Gonzalez and Woods [157].

The structural tensor has been introduced by Bigün *et al.* [42]. The book by Granlund and Knutsson [160] pioneered the subject of tensor operators for representation of local image structures. Discussion of the structural tensor, its construction and application to the detection of features, motion and texture analysis is contained in the works by Jähne [224–227].

A recommended book on matrices and linear algebra is the work by Meyer [308]. It can be used for self-study and also as a reference. However, a shorter and very intuitive approach is provided in the excellent book by Trefethen and Bau [425]. They also give a very in-depth introduction to the singular value decomposition (SVD), starting from its intuitive explanation and basic properties, then presenting some hints on implementation. This book could be recommended for readers not familiar with linear algebra concepts or starting their education in this field. A numerical approach to SVD is also given in the book by Demmel [96]. However, recommended reading that provides concise theory with working computer algorithms in C/C++ is the book by Press *et al.* [352]. This seminal work provides also an in-depth analysis of computation of the numerical derivatives.

Finally, a very good text on data structures and algorithms is the book by Cormen *et al.* [74]. A comprehensive source on fundamental algorithms and seminumerical algorithms can be found in the classic texts by Knuth [248, 249].

4.9.2 Problems and Exercises

1. Verify the decompositions given by formulas (4.12) and (4.13).
2. Find the spectral response of the basic differentiation equation (4.33).
3. Design and implement an *in situ* convolution algorithm, i.e. the procedure *Convolve* which takes only input image and a mask. The result goes back to the input image (for more hints see [226]).
4. Write a 2D convolution procedure which would be tailored for symmetrical 3×3 masks only.

5. Check the separation properties of the filter given by the following mask:

$$\mathbf{B} = \begin{bmatrix} 1 & 4 & 6 & 4 & 1 \\ 4 & 16 & 24 & 16 & 4 \\ 6 & 24 & 36 & 24 & 6 \\ 4 & 16 & 24 & 16 & 4 \\ 1 & 4 & 6 & 4 & 1 \end{bmatrix}$$

   What type of signal processing denotes the above operator **B**?
6. Generate coefficients of the Savitzky–Golay filter of order $N = 2$ and the window span of three pixels using the procedures in Algorithm 4.2. Then test your filters.
7. Repeat problem 6 for $N = 5$ and for the mask of five by five pixels.

5

Scale-space Vision

5.1 Abstract

The properties of image scale and its implications for image matching algorithms are introduced in this chapter. This is followed by a brief explanation of the concept of *scale-space* and how both Gaussian and differential scale-spaces can be constructed. Building on these concepts, we introduce the multi-resolution image pyramid data structure and how this can be parameterized in order to build Gaussian and Laplacian of Gaussians image pyramids. These pyramid structures provide for efficient representation and computation within scale-space and constitute an essential prerequisite to the formation of general purpose image matching algorithms. The notion of subdividing the scale-space within the levels of the pyramid is also presented, such that any specified degree of continuity (in scale) between pyramid levels can be achieved. Examples of image pyramids coded in both the C++ and Matlab programming languages illustrate the practical compromises that must be resolved in practice when building pyramids.

5.2 Basic Concepts

5.2.1 Context

In the preceding chapters we reviewed the geometric process by which stereo-pair images are formed and how, as a consequence, depth information is implicitly encoded within stereo-pairs as relative displacements between the stereo-pair image planes. We then developed the notion of how these relative displacements or *disparities* can be 'decoded' to allow explicit depth information to be recovered via triangulation. This latter process is reliant upon explicit knowledge of the stereo-pair imaging geometry and the disparities themselves which are inferred by solving the *stereo correspondence* problem, i.e. finding the single locations in each of the stereo-pair images that both *correspond* to the same location in the imaged 3D scene. Although we have now considered the basic operations required to tackle the stereo correspondence problem in Chapter 4, there remains an intermediate issue that must first be addressed if we are to develop robust 3D imaging systems, namely that of image scale.

An Introduction to 3D Computer Vision Techniques and Algorithms Bogusław Cyganek and J. Paul Siebert

5.2.2 Image Scale

Why does scale matter? In the most general sense, image scale represents a degree of freedom that the underlying image signal can express; other such degrees of freedom include rotation, contrast and black-level. Scale is simply the manifestation of a change in the spatial size, or *scale*, of a feature, region or complete grouping (e.g. object of interest). In the case of a real camera system, a change in scale of the projected image of an object in the real world is typically caused by the distance from the camera to the object changing. Due to the finite nature in which images projected by a camera lens are sampled in the image plane by a physical sensing device, the amount of information present will also change as the scale of the projected image changes. Hence, an image of a distant object might subtend only a few samples on the image plane and the sensing device will capture the object in terms of a small range of spatial frequencies. As we approach this object, its projection will grow in size on the imaging plane and a correspondingly larger range of spatial frequencies may be present to describe more 'detail' on the surface of the object. Hence, the gross structure of the object is resolved in greater detail as we approach it, but does not transform radically, and new detail emerges as we increase the size of the representation. Evolution of image structure in terms of emerging detail with increasing image size encapsulates the concept of a *multi-resolution scale-space*. The core concept is how to take advantage of the orderly evolution of detail over scale to be able to process image signals in a scale-independent fashion while exploiting the finest levels of detail, i.e. highest acuity information available, for a given task.

5.2.3 Image Matching Over Scale

If we were to develop an algorithm that recognizes a specific object, for example a face, we would like this algorithm to be able to recognize faces regardless of whether they occupy the entire field of view or simply a small region. In this example, a limiting factor, such as the minimum recognizable face size, in pixels, might be part of the recognition algorithm's input parameterization. In the context of the stereo correspondence problem, we would like our *image-matching* algorithm to be able to recover correspondences regardless of how large (or small) the image structures are in our stereo-pair. Furthermore, we would like to be able to *resolve* these correspondences at the finest level of structure available within the stereo-pair images.

This chapter describes a general analysis framework based on scale-space for representing images such that it is possible to devise image matching algorithms that operate over a range of image scales in a consistent manner. As presented in Chapter 4, we shall require certain classes of image operation, such as convolutions, to be able to tackle the stereo correspondence problem. In other words, we are going to need to filter the stereo-pair images in order perhaps to extract image features and then match these features between the stereo-pairs. However, the types of image feature, or, more generally, image structure we extract and match could conceivably span a very large range of spatial scales and, furthermore, the feature scales present within the stereo-pairs are not usually known in advance.

The above issues form the central core of the problem of how to achieve *scale-invariant* analysis; consider the task of detecting an edge using a Laplacian of Gaussians (LoG) operator (4.5.3). The width of edge to which a given LoG kernel is tuned corresponds to

$2\sqrt{2}\sigma$ pixels (Equation (4.73)), where σ is the spread of the Gaussian blur component in the LoG kernel. Accordingly, the LoG kernel should be at least 4σ pixels in both of its spatial dimensions to represent the operator with reasonable accuracy. Given that the size of edges present within captured stereo-pairs will vary continuously, it would be inconceivably cumbersome, and computationally inefficient, if we had to construct a set of LoG filters, each filter corresponding to the (closely approximating) tuning σ required for each width of edge present. Therefore, densely sampling scale-space by constructing LoG filters tuned to each scale present in the scale-space over which we are operating is not a viable approach.

An alternative approach to achieving scale-invariant processing is to sample scale-space with sufficient density such that it is possible to track the evolution of new detail as it emerges from scale to scale. For example, we could utilize a fixed set of LoG filters such that a filter tuned to the scale of the largest edge to be processed is applied first and the locations of these largest edges are labelled. A new LoG filter having a scale tuned to half that of the previous filter would then be used to label all edges present at this finer level of detail and so forth until a stack of edge maps is constructed that range from the coarsest-to-finest sets of edge structure. It is then possible to search, starting with the coarsest edge maps, for the closest corresponding edges in the map containing the next finest level of detail to each edge label in the initial coarsest map. This process can then be repeated until the evolution of each new edge label from every coarser edge label has been traced from scale to scale and hence the term *scale-space tracing* is applied to this analysis. The purpose of this process is to be able to describe the structure within an image in a form that can be compared to similar structures in other images, independently of the scale at which the structure appears in the image. If our structure is compared with a similar structure at lower resolution, there will be a location at a coarser level in the scale-space we constructed for our image that should match (be similar to) the coarser scale version with which it is being compared. Alternatively, if our structure is compared to another structure containing more detail, there will be a coarser level of scale in the compared structure that should match our initial structure.

A computationally more efficient alternative to the above approach to sampling scale-space is to hold constant the size of the kernel, LoG in this example, and resample the input image to generate a set of images, by low-pass filtering and then subsampling, which exhibit progressively lower spatial resolutions. This data structure is referred to as an *image pyramid* (Figures 5.6 and 5.7) and we shall examine in detail how to parameterize the construction of image pyramids in section 5.3. Having deconstructed our input image into an image pyramid, the same size of LoG filter can then be applied to each of the levels of the pyramid to effect edge detection and labelling over a range of edge scales.

The subject of scale-space analysis is central to modern computer vision theory and covers a very broad corpus in the literature. Accordingly, an in-depth treatment of scale-space is beyond the scope of this text and we shall restrict the treatment of scale-space analysis to that required in the context of stereo-pair matching. To conclude the answer to why we need to process over scale, we need to address the following issues.

- Image structure and related features exist over a continuous range of sizes in acquired images.
- The size of features specific to any image is usually not known in advance.

- Basic image operators, such as filters, must be capable of functioning over the same range of feature scales as there are present in the input images.
- It is possible to trace the emergence of structure over scale and thereby embed this technique within a search strategy to achieve scale-independent processing that is also computationally very efficient.

Considering the issue of scale-independent processing listed above in the context of image matching, coarse scale information can be matched first and a local search applied at subsequent scales to refine this search process. Such a coarse-to-fine search strategy is central to solving the correspondence problem successfully and is examined in detail in Chapter 6. The following sections outline in more detail the concepts that underpin scale-space analysis: how to construct a scale-space and how to build the image pyramids on which we can apply scale-independent processing.

5.3 Constructing a Scale-space

5.3.1 Gaussian Scale-space

Lindeberg [277] defines a special form of scale-space that he calls *linear scale-space*:

> Scale-space representation is a special type of multi-scale representation that comprises a *continuous scale parameter* and preserves the *same spatial sampling* at all scales. As Witkin [reference [448] cited] introduced the concept, the scale-space representation of a signal is an embedding of the original signal into a one-parameter family of derived signals constructed by convolution with a one-parameter family of Gaussian kernels of increasing width.

In this view, a signal can be considered to be progressively smoothed by means of a Gaussian kernel. Furthermore, it transpires that only a Gaussian kernel has the form required to produce a family of smoothed signals that meet the specific criteria required of scale-space, namely the orderly emergence of image structure as the scale parameter, σ, decreases. An example of a family of 1D signals derived as a function of σ is illustrated in Figure 5.1 [448]. Note that linear scale-space is not subsampled, as opposed to pyramidal, i.e. multi-resolution, scale-space.

Lindeberg enumerates the following desirable properties enshrined in his definition of linear scale-space.

- Shift invariance: *spatial isotropy*, all spatial positions treated equally.
- Scale invariance: *spatial homogeneity*, all spatial scales treated equally.
- Causality:
  - noncreation of new level curves in scale-space;
  - noncreation of new local extrema (turning points);
  - nonenhancement of local extrema, i.e. no extrema in a given scale becomes larger in the scales above or below.

It also transpires that these criteria can be met by considering the signal to be subject to a first-order diffusion process, as formulated by the standard heat diffusion equation for a heat

Figure 5.1 Family of signals, progressively smoothed by Gaussian convolution. (Reproduced from [448], © 1983 IEEE)

distribution L in an isotropic medium over time t:

$$\partial_t L = \frac{1}{2}\nabla^2 L, \tag{5.1}$$

a solution of which is the (normalized) Gaussian as a function of σ:

$$G(t,\sigma) = \frac{1}{\sqrt{2\pi}\sigma} e^{\left(-\frac{t^2}{2\sigma^2}\right)}. \tag{5.2}$$

Unfortunately, for signal dimensions of 2 and higher, it is not possible to meet the causality requirement using simple kernels, e.g. the 2D Gaussian of:

$$G(x,y,\sigma) = \frac{1}{2\pi\sigma^2} e^{-\left(\frac{x^2+y^2}{2\sigma^2}\right)}, \tag{5.3}$$

such that the numbers of local extrema are guaranteed never to increase with scale.

Figure 5.2 shows the effect of smoothing an image with progressively larger Gaussian kernels. We observe that varying the scale (blurring) parameter σ for a Gaussian filter allows us to compute filters with different degrees of (low-pass) attenuation of high frequencies. Applying this set of filters produces a set of images containing features according to their spatial scale. Intuitively, we can think of the Gaussian kernel as a centre weighted averaging filter that progressively suppresses higher spatial frequencies as the spatial support of the kernel increases with σ.

Input Image

Progressive Gaussian blur with = 8, 6, 4, 2, 1.5, 1, 0.5

Figure 5.2 Family of images generated by convolution using 2D Gaussian functions. (Input image reproduced from Final Year Report, Strathclyde University, Iwan Eising)

Examining the sequence of filtered images in Figure 5.2 from coarse-to-fine scale we see that as high frequencies are introduced (with reducing scale), new levels of complexity appear in the image. However, each new feature that appears in a given scale then persists and evolves into subsequent scales, i.e. each new scale contains all the information of the previous scale. Accordingly, it is possible to track or trace these features from coarse-to-fine images.

The spatial frequency properties of this Gaussian scale-space can be deduced by taking the Fourier transform of the normalized Gaussian:

$$G(\omega) = e^{-\frac{\omega^2\sigma^2}{2}}, \tag{5.4}$$

where ω is the circular frequency, $\omega = 2\pi f$.

High frequencies are attenuated according to the above equation, itself Gaussian (for $\omega > 0$) in form. Figure 5.3 illustrates the effect of varying σ on the attenuation properties of the Gaussian filter for three filters separated by a half-octave and an octave respectively.

5.3.2 Differential Scale-space

The Gaussian scale-space discussed above is redundant in the sense that all the information contained in each scale also includes all of the information of all previous scales. It is often more useful to attempt to isolate new information within the scale at which it appears and this can be accomplished by representing the information *difference* between scales. The simplest method for achieving a differential scale-space is to construct a Gaussian scale-space comprising a set of Gaussian smoothed images and then to subtract images containing neighbouring scales, pixel-by-pixel, to produce a difference of Gaussians (DoG) scale-space, i.e. $DoG(x, y, \sigma) = G(x, y, \sigma_e) - G(x, y, \sigma_i)$, where, using terminology borrowed from biological vision, σ_e and σ_i correspond to the spatial extent of *excitatory* and *inhibitory* Gaussian envelopes respectively. When the ratio $\sigma_e/\sigma_i = 1.6$ then the DoG function, introduced in section 4.5.4, provides a good approximation of the Laplacian of Gaussians (LoG) function, section 4.5.3, adopted here to construct a differential scale-space.

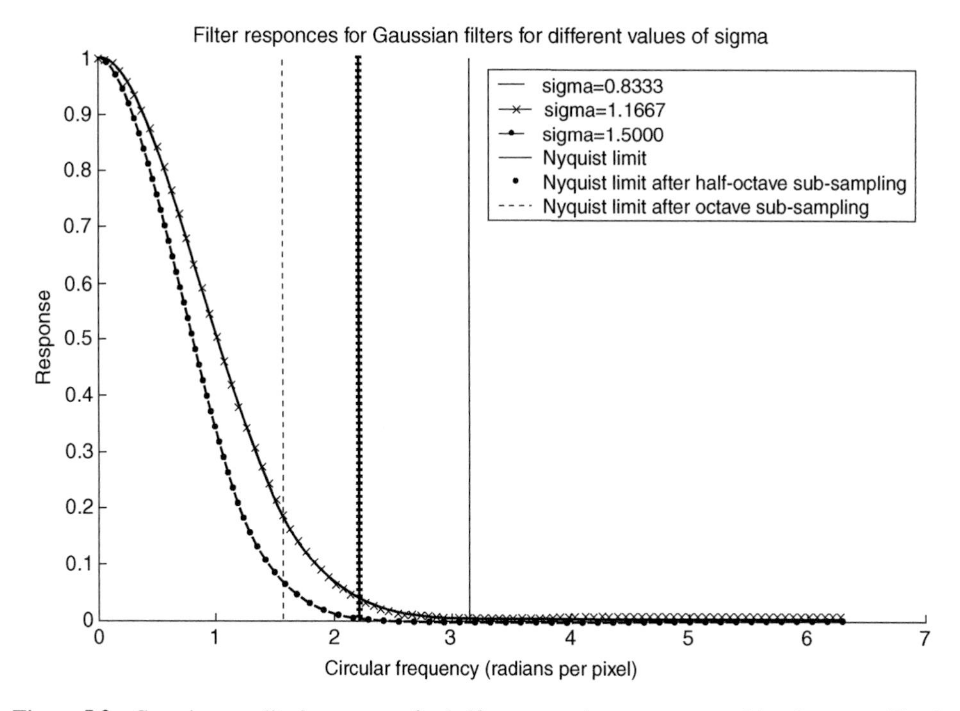

Figure 5.3 Gaussian amplitude response for half-octave and octave separated band passes. (Graphs and Matlab codes kindly supplied by Dr Sumita Balasuriya)

Differentiating the 2D Gaussian function in (5.3) gives the following partial derivatives in x and y:

$$\frac{\partial G(x, y, \sigma)}{\partial x} = \left(\frac{-x}{\sigma^2}\right) \frac{1}{2\pi\sigma^2} e^{-\left(\frac{x^2+y^2}{2\sigma^2}\right)} \tag{5.5}$$

$$\frac{\partial G(x, y, \sigma)}{\partial y} = \left(\frac{-y}{\sigma^2}\right) \frac{1}{2\pi\sigma^2} e^{-\left(\frac{x^2+y^2}{2\sigma^2}\right)} \tag{5.6}$$

The Laplacian of Gaussians function then corresponds to the sum of the second partial derivatives of the Gaussian, i.e. the cross-product terms in the usual definition of the Laplacian are ignored:

$$\nabla^2 G = \frac{\partial^2 G}{\partial x^2} + \frac{\partial^2 G}{\partial y^2} \tag{5.7}$$

$$\nabla^2 G(x, y, \sigma) = -\frac{1}{2\pi\sigma^4} \left(2 - \frac{x^2 + y^2}{\sigma^2}\right) e^{-\left(\frac{x^2+y^2}{2\sigma^2}\right)} \tag{5.8}$$

$$\nabla^2 G(x, y, \sigma) = \left(\frac{(x^2 + y^2) - 2\sigma^2}{\sigma^4}\right) G(x, y; \sigma) \tag{5.9}$$

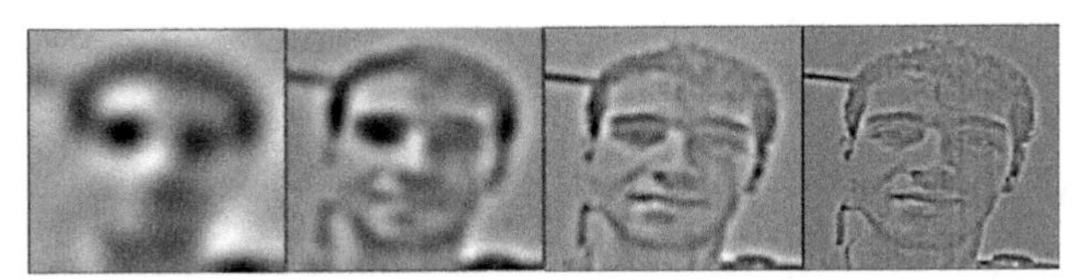

Figure 5.4 Set of images filtered by 2D Laplacian of Gaussian functions. (Input image reproduced from Final Year Report, Strathclyde University, Iwan Eising)

In the frequency domain, the process of isolating information within the scale at which it appears corresponds to deconstructing the original image into a series of *band pass* images. Since each band pass image contains only a range of spatial frequencies, only image features of a certain *characteristic scale* will be 'tuned' in their spatial dimensions to that of a particular band pass (and therefore appear with greatest signal strength within this specific band pass image). Note that such features may also appear in other band pass images, since the pass-band of each scale normally overlaps significantly with that of its neighbouring scales to afford continuity over scale. Figure 5.4 illustrates the emergence of high spatial frequency detail with scale with four examples of an image LoG filtered at progressively finer scales.

Recall in section 4.5.3 taking the Fourier transform of the Laplacian of Gaussian function (4.77); by integration by parts we obtain the following expression that relates frequency, scale and signal amplitude:

$$LoG(\omega) = -\frac{\omega^2 e^{-\frac{\omega^2\sigma^2}{2}}}{\frac{2}{\sigma^2}e^{-1}}, \tag{5.10}$$

where σ is the standard deviation (sigma) of the filter, ω is the circular frequency, $\omega = 2\pi f$ and the denominator is the scaling factor that normalizes the filter's peak response to unity, when $\omega_{\text{peak}} = \frac{\sqrt{2}}{\sigma}$. Figure 5.5 illustrates the circular frequency response for three LoG band passes corresponding to three filters, separated by a half-octave and an octave respectively.

5.4 Multi-resolution Pyramids

5.4.1 Introducing Multi-resolution Pyramids

In the preceding section we constructed both linear and differential scale-spaces with a view to implementing image-matching algorithms that can operate in a scale-independent manner. A severe limitation of these 'pure' scale-spaces is that as their spatial frequency content reduces in (inverse) proportion to scale factor σ, they become not only highly redundant as a

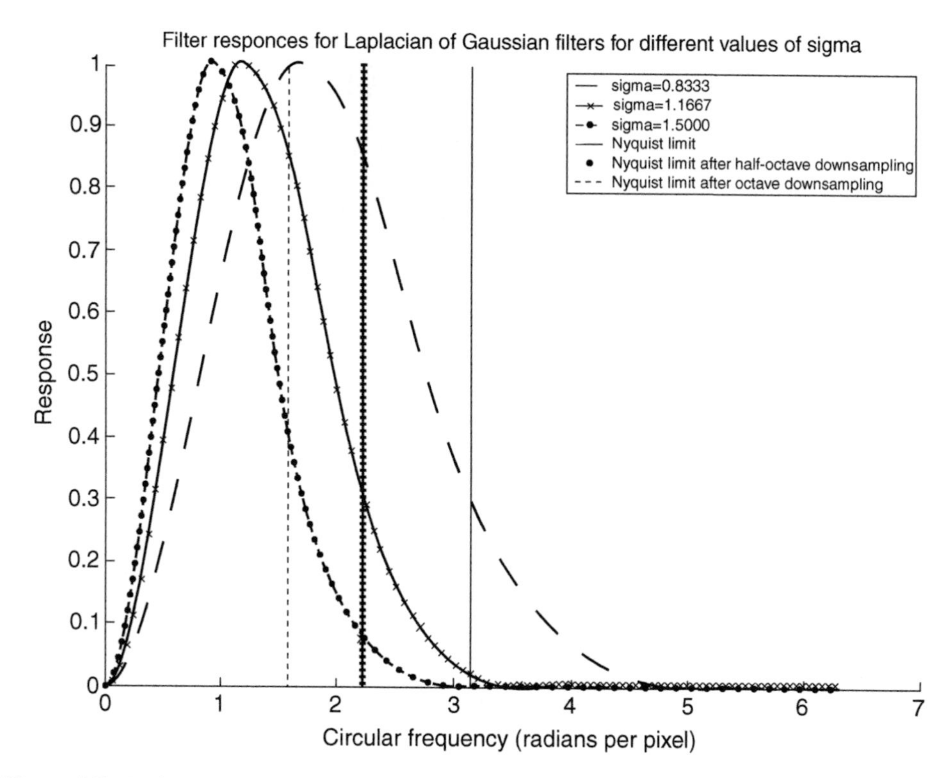

Figure 5.5 LoG amplitude response for half-octave and octave separated band passes. (Graphs and Matlab code kindly supplied by Dr Sumita Balasuriya)

representation but also computationally expensive to generate. The standard method of improving the representational efficiency of pure scale-spaces is to subsample the smoothed signal according to some criterion based on the residual aliasing spectral components present in the signal, at and above the Nyquist limit for each scale in scale-space. Clearly, the larger the scale factor, the lower the spatial frequency band pass and therefore the greater the degree of subsampling that is possible. Hence, if this subsampling operation is performed in a consistent manner, it becomes possible to construct a multi-resolution image data structure commonly referred to as an *image pyramid*.

Figures 5.6 and 5.7 show examples of the DoG and LoG pyramids respectively, for an octave scale-space sampling, i.e. at each level in scale the maximum spatial frequency represented is half that of the previous (finer) scale. In the case of the DoG pyramid, all frequencies are present at the finest scale, while the LoG pyramid contains image band passes in each level of scale. In this latter example, the input image can be reconstructed by expanding and summing each LoG scale to recombine each band pass into a single image containing all the spectral components of the original. Three principal advantages are conferred by the pyramid representation. These are

- improved storage efficiency;
- a uniform basis on which to conduct analysis over scale;

Figure 5.6 Octave scale Gaussian pyramid. (Original image from Final Year Report, Strathclyde University, Iwan Eising)

- improved computational efficiency, both when constructing the representation and also when conducting analysis by means of it.

Issues not yet addressed in this discussion are how densely to sample spatial scale itself and how to select appropriate scale parameters when constructing a scale-space representation. It transpires that these issues are to some degree dependent upon the particular task that is to be conducted using the scale-space representation. However, in the following analysis the above parameterization issues and how to resolve their selection under specific circumstances are discussed.

An often-cited criticism of the image pyramid is that scale-space is overly sparsely sampled, increasing the algorithmic complexity of subsequent analysis (more local search is required at each level of scale if scale-space is not sufficiently continuous between levels). This can be countered in two ways. Firstly we present a method of analysing pyramid construction that is not limited to octave sampling in scale, such that any degree of sampling (in scale) is possible, albeit at a cost. Secondly, the concept of subdivided or semi-pyramids is introduced, where each level of the pyramid is subdivided in scale, but the spatial sampling resolution held constant. Other issues surrounding the image pyramid include the possibility of convolution artefacts being propagated throughout the representation [277], from the finest to coarsest level, particularly when separable filters are employed. To balance these negative aspects, the reader should be aware that many highly successful stereo-pair matching algorithms have been implemented based on pyramid representations and that the use of the pyramid can offer an execution speed advantage of several orders of magnitude over non subsampled scale-space.

Figure 5.7 Octave scale Laplacian of Gaussians pyramid. (Original image from Final Year Report, Strathclyde University, Iwan Eising)

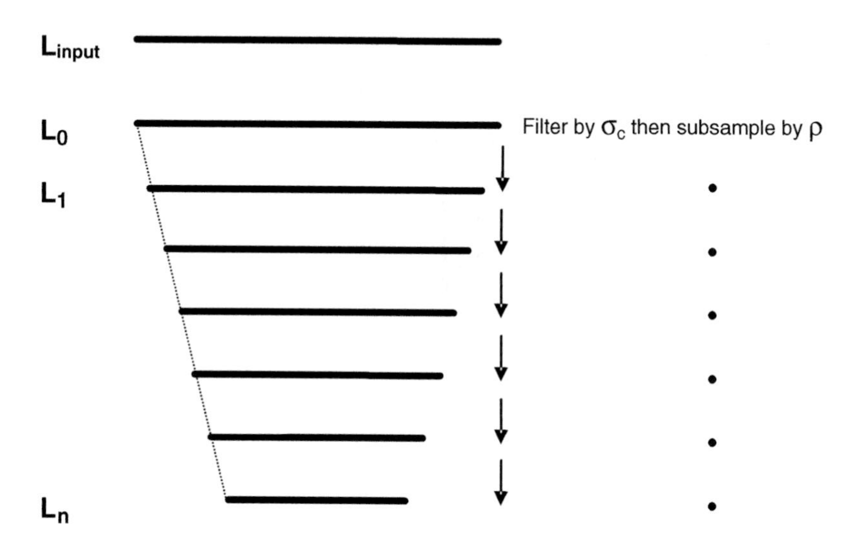

Figure 5.8 Construction of a regular pyramid: apply an initial filter of σ_{init}, filter subsequent levels by the same σ_{c} and subsample by the same ρ factor at each level

5.4.2 How to Build Pyramids

Pyramid construction comprises two basic steps per sampled scale, namely low-pass Gaussian filtering followed by subsampling, i.e. image size reduction. As illustrated in Figure 5.8, an additional, but separate, level of initial filtering can be applied prior to starting the regular pyramid construction process in order to control the degree of apparent blur within each level of the final pyramid, detailed in section 5.4.3. A subsampling factor of $\rho = 2$ (octave reduction factor) is frequently used to serve many applications; $\rho = \sqrt{2}$ (half-octave reduction factor) is also another useful reduction factor that achieves greater interscale continuity than the octave reduction case. It is possible to combine subsampling and convolution into a single operation, e.g. if subsampling by a factor of 2, simply step the kernel over every second pixel of the input image to output a half resolution image (Figure 5.9). What follows is an extended treatment of van Hoff's [197] method for constructing Gaussian and difference of Gaussian pyramids.

5.4.3 Constructing Regular Gaussian Pyramids

The machinery described in section 5.4.2 can be used to construct a regular Gaussian pyramid. Two additional results are required before we can proceed to pyramid construction. Firstly, Gaussian filtering by repeated application of one or more kernels is calculated as follows:

$$\sigma_{\text{total}}^2 = \sigma_{\text{a}}^2 + \sigma_{\text{b}}^2, \tag{5.11}$$

where σ_{a} and σ_{b} denote Gaussian kernels applied in cascade and σ_{total} is the scale (blur) of the equivalent kernel. Secondly, a subsampling factor of ρ reduces the effective size of sigma by ρ. Based on the above, we can construct a Gaussian pyramid as shown in Figure 5.10.

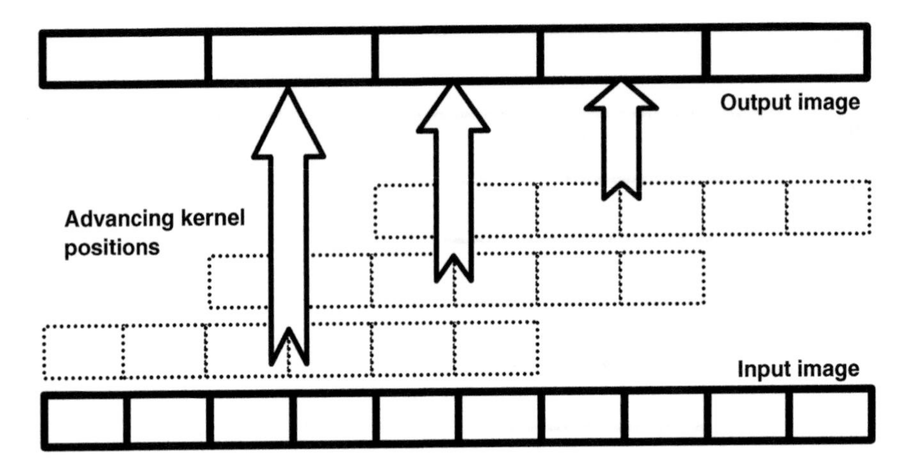

Figure 5.9 Combined filtering and subsampling, e.g. subsample ×2 by stepping filter kernel over every second pixel

In order to ensure that we construct a *regular* Gaussian pyramid, as illustrated in Figure 5.10, we need each level in the pyramid to contain the *same* apparent level of blur, i.e. if we generate an image pyramid of a Dirac impulse image, the impulse response at each level in the pyramid should be the same. Since the response is the same at each level, we can apply the same analysis algorithms at each level, such that these respond in the same manner. In other words, in general:

$$\sigma_0 = \sigma_1 = \sigma_2 \ldots \sigma_i. \tag{5.12}$$

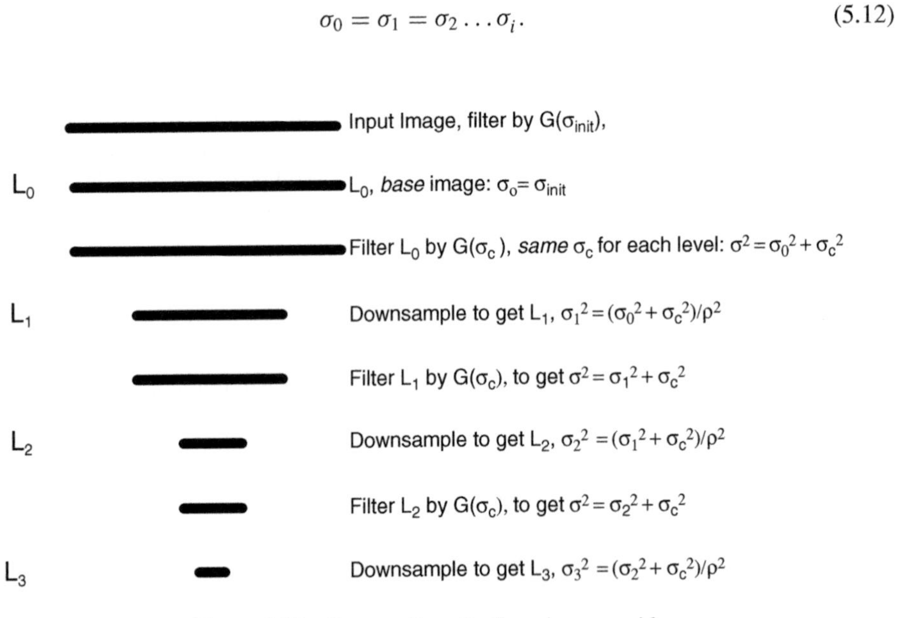

Figure 5.10 Construction of a Gaussian pyramid

All that is changing is the spatial scale of the structures within the original image. To achieve the desired regularity, we need to work out for a given reduction ρ what the relationship is to σ_c and σ_{init}.

Each level of the pyramid illustrated in Figure 5.10 can be implemented efficiently by repeated convolution of a Gaussian filter σ_c, subsampled by a factor ρ. By inspection of Figure 5.10, we can observe that for the i-th level we can find the blur, σ_i, for the current level as follows:

$$\sigma_i = \frac{1}{\rho}\sqrt{\sigma_{i-1}^2 + \sigma_c^2}. \tag{5.13}$$

Substituting $\sigma_i = \sigma_0$ and $\sigma_{i-1} = \sigma_0$ in (5.13), where σ_0 should ideally equal the *intrinsic blur* in the input image for σ_c to remain constant at each level (and thereby avoid the necessity for an additional initial stage of Gaussian convolution σ_{init}), gives:

$$\sigma_c = \sigma_0\sqrt{\rho^2 - 1}. \tag{5.14}$$

The effective σ had they not been reduced is:

$$\sigma_{effective} = \sigma_0\rho^i. \tag{5.15}$$

For an octave pyramid $\rho = 2$ and σ increases in powers of 2 for each level; therefore $\sigma_c = \sigma_0\sqrt{3}$. In the case of a pyramid with $\rho = \sqrt{2}$, $\sigma_c = \sigma_0$. Note that i is numbered such that $i = 0$ corresponds to the finest scale. In Figure 5.10 no intrinsic blur has been assumed, although any image captured using a real (optical) imaging device will typically exhibit a finite point spread function that may well approximate a Gaussian sufficiently for the purposes of pyramid construction. Assuming that we have measured the intrinsic input image blur, σ_{image}, the correct value of σ_0 is simply:

$$\sigma_0 = \sqrt{\sigma_{init}^2 + \sigma_{image}^2}. \tag{5.16}$$

While an initial level of blurring by σ_{init} might seem redundant, it does allow us to tune the remaining filter σ_c parameters to be the same for each level. In effect σ_{init} allows us to add an additional degree of blur to the intrinsic blur in the image to generate a regular pyramid structure of the required blur σ_0 at each level.

5.4.4 Laplacian of Gaussian Pyramids

The straightforward approach to constructing a true Laplacian of Gaussian pyramid (Figure 5.7) is to start by generating a Gaussian pyramid from the input image and then computing and summing second-order partial derivatives in x and y. In this case σ_{init} is chosen to produce a value of σ_0 that produces the desired degree of spatial frequency overlap between pyramid levels, a tradeoff with the degree of aliasing that can be tolerated for an application at hand.

As mentioned, the Laplacians of Gaussians can be approximated by subtracting two difference of Gaussian functions whose σ values take the ratio 1.6. Three methods for generating this ratio can be utilized based on constructing two Gaussian pyramids, such that the blur factor in each level of the (*excitatory*, E) pyramid, σ_{Ei}, is arranged to be 1.6 times the blur, σ_{Ii}, in each level of the (*inhibitory*, I) pyramid. Firstly, the straightforward approach would be to copy the original Gaussian (I) pyramid and then apply a second Gaussian filtering operation σ_{extra} to each of the levels in the copy to generate the E pyramid. For all levels *except* level 0, the E/I σ ratio of 1.6 must hold as follows:

$$\frac{\sigma_{\text{e}}}{\sigma_{\text{i}}} = 1.6 = \frac{\sqrt{\sigma_0^2 + \sigma_{\text{extra}}^2}}{\sigma_0} \Rightarrow \sigma_{\text{extra}} = \sigma_0\sqrt{1.56}. \tag{5.17}$$

Thereafter the I pyramid can be subtracted pixel-wise from the E pyramid to produce a DoG pyramid.

Secondly, two Gaussian pyramids can be constructed, such that the blur factor in each level of the E pyramid, σ_{Ei}, is arranged to be 1.6 times the blur, σ_{Ii}, in each level of the I pyramid, using (5.17) above. Each layer from the I pyramid can be subtracted pixel-wise from the corresponding layer of the E pyramid to form the DoG approximated LoG pyramid.

Finally, it is possible to *expand* the next smaller pyramid level, L_{i+1}, to match the current pyramid level, L_i, in size while arranging that the expanded level, L_{i+1}, contains a spread value 1.6 times that of the current level, L_i. Levels are subtracted as before. The DoG approximated LoG pyramid is computed by subtracting each current level, pixel-wise, from each expanded level, i.e. $L_{i+1} - L_i$.

5.4.5 Expanding Pyramid Levels

Expansion of a level in a pyramid can be accomplished through the convolution operation using Gaussian interpolation. For example, if we wish to expand by a factor of two, we require two (1D) Gaussian kernels; each will be convolved with the input image to be expanded and will be 'centred' on the output pixel location as illustrated in Figure 5.11.

The kernel pair (A, B) generate two output pixels corresponding to the two interpolated values required for 1D interpolation. Hence two passes, one horizontally and one vertically,

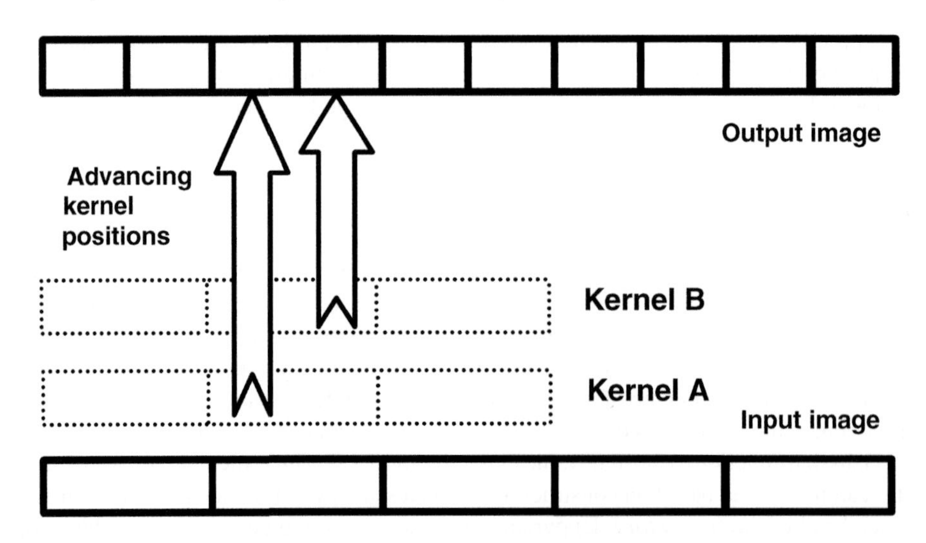

Figure 5.11 Two kernels required for expansion by Gaussian interpolation

will generate the four pixel values required to generate an output image of double the input image dimensions.

While Gaussian interpolation can provide a smoothly expanded image surface, the current level of blur σ_{i} in the image is combined with the expansion blur σ_{e} to give a total blur, σ_{exp}, of:

$$\sigma_{\mathrm{exp}} = \rho\sqrt{\sigma_i^2 + \sigma_e^2}. \tag{5.18}$$

From (5.18) we can now deduce the value of blur σ_{e} required to allow us to construct a difference of Gaussians pyramid that approximates a Laplacian of Gaussians pyramid using the *reduce–expand–subtract* approach. Any level L_n of the LoG pyramid is constructed by subtracting L_n of the DoG pyramid from $L_{n+1}\uparrow\sigma_{\mathrm{e}}$, i.e. where L_{n+1} has been expanded with a Gaussian interpolation expansion filter of blur σ_{e}. To ensure that the ratio $(L_{n+1}\uparrow\sigma_{\mathrm{e}})/L_n =$ 1.6, it follows that:

$$1.6 = \frac{\sigma_{\mathrm{exp}}}{\sigma_0} = \frac{\rho\sqrt{\sigma_0^2 + \sigma_{\mathrm{e}}^2}}{\sigma_0} \tag{5.19}$$

Rearranging (5.18) allows σ_{e} to be computed from (5.19):

$$\sigma_{\mathrm{e}} = \sigma_0\sqrt{1.6^2 - \rho^2}. \tag{5.20}$$

The above equation of course implies that the reduce–expand–subtract approach cannot be used to approximate a LoG pyramid when the expansion factor between levels in the pyramid is greater than 1.6.

5.4.6 *Semi-pyramids*

An often-cited criticism of the pyramid representation is that it samples scale-space too coarsely to achieve adequate continuity between scales for many tasks, including image matching. A compromise between the computational expense of Lindeberg's pure scale-space representation and the typically sparse scale-space sampling of a pyramid can be obtained via a *semi-pyramid*. Figure 5.12 shows a semi-pyramid comprising an octave pyramid that has been subdivided by three to contain two additional interlevel image layers. In this structure, each intermediate layer represents a subdivision of the scale-space between the pyramid levels.

Each level L_i has been subdivided into sections corresponding to a geometric subdivision of scale-space reflecting the scale-space sampling imposed by the gross structure of the pyramid itself. Therefore, as the scale between each pyramid level varies by a factor of ρ, then for a total of N levels of subdivision between each pyramid level the blur factor, σ_n, at the n-th interlevel will correspond to:

$$\sigma_n = \sigma_0\left(\sqrt[N]{\rho}\right)^n, \quad n \in 1\cdots N-1. \tag{5.21}$$

For example, if a pyramid is to be subdivided in two, $N = 2$, there will be a single interlevel at $n = 1$ ($n = 0$ would correspond to the first and $n = N = 2$ would correspond to the second full level of the pyramid).

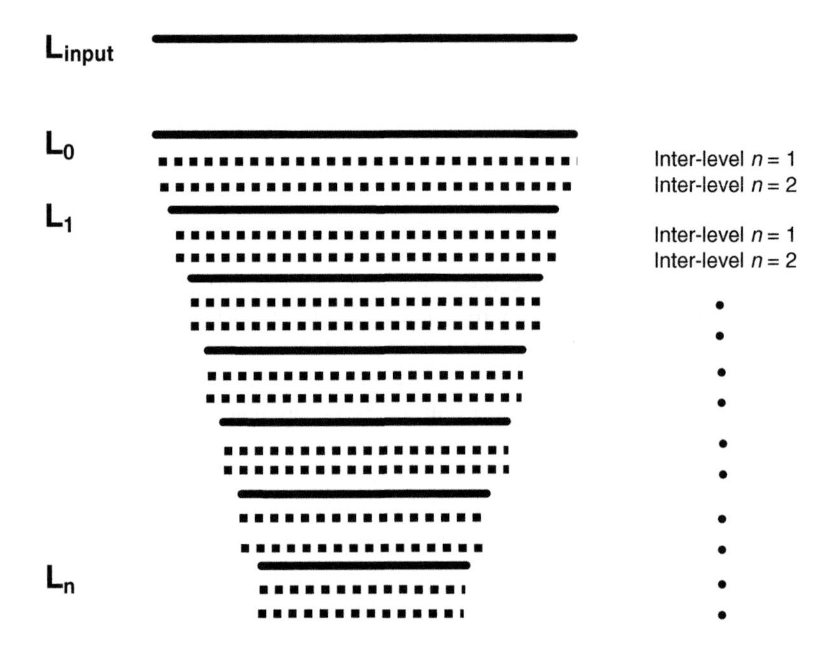

Figure 5.12 Semi-pyramid containing an interlevel subdivision factor of three ($N = 3$) to generate two interlevels ($n = 1, 2$), indicated by dotted lines, per full pyramid level

With only a slight loss of efficiency, the semi-pyramid solves the problem of scale-space continuity and avoids the awkward resampling issues that arise when constructing pyramids with interlayer ratios that are not factors of two. Unfortunately, by definition (5.21), the semi-pyramid layers introduce a significant problem in themselves, namely nonuniformity in the blur factor between these subdivided layers. In turn, this implies that operations on these subdivided layers must be performed using spatial support regions that are themselves a function of σ_n. If a relatively low degree of subdivision is adopted, then the added complexity of requiring one to adapt filters to each sublayer may not be particularly onerous. However, is must be borne in mind that not only filtering algorithms but also search algorithms must be adapted in their parameters to function correctly within each sublayer.

In order to construct a Gaussian semi-pyramid containing N subdivisions of each level, we would like to arrange that the current level, L_i, is copied and then blurred to form the next subdivided layer, $N - 1$ times in all. The additional blur, σ_{sub}, that must be added by convolution with the current subdivision layer blur, σ_n, to provide the correct total blur, σ_{n+1}, is calculated as follows:

$$\sigma_n = \sigma_0 \left(\sqrt[N]{\rho}\right)^n, \quad \sigma_{n+1} = \sigma_0 \left(\sqrt[N]{\rho}\right)^{n+1}, \tag{5.22}$$

$$(\sigma_{n+1})^2 = (\sigma_n)^2 + (\sigma_{\text{sub}})^2, \tag{5.23}$$

$$\sigma_{\text{sub}} = \sigma_0 \left[\sqrt[\frac{N}{2n+2}]{\rho} - \sqrt[\frac{N}{2n}]{\rho} \right]^{1/2}. \tag{5.24}$$

A Laplacian of Gaussians subdivision can be straightforwardly calculated as described in section 5.4.4 by summing the partial derivatives in each subdivided level, or forming a DoG approximation as described before. In this case intersemilevel subtraction can be arranged to give semilevel band pass spatial filters (and corresponding subdivision of scale-space), although only very specific subdivisions will produce good LoG approximations.

5.5 Practical Examples

This section provides codes in both C++ and Matlab that implement functional Gaussian and Laplacian of Gaussians half octave pyramids (reduction factor between levels, ρ, of $\rho = \sqrt{2}$). A reduction factor of $\rho = \sqrt{2}$ was chosen since this value is known to work satisfactorily when matching over scale-space stereo-pair images typically captured in close-range photogrammetry applications. Since $\sqrt{2}$ is not a particularly convenient reduction factor to implement within a pyramid, the following examples are therefore particularly illustrative, in terms of implementation compromises that must be reached in practice. In both C++ and Matlab implementations given here separable 1D kernels have been used to compute Gaussian filtering.

5.5.1 C++ Examples

In this example of constructing a half octave pyramid using the C++ programming language, a reduction factor of 1.5 has been selected to ease the task of subsampling each level. However, the filter coefficients adopted assume that the scale-space is being sampled in half octave increments between levels. Direct decimation of the previous level by subsampling and filtering combined into a single operation has been adopted.

5.5.1.1 Building the Laplacian and Gaussian Pyramids in C++

Figure 5.13 presents class hierarchy for the three types of image pyramids: the Gaussian, DoG and Laplacian pyramids. The *TImagePyramids* base class defines a common interface of this family of classes. Its only derived class is *TGaussianImagePyramids* from which the *TDOGImagePyramids* and *TLaplacianImagePyramids* are derived in turn. The reason for such organization is that the last two types of pyramids are built upon the Gaussian pyramids. This is well visible in Algorithm 5.1(b) as well.

The auxiliary class *TRealLinearFilter_Factory* in Figure 5.13 implements the factory pattern the role of which is to supply different types of linear filter objects such as *TBinomialFilter* or *TGaussianFilter* objects. These are discussed in section 13.3.8.

Algorithm 5.1(a) presents the flow chart of the algorithm that produces the Gaussian pyramid. Its input consists of a single image, while an output is composed of a set of images, each being a copy of an original one but at different scale. Such a set of images forms a pyramid of images. The implementation can be simplified by substitution of the Gaussian filter (section 4.3.1) with the binomial one (section 4.3.2).

The input parameters of Algorithm 5.1(a) are as follows.

1. A value of σ for the Gaussian filter (expressed in pixels).
2. An input image.
3. A value M of the required levels of the Gaussian pyramid.

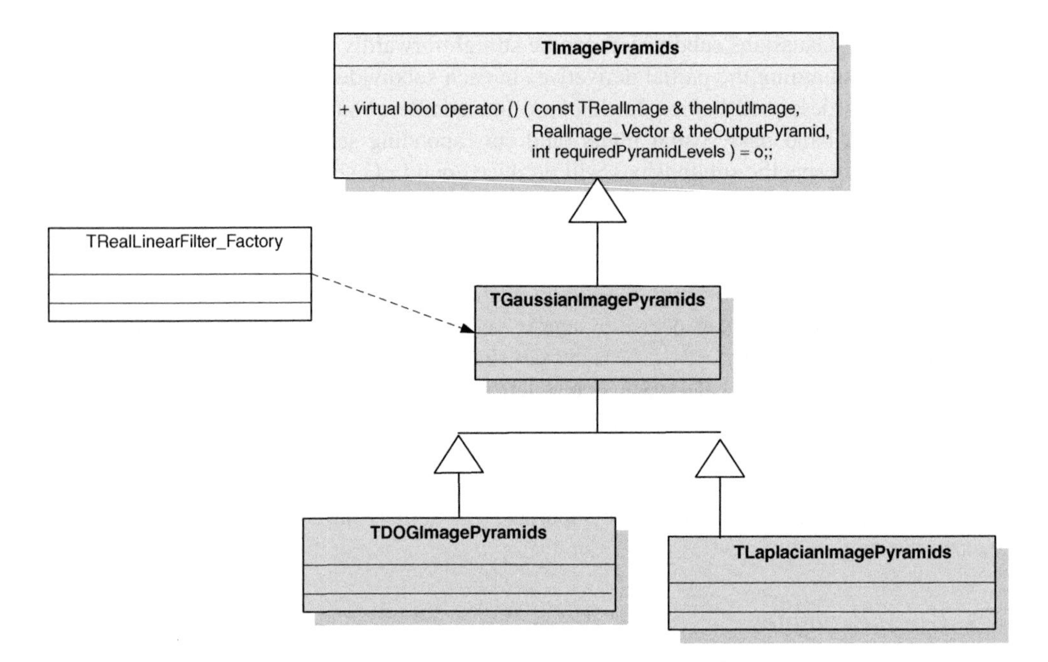

Figure 5.13 Hierarchy of the scale-space pyramids

The output is as follows.

1. A series (a vector) of M images of the Gaussian pyramid.

Algorithm 5.1(a) consists of the following steps.

1. Creation of the Gaussian mask. The 2D Gaussian mask is created based on the supplied σ value and (optionally) mask size, given in pixels. The mask is separable which means that instead of a single 2D mask there are two 1D masks, one horizontal and the second vertical. This speeds up the convolution (section 4.2).
2. The algorithm assumes a temporary image *TMP_IM* (array of pixels) that stores intermediate results.
3. The loop starts at this point. The loop will be performed M times, with an iteration variable k going from 0 up to $M - 1$ (inclusive).
4. *TMP_IM* is an image at a scale k in the computed pyramid. Store the *TMP_IM* in the output data structure after all previously stored images.
5. Filter the *TMP_IM* image with the mask created in step 1. The result is assumed to go to the *TMP_IM*, although there can be a necessity for an additional storage for intermediate results of convolution.

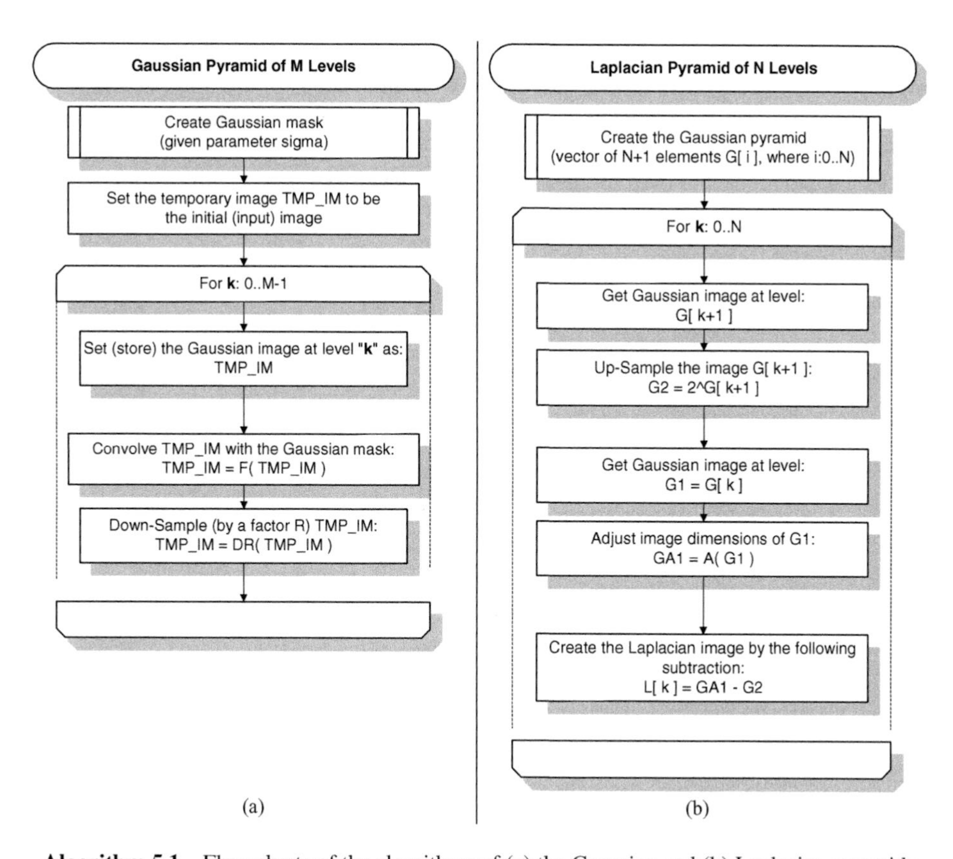

Algorithm 5.1 Flow charts of the algorithms of (a) the Gaussian and (b) Laplacian pyramids

6. Down sample the image just obtained (i.e. after low-pass filtering). The result is assumed again to go to the *TMP_IM*, although there can be a necessity for an additional storage for intermediate results of this operation.
7. Go to step 4 and repeat the loop for the next pyramid levels ($k \leftarrow k + 1$), unless last level has been just processed.

An exemplary implementation is presented in Algorithm 5.2.

The flowchart of the Laplacian pyramid is presented in Algorithm 5.1(b). It utilizes the Gaussian pyramid from Algorithm 5.1(a) as its subroutine. The format of the input and output parameters is the same as for the Gaussian pyramid. The method proceeds as follows.

1. Create the Gaussian pyramid with $N + 1$ levels, in accordance with Algorithm 5.1(a).
2. The loop starts at this point and embraces steps 3–6. The loop will be performed N times, with a variable k going from 0 up to $N - 1$ (inclusive).
3. Get an image from the Gaussian pyramid at level $k + 1$ and up-sample. The result goes to the temporary image G2.
4. Get an image G1 from the Gaussian pyramid at level k and adjust its dimensions to fit exactly the dimensions of the just-computed image G2. The adjusted image becomes GA1.

This step is necessary because the consecutive processes, first of down sampling then up sampling, do not necessarily result in an image with dimensions of the original image.

5. The Laplacian image L[k] at the k-th level of the pyramid is obtained simply by subtraction: GA1 − G2.
6. Go to step 3 and repeat the loop for the next pyramid levels ($k \leftarrow k + 1$), unless the last level has been just processed.

A function operator which implements the Laplacian pyramid is presented in Algorithm 5.3.

```
///////////////////////////////////////////////////////////////////
// This function creates an image pyramid from the input
// image.
///////////////////////////////////////////////////////////////////
//
// INPUT:
//      theInputImage - reference to the input image
//      theOutputPyramid - vector of output images
//      requiredPyramidLevels - required number of
//         the pyramid levels
//
// OUTPUT:
//      true - if operation successful
//      false - otherwise
//
// REMARKS:
//
//
bool TGaussianImagePyramids::operator () (
                        const TRealImage & theInputImage,
                        RealImage_Vector & theOutputPyramid,
                        int requiredPyramidLevels )
{
   // At first we need a Gaussian filter.
   // Get the smoothing filter through the current factory.
   RealLinFilter_AP theSmoothingFilter(
                        fRealLinearFilter_Factory( fMaskWidth, fSigma ) );

   TImage_SimpleByTwo_DownSampler theDecimator;

   REQUIRE( theOutputPyramid.size() == 0 ); // require no prior images

   // save the first image of the pyramid (which is the original image)
   theOutputPyramid.push_back( theInputImage );

   // prepare the starting smoothed version
   RIAP smoothedImageAtLevel( ( * theSmoothingFilter )(theInputImage) );

   // Go through all remaining levels
   for( int theLevel = 1; theLevel < requiredPyramidLevels; theLevel ++ )
   {
      // decimate the smoothed version
      RIAP decimatedImage( theDecimator( * smoothedImageAtLevel ) );

      // save that level of images
      theOutputPyramid.push_back( * decimatedImage );

      // the starting image for the next level
      smoothedImageAtLevel = RIAP( ( * theSmoothingFilter )(
                        * (const TRealImage *) decimatedImage.get() ) );
   }

   return true;
}
```

Algorithm 5.2 Implementation of the functor which builds the Gaussian pyramid. (Reproduced with permission of Pandora Int. Inc., London)

```
bool TLaplacianImagePyramids::operator () (
                        const TRealImage & theInputImage,
                        RealImage_Vector & theOutputPyramid,
                        int requiredPyramidLevels )
{
    // At first create the pure Gaussian pyramid
    RealImage_Vector thePureGaussianPyramid;
    if( TGaussianImagePyramids::operator ()(
        theInputImage, thePureGaussianPyramid, requiredPyramidLevels+1 )
            == false )
        return false;

    // Now take two neighbouring images from the pure Gaussian pyramid,
    // up-sample the smaller one, and subtract the two.
    // The result put into the output pyramid.

    TImage_TwoNeighborInterpol_UpSampler      theImageUpSampler;

    // Go through all remaining levels
    REQUIRE( thePureGaussianPyramid.size() == requiredPyramidLevels + 1 );
    for( int theLevel=0; theLevel<requiredPyramidLevels-1; theLevel++ )
    {
        REQUIRE( theLevel + 1 < thePureGaussianPyramid.size() );
        // interpolate the smoothed and down-sampled version
        RIAP upSampledImage(
            theImageUpSampler( thePureGaussianPyramid[ theLevel + 1 ] ) );

        // It can happen that the two images we wish to subtract can have
        // different dimensions. This happens because when down sampling,
        // then up sampling, image dimensions are adjusted to the nearest
        // even value. At the same time the second image, the one
        // from the Gaussian pyramid that has not been downsampled,
        // can have an odd dimension.
        // Thus a necessity to adjust image dimensions before subtracting.

        RIAP adjustedGaussianImage(
                thePureGaussianPyramid[ theLevel].
                    OrphanBitBlit(
                            0, 0,
                            upSampledImage->GetCol(),
                            upSampledImage->GetRow() ) );

        theOutputPyramid.push_back(
                    * adjustedGaussianImage - * upSampledImage );
    }

    // The last image in the Laplacian pyramid is a low-pass filtered
    // Gaussian image containing only the coarsest structures.
    REQUIRE( theLevel == requiredPyramidLevels - 1 );
    theOutputPyramid.push_back( thePureGaussianPyramid[ theLevel ] );

    return true;
}
```

Algorithm 5.3 Implementation of the functor which builds the Laplacian pyramid. (Reproduced with permission of Pandora Int. Inc., London)

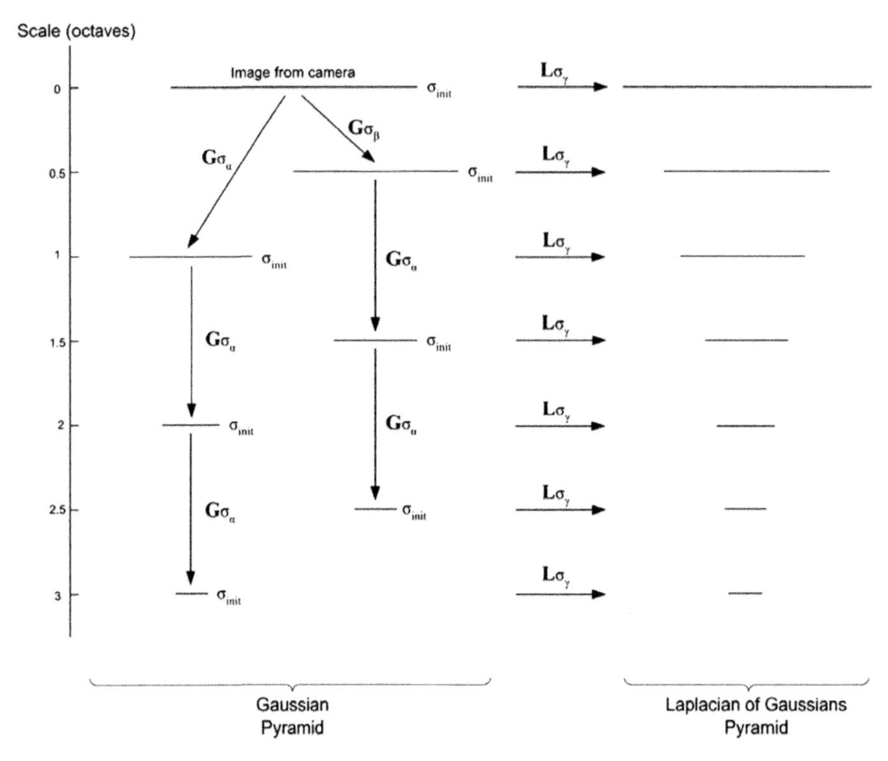

Figure 5.14 Constructing half octave DoG and LoG pyramids using two octave pyramids, the second offset by a half octave to the first. (Graphs and codes kindly supplied by Dr Sumita Balasuriya)

The DoG pyramid is built in an analogous way. Details can be found in the source code accompanying the book [216].

5.5.2 Matlab Examples

In the following Matlab codes (kindly supplied by Dr Sumitha L. Balasuriya who also supplied Figures 5.3, 5.5 and 5.14) to implement a half octave pyramid, an alternative approach has been adopted to achieving a half octave pyramid reduction factor. Rather than perform direct subsampling of each pyramid level with an awkward 1.5 reduction factor, which mismatches slightly with the desired scale-space subdivision of 2, two octave separated pyramids have been constructed. In this case, one of the octave pyramids has been constructed from a copy of the input image which has been subsampled by a half octave, as illustrated in Figure 5.14.

Below are the Matlab codes to implement the half octave pyramid of Figure 5.14. A Gaussian half octave pyramid is first constructed and the Laplacians of Gaussians pyramid is then constructed by Laplacian filtering each level of the LoG pyramid, as follows:

```
function Pyr=half_octave(I,levels,sigma_init);
% rho=1.5 aproximation to half-octave
% rho=2 octave

% initialise
masksize=7; % size of mask used in pyramid
Pyr{1}=double(I);

% sigma_alpha, subsample down to first octave layer = sigma_c
sigma_alpha=sigma_init*sqrt(3);

M_alpha=mask(masksize,sigma_alpha,0.5,0.5,'gau',sigma_alpha);
M_alpha=M_alpha/sum(sum(M_alpha));

% sigma_beta, subsample down to first half-octave layer
sigma_beta=sigma_init*1.1180;

M_beta{1,1}=mask(masksize,sigma_beta,0.25,0.25,'gau',sigma_beta);
M_beta{1,1}=M_beta{1,1}/sum(sum(M_beta{1,1}));
M_beta{3,1}=mask(masksize,sigma_beta,-0.25,0.25,'gau',sigma_beta);
M_beta{3,1}=M_beta{3,1}/sum(sum(M_beta{3,1}));
M_beta{1,3}=mask(masksize,sigma_beta,0.25,-0.25,'gau',sigma_beta);
M_beta{1,3}=M_beta{1,3}/sum(sum(M_beta{1,3}));
M_beta{3,3}=mask(masksize,sigma_beta,-0.25,-0.25,'gau',sigma_beta);
M_beta{3,3}=M_beta{3,3}/sum(sum(M_beta{3,3}));

[height,width]=size(I);

%%%


% HALF-OCTAVE DOWN FROM ORIGINAL IMAGE

Coarser=Pyr{1};

b=1;
for j=1+(masksize-1)/2:height-(masksize-1)/2,
    jj=rem(j,3);
    if jj==0, jj=3; end;  % Matlab indexes from 1

    if jj~=2,
        a=1;
        for i=1+(masksize-1)/2:width-(masksize-1)/2,
            ii=rem(i,3);
            if ii==0, ii=3; end;  % Matlab indexes from 1

            if ii~=2,
                try
                    Out(b,a)=sum(sum(Coarser(j-(masksize-1)/2:j+(masksize-1)/2,i-
(masksize-1)/2:i+(masksize-1)/2).*M_beta{ii,jj}));
                    a=a+1;
                catch

                end

            end
        end
        b=b+1
    end
end

Pyr{2}=Out;


% OCTAVE DOWN FROM ORIGINAL IMAGE

clear Out;
```

```
Coarser=Pyr{1};

b=1;
for j=1+(masksize-1)/2:height-(masksize-1)/2,
    if rem(j,2)~=0,
        a=1;
        for i=1+(masksize-1)/2:width-(masksize-1)/2,
            if rem(i,2)~=0,
                try
                    Out(b,a)=sum(sum(Coarser(j-(masksize-1)/2:j+(masksize-1)/2,i-
(masksize-1)/2:i+(masksize-1)/2).*M_alpha));
                    a=a+1;
                catch

                end

            end
        end
        b=b+1
    end
end

Pyr{3}=Out;


% OCTAVE SEPERATED CONVOLVE AND REDUCE

% recursive subsampling


for level=4:levels,
    Coarser=Pyr{level-2};
    clear Out;
    [height,width]=size(Coarser);
    b=1;
    for j=1+(masksize-1)/2:height-(masksize-1)/2,
        if rem(j,2)~=0,
            a=1;
            for i=1+(masksize-1)/2:width-(masksize-1)/2,
                if rem(i,2)~=0,
                    try
                        Out(b,a)=sum(sum(Coarser(j-(masksize-1)/2:j+(masksize-1)/2,i-
(masksize-1)/2:i+(masksize-1)/2).*M_alpha));
                        a=a+1;
                    catch

                    end

                end
            end
            b=b+1
        end
    end

    Pyr{level}=Out;

end

function M=mask(siz,sigma,X,Y,type,sigma2);
% siz can be even or odd
% X,Y denotes the subpixel distance of the centre of the
% gaussian from the centre of the mask
% M=mask(siz,sigma,X,Y,type,sigma2);

M=zeros(siz);

[x,y] = meshgrid(-(siz-1)/2:(siz-1)/2,-(siz-1)/2:(siz-1)/2);
x=x+X;
y=y+Y;
```

```
if type=='gau',
    h = (1/(2*pi*sigma2^2))*exp(-(x.*x + y.*y)/(2*sigma2*sigma2));
    %h = h/sum(sum(h)); % can keep this line if want to be accurate

elseif type=='dog',
    h1 = (1/(2*pi*sigma^2))*exp(-(x.*x + y.*y)/(2*sigma*sigma));
    %h1 = h1/sum(sum(h1)); % can keep this line if want to be accurate
    h2 = (1/(2*pi*sigma2^2))*exp(-(x.*x + y.*y)/(2*sigma2*sigma2));
    %h2 = h2/sum(sum(h2)); % can keep this line if want to be accurate

    h=h1-h2;

elseif type=='cir',
    h1 = exp(-(x.*x + y.*y)/(2*sigma*sigma));
    h1 = h1/sum(sum(h1));
    h2 = exp(-(x.*x + y.*y)/(2*sigma2*sigma2));
    h2 = h2/sum(sum(h2));

    h=h1-h2;
    h=-(h.*(h>0));

elseif type=='log',
    std = sigma*sigma;
    h1 = exp(-(x.*x + y.*y)/(2*std));
    h = h1.*(x.*x + y.*y - 2*std)/(2*pi*(sigma^6));
    h = h - sum(h(:))/prod(size(h));

elseif type=='gab',
    %[C,S] = gabormask(size,sigma,[],sigma2);
    %h=C;

    orient = sigma2;
    period = [];
    Size = siz;

    sy=Size-1; sx=Size-1;
    period = sigma*2*sqrt(2);

    % Basic grid
    hy = sy/2; hx = sx/2;
    [x, y] = meshgrid(-hx:sx-hx, -hy:sy-hy);
    x=x+X;
    y=y+Y;

    % Parameters
    omega = 2*pi/period;
    cs = omega * cos(orient);
    sn = omega * sin(orient);
    k = -1/(2*sigma*sigma);

    % Main computations
    g = exp(k * (x.*x + y.*y));         % Gaussian mask
    xp = x * cs + y * sn;               % Rotated x coords, phase
    cx = cos(xp);                       % cos grating
    cmask = g .* cx;                    % modulated cos grating
    sx = sin(xp);                       % sin grating
    smask = g .* sx;                    % modulated sin grating

    % Normalise so that convolution of mask with a harmonic curve of the
    % matching frequency gives unity peaks
    cnorm = sum(sum(cmask.*cx));
    cmask = cmask/cnorm;
    snorm = sum(sum(smask.*sx));
    smask = smask/snorm;

    C=cmask; S=smask;
    h{1}=C;
    h{2}=S;
```

```
end

%  [sy,sx]=size(h);
%  element=h(round(sy/2),1);
%  M=h.*(h>=element);
%
%  M=M/sum(sum(M));

M=h;
```

5.5.2.1 Building the Gaussian Pyramid in Matlab

The Gaussian pyramid is constructed using the Matlab code in the previous section as follows. Invoke the construction of a half-octave Gaussian pyramid using:

```
Pyr=half_octave(Image,levels,init)
```

The variable definitions are

Image = input image
levels = no. of levels of pyramid
init = Intrinsic blurring in Image
Pyr = Cell data structure with Gaussian pyramid levels.

In this example we construct two pyramids, for the 'right' and 'left' images of a stereo-pair, as follows:

```
Pyr_L=half_octave(Left,7,0.5);
Pyr_R=half_octave(Right,7,0.5);
```

5.5.2.2 Building the Laplacian of Gaussians Pyramid in Matlab

The code to invoke the construction of the Laplacian of Gaussians pyramid is slightly more involved than that above, since this is constructed here by convolving each level of a Gaussian pyramid with a Laplacian filter. A mask generator function is provided with the appropriate parameters to generate the required Laplacian kernel:

```
M=mask(siz,sigma,X,Y,type,sigma2);
```

The variable definitions are

siz = Size of filter
sigma = Sigma of filter
X = Horizontal subpixel offset
Y = Vertical subpixel offset
type = type of filter ('lap' in this case)
sigma2 = Sigma of excitatory DoG subfield if generating DoG filter.

```
Pyr_L=half_octave(Left,7,0.5); Pyr_R=half_octave(Right,7,0.5);
patchsize=5; sigma=5/6;
Lap_Pyr_L=laplacian(Pyr_L,patchsize,sigma);
Lap_Pyr_R=laplacian(Pyr_R,patchsize,sigma);
```

5.6 Closure

5.6.1 Chapter Summary

The key concept introduced in this chapter is that of image scale and how this must be accommodated to allow matching and analysis algorithms to operate independently of the scale of image structures represented within images. Scale-space provides an embedded decomposition of the image signal into a family of signals that provide a structure on which analysis over scale can be applied. Differential scale-space takes this notion a step further by decomposing the input image into a set of overlapping band pass images, each band pass representing a portion of the spatial frequency content of the original image. The image pyramid allows both Gaussian (low-pass) and Laplacians of Gaussians (band pass) decompositions to be represented efficiently and these data structures form the foundations of a very substantial fraction of modern image matching and analysis algorithms. It is possible to parameterize the construction of the image pyramid to accommodate specific tasks and to achieve specific degrees of continuity between scales based on careful analysis of the reduction factor between pyramid levels and the degree of blurring applied. The semi-pyramid also provides a mechanism for simplifying the construction of pyramid structures with increased continuity between scales over the basic pyramid structure.

5.6.2 Further Reading

The two key introductory papers to scale-space concepts used here are by Lindeberg [277] and Florack *et al.* [130]. Witkin's seminal paper on search over scale [448] should be read in the context of image matching (Chapter 6). A number of textbooks have emerged on the subject of scale-space. Some examples are as follows.

- *Front-End Vision and Multi-Scale Image Analysis: Computer Vision Theory and Applications written in Mathematica* by Bart M. Ter Haar Romeny.
- *Gaussian Scale-Space Theory* by Jon Sporring.
- *Scale-space Theory in Computer Vision* by Tony Lindeberg.
- *Curvature Scale Space Representation: Theory, Applications and MPEG-7 Standardization* by Farzin Mokhtarian and Miroslav Bober.

The final text quoted above is somewhat different in that it deals with curvature embedded in scale-space, but is none the less highly relevant to understanding certain classes of matching algorithm.

The state-of-the-art in scale-space theory and techniques has been presented every two years at the Scale-Space International Conference since 1997. The proceedings published to date are listed below.

- Scale-Space Theory in Computer Vision: First International Conference, Scale-Space 1997, Utrecht, The Netherlands.
- Scale-Space Theories in Computer Vision: Second International Conference, Scale-Space 1999, Corfu, Greece.

- Scale-Space and Morphology in Computer Vision: Third International Conference, Scale-Space 2001, Vancouver, Canada.
- Scale Space Methods in Computer Vision: Fourth International Conference, Scale-Space 2003, Isle of Skye, UK.
- Scale Space and PDE Methods in Computer Vision: Fifth International Conference, Scale-Space 2005, Hofgeismar, Germany.
- Scale Space and Variational Methods in Computer Vision, First International Conference, SSVM 2007, Ischia, Italy.

5.6.3 Problems and Exercises

1. Plot the PSNR for a Gaussian filter: at the Nyquist limit, subsampled a half octave below this limit and subsampled one octave below.
2. What value of σ is required to ensure a PSNR of 150 dB at each level of a regular Gaussian pyramid with an octave reduction factor?
3. Plot the PSNR for a Laplacian of Gaussians filter: at the Nyquist limit, subsampled a half octave below this limit and subsampled one octave below.
4. What value of σ is required to ensure a PSNR of 150 dB at each level of a regular Laplacian of Gaussians pyramid with an octave reduction factor?
5. How would you construct a Laplacian of Gaussians pyramid such that the spectral overlap between levels resulted in the half-power points of adjacent band passes (pyramid levels) being congruent? Plot the frequency response of each level of the resultant pyramid.

6

Image Matching Algorithms

6.1 Abstract

In this chapter we discuss some basic aspects of image matching algorithms. Matching can be viewed as a process of finding the degree of correlation between two groups of data. This area belongs to the one of the most explored topics in computer science. Therefore the key issue of this chapter is to provide the basic concepts followed by some of the most common image matching strategies.

We begin with an overview of matching measures, starting with measures operating on scalar intensity signals, bit strings, progressing then to vector and matrix data, as well as to the statistical and information theory-based methods. The algorithmic aspects are then discussed with particular emphasis on techniques for increasing the effectiveness of the methods.

Matching is sometimes more efficient when operating on image signals that have been transformed in some way, rather than operating on the pure intensity values themselves. Examples here are the nonparametric *Census* and *Rank*. Another type of transformation discussed is the nonlinear log-polar transform, which allows more reliable image matching. Its application to area matching around salient points is also presented, where it can be used to find point correspondences necessary for computation of the fundamental matrix.

The rest of the chapter is devoted to the broad group of stereo matching methods, i.e. the computer algorithms by means of which disparity information can be extracted from stereo-pair images of a scene. We discuss some algorithmic problems encountered in stereo matching; different stereo methods are then described with software implementations of some of them. Finally, gradient-based matching, dynamic programming, graph cuts and optical flow methods are discussed.

6.2 Basic Concepts

Comparing different things is one of the most common actions performed by humans. We often compare prices for the best deal, maximum speed of a car with a speed limit sign, people's heights, but also we are able to compare meanings of words, or compare chances of politicians in elections, for instance. Each comparison is based on some prerequisites – or *a priori* knowledge – that sometimes can be expressed by a simple mathematical formula. Otherwise, people use their own 'fuzzy' interpretation of things, sometimes with various meanings for different persons. Therefore the meanings of a 'distance' and a 'length' are very important in science. In mathematics these concepts are known as a metric and a norm, respectively.

In computer vision, and also in other disciplines of science, data matching belongs to one of the fundamental processing methods. There are also many types of 'data' that are to be matched. For instance, these can be intensity signals, detected shape contours, graphs, etc.

6.3 Match Measures

In this section we analyse the most common and practical matching measures that can be used to tell which areas in different images fit together and how to assign a scalar value to describe 'goodness' of a match.

6.3.1 Distances of Image Regions

Table 6.1 contains a list of the most common area or region matching measures operating directly on pixels. In all of the following definitions we assume that two compatible image regions I_1 and I_2 are compared. Both can belong to the same or different image spaces. I_1 is built around a reference point (x, y), expressed in its local coordinate space; I_2 is built around a point $(x + d_x, y + d_y)$ in its local coordinate space. For both, the matching regions are defined by a set U of offset values, measured from their reference points, i.e. (x, y) and $(x + d_x, y + d_y)$, respectively. Thus, the matched regions are not necessarily compact. We assume also that all indices defined by U fall into ranges of valid pixel location for I_1 and I_2, respectively.

Pixel values can be scalars, vectors, matrices or even tensors, i.e. for pixels we allow all mathematical objects for which the involved operators and norms are defined. For instance D_{SAD} can be defined for all such objects (pixels) for which the subtraction operator '$-$' and the norm $||$ are defined. However, the following measures are uniquely defined only for scalars. Usually, for higher dimensional objects a result can be obtained in many different ways. Moreover, we are usually interested to get a scalar value as a result of matching.

In the task of region matching we are usually interested in finding the central points (x, y) and $(x + d_x, y + d_y)$, and/or values of d_x and d_y, for which a matching measure obtains its extreme value. Table 6.2 presents an explanation of the symbols used in Table 6.1.

Although Table 6.1 gives many possibilities, a practical choice of a proper distance measure for a given application is not an easy one. The way to overcome this problem is to get some more in-depth knowledge on the distances and to experimentally verify their behaviour.

The most commonly known and used are D_{SAD} and D_{SSD}, although the first one usually requires the least computations. This is especially important when the speed of computations is a priority, although bit matching in the nonparametric domain can be an alternative (section 6.3.7).

Sebe *et al.* [378] examined relations of the common matching measures with respect to the noise distribution encountered in real images and different applications of computer vision. Their results show that the usual assumption about the Gaussian noise distribution and, as a consequence, choice of D_{SSD} are not well justified in many cases. A better approximation gives the Cauchy distribution. To cope with real situations Sebe *et al.* propose either to precondition image statistical properties, so the inner noise is more Gaussian like, or use a

Table 6.1 The most common matching measures for intensity signals

D_{SAD} – *sum of absolute differences* (6.1)

$$D_{SAD} = \sum_{(i,j)\in U} \left| I_1(x+i, y+j) - I_2(x+d_x+i, y+d_y+j) \right|$$

D_{ZSAD} – *zero mean sum of absolute differences* (6.2)

$$D_{ZSAD} = \sum_{(i,j)\in U} \left| \left(I_1(x+i, y+j) - \overline{I_1(x,y)}\right) - \left(I_2(x+d_x+i, y+d_y+j) - \overline{I_2(x+d_x, y+d_y)}\right) \right|$$

D_{SSD} – *sum of squared differences* (6.3)

$$D_{SSD} = \sum_{(i,j)\in U} \left(I_1(x+i, y+j) - I_2(x+d_x+i, y+d_y+j)\right)^2$$

D_{ZSSD} – *zero mean sum of squared differences* (6.4)

$$D_{ZSSD} = \sum_{(i,j)\in U} \left[\left[I_1(x+i, y+j) - \overline{I_1(x,y)}\right] - \left[I_2(x+d_x+i, y+d_y+j) - \overline{I_2(x+d_x, y+d_y)}\right]\right]^2$$

$D_{\mathrm{SSD-N}}$ – *normalized sum of squared differences* (6.5)

$$D_{SSD-N} = \frac{\sum_{(i,j)\in U} \left[I_1(x+i, y+j) - I_2(x+d_x+i, y+d_y+j)\right]^2}{\sqrt{\sum_{(i,j)\in U} I_1(x+i, y+j)^2 \cdot \sum_{(i,j)\in U} I_2(x+d_x+i, y+d_y+j)^2}}$$

$D_{\mathrm{ZSSD-N}}$ – *zero mean normalized sum of squared differences* (6.6)

$$D_{ZSSD-N} = \frac{\sum_{(i,j)\in U} \left[\left(I_1(x+i, y+j) - \overline{I_1(x,y)}\right) - \left(I_2(x+d_x+i, y+d_y+j) - \overline{I_2(x+d_x, y+d_y)}\right)\right]^2}{\sqrt{\sum_{(i,j)\in U} \left(I_1(x+i, y+j) - \overline{I_1(x,y)}\right)^2 \cdot \sum_{(i,j)\in U} \left(I_2(x+d_x+i, y+d_y+j) - \overline{I_2(x+d_x, y+d_y)}\right)^2}}$$

D_{CV} – *covariance-variance* (6.7)

$$D_{CV} = \frac{\sum_{(i,j)\in U} \left(I_1(x+i, y+j) - \overline{I_1(x,y)}\right) \cdot \left(I_2(x+d_x+i, y+d_y+j) - \overline{I_2(x+d_x, y+d_y)}\right)}{\sqrt{\sum_{(i,j)\in U} \left(I_1(x+i, y+j) - \overline{I_1(x,y)}\right)^2 \cdot \sum_{(i,j)\in U} \left(I_2(x+d_x+i, y+d_y+j) - \overline{I_2(x+d_x, y+d_y)}\right)^2}}$$

(continued)

Table 6.1 The most common matching measures for intensity signals (*Continued*)

D_{SCP} – *sum of cross products* (6.8)

$$D_{SCP} = \sum_{(i,j)\in U} I_1(x+i, y+j) \cdot I_2(x+d_x+i, y+d_y+j)$$

$D_{\text{SCP-N}}$ – *normalized sum of cross products* (6.9)

$$D_{SCP-N} = \frac{\sum_{(i,j)\in U} I_1(x+i, y+j) \cdot I_2(x+d_x+i, y+d_y+j)}{\sqrt{\sum_{(i,j)\in U} I_1(x+i, y+j)^2 \cdot \sum_{(i,j)\in U} I_2(x+d_x+i, y+d_y+j)^2}}$$

metric which better fits real situations. For the latter solution, a good choice is the Cauchy metric which in their opinion is better than the Kullback measure (section 6.3.4) and much better than D_{SSD} or D_{SAD}. However, practical application of this idea should be preceded by a statistical test to verify a hypothesis of a distribution encountered in the images to be processed. The second problem is determination of a height–tail parameter of the Cauchy distribution – a suitable algorithm is outlined in [378]. The Cauchy distance D_{CHY} among two regions can be computed in accordance with (6.26).

Similarly Bhat and Nayar [41] showed that D_{SAD} and D_{SSD} are very sensitive to outliers, and suggested that other measures should be used for image matching.

Table 6.2 Explanation of symbols used in Table 6.1

Expression in Table 6.1	Description
$U = U(x, y)$	A set of points (in practice, point coordinates) located around a point with local image coordinates (x, y).
$I_k(x, y)$	An intensity value of the k-th image at a point with local image coordinates (x, y).
$\overline{I_k(x, y)} = \frac{1}{N} \sum_{(i,j)\in U} I_k(x+i, y+j)$ where $N = \#U$	An average intensity value of the k-th image at a certain neighbourhood U around a point with local image coordinates (x,y). N denotes the number of points that were taken into computations or, in other words, the number of distinctive pairs (i, j) which denote relative displacements around (x, y). They can be positive or negative with only one assumption that $I_k(x+i, y+j)$ belongs to the domain of an image I_k, otherwise a value would not be defined.
d_x, d_y	Parameters that denote relative horizontal and vertical displacements of the two image blocks being compared. Note that in many computer vision tasks these are just the values we are looking for, under a constraint that a distance measure $D_{()}$ from Table 6.1 attains its extreme value (this leads to the optimization problems).

Thus, to cope with local inaccuracies caused by noise, different lighting conditions or camera characteristics, other distances, such as D_{ZSAD}, D_{ZSSD}, $D_{ZSSD\text{-}N}$ and D_{CV}, can be considered. The mentioned measures have a common feature – they assume signal preconditioning to obtain zero mean values in compared regions. In some applications, such as stereo matching [118] or image registration in the log-polar space [465], D_{CV} gives the best results. However, its computational complexity is also the highest from the distances presented in Table 6.1.

The nice feature about D_{CV} is that it is invariant to the linear transformation of the two matched signals, i.e. it holds that

$$D_{CV}(I_1, I_2) = D_{CV}(a_1 I_1 + b_1, a_2 I_2 + b_2), \tag{6.10}$$

where $a_{1,2}$ and $b_{1,2}$ are constants ($a_{1,2}$ needs to have the same sign). Therefore D_{CV} is frequently used in template matching, where one image, say I_1, describes a pattern. Then, each possible region in I_2 is matched against the pattern I_1 by means of D_{CV}. The best match can indicate a region in I_2 where the pattern is found.

The other subgroup of distances presented in Table 6.1 constitutes D_{SCP} and $D_{SCP\text{-}N}$. Both follow the idea of a scalar product between two vectors. In this case components of the vectors are created from the intensity signals of matched blocks of images. The scalar product can be used to measure *the phase difference* between vectors,[1] assuming however that the vectors are normalized (i.e. their lengths are set to one). Otherwise, a vector with components close to zero would match all other vectors, which obviously is not what we are interested in. The postulate of normalization has been expressed in the $D_{SCP\text{-}N}$ measure – at a cost of additional computations, however. Unfortunately for simple D_{SCP} such normalization conditions are not met directly. Therefore, in this case, even D_{SSD} does better than D_{SCP} since it takes into account local signal energies. The two distances are related as follows:

$$\begin{aligned}
D_{SSD}(I_1, I_2) &= \sum_{(i,j)\in U} \left(I_1(x+i, y+j) - I_2(x+d_x+i, y+d_y+j)\right)^2 \\
&= \sum_{(i,j)\in U} (I_1^2(x+i, y+j) - 2I_1(x+i, y+j)I_2(x+d_x+i, y+d_y+j) \\
&\quad + I_2^2(x+d_x+i, y+d_y+j)) = \sum_{(i,j)\in U} \left(I_1^2(x+i, y+j) + I_2^2(x+d_x+i, y+d_y+j)\right) \\
&\quad - 2\sum_{(i,j)\in U} I_1(x+i, y+j)I_2(x+d_x+i, y+d_y+j) \\
&= 2\left[\frac{1}{2}\sum_{(i,j)\in U} \left(I_1^2(x+i, y+j) + I_2^2(x+d_x+i, y+d_y+j)\right) - D_{SCP}(I_1, I_2)\right].
\end{aligned} \tag{6.11}$$

The first factor in the last expression in brackets of (6.11) conveys information proportional to the cumulative signal energy in the matched image regions. Thus, D_{SSD} can be seen as the average cumulative signal energy reduced by the dot product D_{SCP} between matched 'vectors'.

[1] A scalar product between two vectors, say **a** and **b**, is defined as $\mathbf{a}\cdot\mathbf{b}=|\mathbf{a}||\mathbf{b}|\cos(\mathbf{a},\mathbf{b})$.

A choice of the matching window U in (6.1)–(6.7) is even more cumbersome than a choice of the matching distance. There are no strict rules to define U and a choice usually depends on the application and image contents.

6.3.2 Matching Distances for Bit Strings

In some cases we can treat data as simple series of bits, each bit conveying some specific information. For instance an intensity signal can be preprocessed by a nonparametric *Census* transformation (section 6.3.7) – in this case each bit reads whether a given intensity value is less, or is not, than a reference value. In such cases it is better to compare bit strings with binary measures; some of the most common are presented in Table 6.3.

Table 6.3 The most common matching measures for bit streams

D_H – *Hamming distance* (6.12)

$$D_H(\mathbf{a}, \mathbf{b}) = \frac{1}{N}\sum_{i=1}^{N} a_i \otimes b_i$$

D_T – *Tanimoto distance* (6.13)

$$D_T(\mathbf{a}, \mathbf{b}) = \begin{cases} 1 & \text{if } \mathbf{a} = \mathbf{b} = \mathbf{0} \\ 1 - \dfrac{\mathbf{a}^T\mathbf{b}}{\mathbf{a}^T\mathbf{a} + \mathbf{b}^T\mathbf{b} - \mathbf{a}^T\mathbf{b}} & \text{otherwise} \end{cases}$$

D_{DK} – *Dixon–Koehler distance* (6.14)

$$D_{DK}(\mathbf{a}, \mathbf{b}) = D_H(\mathbf{a}, \mathbf{b})\, D_T(\mathbf{a}, \mathbf{b})$$

D_{WT} – *weighted Tanimoto distance* (6.15)

$$D_{WT}(\mathbf{a}, \mathbf{b}) = \eta D_T(\mathbf{a}, \mathbf{b}) + (1 - \eta)\, D_T(\neg\mathbf{a}, \neg\mathbf{b})$$

In Table 6.3 **a**, **b** are the compared vectors of the same length N, $-$ denotes bit negation and $\otimes$ denotes the exclusive-or (XOR) operation which is simply a number of mismatched bits when comparing the aligned vectors bit by bit.

The Hamming measure D_H (6.12) treats all compared bits (0 or 1) with the same weight. The other three metrics, in Table 6.3, originated in biological and chemical sciences. Contrary to D_H, however, D_T and D_{DK} stress more matches on '1s' than on '0s' [105]. They can present some advantage if bits '0' are less important, e.g. denote image areas with the same intensity. Such areas with uniform intensity, which usually cause problems in matching, will be treated with a slightly lower weight. In the D_{WT} measure we can control the influence of the matches on '1s' and '0s' at the same time. The first term in D_{WT} is simply the weighted D_T whereas the second term presents a reversely weighted complement of D_T. The weight parameter η

Table 6.4 Comparison of different matching strategies for bit strings: Hamming D_H, Tanimoto D_T, Dixon–Koehler D_{DK} and weighted Tanimoto D_{WT}. The match values are in the range [0, 1] with 0 for the best and 1 for the worst match

a	**b**	D_H	D_T	D_{DK}	D_{WT}	
					η in (6.16)	η in (6.17)
10010	01101	1	1	1	0.836	0.877
01010	00110	0.4	0.667	0.267	0.508	0.548
01100	11100	0.2	0.333	0.067	0.26	0.279
01101	11111	0.4	0.4	0.16	0.4	0.4
11111	11101	0.2	0.2	0.04	0.226	0.219
11111	11111	0	0	0	0	0
00000	00000	0	1	0	0.667	0.75

stabilizes situations of strong correlations exclusively on ‘1s’ or only on ‘0s’. Usually this parameter is given as follows [129]:

$$\eta = \frac{2-p}{3}, \quad \text{where } p = \frac{\mathbf{a}^T\mathbf{a} + \mathbf{b}^T\mathbf{b}}{2N}. \tag{6.16}$$

Certainly, $p \in [0, 1]$ and $\mathbf{a}^T\mathbf{a}$ is a number of ‘1s’ in **a**, while $\mathbf{b}^T\mathbf{b}$ in **b**. However, to favour all matches on ‘1s’ we can set for example

$$\eta = \frac{3-p}{4}. \tag{6.17}$$

Table 6.4 contains some examples of comparison of bit vectors **a** and **b** with the presented distances.

When observing the consecutive rows in Table 6.4 from top to bottom we see that the vectors **a** and **b** have 0, 1, 2, 3, 4 or 5 matches on ‘1s’, respectively. The last row has no matches on ‘1s’ but a maximum number of matches on ‘0s’. Notice also that, in contrast to the distances in Table 6.1, the relative order of bits (i.e. their permutation) does not influence the results of this group.

In practice, for block matching (section 6.6) the best results in quality and speed are obtained with the D_H measure [87]. The other binary distances are more suitable for matching of binary patterns, such as hand-drawn images or digits [240].

6.3.3 Matching Distances for Multichannel Images

Sometimes we can have images the pixels of which are not scalars (e.g. colour images). In this case it is also possible to define measures that can be used to compare relative distances among pixels.

The most popular metrics for vector data follow the Minkowski metric $D_{M\alpha}$ and for 2D mathematical objects (like matrices or tensors) the Frobenius metric D_F. The latter is characterized by its desirable rotation invariance property. Table 6.5 summarizes some of the most common distances for this type of data.

Table 6.5 The most common matching measures for vectors and matrices (or tensors)

D_{M} – *Minkowsky distance for vectors* **a** *and* **b** *with parameter* α

$$D_{\mathrm{M}\alpha}(\mathbf{a}, \mathbf{b}) = \left(\sum_{k=1}^{S} |a_k - b_k|^{\alpha} \right)^{1/\alpha} \tag{6.18}$$

$D_{\mathrm{M}1}$ – *Minkowsky distance between matrices* **A** *and* **B** *with* $\alpha = 1$

$$D_{\mathrm{M}1}(\mathbf{A}, \mathbf{B}) = \max_{1 \le i \le m} \sum_{j=1}^{n} \left| a_{ij} - b_{ij} \right| \tag{6.19}$$

$D_{\mathrm{M}\infty}$ – *Minkowsky distance between matrices* **A** *and* **B** *with* $\alpha \to \infty$

$$D_{\mathrm{M}\infty}(\mathbf{A}, \mathbf{B}) = \max_{1 \le j \le n} \sum_{i=1}^{m} \left| a_{ij} - b_{ij} \right| \tag{6.20}$$

D_{F} – *Frobenius distance between matrices* **A** *and* **B**

$$D_{\mathrm{F}}(\mathbf{A}, \mathbf{B}) = \sqrt{\sum_{i=1}^{m} \sum_{j=1}^{n} \left| a_{ij} - b_{ij} \right|^2} \tag{6.21}$$

In Table 6.5 a_k and b_k stand for the k-th component of the vectors **a** and **b** from S-dimensional space, respectively, and α is a parameter. Similarly, **A** and **B** are $m \times n$ matrices (or tensors) with scalar elements a and b, respectively.

A common topological question is: what is *a unit distance* for a given metric? To visualize the behaviour of the Minkowsky measure with change of the parameter $\alpha > 1$, without loss of generality let us assume that the vector **a** is placed in the centre of the 2D coordinates system, i.e. $\mathbf{a} = [0, 0]$. Now, the above question can be formulated mathematically as follows:

$$D_{\mathrm{M}\alpha}(0, \mathbf{b}) = \left(\sum_{k=1}^{S} |b_k|^{\alpha} \right)^{1/\alpha} \stackrel{?}{=} 1, \tag{6.22}$$

or in the 2D case

$$|b_1|^{\alpha} + |b_2|^{\alpha} = 1. \tag{6.23}$$

Thus, for $\alpha = 1$, we have $|b_1| + |b_2| = 1$; for $\alpha = 2$, $b^2{}_1 + b^2{}_2 = 1$; and so on. The solution is depicted in Figure 6.1.

Similar plots can be drawn for N-dimensional space. D_{M} with $\alpha = 2$ (i.e. the SSD measure) seems to comply with our everyday intuition on geometric distance, which explains the great popularity of this measure in many applications.

Table 6.6 The most common statistical distances

D_G – *Gaussian distance between vectors* **a** *and* **b** *with parameter* σ

$$D_G(\mathbf{a}, \mathbf{b}) = e^{-\left(\frac{D_X(\mathbf{a},\mathbf{b})}{\sigma}\right)^2} \tag{6.24}$$

where σ is a parameter that controls the width of the distribution [341]; D_x usually denotes the Euclidean norm on the difference between **a** and **b**, i.e. ${D_x}^2 = D_{SSD}$ given by (6.3), or the Mahalanobis distance given below (in this case we set $\sigma = 1$).

D_{MAH} – *Mahalanobis distance between vectors* **a** *and* **b**

$$D_{MAH}(\mathbf{a}, \mathbf{b}) = (\mathbf{a} - \mathbf{b})^T \mathbf{A}^{-1} (\mathbf{a} - \mathbf{b}) \tag{6.25}$$

where **A** is the covariance matrix which is computed for a given population of data points, from which **a** and **b** are drawn.

D_{CHY} – *Cauchy distance between vectors* **a** *and* **b** *with parameter* τ

$$D_{CHY}(\mathbf{a}, \mathbf{b}) = \log\left[1 + \left(\frac{D_x(\mathbf{a}, \mathbf{b})}{\tau}\right)^2\right] \tag{6.26}$$

where τ is a parameter that controls height and tails of the Cauchy distribution. D_x is a norm on the difference between **a** and **b**; usually computed as the Euclidean distance.

6.3.3.1 Statistical Distances

There are some distances which come from the domain of mathematical statistics. These are the Gaussian and the Mahalanobis distances defined for scalar or N-dimensional vector data. Additionally, the Mahalanobis distance requires knowledge of the covariance matrix computed for a population of data. The most common statistical distances are summarized in Table 6.6.

Computation of the Mahalanobis distance D_{MAH} requires computation of the covariance matrix **A**, then finding its inverse. However, this means that we need to know a population of data, say $\{\mathbf{x}\}$, in which we then try to compute a distance between two vectors **a**, **b**, which do not necessarily belong to this population. Computation of $\mathbf{A}^{-1}$ may be time consuming for a large number of data. However, it has to be done once for the whole population.

The covariance matrix **A** for a population $\{\mathbf{x}\}$ is given as [157, 163, 341]

$$\mathbf{A}_{\{x\}} = E\left\{(\mathbf{x} - \mathbf{m}_x)(\mathbf{x} - \mathbf{m}_x)^T\right\}, \tag{6.27}$$

where **x** are assumed to be $N \times 1$ column vectors, N being the dimension of the data space, $E\{\cdot\}$ denotes the expectation value and $\mathbf{m}_x$ is a mean vector of the population $\{\mathbf{x}\}$. The mean vector can be estimated by the following expression:

$$\mathbf{m}_x = \frac{1}{N}\sum_{i=1}^{N} \mathbf{x}_i. \tag{6.28}$$

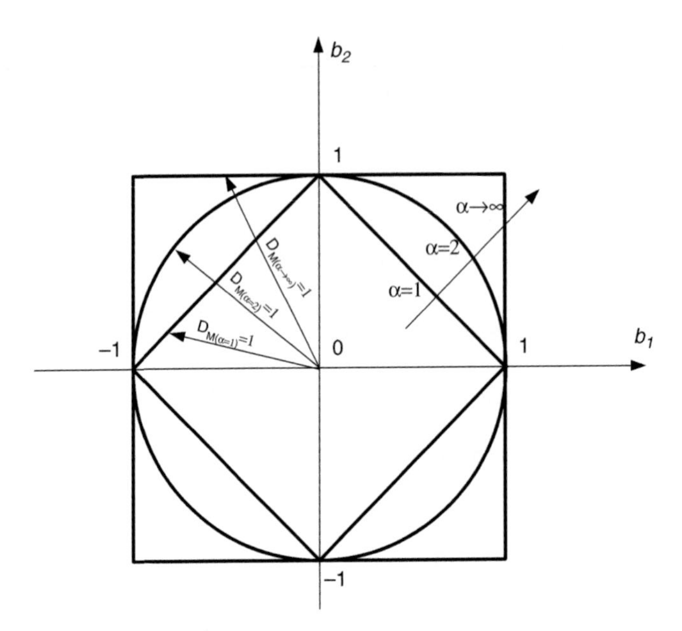

Figure 6.1 Plots of a unit distance from the origin of the 2D coordinate system in the sense of the Minkowsky metric for different parameters α

After some rearrangements (6.27) transforms into

$$\mathbf{A}_{\{x\}} = \frac{1}{N} \sum_{i=1}^{N} \mathbf{x}_i \mathbf{x}_x^{\mathrm{T}} - \mathbf{m}_x \mathbf{m}_x^{\mathrm{T}}, \tag{6.29}$$

which can be directly used for computation of $\mathbf{A}$. Discussion of numerical computations of the inverse matrix can be found, for example, in [352]. Let us note also that if $\mathbf{A} = \mathbf{I}$ then D_{MAH} reduces to D_{SSD}.

Some simple examples of colour image segmentation based on the Mahalanobis distance computed in the RGB space can be found in [157]. More information on the Mahalanobis distance in the light of statistics for geometric computations can be found in the monograph by Kanatani [239].

6.3.4 Measures Based on Theory of Information

Very important knowledge about data can be measured by means of the entropy, relative entropy – known also as the Kullback–Leibler distance – and, as a consequence, by the conditional entropy and maximum mutual information principle [75, 184]. These concepts can be used to build practical and very effective methods of image segmentation, matching, etc. Table 6.7 presents an overview of the most basic concepts of information theory that can also be applied to image processing.

Table 6.7 The most common concepts of information theory

Entropy, a measure of uncertainty of a random variable

$$H(X) = -\sum_{x \in A} p(x) \log_B p(x), \tag{6.30}$$

where X is a random variable for which a set A of its allowable discrete values is given (called also an alphabet). B in this and the following formulas denotes the base of the logarithm. Usually it is chosen to be 2, 10 or $e \approx 2.71$. The probability of X taking one of the values is given by the probability mass function $p(x) = \Pr\{X = x\}, x \in A$ [75].

Entropy describes an 'amount' of information conveyed by a random variable (or, in other words, required to describe such a variable).

Joint entropy, a measure of uncertainty of two random variables X and Y

$$H(X, Y) = -\sum_{x,y \in A} p(x, y) \log_B p(x, y), \tag{6.31}$$

where X, Y are two random variables with the same alphabet A.

Conditional entropy, a measure of uncertainty of one random variable given the second one

$$H(Y|X) = -\sum_{x,y \in A} p(x, y) \log_B p(y|x), \tag{6.32}$$

where X, Y are two random variables with the same alphabet A.

Kullback–Leibler distance between two distributions p and q (also called relative entropy)

$$D_{\mathrm{KL}}(p, q) = \sum_{x \in A} p(x) \log_B \frac{p(x)}{q(x)}, \tag{6.33}$$

where p and q are two probability functions. It is assumed that $0\log(0/0) = 0$, $0\log(0/q) = 0$, and $p\log(p/0) = \infty$.

D_{KL} is a measure of discrepancy between two distributions p and q. The further the two, the less justified is the assumption of a distribution q when the true distribution is just p.

Symmetric Kullback–Leibler distance between two distributions p and q (also called J-divergence or J-distance)

$$D_J(p, q) = \frac{1}{2}\left[D_{\mathrm{KL}}(p, q) + D_{\mathrm{KL}}(q, p)\right], \tag{6.34}$$

with the same meaning of symbols as described for the D_{KL} case. The nice feature of D_J is that contrary to D_{KL} it is a true metric. Thus, a distance from p to q is the same as from q to p. Similarly, the triangle inequality holds.

Mutual information

This is a relative entropy between the joint distribution $p(x, y)$, and the product of distributions $p(x)p(y)$, as follows:

$$I(X, Y) = \sum_{x,y \in A} p(x, y) \log_B \frac{p(x, y)}{p(x)\, p(y)}. \tag{6.35}$$

(continued)

Table 6.7 The most common concepts of information theory (*Continued*)

Mutual information gives a measure of the amount of information that one random variable conveys about the second one.

It can be shown that the mutual information can be expressed in terms of entropy and joint entropy, as follows [75]:

$$I(X, Y) = H(X) + H(Y) - H(X, Y) \tag{6.36}$$

and

$$I(X, Y) = H(X) - H(X|Y) = H(Y) - H(Y|X). \tag{6.37}$$

From the above one can easily notice that the mutual information amounts to the uncertainty of X, given by entropy $H(X)$, reduced by the uncertainty of X due to knowledge of Y, or vice versa.

Entropy can be used to measure the amount of information conveyed in an image by a certain local region around a pixel. Such information can be used to decide whether this amount is sufficient for subsequent matching of such regions. If not, then this location can be skipped or the size of the region needs to be increased. Such a simple concept was employed for adaptive window growing in the nonparametric representation of images (section 6.3.7). Matching is done with areas of a minimum size which convey sufficient information for match discrimination, however. Because of this technique it was possible to increase the accuracy of the disparity map compared to basic area matching [88].

It is worth noting that entropy of a discrete random variable is invariant to its rotations and translations. This feature of entropy can be employed for template matching. Sometimes it is more convenient to consider the entropy (6.30) in terms of the probabilistic expected value:

$$H(X) = E_p\left\{\log \frac{1}{p(x)}\right\},$$

where[2] the expected value $E_q\{p(X)\}$ of $p(X)$, with X having the probability distribution function $q(X)$, is given as [341]

$$E_q\{p(X)\} = \sum_{x \in X} q(x)\, p(x). \tag{6.38}$$

The Kullback–Leibler distance D_{KL} (i.e. the relative entropy) and its symmetries version D_J are used when comparing two distributions of probability. D_{KL} can be used immediately for matching of histograms, as discussed in section 6.3.5. In image processing D_{KL} is employed

[2]Henceforth we skip the base B in the logarithms.

most frequently for pattern recognition. Let us observe that (6.33) can be expressed as

$$\begin{aligned} D_{\mathrm{KL}}(p, q) &= \sum_{x \in A} p(x) \log [p(x)] - \sum_{x \in A} p(x) \log [q(x)] \\ &= E_p \{\log [p(x)]\} - E_p \{\log [q(x)]\} \\ &= S - E_p \{\log [q(x)]\}. \end{aligned} \tag{6.39}$$

Thus, if we consider $p(X)$ as a model which does not change and $q(X)$ as a test pattern which changes from instant to instant, then pattern matching with D_{KL} can be seen as a search for the expected value (6.38) of $\log[q(X)]$ in respect to the probability distribution $p(X)$ of a model. The constant S in the above formula does not depend on the test pattern. Interpretation of a model and test pattern can be exchanged, however. Thus, we can match many test patterns against a model, or a test pattern with a database of prototypes. In the former case the matching can be seen as an optimization problem of the form

$$\arg\min_j \left[D_{\mathrm{KL}}\left(p, q_j\right) \right] = \arg\min_j \left[E_p \left\{ \log \left[q_j(X) \right] \right\} \right]. \tag{6.40}$$

Joint (6.31) and conditional entropies (6.32) are side products when computing the mutual information in accordance with (6.36) or (6.37), respectively.

Image matching in terms of their mutual information as a similarity measure has attracted great interest among researchers. This is especially so in the areas of medical image registration and object recognition, since it is independent of translation and rotation, as well as being robust to outliers and noise [365, 439]. Maximization of the mutual information between images can be thought of as finding their largest overlapping regions such that they explain each other well in the information theoretic terms, i.e. by minimizing their joint entropy [75].

6.3.5 Histogram Matching

Histograms are 2D diagrams in which the ordinate depicts frequencies of occurrences of values from the abscissa. More often than not histograms are represented as linear arrays (or vectors), which belong to the 1D data structures. These data structures have found vast application in image processing, mostly to acquire information on the frequency of occurrence of different features in images. Indeed, when properly normalized, histograms can be thought of as estimations of the probabilistic density function of a random variable (image features, etc.) [351].

Approaching the problem of histogram matching we can go two ways, depending on how we treat these structures. If we look at histograms as vectors of data [145], we can apply any of the already presented methods for vector matching (Table 6.5; see also Table 6.1). This is a quite obvious approach when two vectors are of the same length. Otherwise, the partial matching techniques can be used.

The probabilistic approach is the second way that can be undertaken for histogram matching. In this case, we can treat each entry of the histogram as a discrete value of a probabilistic density function (pdf) [75, 341, 351]. Thus, matching two histograms is equivalent to matching two probabilistic densities, for which very common is application of the already presented

Kullback relative information measure. Such a strategy has been suggested by many authors, for instance Sebe *et al.* [378] or Pratt [351].

Let us assume that we have two histograms, represented as vectors **a** and **b**, each consisting of N data, i.e. $\mathbf{a} = \{a_i\}_{1 \leq i \leq N}$ and $\mathbf{b} = \{b_i\}_{1 \leq i \leq N}$. The two probabilities associated with **a** and **b** can be approximated respectively as [351]

$$P(a_i) \approx \frac{a_i}{\sum_{k=1}^{N} a_k} = \frac{a_i}{A} \quad \text{and} \quad P(b_i) \approx \frac{b_i}{\sum_{k=1}^{N} b_k} = \frac{b_i}{B}, \tag{6.41}$$

assuming that A and B are different from zero. Then, the Kullback–Leibler measure D_{KL} takes on the form

$$D_{\mathrm{KL}}(\mathbf{a}, \mathbf{b}) = \sum_{i=1}^{N} P(a_i) \log \frac{P(a_i)}{P(b_i)}, \tag{6.42}$$

where it is assumed that $\forall\, i$: $P(a_i) \neq 0$, $P(b_i) \neq 0$. Entering (6.41) into (6.42) we obtain the following formula which can simplify computation of the D_{K}:

$$D_{\mathrm{KL}}(\mathbf{a}, \mathbf{b}) = \log \frac{B}{A} + \frac{1}{A} \sum_{i=1}^{N} a_i \log \frac{a_i}{b_i}, \quad \text{assuming A, B} \neq 0 \text{ and } \forall i: a_i,\ b_i \neq 0, \tag{6.43}$$

where $A = \Sigma_i a_i$ and $B = \Sigma_i b_i$, as already denoted in (6.41).

The other distance that can be used to match histograms is an approximation of the statistical χ^2 functions, given as

$$D_{\chi^2}(\mathbf{a}, \mathbf{b}) = \sum_{i=1}^{N} \frac{[P(a_i) - P(b_i)]^2}{P(a_i) + P(b_i)}, \quad \text{assuming } P(a_i) + P(b_i) \neq 0. \tag{6.44}$$

The last measure awards matches on larger values of $P(a_i)$ and $P(b_i)$ which can be advantageous when matching histograms of some image features.

There are many methods of image matching with histograms that measure frequency of occurrences of different image features, starting from bare intensities up to local orientations [137]. The latter can be easily obtained with the structural tensor presented in section 4.6.

6.3.6 *Efficient Computations of Distances*

When computing match measures for successive pixels usually we place a square window around a given pixel. Then each pixel from that window is taken into the computation of a match value. The situation is depicted in Figure 6.2 for a 3×3 square window W_i.

Additionally for some measures, such as $D_{\mathrm{ZSSD\text{-}N}}$ or D_{CV}, we traverse the window twice to compute the mean value. However, when moving computations to the next pixel position, the new window W_{i+1} overlaps with the previous window W_i. In Figure 6.2, pixel nos. 1, 4,

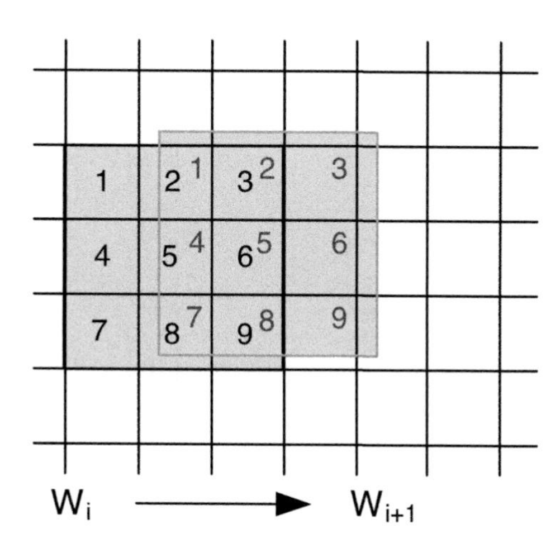

Figure 6.2 Efficient computations of the distance among pixel blocks

7 from window W_i are not taken into computation of a new value in the new window W_{i+1}. Instead, new pixel nos. 3, 6, 9 are acquired. However, pixels 2, 3, 5, 6, 8 and 9 in W_i remain, now numbered respectively as 1, 2, 4, 5, 7 and 8 in W_{i+1}. For shift invariant measures local numbering of pixels is not important. Thus, we easily see that when computing mean value in W_{i+1} we can *reuse* the mean value previously found for W_i. The only thing to do is to subtract values from positions 1, 4, 7 in W_i, and add new values at pixel locations 3, 6, 9 in W_{i+1}. The same can be done when computing match value for comparison of two windows in different images. Such techniques which save computations by reusing previously computed values are well known in the computer vision or computer graphics community. It was suggested for example by Faugeras *et al.* [118] in their real-time matching system. This method resembles also the moving histogram algorithm, used for fast update of histograms in the progressing windows. This technique finds application when computing arbitrary rank filters [396].

The other improvement to the simple window matching technique was proposed by Chen *et al.* [67]. It is called a *winner-update* technique since only the best (winning) match is checked again and again until another match gets the best value. This method can be applied when looking for the best match among a number of potential comparisons, such as in stereo matching or motion analysis. Let us analyse the simple procedure of finding a best match $m(W_j)$ for a window W_j:

$$m\left(W_j\right) = \underset{1 \leq k \leq N}{\arg\min} \left\{ D_x\left(W_j, W_k\right) \right\}, \tag{6.45}$$

where W_k denotes a series of check windows and D_x is a match measure between pairs of windows, for which it is assumed that the best match is given for minimum value of D_x (section 6.3.1). In the same way we can also search for maximal value. It is very important to realize that when solving (6.45) basically we are not interested in finding *all* possible values, from which the best one is chosen. Instead, we are interested in finding the best value in a minimum number of steps. Here the winner-update strategy can be of help. The best way to explain this methodology is to create a simple card game, as proposed in [67] (Figure 6.3).

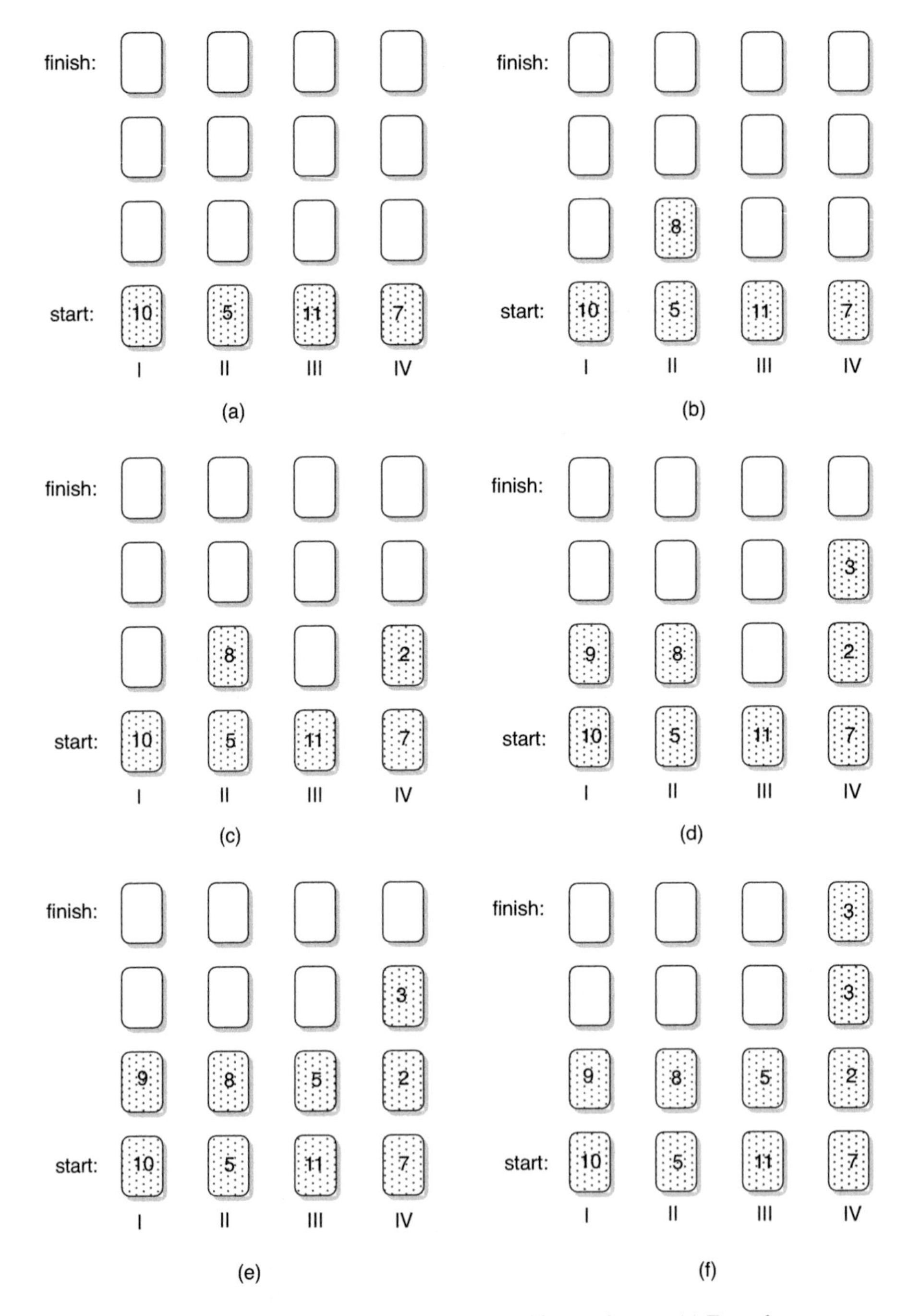

Figure 6.3 Explanation of the winner-update technique with a card game. (a) Four players start opening their cards. Only one player with the lowest score at a moment is allowed to draw. After a few steps (b–e) the winner (f) is the player with the lowest cumulative sum

At the first step the bottom row of cards is opened. Then, the rule is that a new card can be opened only in the column that has a minimum total score (if there are two equal values, then the left one has priority, for example). The winning column is the one which first reaches the last row named 'finish'. In the first step in Figure 6.3(a) the last row is opened. Since column II has a minimal value then this row has a right to draw again – the results are visible in Figure 6.3(b). Now, column IV has a minimum value however. Thus, column IV can be drawn again, as in Figure 6.3(c), and then once again, as depicted in Figure 6.3(d). At this step, however, column III reaches a global minimum value, so it is allowed to draw a card. The result of this is visible in Figure 6.3(e). Now again, column IV is allowed to draw – after this step it reaches the final row and thus column IV is a winner (Figure 6.3(f)). Notice that to find the best match, which is column IV in our simple case, we do not need to compute all partial sums; that is, we do not open all the cards. Thus, we save on computations. In practice we found however that selection of a current winner in each step can take some time [85], so it is important to implement a fast technique of best match selection such as a hashing table [74].

The third technique that can speed up computations of the match values can be applied if sums of pixels are computed many times for different rectangular regions within an image. This technique, called the *cumulative image method* (or integral image), is well known in the computer graphics community [440]. It starts with preparation of an image of cumulative sums. Each pixel of that image contains partial sums of all pixels from the original image, whose positions are above and to the left of a current pixel, as depicted in Figure 6.4(a). Thus, pixels of the cumulative image have to be able to store such sums. For instance, for a HDTV image (1k $\times$ 2k $\times$ 10 bits) the cumulative sum has to have width of at least 31 bits.

However, after the cumulative image is ready, computation of the sum of pixel values within any rectangular window of the original image can be done in linear time equal to reading four values and performing two subtractions and one addition:

$$\Sigma_1 = P_1 - P_2 - P_3 + P_4, \tag{6.46}$$

where Σ_1 is a cumulative sum of pixels in the bold rectangle and P_1–P_4 are cumulative values taken from positions presented in Figure 6.4(b). This technique can also be used for efficient computation of histograms in the selected regions of the original image. Such a method is called an *integral histogram* [350]. The cumulative image method has been also used by Veksler for stereo matching with variable windows [435].

6.3.7 Nonparametric Image Transformations

The nonparametric measures transform intensity values of pixels into mutual relations of those values. These relations can be the number of permutations that are necessary to sort the pixels, the number of pixels whose values are greater than the chosen one, a stream of bits that convey relations of a chosen pixel with its neighbours, etc. By this operation the statistical parameters of the signal are changed. The input intensity values, usually with unknown statistical distribution, are transformed into data characteristic of the uniform distribution, such as random value drawn from the set of N integers. The local neighbourhood for computation of the nonparametric transformation can be set arbitrarily, although the most common is an odd size square. Moreover, such neighbourhood does not necessarily need to be compact.

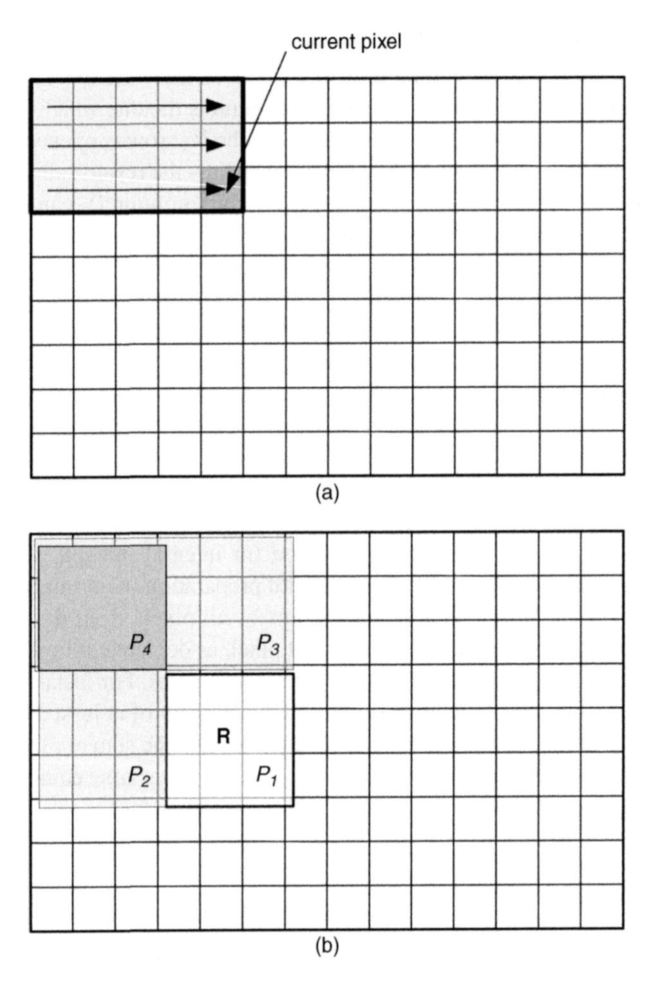

Figure 6.4 Cumulative image method. (a) The image (matrix) of the cumulative sums – each element stores a cumulative sum of the pixel values from the original image with position to the left and up from the current pixel position (grey area). (b) For a rectangle R, a sum of its pixels can be computed in two subtractions and one addition: $\Sigma_R = P_1 - P_2 - P_3 + P_4$

The two most common nonparametric transformations are *Census* and *Rank*. Both were proposed by Zabih and Woodfill for hardware computation of correspondences in stereo matching [455]. Nevertheless, they can be used in other computer vision tasks such as object recognition or optical flow. They have also been shown to be useful for signal conditioning before application to the input layer of neural networks. Finally, their software or hardware implementation is also straightforward.

Let us assume that a region around a central pixel was selected in an image. The *Rank* transform is defined as the number of pixels in that region for which the intensity signal is greater than or equal to the central one. The *Census* transform is an ordered stream of bits

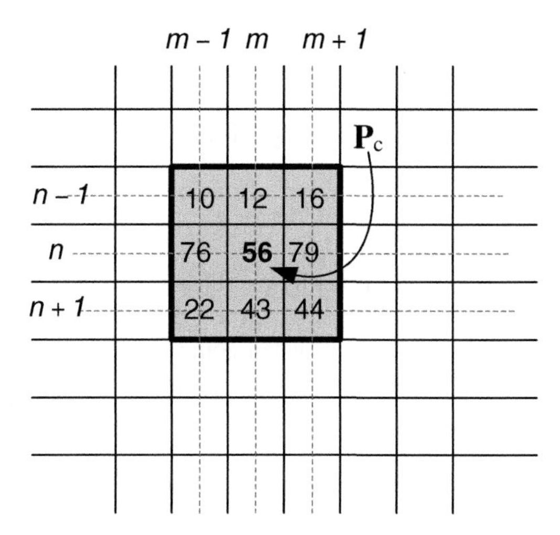

Figure 6.5 Computation of the nonparametric transformations in a square 3 × 3 neighbourhood. The central pixel at (*m*, *n*) is compared with each other pixel except itself. This results in 1 or 0 depending on the mutual relation of the compared pixels

where each bit conveys information on whether the intensity of a central pixel is greater than a pixel from its neighbourhood or not.

Figure 6.5 explains the computation of the *Rank* and *Census* transformations in a 3 × 3 window. The central pixel $\mathbf{P}_c$ at position (*m*, *n*) is compared with each other pixel **P** in the window, except itself. Each comparison results in one bit 0 or 1 depending on whether the central pixel is less than a neighbouring pixel **P** or not. Assumed is natural bit ordering, i.e. from left to right, from top to bottom. Thus, in this case, we compare 56 with 10, 56 with 12. . ., and finally 56 with 44, which results in a bit stream of 00011000. From this we easily find its *Rank* value which is 2, i.e. there are two '1s' in the bit stream.

An interesting observation for *Census* is that for 3 × 3 and 5 × 5 windows the length of the output bit stream is 8 or 24 bits, i.e. one or three bytes, respectively. These fit nicely into computer memory on bytes boundary.

The *Census* transformation *T* for a pixel $\mathbf{P}_c$ in the image **I** is defined as [455]

$$T\left[\mathbf{I},\ \mathbf{P}_c\right] = \underset{\mathbf{P} \in W(\mathbf{P}_c, \beta)}{\otimes} \xi\left(\mathbf{I},\ \mathbf{P}_c,\ \mathbf{P}\right), \tag{6.47}$$

where **I** denotes the space of input image with pixels of scalar values, $\mathbf{P}_c$ is a central pixel, ⊗ is a bit concatenation operation, $W(\mathbf{P}_c, \beta)$ is a local pixel neighbourhood around a pixel $\mathbf{P}_c$ with a radius β and **P** denotes pixels belonging to *W*; ξ is given by the formula

$$\xi\left(\mathbf{I}, \mathbf{P}_c, \mathbf{P}\right) = \begin{cases} 0 & \text{if } I\left(\mathbf{P}\right) \le I\left(\mathbf{P}_c\right) \\ 1 & \text{otherwise} \end{cases}, \tag{6.48}$$

where $I(\mathbf{P})$ is a scalar pixel value in image **I** at position **P**. Application of (6.47) and (6.48) to all pixels in *W* produces a stream of bits.

The important feature of the nonparametric representations is their resistance to noise and local image distortions. This is a direct result of a change of an input signal, which can be affected by noise, by the stream of bits reflecting mutual relations in local neighbourhoods. The output bit stream has different statistical properties from the input signal.

6.3.7.1 Reduced *Census* Coding

Image matching can be done solely on a pixel-by-pixel basis, although if taking only intensity values, such a strategy leads to many errors which are due to the limited dynamics of the intensity representation and noise. However, when computing *Census* representation wider neighbourhoods are visited and hence information is gathered on more than one pixel. Thus with sufficiently large windows W in (6.47) it is even possible to perform reliable pixel-by-pixel matching [87, 90].

Nevertheless, in many cases pixel-by-pixel matching even with the *Census* representation is not sufficient (e.g. for large baseline stereo). Therefore larger support regions are necessary for reliable area-based matching. Then matching is usually done in the corresponding rectangular windows placed in the source and destination images (section 6.6). However, if the *Census* values are computed in square windows, as presented in Figure 6.5, such matching methods lead to some data redundancy. This is explained in Figure 6.6 for a 3×3 match window with each pixel already converted to the 3×3 *Census* representation (i.e. although having different meaning, the two windows are of the same size).

In Figure 6.6 it can be seen that if *Census* was computed for all pixels from the local neighbourhoods (as in Figure 6.5), then when such *Census* values are cumulated in a bigger region some comparisons are done twice. In Figure 6.6 comparisons 0–4 and 4–0 are done twice. Such repeated bits do not convey useful information since they are highly correlated. Thus, one comparison can be simply omitted to save on bits in the representation. So, if computing *Census* for pixels that will be gathered into aggregation blocks (e.g. very common

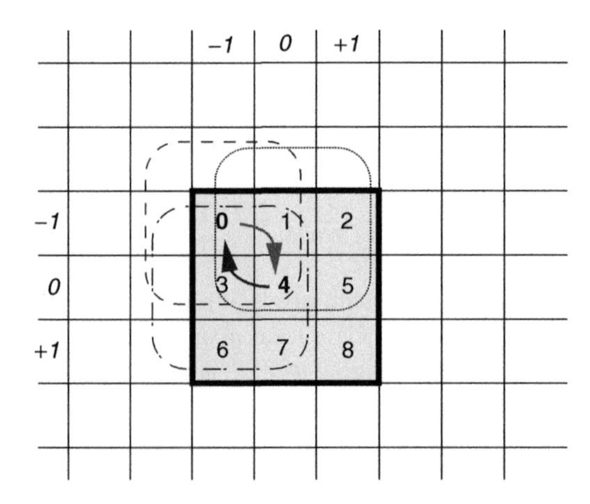

Figure 6.6 Data redundancy in blocks of pixels in *Census* representation. If *Census* was computed in full square windows then comparison 0–4 and 4–0 is done twice. (From [90] with kind permission of Springer Science and Business Media)

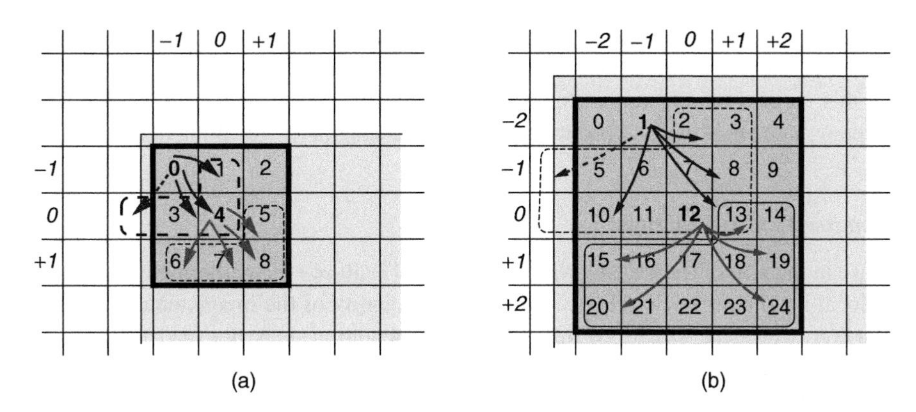

Figure 6.7 Reduced *Census* coding for blocks of pixels. (a) 3 × 3 and (b) 5 × 5 neighbourhoods. Larger neighbourhoods can be encoded in a similar way. (From [90] with kind permission of Springer Science and Business Media)

in stereo matching) we need only to compute half the number of comparisons (6.48). A simple modification is proposed in Figure 6.7.

Formally, the reduction of redundant comparisons can be obtained by changing in (6.47) window W of neighbouring pixels that are used in (6.48). Assuming the top–down and left–right bit numbering in W, and assuming that a central pixel $\mathbf{P}_c$ has an ordinal number p_c, then only pixels with numbers greater than p_c are taken into the representation. This adds to savings in terms of computation time and memory occupation.

The aforementioned reduction technique for blocks of pixels in *Census* domain can also be obtained by taking each i-th sample in a block, assuming however that *Census* has been computed from the nonreduced windows W. This idea is illustrated in Figure 6.8.

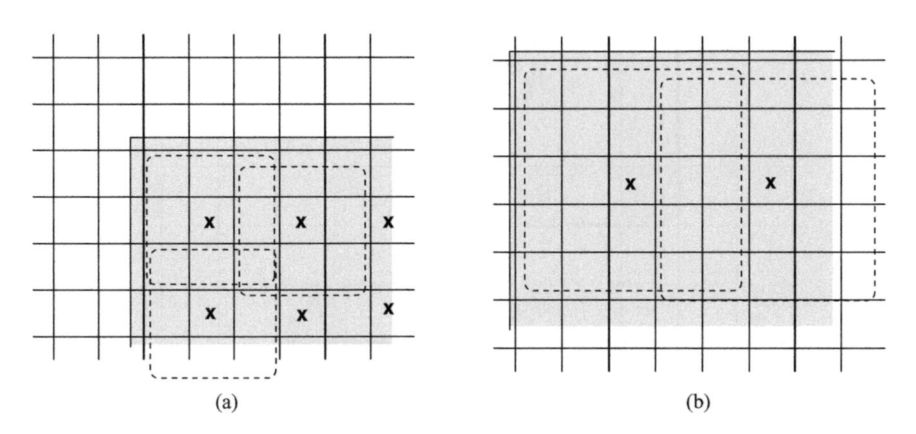

Figure 6.8 Taking each i-th pixel in matching blocks of nonreduced *Census* pixels: (a) 3 × 3 and (b) 5 × 5 neighbourhoods. Central pixels are denoted with '×'. (From [90] with kind permission of Springer Science and Business Media)

However, the drawback of this approach is that it requires a wider representation of the matching blocks. For example, the 5×5 *Census* transformation and the 3×3 matching block, in this method, would require comparisons within a block of 11×11 pixels. In the case of stereo matching this can produce excessive smearing in the resulting disparity map.

6.3.7.2 Sparse *Census* Relations

An increase in the size of the *Census* window W in (6.47) allows the gathering of information in wider local neighbourhoods. This can improve the quality of the image matching (section 6.6). However, excessive size of W in *Census* representation does not lead to further improvements, since in bigger matching blocks, even if the blocks correspond to each other, the differences in pixel values are frequent due to different projective transformations of the two images. From the computational point of view, *Census* windows W that are too big result in much slower computations and high memory occupation. For example W of size 7×7 results in 6 bytes per pixel. Thus, instead of increasing W in (6.47) a better idea is to make it not compact and compute *Census* from a *sparsely sampled* neighbourhood. Thus, the mutual relations are computed among the central pixel and its neighbours *separated* by a certain distance. This technique is visualized in Figure 6.9. Notice that in this case the reduced aggregation scheme is also assumed, so only the pixels to the right and down from the central one are considered to be taken into the relations. This can be seen as a special definition of the window $W(\mathbf{P}_c, \beta)$ in (6.47).

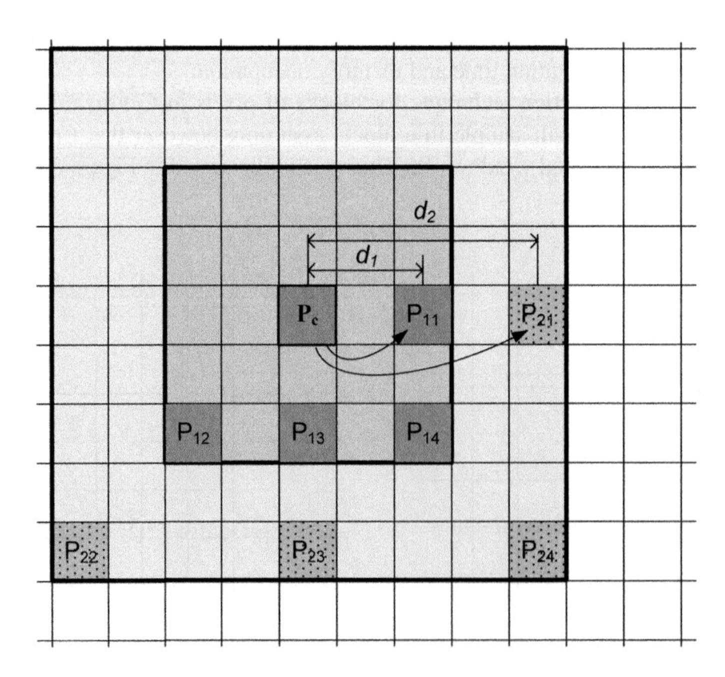

Figure 6.9 Sparse relations for *Census* matching. The inner window size is 5×5 but we compute relations with only 4 neighbours distant by d_1. In the outer window we also compute only 4 relations among pixels distant by d_2. (From [90] with kind permission of Springer Science and Business Media)

In Figure 6.9 the two *Census* windows are shown: the inner 5×5 and the outer 9×9, respectively. For the inner representation $\mathbf{P}_c$ is compared with only four of its neighbours, $\mathbf{P}_{11}$–$\mathbf{P}_{14}$, separated from each other by d_1 pixels.

Such a four-bit representation produces almost the same matching results as the full 5×5 representation with 24 bits [90]. However, we can increase discriminative properties even further by addition of the next four bits from the outer window given by pixels P_{21}–P_{22} separated by a distance of d_2 pixels. This results in eight comparisons which can be efficiently encoded into a single byte. As a payoff we obtain representation of the whole 9×9 neighbourhood. Recall that such sparse sampling is done for *each* pixel in the matching block.

The sparse *Census* coding can also be justified in terms of the probabilistic dependence among pixels and their nearest neighbours. This is a result of some physical phenomena encountered in digital cameras, e.g. charge leaking in neighbouring cells of CCD devices.

6.3.7.3 Fuzzy Relationships Among Pixels

Let us enhance the concept of the *Census* measure to convey more detailed information on pixel relations. This can be done by assigning *more than one bit* for the relation between two pixels, whose relation D_k can now be defined as [86]

$$D_k = I(\mathbf{P}_c) - I(\mathbf{P}_k), \tag{6.49}$$

where $\mathbf{P}_c$ is a central pixel and $\mathbf{P}_k$ is a pixel from the neighbourhood of the central one.

Table 6.8 presents a proposition of fuzzy rules F for the relation between pixels based on their relative difference in intensity. Fuzzy rules have found broad applications in computations with imprecision, such as in common expressions of a spoken language [242, 456]. Here they allow a unified description of some more or less precise relations between pairs of intensities. For computer representation each fuzzy relation is encoded on three bits. A proposition of such an encoding is given in the third column of Table 6.8.

What is still missing is a relation between the fuzzy rules F and the actual value of D_k in (6.49). This can be obtained directly from a value of D_k, put into the sigmoidal or hyperbolic functions. However, much simpler is a piecewise linear approximation which can be easily computed providing six threshold values on D_k. Nonetheless, thresholds can be a little cumbersome in practice. Therefore a binary partitioning of D_k has been chosen to facilitate

Table 6.8 Fuzzy rules for the relation between pixels based on their relative difference in intensity

	Relation type F	Bit encoding $B(F)$
1	*much smaller*	011
2	*smaller*	010
3	*slightly smaller*	001
4	*equal*	000
5	*slightly greater*	101
6	*greater*	110
7	*much greater*	111

```
int BF = 0x00; // Initial value of the encoding
// Dk is an integer difference of pixel values
if( Dk < 0 )
{
  Dk = - Dk;                // Make Dk positive
  BF = 0x04;                // Set a sign
}
for( int i = 0; i <= 2; ++ i )
{
  if( ( Dk >>= 2 ) != 0 )             // Shift R, and if not 0 then
     ++ BF ;                          // increment E by 1
}
```

Algorithm 6.1 Bit encoding algorithm for fuzzy relations of pixels. Only operations used are bit shifting and an increment by one

implementation requiring only integer arithmetic. That is, at each iteration the positive value D_k is shifted right by two bits and if the result is still different from zero, the encoding value is incremented by 1. Finally, the first bit (i.e. the most significant one) in $B(F)$ conveys information on the sign of comparison. This scheme is presented in Algorithm 6.1. Algorithm 6.1 leads to the following discrete thresholds of D_k.

1. If $|D_k| \in [0, 3]$ then F is '*equal*'.
2. If $|D_k| \in (3, 15]$ then F is '*slightly smaller/greater*'.
3. If $|D_k| \in (15, 63]$ then F is '*smaller/greater*'.
4. If $|D_k| \geq 64$ then we classify as F is '*much smaller/greater*'.

Smaller/greater is resolved by a sign. The proposed algorithm can be easily implemented in assembly or in hardware logic since only operations of bit shifting and an increment by one are necessary.

Finally, let us notice that the described procedure can be used not only for neighbourhood encoding but also it defines a *fuzzy subtraction* of images in which each difference of pixels is given by a fuzzy rule.

6.3.7.4 Implementation of Nonparametric Image Transformations

In the simplest case, the *Census* and *Rank* transformations, computed in dense 3×3 neighbourhoods, result in the same number of bits as required for monochrome images, which is eight bits per pixel (section 3.7.1.2). Thus, in this case the nonparametric transformation does not change the number of bits of pixels, although they belong to different domains. Therefore it is quite easy to write simple procedures which transform one *MonochromeImage* with grey-value pixels into the other *MonochromeImage* with nonparametric pixels in the *Census* or *Rank* representations. Two exemplary procedures are presented in Algorithms 6.2 and 6.3. It has to be remembered that the returned image is orphaned, which means that the caller is responsible of its final disposal. An alternative is to use the *auto_ptr<>* pattern (section 13.4).

In Algorithm 6.2 the first two loops *L[33–67]* and *L[38–65]* organize iteration through all pixels in the input image, given by the input *image* reference. Then the boundary values for the inner loops are prepared. These, organized around lines *L[48–61]* and *L[50–60]*, are responsible for accessing each pixel in the local neighbourhoods of pixels, which are square

```
1  ////////////////////////////////////////////////////////////
2  // This function creates a non-parametric 3x3 Census
3  // image from the supplied monochrome image.
4  ////////////////////////////////////////////////////////////
5  //
6  // INPUT:
7  //                image - a reference to the input monochrome image
8  //
9  // OUTPUT:
10 //                pointer to the orphaned object - image of which
11 //                       pixels are in the 3x3 Census format
12 //
13 // REMARKS:
14 //                The returned object is orphaned which means
15 //                       that the caller is responsible for destroying
16 //                       this object!
17 //
18 MonochromeImage * Orphan_3x3_Census( const MonochromeImage& image )
19 {
20     register int i, j, m, n;
21     register int h_from, h_to, v_from, v_to;
22
23     const int kWinSwing = 1; // = 3 / 2;
24
25     const int row = image.GetRow();
26     const int col = image.GetCol();
27
28     MonochromeImage* nonParamImage = new MonochromeImage(col,row,0);
29
30     unsigned char central_pixel, non_param_pixel;       // exactly 8 bits
31
32     // For each pixel ...
33     for( i = 0; i < row; i ++ )
34     {
35             v_from = ( i >= kWinSwing ? i - kWinSwing : 0 );
36             v_to = ( i + kWinSwing >= row ? row - 1 : i + kWinSwing );
37
38             for( j = 0; j < col; j ++ )
39             {
40                     h_from = ( j >= kWinSwing ? j - kWinSwing : 0 );
41                     h_to = ( j + kWinSwing >= col ? col - 1 : j + kWinSwing );
42
43                     central_pixel = image.GetPixel( j, i );
44
45                     non_param_pixel = 0;
46
47                     // Now move in the census window
48                     for( m = h_from; m <= h_to; m ++ )
49                     {
50                             for( n = v_from; n <= v_to; n ++ )
51                             {
52                                     if( m == j && n == i )
53                                       continue;      // skip the central pixel
54                                       non_param_pixel <<= 1; // shift left the already
55                                                             // acquired series of bits
56
57                                        if( image.GetPixel( m, n ) > central_pixel )
58                                                non_param_pixel |= 0x01;  // set the least
```

Algorithm 6.2 Listing of the *Orphan_3x3_Census* function for a computation of the 3×3 nonparametric *Census* representation

```
59                                                                         // significant bit
60              }
61           }
62
63           // write out the non-param pixel
64           nonParamImage->SetPixel( j, i, non_param_pixel );
65        }
66
67     }
68
69     return nonParamImage;
70  }
```

Algorithm 6.2 (*Continued*)

3×3 panes in this particular implementation. Finally, the first 'if' excludes comparison of a central pixel to itself, whereas the second 'if' checks the pixel relation which determines the bit value.

The *Orphan_3x3_Rank* procedure in Algorithm 6.3 is organized in the same manner, except for the second 'if' in *L[54–55]* in the innermost loops which simply count the number of bits greater than the central one.

In the general case of $n \times m$ nonparametric neighbourhoods the input and output pixels are two different structures with different number of bits. For this purpose a special bit stream needs to be defined. Then, having defined this new data type, an appropriate image can be created almost immediately due to the template definition of the *TImageFor<>* (section 3.7.1.2).

Implementations of the sparse *Census* (section 6.3.7.2) and fuzzy encoding (section 6.3.7.3) representations require only different organization of the inner loops.

6.3.8 Log-polar Transformation for Image Matching

The log-polar transformation takes points (x, y) from the Euclidean space into the (r, φ) points in the polar space defined as [465]

$$r = \log_B\left(\sqrt{(x-x_0)^2+(y-y_0)^2}\right), \tag{6.50}$$

$$\varphi = \arctan\frac{y-y_0}{x-x_0}, \quad \text{for} \quad x \neq x_0, \tag{6.51}$$

for a point (x, y), where $O = (x_0, y_0)$ is a centre of transformation, and B denotes the base of the logarithm which can be any positive value different from 1. Usually it is chosen to fit the maximal expected distance r_{max} from the centre O in a local coordinate space of a given image.

In many applications it is necessary to find an inverse transformation. For instance, in the inverse image warping scheme the output pixel grid is given *a priori*. Then the coordinates in the input image space have to be found (section 12.5). An inverse log-polar transformation is given as

$$x = B^r \cdot \cos(\varphi) + x_0, \quad y = B^r \cdot \sin(\varphi) + y_0, \tag{6.52}$$

```
1     ////////////////////////////////////////////////////////////
2     // This function creates a non-parametric 3x3 Rank
3     // image from the supplied monochrome image.
4     ////////////////////////////////////////////////////////////
5     //
6     // INPUT:
7     //                      image - a reference to the input monochrome image
8     //
9     // OUTPUT:
10    //                      pointer to the orphaned object - image of which
11    //                              pixels are in the 3x3 Rank format
12    //
13    // REMARKS:
14    //                      The returned object is orphaned which means
15    //                              that the caller is responsible for destroying
16    //                              this object!
17    //
18    MonochromeImage * Orphan_3x3_Rank( const MonochromeImage & image )
19    {
20      register int i, j, m, n;
21      register int h_from, h_to, v_from, v_to;
22
23      const int kWinSwing = 1; // = 3 / 2;
24
25      const int row = image.GetRow();
26      const int col = image.GetCol();
27
28      MonochromeImage* nonParamImage = new MonochromeImage(col,row,0);
29
30      unsigned char central_pixel, non_param_pixel;
31
32      // For each pixel ...
33      for( i = 0; i < row; i ++ )
34      {
35              v_from = ( i >= kWinSwing ? i - kWinSwing : 0 );
36              v_to = ( i + kWinSwing >= row ? row - 1 : i + kWinSwing );
37
38              for( j = 0; j < col; j ++ )
39              {
40                      h_from = ( j >= kWinSwing ? j - kWinSwing : 0 );
41                      h_to = ( j + kWinSwing >= col ? col - 1 : j + kWinSwing );
42
43                      central_pixel = image.GetPixel( j, i );
44
45                      non_param_pixel = 0;
46
47                      // Now move in the rank window
48                      for( m = h_from; m <= h_to; m ++ )
49                      {
50                              for( n = v_from; n <= v_to; n ++ )
51                              {
52                                      if( m == j && n == i )
53                                              continue;       // skip the central pixel

54                                      if( image.GetPixel( m, n ) > central_pixel )
55                                              ++ non_param_pixel;   // increase the counter

56                              }
57                      }
58
59                      // write out the non-param pixel
60                      nonParamImage->SetPixel( j, i, non_param_pixel );
61              }
62
63      }
64
65      return nonParamImage;
66    }
```

Algorithm 6.3 Listing of the *Orphan_3x3_Rank* function for a computation of the 3×3 nonparametric *Rank* representation

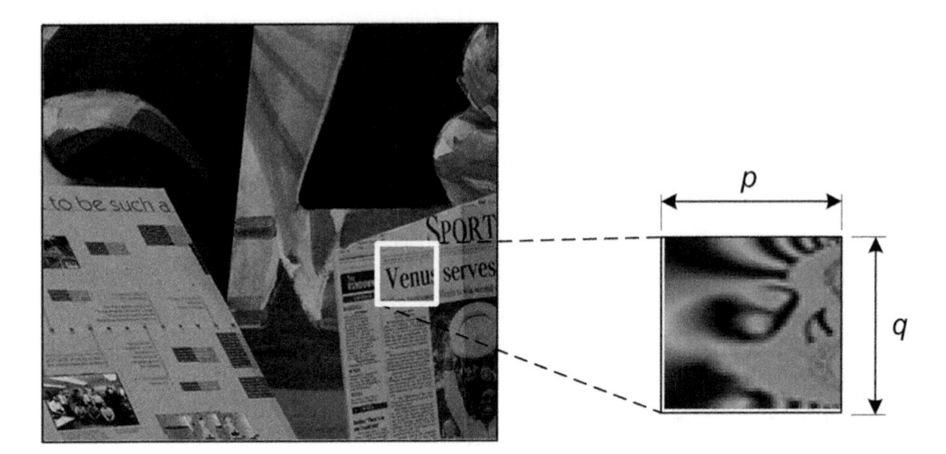

Figure 6.10 Template cropped from the left 'Venus' and view of its log-polar transformation. (Image on left courtesy of Prof. Rick Szeliski, Middlebury University, www.middlebury.edu)

assuming $B > 1$, and $0 \leq r \leq r_{\max}$, $0 \leq \varphi < 2\pi$. We can assume that r is a vertical and φ a horizontal coordinate of the output space. Although the choice of B is arbitrary, in practice B is chosen to fit all possible values of r, but not to exceed $r_{\max}$. To fulfil this requirement it should be set as follows:

$$B = \sqrt[r_{\max}]{d_{\max}}, \; d_{\max} > 1, \; r_{\max} > 1, \tag{6.53}$$

where $d^2{}_{\max} = (x_{\max} - x_0)^2 + (y_{\max} - y_0)^2$ is the maximal distance of a point in the image from the centre O. Finally, for discrete images the values of r and φ should be quantized.

Image registration is a process that relies on image matching [159]. An image registration method which employs matching in the log-polar space is proposed by Zokai and Wolberg [465]. It requires a 4D search, however. Let us assume that a single template is to be matched in an input image. Then, for each position (x_i, y_i) in the input image a region of the exact size of the template is selected and transformed into its log-polar representation. To exemplify our discussion, Figure 6.10 depicts a left image from the 'Venus' stereo-pair[3] with a selected region of size 30×30 pixels and its log-polar transformed version. This region of size $p \times q$ pixels constitutes a template that we will try to match in the right image of 'Venus'.

Matching of the log-polar transformed signals is done in an extended search space depicted in Figure 6.11. The test pattern is wrapped around to $2q$ to allow full search of the rotation value. The range of the scale search is also extended by a distance u, which can be up to width p of the template. Each position in the extended space is then matched pixel-by-pixel. For this purpose Zokai and Wolberg propose the D_{CV} measure (section 6.3.1). Alternatively the template and the extended space can be transformed into the Census representation and the matching is done with the Hamming measure D_{H}.

The position found of the best match in the extended space reflects the internal change of scale and rotation (r, φ) between the image and the template. Thus, for each checked position in the input image, the four parameters (x_i, y_i, r, φ) of the best matches are stored. Two techniques of selecting the best matches can be proposed. The first one relies on setting a

[3]This and other stereo-pairs can be seen in Table 3.4.

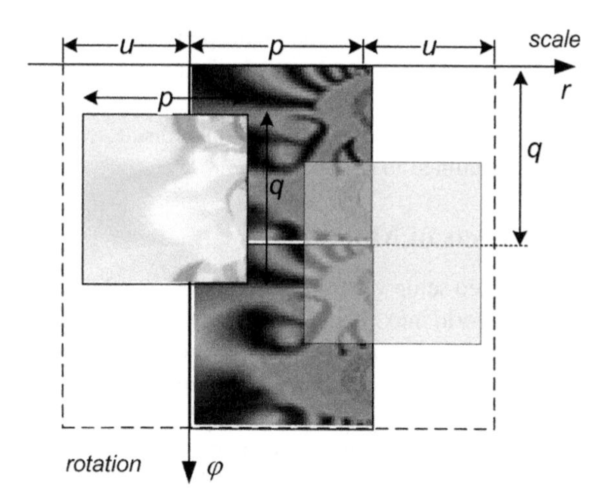

Figure 6.11 Matching a pattern in the extended log-polar space. For each position a search is done in two dimensions to account for scale and rotation. Thus, the matching requires a 4D search

fixed threshold value and acceptance of only correlation measures above this threshold. The second technique is to build a priority queue of a fixed length L, which stores L positions of the best matches.

Figure 6.12 shows template matching results in the right image from the 'Venus' stereo-pair, as well as its affinely transformed versions. The latter was obtained by an affine transformation consisting of scaling ($s_x = 0.77$, $s_y = 0.88$) and rotation ($-13°$). In both test images the template pattern has been found correctly, although the second version was additionally deformed.

Because of the 4D search space the method is not practical for dense image matching, unless a hardware acceleration is employed. Nevertheless the method is very useful if only

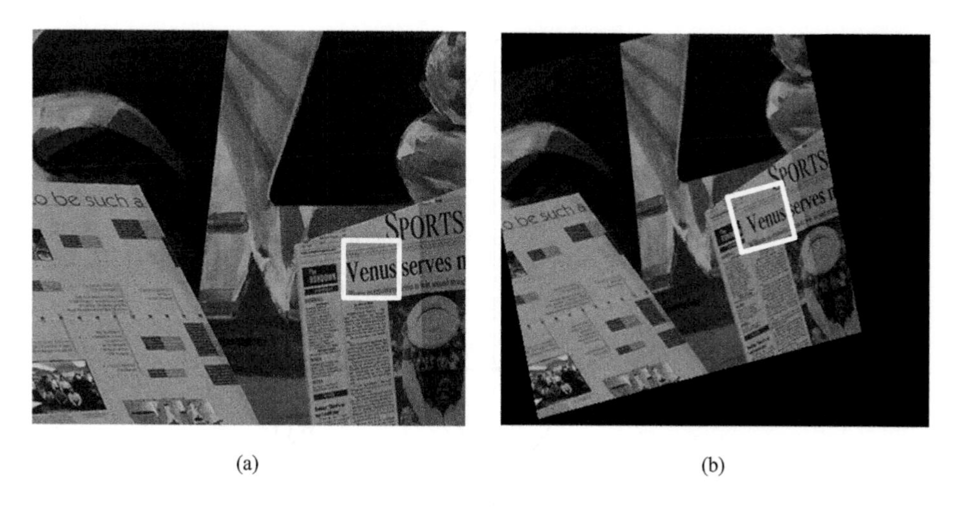

Figure 6.12 (a) Template matched in the original right 'Venus' image, and (b) in its affinely transformed version. (Figure 6.12(a) courtesy of Prof. Rick Szeliski, Middlebury University, www.middlebury.edu)

selected points from one image are to be matched with points in another image. Hence, the method is suitable for finding point correspondences for image rectification (section 3.4.8) or the fundamental matrix (section 3.4.5).

Simple implementation of the log-polar transformation is discussed in section 6.13.2. It is based on the warping module outlined in Chapter 12.

6.4 Computational Aspects of Matching

Geometrical properties of a stereo setup were discussed in section 3.4. We know that cameras convert information of the 3D world into 2D images, from which one tries to recover information on 3D space. However, this inverse process might not be unique. For instance, a 3D point can be visible only to one camera, being occluded to the second one at the same time. Further, digital images convey information only with highly limited resolution and with quantized luminance values subject to noise and distortions. As a consequence, there are no guarantees that the best matching points are images of the same real point from the 3D space. Finally, even if true corresponding points are correctly found their positions are discrete which results in only integer values of disparity.

In this section we discuss the problems of occlusions, discrete values of the disparity field and methods for evaluation of the quality of stereo methods.

6.4.1 Occlusions

When observing 3D objects from two different view positions some of their details are visible for all view positions, some are visible only for one, and some are totally invisible. The problem of invisible points is due to scene geometry and occlusions of objects in the scene. Such a situation is depicted in Figure 6.13 where an object B_1 is partially occluded by an object B_2. A point m_l, which is a left image of a 3D point M on a surface of the object B_1, cannot be matched with any other point on the right image since it is not visible to the right camera. This is an

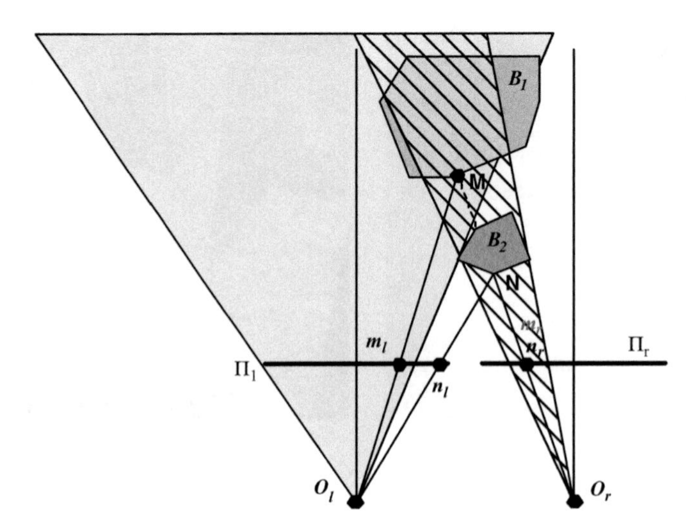

Figure 6.13 Mutual occlusions of objects

example of a binocular half-occlusion. Moreover, the virtual point m_r is substituted by a point n_r, which actually is an image of a 3D point N which belongs to the object B_2. Conversely, the point N can be correctly matched since both its image points are visible to the two cameras.

In real systems the problem of occlusions is inevitable. Thus, it is necessary to consider this fact in the development of matching methods. Especially it is important to find the occluded points and provide a means of interpolating depth values in those places. Based on [114], Table 6.9 summarizes the most common methods that help in the detection of binocular half-occlusions.

Table 6.9 Methods for detection of binocular half-occlusions

Occlusion method	Description
Cross-checking = left–right checking (LRC)	In this method the matching process is performed twice: at first with the left image kept as a reference, then the process is reversed and the right image of a stereo-pair constitutes a reference image. Then, the two disparity maps are checked as follows. 1. Given are two input images $I_L(x, y)$, $I_R(x, y)$ in the canonical setup, as well as two disparity maps, $D_L(x, y)$ and $D_R(x, y)$, which are computed with the left or right reference image, respectively. 2. For each allowable location (x, y) in the left image, take $d_L = D_L(x, y)$. 3. Compute its corresponding position in the right disparity map: $d_R = D_R(x + d_L, y)$. 4. If $\lvert d_L \rvert = \lvert d_R \rvert$ then the (x, y) location in the left-referenced disparity map, and $(x + d_L, y)$ location in the right-referenced disparity map have correct disparities. Otherwise, the location can be an occluded point. This method has been applied by many researchers: for example [44, 139, 281, 369, 464, 465].
Occlusion constraint (OCC)	In this approach it is assumed that a discontinuity found in the disparity map corresponds to an occlusion area. Therefore to find occluded areas it suffices to find discontinuities in the disparity map. This process is performed twice: once with the left image being a reference, then with the right one (similarly to the LRC method).
Point ordering constraint (ORD)	This follows the ordering constraint presented in Figure 6.13. This method assumes that if the order of matched points is different in the two images then the matched point in the scene is an occlusion point.
Bimodality (BMD) – occlusion borders only	Bimodality rule says that points near an occlusion have in their close location disparities that come from the occluding and occluded areas. More precisely, the disparity histogram in this area shows two close extrema. Detection of such a situation leads to a potential conclusion of occluded points.
Match goodness jumps (MGJ) – occlusion borders only	This method assumes that if matching the occluded points, their matching measure will be worse than in the case of not occluded points. Thus, it would be possible to detect the occlusions directly in the matching method.
Null method (NM)	In this approach the occlusion problem is ignored, which follows an idea that the occluded points are not so numerous compared to the total amount of matched points. Such an approach is sometimes justified for aerial images.

The empirical results presented in [114] are as follows.

1. The ORD method has the lowest overall false positive rate and the lowest hit rate, at the same time.
2. The OCC is usually the best method, having the highest hit rate and the lowest false positive rate.
3. LRC is almost as good as OCC. However, it shows bad results in areas of the scene with low spatial frequency structure.

It is interesting to notice that half-occlusions play a very important role in the recognition process of the human visual system since they provide information on scene structure [7, 201].

6.4.2 Disparity Estimation with Subpixel Accuracy

Disparities are computed as differences between the positions of the corresponding points in the matched images. Since these positions are restricted to lie on the integer grid of pixels, then computed disparities can have only integer values. In some situations, for instance in the case of low image resolution, this can excessive errors during space reconstruction.

One way to alleviate this problem is to take advantage of the shape of matching measure in a wider range rather than in a single pixel position. This way we can infer a more precise position of a minimum of a matching measure (a cost function) which, because of continuous support, does not need to fall under the integer pixel position. The most common technique is to fit a third-order curve, a parabola, to the three values of a matching measure with a point of interest being in a centre of the chosen window. Then, the position of a minimum of this parabola is found, which indicates a new disparity value, now with a subpixel resolution, however. Certainly, it is also possible to fit higher order polynomials and/or to a larger number of points. Nevertheless, in practice fitting a parabola is the most efficient method in terms of accuracy achieved versus computational effort. In this section we present details of this technique [323, 369].

Figure 6.14 depicts a matching cost function (values denoted by rectangles) – this can be one of the matching measures presented in Table 6.1 – for which a minimal integer value was found, denoted as d_i. Our task now is to take match values at two nearest neighbours of d_i, i.e. at d_{i-1} and d_{i+1}, and fit a third-order polynomial to them (shown in Figure 6.14). Then a new minimal value d_x can be found which no longer is restricted to lie on an integer grid. More precisely, we have the following three pairs of data:

$$\{d_{i-1}, m_{i-1}\}, \{d_i, m_i\}, \{d_{i+1}, m_{i+1}\}, \tag{6.54}$$

where $m_i = m(d_i)$ is a match value for the displacement d_i. Based on them we wish to determine coefficients of the third-order polynomial, given as

$$ad_i^2 + bd_i + c = m_i, \quad (a \neq 0), \tag{6.55}$$

which is a well-known quadratic equation [259]. The point d_x of the minimum can be easily found by doubly differentiating (6.55) with respect to d_i, and then equating the first derivative

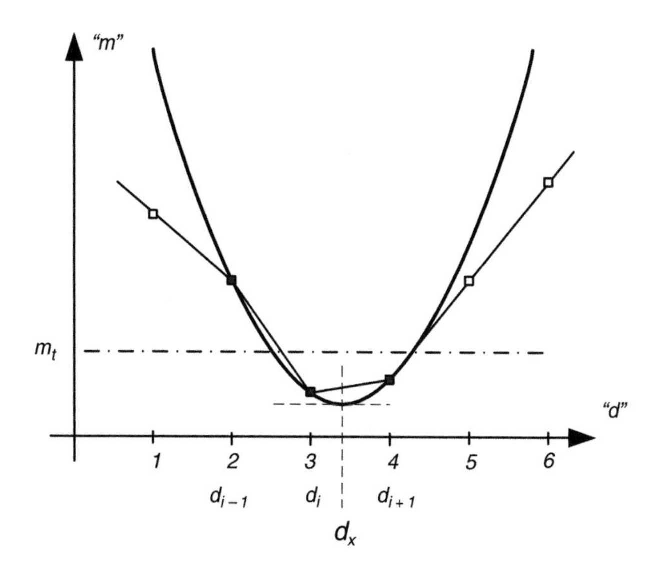

Figure 6.14 Subpixel estimation of a disparity value in a local neighbourhood of pixels

to zero while requiring the second to be positive:

$$2ad_x + b = 0 \quad \text{and} \quad 2a > 0,$$

thus

$$d_x = -\frac{b}{2a} \quad \text{and} \quad a > 0. \tag{6.56}$$

Inserting (6.54) into (6.55) we obtain a set of three equations

$$\begin{cases} ad_{i-1}^2 + bd_{i-1} + c = m_{i-1} \\ ad_i^2 + bd_i + c = m_i \\ ad_{i+1}^2 + bd_{i+1} + c = m_{i+1} \end{cases},$$

which we solve for a and b. However, it is easier to directly compute b/a using Cramer's rule [259]. Prior to this, we can simplify things by assuming that $d_{i-1} = -1$, $d_i = 0$ and $d_{i+1} = 1$. This is equivalent to shifting the origin of the coordinate system to the point $(d_i, 0)$ which we will account for at the end of our computations. Thus

$$\frac{b'}{a'} = -\frac{\begin{vmatrix} 1 & m_{i-1} & 1 \\ 0 & m_i & 1 \\ 1 & m_{i+1} & 1 \end{vmatrix}}{\begin{vmatrix} -1 & m_{i-1} & 1 \\ 0 & m_i & 1 \\ 1 & m_{i+1} & 1 \end{vmatrix}} = -\frac{m_{i-1} - m_{i+1}}{m_{i-1} - 2m_i + m_{i+1}}, \tag{6.57}$$

where a' and b' are in the shifted coordinate system. Now, we insert (6.57) into (6.56) and shift by d_i, to obtain the subpixel disparity d_x, as follows:

$$d_x = d_i + \frac{m_{i-1} - m_{i+1}}{2\left(m_{i-1} - 2m_i + m_{i+1}\right)}, \tag{6.58}$$

for which we assume that

$$(m_{i-1} - m_i) + (m_{i+1} - m_i) > 0, \tag{6.59}$$

which simply is a consequence of the assumption made that $a > 0$. From Figure 6.14 we see that the above means that d_i indeed is a local minimum. This feature will be used further for match clarification. For example, in [323] it is proposed to consider a match as unique if only one value, or at most two values, lie below a threshold m_τ. However, if a third smallest value does not lie above this threshold, then such a match is considered as bad. Thus the authors of [323] propose to keep track of the three smallest values. The smallest of them sets a threshold above which the third smallest value should lie to be considered as a valid match. They propose to set this threshold in the range 5–20% above the minimum value of a matching measure. In [323] only the D_{SAD} (Equation (6.1)) was used; however, we can extend this technique to other matching measures as well. In practice this technique has been shown to be very efficient.

6.4.3 *Evaluation Methods for Stereo Algorithms*

To evaluate the quality of computed disparity maps by different stereo matching methods we need special measures. The most obvious is comparison with the truth data, i.e. true disparity or depth values for a scene, usually acquired by other methods, such as laser scanner or created for artificial images. However, such reference data are not easy to gather and only a few artificial test images are available (section 3.4.10). This and other evaluation methods are discussed in this section (Table 6.10).

There are many advantages and disadvantages of the presented comparison methods. On the one hand, the ground-truth approach seems to be very appealing. However, in practice only a few images are equipped with the ground-truth data. Usually this is available for only artificial images (see section 3.4.10) or when a precise laser range scanner is available. It happens also that because we compute the measure for the whole image, a visually worse disparity map can have a higher *GT_RMS* than other visually better maps. A similar problem can be encountered with the method of comparison of the synthesized view.

As discussed by Gong and Yang, the *G*(*I*) measure penalizes mostly the regions with large dissimilarities in disparity. As a consequence, it does not produce the best evaluation for the ground-truth data. Nevertheless, both measures *F*(*I*) and *G*(*I*) are suitable alternatives for evaluation of disparity maps, although sometimes their best results are not what we call visually the best.

The measure relying on the number of rejected points by the LRC method can be used if no ground-truth data is available. However, it is obvious that this method depends

Table 6.10 The most popular quality measures for evaluating performance of matching algorithms

Quality measure	Description
RMS on ground-truth data GT_RMS	This quality measure assumes *a priori* knowledge of the true values of disparities in each pixel position. In practice, this can be fulfilled for artificial stereo-pairs, such as the ones presented in section 3.4.10. Further, the measure can be computed for all pixels belonging to an image or only for its area A, i.e. we assume that point coordinates $(x, y) \in A$. The root-mean-square measure *GT_RMS* of disparities $d(x, y)$ in respect to the ground-truth disparities $d_{GT}(x, y)$, can be defined as $$GT\_RMS_A = \frac{1}{N}\sqrt{\sum_{(x,y)\in A} [d(x, y) - d_{GT}(x, y)]^2} \quad (6.60)$$ where N denotes the number of points that belong to the area A. In [370] the special areas A were divided into the following three groups. 1) Texturless areas (T-A) – regions in which, according to some texture-content measure, there is no texture. 2) Occluded areas (O-A) – image regions which are only visible by one camera. 3) Depth discontinuity areas (D-A) – disparity map regions for which their gradient value is excessive, according to some function (or simple threshold value, etc.).
Percentage of incorrect matches on the ground-truth GT_BP	Similarly to the previous measure, the percentage of bad matched pixels *GT_BP* over an area A can be defined as $$GT\_BP_A = \frac{1}{N}\sum_{(x,y)\in A} [\lvert d(x, y) - d_{GT}(x, y)\rvert > \delta_d] \quad (6.61)$$ where, as before, N denotes the number of points that belong to the area A. We see that GT_BP sums up all those points for which their disparity value is different from the ground-truth data of more than a threshold δ_d. Then the sum is normalized by a total number of points.
Synthesized view prediction errors	This comparison method is performed in two steps. At first a new view is synthesized based on a base view and the computed disparity map. Then, at the second step, this synthesized view is compared with another base view. The authors of [370] distinguish two possibilities for this method: 1) The forward warp – here we take a reference view, then we warp it based on the computed disparity map. Finally, the synthesized new view is compared with the reference view taken at the beginning of this process. This way we obtain the forward prediction error. 2) The inverse warp – a view is inversely warped by the disparity map, then compared against the reference image. This way we obtain the inverse prediction error. For comparison of the views (i.e. the synthesized and the reference ones) we can use one of the already presented matching measures (Table 6.1). Further discussion on the two warping schemes can be found in [370].

(*continued*)

Table 6.10 (*Continued*)

Quality measure	Description
Parameter-free measures	Gong and Yang noticed a similarity between evaluation problems encountered in image segmentation and stereovision [155]. In both, the areas selected for comparison should be smooth and exhibit low colour errors among compared pixels. Thanks to this observation Gong and Yang adopted the parameter-free measure from the segmentation domain to be used in stereo assessment, as follows [155]: $$F(I)=\frac{\sqrt{\#A}}{10^3 MN}\sum_{i=1}^{\#A}\frac{e_i^2}{\sqrt{A_i}} \quad (6.62)$$ where A_i is the i-th area of consideration, e_i is an error computed for A_i and $\#A$ denotes the number of regions. Image size is $M \times N$ pixels. For each area A_i, the error e_i is computed as SSD (6.18) between corresponding colour pixels of the original and the segmented images, respectively. $F(I)$ was extended in [155] to penalize regions characterized by large error. This extension $G(I)$ is as follows: $$G(I)=\frac{\sqrt{\#A}}{10^6 MN}\sum_{i=1}^{\#A}\frac{E_i^2}{\sqrt{A_i}} \quad (6.63)$$ where the only difference compared to $F(I)$, despite the larger constant in denominator, is that errors E_i are *squared* Euclidean distances now. Both measures, $F(I)$ and $G(I)$, were devised for segmentation but can be used to assess stereo methods as well. The most cumbersome part however is finding the proper regions A_i. Gong and Yang proposed classifying two pixels into the same region if they have the same disparities, as well as there being a four-connected path on which all the pixels have the same or higher disparities. The parameter-free measures $F(I)$ and $G(I)$ are evaluated in accordance with the following postulates [155]. • The measures should give the best results for the ground truth data. • For disparity maps with similar number of regions A_i, the measures should promote the one with higher rate of correct matches. • For disparity maps with similar matching rate, the measure should promote the one with lower noise, i.e. which has fewer number of regions A_i.
Number of pixels rejected by the left–right consistency check	If the ground-truth data is not available, which is the case for most stereo images, some intuition on the matching quality can be gained by examining the number of pixels rejected by the left–right consistency check (LRC; Table 6.9). As already alluded to, the main purpose of LRC is detection and correction of the problem arising from the occlusions. In such a case, LRC detects inconsistencies in the left–right and right–left matches. However, LRC can also detect match inconsistencies due to poor behaviour of a matching method. Thus, this feature of LRC has been used to assess the quality of the matching methods [25, 90].

heavily on the image contents since it sums up all match inconsistencies which are due to half occlusions, but can also arise from poor match clarification abilities of a matching method. Assuming that the number of half-occlusions for a given stereo-pair is constant we can use this method to comparatively qualify different matching strategies applied to this stereo-pair [25, 90].

6.5 Diversity of Stereo Matching Methods

The stereo matching process is a key method of recovering information on the 3D environment based on two simultaneously acquired images of the same scene, taken however from slightly different positions. Needless to say, the human visual system is doing the same for full space orientation. Section 3.4 presents the mathematical background behind this process. In this and subsequent sections we present some of the basic matching algorithms, discuss their advantages and disadvantages, and provide information on computational complexity, practical realizations and applications. Since stereo is one of the key topics of computer vision there is ample literature on this subject as well. For a general overview the paper by Brown *et al.* [57] or the report by Scharstein and Szeliski [370] are highly recommended. The latter provides a taxonomy on the vast realm of dense two frame stereo matching methods. The mentioned works were very influential to the synopsis presented in this chapter.

The reason for presenting such a review of stereo methods is to give an insight into different approaches to the same problem. Based on this, further improvements can be created. More literature references are provided and are discussed at the end of this chapter (section 6.14.1).

The first and the simplest division of the stereo methods is based on the type of output disparity map (Figure 6.15). The most desirable are dense disparity maps, in which all or almost all pixels have determined disparity values. Such maps are very useful and can be used, for instance, in image synthesis [369]. On the other hand, sparse disparity maps have disparity values determined only for selected image points (usually these are features, such as corners or edges). More often than not, they are faster in computation but have limited applications since missing values have to be interpolated.

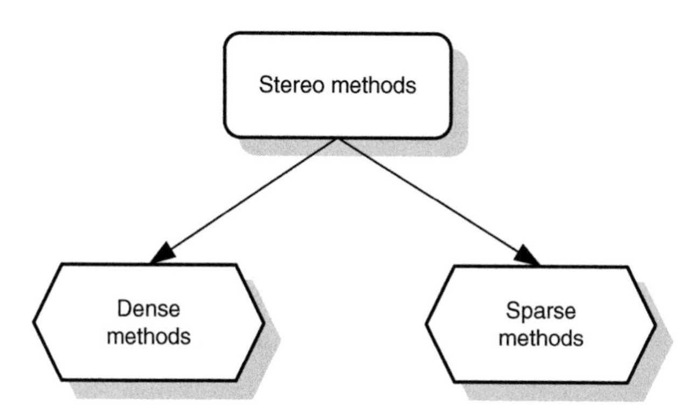

Figure 6.15 Diversity of stereo methods according to the output disparity map

It is interesting to take a look at the format of the output data, i.e. the ways of representing the resulting disparity maps. The most common way is to provide a 2D disparity map, which has to be considered with respect to a certain reference image (left, right or 'cyclopean') [180, 246]. In the case of a rectified stereo setup, this disparity map contains only horizontal displacements from the reference image. Otherwise the vertical displacements are also computed. This way we come to the multivalued representation of disparities [412] which is common to the stereo methods that are based on multiple views. Other representations come in the form of 3D models such as deformation fields, level-sets or triangulated meshes.

In the multiple baseline stereo methods a reference image is matched against more than one image. This way, many costs maps are obtained which are then summed up since there is a common reference frame [335]. This concept can be further extended to the arbitrary camera configurations with a plane sweep method [412]. The plane sweep algorithm transforms each image on to a common plane with a projective transformation. This is done for each disparity value. This way, disparity is defined as a 3D projective homography of an original camera space. This nicely connects the geometry of the camera setup with the disparity space.

Another possible classification of the stereo methods is based on the format of the signal taken for computation of match values (Figure 6.16). Many methods rely directly on the intensity values, whereas others first transform intensity into other domains (section 6.3.7) or compute some characteristic features which are then used for matching.

Figure 6.17 depicts a hierarchy of stereo matching methods divided into two groups: local and global methods. Methods of the first group compute disparity values based solely on the local information around certain positions of pixels. Local methods are discussed in sections 6.6–6.8.

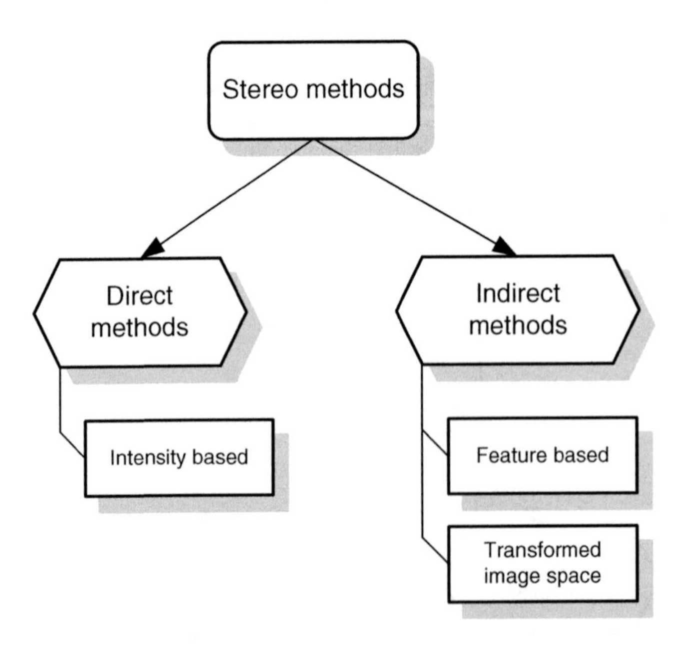

Figure 6.16 Diversity of stereo methods – division into direct methods that are based on bare intensities and indirect ones which operate on transformed space

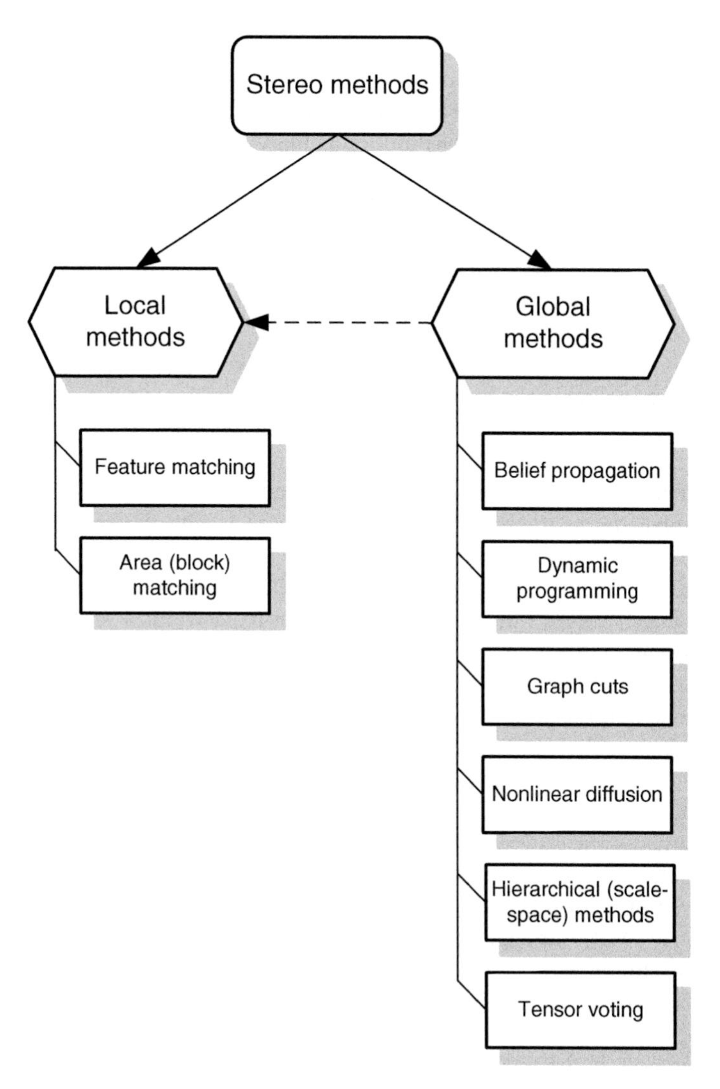

Figure 6.17 Diversity of stereo methods – division into local and global methods

Global methods use all cost values in the optimization process to determine disparity *and* occlusions. There are many different methods in this group, although the most characteristic are depicted in Figure 6.17. We can briefly characterize them as follows.

- Belief propagation – in this approach the stereo problem is formulated in the probabilistic way by means of Markov random fields. From this the maximum *a posteriori* estimation is obtained by applying a Bayesian belief propagation (BP) algorithm. BP performs a kind of message passing, where the message is meant as a probability that a receiver (a node in MRF) should exhibit disparity which is congruent with all information already passed to it by a sender. The nodes are divided into high-confidence and low-confidence ones. The entropy of a message from high-confidence nodes to low-confidence nodes is smaller than in the opposite direction. A formulation with three MRFs was proposed by Sun *et al.* [405]. The first MRF is responsible for modelling of a smooth field for disparity, the second for line process for depth discontinuity and the third for a binary process for occlusions. The model is further extended to incorporate other visual cues which are not taken by the three MRFs. Formulation of the stereo problem in terms of nonparametric belief propagation (NBP) was done by Sudderth *et al.* [404]. They propose an algorithm which uses stochastic methods to propagate kernel-based approximations to the true continuous messages. Each message update in the nonparametric formulation of NBP is based on a sampling procedure which allows other than Gaussians distributions.

- Dynamic programming – the main idea of the methods from this group lies in division of the 2D search problem into a series of *separate* 1D search problems on each pair of epipolar lines. Further discussion on this technique is provided in section 6.10.
- Graph cuts – methods from this group assume solution to the stereo problem, formulated as an energy functional, as computation of a maximum flow in graphs (discussed in section 6.11). Comparison of the belief propagation approach with the graph cuts method is provided in a paper by Tappen and Freeman [416].
- Nonlinear diffusion – in this approach the nonlinear diffusion is employed in the aggregation step of partial match values (this is in contrast to the fixed window encountered in local methods, for instance). Such an approach was proposed by Scharstein and Szeliski [368], and also discussed in the book by Scharstein [369]. The method derives from a version of the diffusion equation, for instance in the formulation (4.85), called a membrane equation with a local stopping policy. In this formulation a diffusion equation is endowed with a term that controls the amount by which current energy values had diverged from the original value. Thus, the diffusion can progress only to a certain degree. The second term which is built upon a certainty measure ensures that the diffusion takes places only in locations of ambiguous matches. Two certainty measures are proposed, the first called a winner margin, the second based on entropy. Then, the Bayesian model for stereo matching is proposed which incorporates MRF for aggregation step and robust non-Gaussian statistics to handle outliers and discontinuities.
- Hierarchical (scale-space) – in this formulation the stereo problem is computed based on the Gaussian scale-space, in which information on matches from the coarsest level controls the matching process in the finer (detail) levels, and so on. Thus, in each level the search space is greatly reduced to only local deformations in respect to a given matching level. This, in turn, leads to improvements in quality and run time, although the matching process is repeated at each level of the scale pyramid. A version of the hierarchical matching, called elastic matching, is presented in section 6.7.
- Tensor voting – the method originally proposed by Mordohai and Medioni [319] relies on a concept of perceptual organization postulated in the famous Gestalt theory. This is implemented by the tensor voting method which constitutes a computational framework of perceptual organization of salient local structures in images. This bears some resemblance to the structural tensor (discussed in section 4.6). It appears that such a second-order symmetric nonnegative tensor can be factored out into the stick and ball components which constitute tokens that are further used for image analysis. The tokens form a voting field in which votes propagate, separately for stick and ball components. In this framework Mordohai and Medioni propose an efficient stereo method which is composed of four steps: initial

matching (using a local method), detection of the correct matches, surface grouping and refinement, and finally estimation of disparity values for unmatched pixels. Thus, the method results in a dense disparity map. More details are provided in [320]. Some experimental results can be found on the Middlebury web page [209].

As indicated in Figure 6.17 there is a link between global and local methods since there are many variants in which some of the global methods utilize a local approach to the computation of disparity maps. For instance some of the global methods are initialized with a disparity map obtained by a local one, etc.

Usually local methods are more straightforward in implementation and exhibit faster run time, although many new optimization algorithms allow sufficient execution time for global realizations. The main benefit of the global methods is usually qualitatively better disparity maps (i.e. fewer errors) which takes into account occlusions (i.e. discontinuities). This comes from the fact that global methods usually perform better in areas with insufficient texture for errorless matching. Further details and properties of the outlined methods are discussed in the following subsections.

6.5.1 Structure of Stereo Matching Algorithms

In recent years dozens of algorithms for stereo correspondence have been developed, and, although they differ in many aspects, for many of them it is possible to distinguish common characteristic steps [369, 370] which are presented in Figure 6.18.

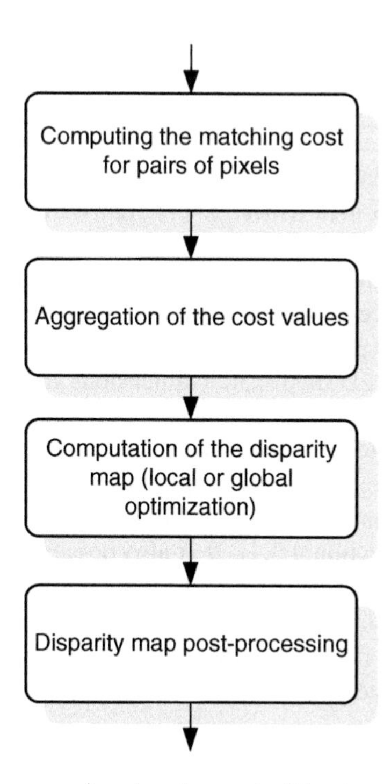

Figure 6.18 Basic processing steps for most of the stereo correlation methods

The first step in Figure 6.18, i.e. computation of the matching cost for pairs of pixels, is relatively straightforward. It can be done with one of the matching measures, already discussed in section 6.3. Other steps of computation are discussed below.

6.5.1.1 Aggregation of the Cost Values

Computation of the matching costs for pairs of pixels is the most basic step of stereo correlation. It consists in computation of a fit measure for pairs of pixels, where each pixel in a pair comes from a different image. The most common measures for this purpose are presented in sections 6.3.1–6.3.5. However, taking only single intensity values has many limitations. The first comes from the fact that in practice intensities are represented on a limited number of bits (usually 8–10 bits per pixel). Therefore, their discriminative power is very limited. The next problem is noise superimposed on the intensity signal. There are different types of noise (discussed in Chapter 11) which add additional error on match values. Also, we have to remember that more often than not images are taken by different cameras, which can differ in some parameters of their image processing path. The most common are variations of the bias-gain factor in cameras' transmission channels. For these reasons image preprocessing steps or/and fit measures that account for the mentioned problems can be of help. Usually such solutions rely on some information gathered in a local neighbourhood of a pixel, rather than its single intensity value. The most common preprocessing step is transformation from the intensity to the nonparametric space (section 6.3.7), where each (single) pixel is augmented with some information on its closest neighbours. Then matching costs can be done on pairs of single pixel values. Log-polar transformation is another example of a nonlinear transformation of the input local neighbourhoods which aims at more reliable matching (section 6.3.8). A second possibility is to use a measure that accounts for some intensity variations in local image areas. These are, for instance, measures that from each matched pixel subtract a mean value, which is usually computed in small image areas (section 6.3.1). Such techniques also assume information gathered from wider areas (local pixel neighbourhoods) than single pixels, even if this is only computation of a mean etc.

Since comparison of only single pixels has a limited discriminative power it is necessary to count on the cooperative influence of neighbouring pixels. This is called aggregation of single cost values, while neighbouring pixels involved in this process are called a support region. The support region can be either 2D (two spatial dimensions and a fixed disparity) or 3D (two spatial dimensions and disparity as the third dimension). For the former, an observed scene is assumed to be composed of frontoparallel surfaces.

The simplest aggregation scheme follows simple low-pass filtering in the support region. This can be done by convolution with a uniform (box filters), binomial, Gaussian or other fixed-size convolution kernel (section 4.3.1). It is also possible to adjust the support region to fit image contents. The simplest way is for each location to use multiple supported regions of different size [141, 195]. Its modification consists of many windows, placed however at slightly different positions (also called shiftable windows) [47]. The other idea is to build a variable size window which is adjusted to the image content [237, 281, 435, 452] or local signal statistics [88].

A separate group constitutes methods that employ the iterative diffusion scheme for the aggregation step [271, 368].

6.5.1.2 Computation of the Disparity Map

This step can be organized either as a local or a global optimization. The basic equation that governs matching of two images I_1 and I_2 can be written as

$$I_1(x, y) = I_2(\Psi(x, y)), \tag{6.64}$$

where $\Psi(x, y)$ is a function that defines a local deformation model, i.e. a deformation of the initial coordinate grid for which (6.64) holds. Thus, computational stereo is about computing $\Psi(x, y)$. For the linear case of horizontal disparities, the above can be reduced to the following simpler form:

$$I_1(x, y) = I_2(x + D(x, y), y), \tag{6.65}$$

where $D(x, y)$ denotes a disparity field.

Further, a similar equation can be superimposed on some functions of the intensity signals:

$$\Xi\{I_1(x, y)\} = \Xi\{I_2(\Psi(x, y))\}, \tag{6.66}$$

where $\Xi(I)$ denotes a transformation of the intensity signal. The most common here is computation of the gradient from intensity, i.e. $\Xi(I) \equiv \nabla(I)$. However, (6.66) is very general, i.e. it can comprise any combination of signals and their transformed (derived) versions. The majority of the simple stereo correspondence methods rely on intensity matching. However, other versions rely on intensity and/or other signal representations. For instance, different versions of matching methods operating on the gradient fields are presented in section 6.9.

In local optimization, disparity values are computed locally, i.e. based on the aggregation values constrained in certain local match regions. The local regions are shifted in a predefined range to find the best match value. The shift value of the winning match becomes a disparity. Therefore such a strategy is commonly known as a winner-takes-all (WTA) approach. The limitation of this approach is that uniqueness of matches is achieved only for a reference image whereas points in the matched images can be matched to multiple positions [369]. This can be resolved by the cross-check procedure (section 6.6.6). Sometimes selection of the best match is very problematic, since there is no unique strong extreme in the cost function. An interesting idea to overcome this problem is proposed in the work by Mühlmann *et al.* [323] which consists of analysing not a single minimum but at least three consecutive minimal values. If they are close to a certain threshold, which in [323] is reported to be about 5–20% between the best and next value, then such a match is rejected. This is a very useful technique in many other computer vision methods that require selection of the best fit.

Global optimization is usually a more powerful technique than local matching since all local cost values and other constraints can be taken simultaneously to find the disparity values that fit best into this optimization task. Because all the cost values are involved in the optimization task then the aggregation step is usually omitted. The common approach to formulate the global optimization process is to design an energy functional involving the disparity function that is to be evaluated during the energy optimization process. The energy functional for stereo matching can be stated as

$$E(\theta) = E_{\text{data}}(\theta) + E_{\text{smooth}}(\theta), \tag{6.67}$$

where θ denotes a set parameters that affect the energy value. $E_{\text{data}}(\theta)$ relates disparity values with values of the matched pixels. This can be written for instance as a sum of the local cost values [370]:

$$E_{\text{data}}(\theta) = \sum_{(x,y)\in I_{1,2}} S(x, y, D), \tag{6.68}$$

where $S(x, y, D)$ denotes a cost value. More generally, $E_{\text{data}}(\theta)$ conveys a level of disagreement between θ and the input (observed) data.

The term $E_{\text{smooth}}(\theta)$ is introduced to enforce a smoothness of the solution, i.e. an additional constraint on the resulting disparity map. Usually it is a function of disparities, sometimes additionally related with a function of image intensity. For instance, this can be stated as

$$E_{\text{smooth}}(\theta) = \sum_{(x,y)\in I_{1,2}} \Phi(x, y, |\nabla D|), \tag{6.69}$$

where $\Phi(x, y, |\nabla D|)$ is a certain functional of the disparity gradient. The above can take on a form proposed by Scharstein and Szelisky [368]:

$$E_{\text{smooth}}(\theta) = \sum_{(x,y)\in I_{1,2}} f(D(x, y) - D(x + 1, y)) + f(D(x, y) - D(x, y + 1)), \tag{6.70}$$

where $f(\,)$ denotes a monotonically increasing function on its argument [411]. Choice of f influences the quality of the output disparity map. For instance, if f is a quadratic function, then disparity is smoothed across object boundaries in the input images. Therefore some robust $f(\,)$ have been also proposed [368]. Additionally, the term of disparity gradient can be augmented with a condition for intensity values:

$$\begin{aligned} &E_{\text{smooth}}(\theta) \\ &= \sum_{(x,y)\in I_{1,2}} \{f\left(D(x, y) - D(x + 1, y)\right) \cdot g\left(\|I(x, y) - I(x + 1, y)\|\right) \\ &+ f\left(D(x, y) - D(x, y + 1)\right) \cdot g\left(\|I(x, y) - I(x, y + 1)\|\right)\}, \end{aligned} \tag{6.71}$$

where the new function g is a monotonically decreasing function of its argument. It lowers the smoothness costs for large intensity gradients. Choice of the smoothness penalty term Φ in (6.69) is sometimes referred to as the Potts model [51, 252, 253].

Then, finding disparity map D is equivalent to solving the following optimization problem:

$$D = \arg\min_{\theta}\{E(\theta)\}. \tag{6.72}$$

Many algorithms exist that help approach this optimization problem [36, 37, 127, 184, 352]. Usually the scheme of this process is to change in some way θ_i to θ_{i+1}, then compute a new energy value and if it is smaller than the previous one, that is

$$E\left(\theta_{i+1}\right) < E\left(\theta_i\right), \tag{6.73}$$

then θ_{i+1} is chosen as a new state. Finally, a state θ_o,

$$E(\theta_o) \leq E(\theta_j), \tag{6.74}$$

is a (local) minimum of energy E, if the above is fulfilled for all θ_j.

Many optimization methods suffer from falling into local minima whereas we are mostly interested in global minima values. However, finding a global minimum does not belong to simple tasks and in general the problem is NP hard, i.e. it cannot be solved even in a polynomial time.

Some examples for construction of the energy functional follow. For instance Robert and Deriche propose the following continuous version of the energy functional for direct evaluation of the depth map [359]:

$$E(\theta) = \int\int [I_1(x, y) - I_2(\Psi(x, y,\ D(x, y)))]^2 \mathrm{d}x\mathrm{d}y + \upsilon \int\int \Phi(\|\nabla \mathrm{D}\|)\, \mathrm{d}x\mathrm{d}y, \tag{6.75}$$

where I_1 and I_2 are left and right views of a stereo-pair, respectively, $D(x, y)$ is a disparity map and $\Psi()$ denotes a local displacement model which is dependent on the extrinsic and intrinsic parameters of the stereo setup. The minimum of the above $E(\theta)$ is calculated by methods pertinent to the Euler–Lagrange equations.

Another energy function which is then solved by the radial-based neural network was proposed by Wei *et al.* [444]. We outline the principles of this method in section 6.9.

Gong and Yang propose using genetic optimization for stereo matching [155]. This is known as a method which is able to find a global solution to an optimization problem. However, at first the problem needs to be encoded into the domain of genes. Then the iterative gene processing mechanism starts. It consists of gene mutations, inversions and crossover. At each step only the optimal solutions are left with all others being disposed of. The process is governed by a chosen optimality condition; such as (6.67) for stereo problem. A drawback of the method is computation time.

Finally, simulated annealing is an optimization method that, under certain conditions, can attain global-extrema [184]. At each step an energy functional is assessed and a direction of maximal energy decrease is chosen for the next step. However, with a certain probability a direction different from maximally optimal at a moment can be also chosen. In this way traps of local minima can be avoided.

6.5.1.3 Disparity Map Postprocessing

This stage is aimed at improving the quality of the output disparity map. The most common methods are as follows.

1. Subpixel estimation of disparity values (section 6.4.2) – a process used to overcome the problem of integral grid of input images. Often it is done by fitting a polynomial to the discrete values of match values in a local neighbourhood. Then, the disparity with a subpixel accuracy can be found by interpolation.
2. Disparity verification (section 6.4.1) – more often than not performed with the cross-checking method.

3. Filtering of the disparity values – the most popular here is application of the median filter to get rid of spikes in disparity space.
4. Interpolation of missing disparity values – this is a process of filling up those places for which the disparity could not be determined or the disparity value was rejected during the verification stage.

6.6 Area-based Matching

Area-based matching consists of measuring the degree of correlation between pixels in matched images. However, instead of comparing single pixels for best match evaluation, groups of pixels – usually gathered in a fixed sized image patch, i.e. *neighbourhood N* – are taken *simultaneously* for comparison (Figure 6.19). This comes from the small discriminative power of comparison which would be based solely on a single pair of pixels. This is a result of the very local information conveyed by a pixel, whose value in most cases is represented by a limited number of bits. The small dynamic range of values leads in consequence to ambiguities when only single pixels are compared.

The situation is different if a group of pixels is compared with another group of pixels. In this case, not only are their values important, but also their spatial positions can be taken into consideration since pixels on corresponding positions in the two groups are now being compared.

Area-based matching algorithms are generally designed to recover *dense* disparity fields between pairs of corresponding images. This property in itself is useful for applications where continuous disparity or surface measurement fields are required for scene synthesis [369] or 3D scanning applications, where we wish to reconstruct virtual models of the scene. If we intend to match a stereo-pair of images, e.g. as shown in Figure 6.19, for each pixel in the left, I_l, image, the goal of the area-based algorithm is to find *every* corresponding location in the right, I_r, image of the stereo-pair. We define a displacement field, or disparity map, $D_{xy}(x, y)$, such that the $D_{xy}(x, y)$ field maps each pixel of I_l to a *single*, i.e. unique, corresponding location

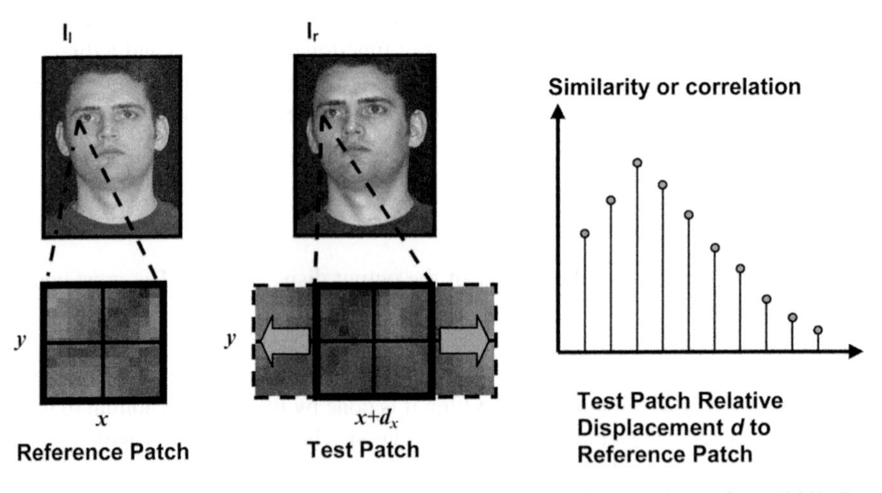

Figure 6.19 Basic 1D area-based search along a horizontal scanline. (Redrawn from [389], Emerald Group Publishing Limited)

in I_r. While the disparity map should ideally comprise such a *bijective* mapping, there are circumstances in practice that may result in this not being the case, e.g. match failure or surface features projecting to a singularity in one of the images to be matched (consider a plane viewed edge-on in one camera but not the other)

$$I_l(x, y) \to I_r(x', y'). \tag{6.76}$$

Usually the disparity field is structured as two maps $D_x(x, y)$ and $D_y(x, y)$ which store horizontal and vertical displacements, respectively. In general these displacements are real-valued and hence represent subpixel correspondences between matched images. In this case (6.64) takes on the form

$$I_r(x', y') = I_l(x + D_x(x, y), y + D_y(x, y)), \tag{6.77}$$

where $D_x(x, y)$ and $D_y(x, y)$ are horizontal and vertical disparity values, respectively.

6.6.1 Basic Search Approach

In order to recover dense disparity fields, i.e. recover a disparity value $D(x, y)$ at *every* x, y location in the reference image, we must find the corresponding locations of the compared image patches referenced at these locations. We have met in section 6.3 a variety of metrics that allow us to compare image patches in terms of their similarity. Using such a metric we can place a local neighbourhood (reference patch) N over I_l at pixel location x, y and search about the corresponding x, y location in I_r using the same (test patch) neighbourhood. This local search process is applied to find the local test neighbourhood in I_r that is most similar to the reference neighbourhood placed in the left image. We simply repeat this process for every pixel in I_l to find the most similar corresponding neighbourhoods in I_r. Accordingly we can obtain a disparity estimate at every location in I_l. Figure 6.19 illustrates a simple 1D search and Figure 6.20 illustrates a full 2D search.

Usually, for the sake of simplicity in implementation, matching areas are rectangles or even squares of the same size, depicted in Figure 6.20. The task now is to find pairs of corresponding areas in the two images which fit the best. However, even taking areas of pixels for comparison it is not always guaranteed to find unique matches between these areas. This is a result of the already mentioned nonuniqueness of the inverse projective problem (section 3.4).

In practice, when employing the area-based method the following questions have to be answered.

1. What should be the shape and size of the matching area?
2. How does one measure the 'goodness' of a match?
3. How does one find the best match?

Increasing the size of a matching area increases its discriminative power. Thus, less ambiguous matches are computed. However, when increasing the size of the matching window we encounter much smearing in the disparity map. This is a result of the reduced influence of local feature regions inside the matched areas. Examples of this effect are presented in the next sections.

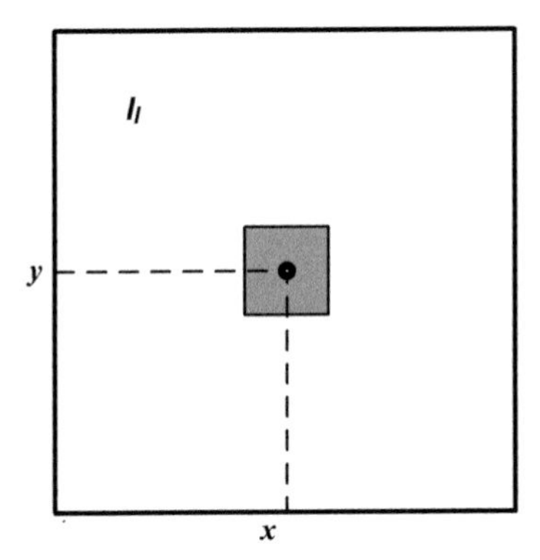

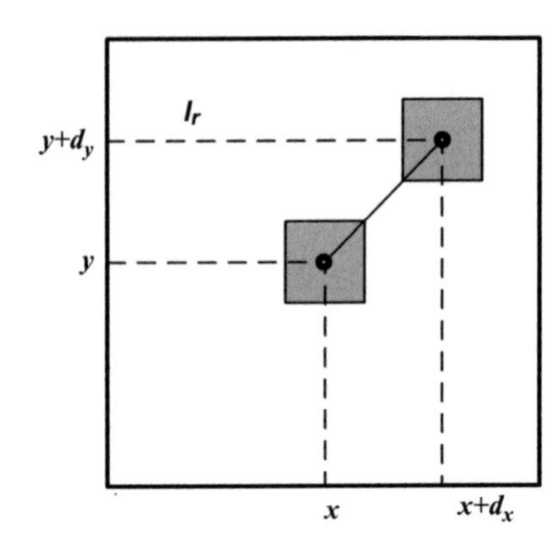

Figure 6.20 General 2D search. A patch in the left-hand image is compared with a series of trial patches in the right-hand image and the best position selected by interpolation of the matched values as before. The figure illustrates a 2D search in the right image to find the relative displacement of the patch located in the left image

Nevertheless, the area-based matching methods are very versatile and useful in practical realizations. They can use diverse matching metrics (section 6.3), different image transformations (section 6.3.7) and they can be implemented in different ways depending on the application (sections 6.6.3 and 6.6.4).

An interesting feature of these methods is that all pixels are treated in the same way and all are taken for matching. This is different from the feature-based methods for instance, in which only a *selected* group of characteristic pixels (features) are matched (section 6.8). Thus, these methods are called sometimes 'direct' methods [220]. However, the pixel values can be any values suitable for comparison, i.e. these are not necessarily image intensities but can be any other information computed from intensities. For instance, prior to matching, the input intensities can be low-pass filtered, processed to extract the statistical information parameters (not necessarily section 6.3.4) or transformed into a nonparametric representation (section 6.3.7). At this stage it is also possible to change the dimensionality of the input data. For instance colour images can be converted into a monochrome representation and then matched [91]. On the other hand, the monochrome (or colour) input images can be transformed by the structural tensor operator which results in 3D data. Then, the matching takes place in the tensor space (section 6.6.7.3).

Figure 6.21 presents a block diagram of a typical area-based matching method. The acquisition modules (1) supply a pair of images. Usually, these are RGB colour or monochrome images and are of the same size. The input images are then optionally transformed, which is done in stage (2). The actual area-based matching is performed in stage (3). There are different ways of organising this module, which we will explore in more detail. Implementation of (3) can follow the point-oriented or disparity-oriented strategy. In the former case, the local matching method is obtained (section 6.5). The latter approach allows a global optimization for computation of a disparity map. Both solutions have their advantages and drawbacks, which are discussed in the next sections.

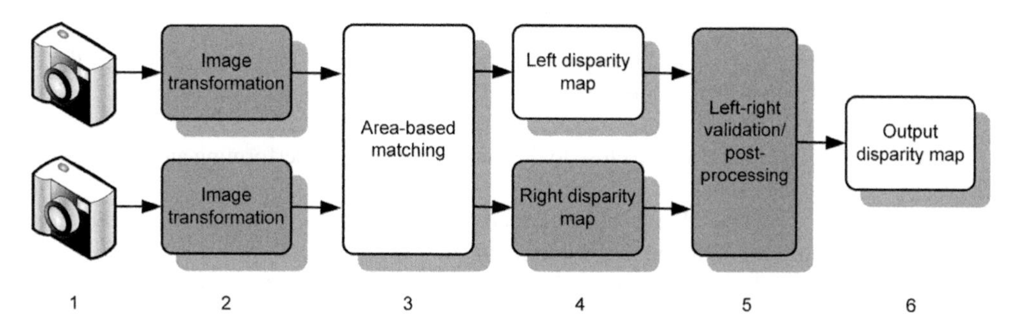

Figure 6.21 Area-based matching algorithms for stereovision. Shaded modules are optional

The other important distinction of module (3) is that if the two cameras are in the canonical stereo setup, then the matching can be done in a 1D fashion, alongside the scanlines (section 3.4.2). In other case, the images should be rectified (section 3.4.8) or the matching has to be done in two dimensions.

The matching results in a dense disparity map. For an optional validation (section 6.4.1), the matching process can be run again (or in parallel) with input images reversed, however. This produces a second disparity map. If the two maps are available then they can be verified by the left–right validation module (5). Finally we obtain the output disparity map (6), potentially with a subpixel accuracy (section 6.4.2).

6.6.2 *Interpreting Match Cost*

In order to determine the relative displacement to achieve best match between the *reference* and *test* images patches, $P_r(i, j)$ and $P_t(i, j)$, indexed by (i, j), we require a metric to determine their similarity, or *correlation* coefficient, c_{rt} and a selection of metrics were introduced in section 6.3, discussed in section 6.3.1 and listed in Table 6.1. As suggested in Equation (6.3) we could subtract the pixel values of each patch, take the square and add the values together:

$$c_{rt} = \sum_{1}^{i} \sum_{1}^{j} (P_r(i, j) - P_t(i, j))^2. \tag{6.78}$$

Notice that this formula is in effect treating P_r and P_t as vectors and computing the (square) of the modulus of their vector difference:

$$c_{rt} = |P_r - P_t|^2 . \tag{6.79}$$

If we wanted to make the comparison measure independent of the size of the patches, we could normalize this measure by the number of elements in the patch of dimensions $I \times J$:

$$c_{norm} = \frac{1}{I \times J} c. \tag{6.80}$$

While the above measure for comparing patches is simple, it suffers from a number of limitations: If one signal has a much larger gain, or DC offset, than the other, it swamps the correlation value. The correlation measures have *arbitrary* output values, consequently:

- it is difficult to tell if we are getting close to the best match without performing an *exhaustive* search;
- it would be efficient to be able to halt the search when we passed through a correlation valley, in the belief we had found an optimum match;
- it would also be useful to be able to use the score as a *confidence* measure to evaluate how good our match is;
- confidence metrics computed using the above measures are dependent on the images being matched;
- it would also be more useful if the measure *increased* with similarity.

The traditional solution to the above is the use of *statistical correlation* (Table 6.1). Equation (6.7) is now derived as follows. To afford a different basic *similarity* measure c_{rt} similar to convolution, without loss of generality we can drop the 2D indexing notation for clarity, where $N = i \times j$, and treat the 2D patches as 1D matrices:

$$c_{fg} = \frac{1}{N} \sum_{n=1}^{n=N} P_{\mathrm{r}}(n) P_{\mathrm{t}}(n). \tag{6.81}$$

We are now summing the *product* of the two signals (each of N elements) to find their best match (Table 6.1; Equation (6.8)). It is possible to obtain invariance to DC offsets by subtracting the mean μ (average value of each signal in the current correlation window) from each signal:

$$c_{fg} = \frac{1}{N} \sum_{n=1}^{n=N} (P_{\mathrm{r}}(n) - \mu_{\mathrm{r}})(P_{\mathrm{t}}(n) - \mu_{\mathrm{t}}) \tag{6.82}$$

where c_{rt} is termed the *covariance* of P_{r} and P_{t}, and indicates if these signals change together, i.e. are correlated. In practice, we remove the mean from each image patch by prefiltering the input images prior to correlation. A zero-mean filter such as *DoG* or *LoG* filter is used and the spatial scale of this filter is selected according to criteria set out in sections 4.5.3 and 4.5.4.

Recall that the standard deviation of a signal sample is equivalent to the RMS signal amplitude of the varying (AC) component. Therefore, we can normalize the correlation score in terms of signal amplitude by dividing by the standard deviations of each patch, σ_{f} and σ_{g}, as follows:

$$c_{\mathrm{rt}} = \frac{\frac{1}{N} \sum_{n=1}^{n=N} (P_{\mathrm{r}}(n) - \mu_{\mathrm{r}})(P_{\mathrm{t}}(n) - \mu_{\mathrm{t}})}{\sigma_{\mathrm{r}} \sigma_{\mathrm{t}}}. \tag{6.83}$$

It is now possible to use the correlation score as a confidence measure since the above similarity function gives output values in the range $[-1\ldots 1]$:

- $1.0 =$ maximum similarity and greatest confidence;
- $0.0 =$ neutral;
- $-1.0 =$ maximum *anti correlation*, the correlated patches are inverses of each other.

It is also possible to consider the statistical correlation process in terms of vectors (see also section 6.3.3). Let us consider our image signals (i.e. patches) P_{r} and P_{t} to be correlated, having first been passed through the *DoG* or *LoG* filter such that they have zero mean to start with:

$$P'_{\mathrm{r}} = P_{\mathrm{r}} - \mu_{\mathrm{r}}, \quad P'_{\mathrm{t}} = P_{\mathrm{t}} - \mu_{\mathrm{t}}.$$

The correlation equation now simplifies to

$$c_{\mathrm{rt}} = \frac{\frac{1}{N}\sum_{n=1}^{n=N} P'_{\mathrm{r}}(n)P'_{\mathrm{t}}(n)}{\sigma'_{\mathrm{r}}\sigma'_{\mathrm{t}}}. \tag{6.84}$$

Treating patches P_{r} and P_{t} as vectors, the equation for standard deviation, σ', now becomes

$$v_{\mathrm{r}} = \sigma'^{2}_{\mathrm{r}} = \left[\frac{1}{N}\left|P'_{\mathrm{r}}\right|^2\right], \quad v_{\mathrm{t}} = \sigma'^{2}_{\mathrm{t}} = \left[\frac{1}{N}\left|P'_{\mathrm{t}}\right|^2\right]. \tag{6.85}$$

$$\sigma'_{\mathrm{r}} = \frac{1}{\sqrt{N}}\left|P'_{\mathrm{r}}\right|, \quad \sigma'_{\mathrm{t}} = \frac{1}{\sqrt{N}}\left|P'_{\mathrm{t}}\right|$$

Furthermore correlation for signals P'_{r} and P'_{t} (for zero mean) is equivalent to their vector dot product:

$$c_{\mathrm{rt}} = \frac{\frac{1}{N}\sum_{n=1}^{n=N} P'_{\mathrm{r}}(n)P'_{\mathrm{t}}(n)}{\sigma'_{\mathrm{r}}\sigma'_{\mathrm{t}}} = \frac{\frac{1}{N}\sum_{n=1}^{n=N} P'_{\mathrm{r}}(n)P'_{\mathrm{t}}(n)}{\frac{1}{N}\left|P'_{\mathrm{r}}\right|\left|P'_{\mathrm{t}}\right|}. \tag{6.86}$$

Finally, substituting in the normalization terms and the correlation simplifies to

$$c_{\mathrm{rt}} = \frac{\left|P'_{\mathrm{r}}\right|\left|P'_{\mathrm{t}}\right|\cos(\theta)}{\left|P'_{\mathrm{r}}\right|\left|P'_{\mathrm{t}}\right|} = \cos(\theta). \tag{6.87}$$

Since $\cos(\theta)$ takes up the range $[-1\ldots+1]$ depending on whether the vectors are in exact opposition or exact alignment respectively, we once more have a similarity metric that can be thought of as the *cosine angle* between our image patches expressed as vectors, as discussed also in section 6.3.1. This property gives us a uniform metric for comparing two image patches (or vectors) independently of their gains or black levels as before. Hence, for zero-mean signals, statistical correlation and the normalized vector dot product are identical.

A potential flaw with the above statistical correlation metric is a lack of stability that can arise if a strong signal appears at the edge of the correlation window to dominate the correlation score. In other words, when searching for a similar patch it is desirable if the correlation score rises and falls monotonically as the 'best match' position is reached and then passed. If the match-window 'just clips' another high-contrast structure, then this match continuity will be disturbed. To combat this boundary effect, Jin [229] describes how the correlation window can be weighted with a Gaussian function as follows. The correlation coefficient, c_{rt}, between a reference window situated at the left band-pass image and the test search window situated at a trial match point in the right band-pass image is calculated as follows:

$$c_{\mathrm{rt}} = \frac{\rho_{\mathrm{rt}}}{\sqrt{\sigma_{\mathrm{r}}^2 \sigma_{\mathrm{t}}^2}}. \tag{6.88}$$

The above formulation assumes that the input images I_{L} and I_{R} have been prefiltered with a *LoG* filter of scale σ_{s} (not to be confused with σ, standard deviation, of the image patch signal) such that reference and test patches P'_{r} and P'_{t} with zero mean can be extracted for comparison. The Gaussian windowed covariance is computed as

$$\rho_{\mathrm{rt}} = \sum_{u=-\frac{I}{2}}^{\frac{I}{2}} \sum_{v=-\frac{J}{2}}^{\frac{J}{2}} G(u, v) \cdot P'_{\mathrm{r}}\left(\frac{I}{2} + u, \frac{J}{2} + v\right) \cdot P'_{\mathrm{t}}\left(\frac{I}{2} + u, \frac{J}{2} + v\right). \tag{6.89}$$

The local Gaussian windowed standard deviations are similarly computed as

$$\sigma_{\mathrm{r}}^2 = \sum_{u=-\frac{I}{2}}^{\frac{I}{2}} \sum_{v=-\frac{J}{2}}^{\frac{J}{2}} G(u, v) \cdot P'_{\mathrm{r}}\left(\frac{I}{2} + u, \frac{J}{2} + v\right) \cdot P'_{\mathrm{r}}\left(\frac{I}{2} + u, \frac{J}{2} + v\right), \tag{6.90}$$

$$\sigma_{\mathrm{t}}^2 = \sum_{u=-\frac{I}{2}}^{\frac{I}{2}} \sum_{v=-\frac{J}{2}}^{\frac{J}{2}} G(u, v) \cdot P'_{\mathrm{t}}\left(\frac{I}{2} + u, \frac{J}{2} + v\right) \cdot P'_{\mathrm{t}}\left(\frac{I}{2} + u, \frac{J}{2} + v\right). \tag{6.91}$$

Note that to maintain stability the Gaussian window must have a scale factor $\sigma_{\mathrm{w}} \geq \sigma_{\mathrm{s}}$, i.e. have a standard deviation (blur factor) equal to or greater than that of the *LoG* kernel used for zero-mean filtering. A factor of 1.3 has been found to work well in practice [72, 229]. The size of the correlation window can be set to $I = 4\sigma_{\mathrm{w}}$ rounded (up) to the nearest odd integer. The Gaussian weight has the effect of enforcing a *continuity constraint* on the matching process, i.e. the reference patch is correlated at intervals smaller than the Gaussian envelope on the search image; the resulting match scores are themselves correlated since the image patches being compared overlap within the Gaussian weighted support window. As will be discussed (section 6.7.1.1) the resulting matching search continuity allows the search scale parameters to be deduced such that we can determine the relations between the size of correlation window required to ensure that a given magnitude of disparity can be matched.

Despite the above refinements to the basic statistical correlation process, limitations remain when conducting matching search, including: lack of scale and rotation invariance, *false target* problems and the ability to correlate only highly textured patterns. The statistical correlation match metric is also comparatively computationally expensive, especially when applied exhaustively. Section 6.7.1 investigates the matching range attainable at a single scale while section 6.7.3 shows how the dynamic range of the matching algorithm can be extended by matching over the multi-resolution pyramid data structure (section 6.7.4).

An interesting improvement to discrete pixel matching was proposed by Birchfield and Tomasi [44]. Their idea consists of comparing each pixel in the reference frame with a linearly interpolated value of a pixel, rather than an original discrete value, from the second image. By this method, matching is more tolerant of different image sampling schemes, applied to the original image signal.

6.6.3 Point-oriented Implementation

In this section we present a complete implementation of an area-based stereo matching method. It follows the point-oriented organization of the inner loops (Algorithm 6.4). Its purpose is to give an in-depth view of the inner structure of the algorithm. However, it must be remembered that this is a very simple, nonoptimal and strictly procedural implementation with a main didactical purpose.

Being point oriented, the external loops traverse each pixel position in the reference image. The innermost loop traverses all possible disparity values and for each of them a match is computed between an area in the reference and an area in the second image. This follows the idea listed in Algorithm 6.4.

There are three functions that constitute the point-oriented matching algorithm. Their mutual relation is presented in Figure 6.23. The external function *ComputeDisparity_Local* takes on six parameters which are the two input monochrome images, the output image which upon

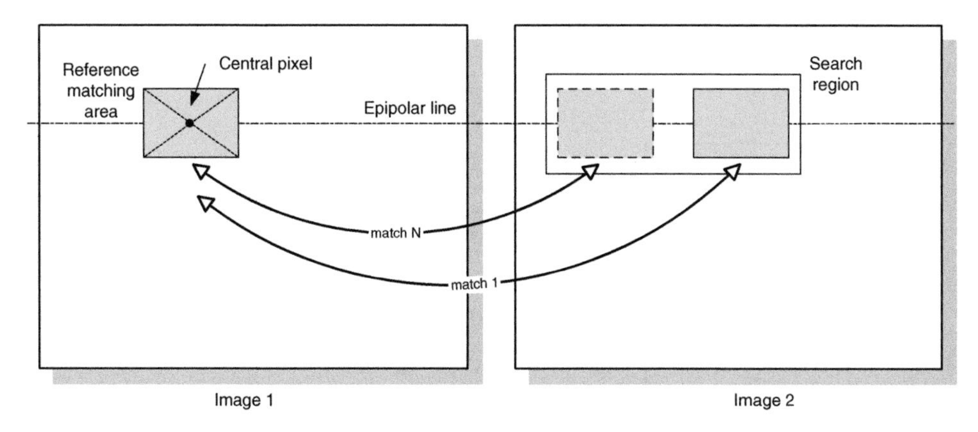

Figure 6.22 Local matching strategy in the point-oriented scheme and rectified stereo system. A reference matching area from the first image is matched against all possible areas on the epipolar line in the second image. A position of the 'best' match is saved

```
for(int i = 0; i < max_image_rows; i ++)
{
      for(int j = 0; j < max_image_cols; j ++)
      {
            for(int d = 0; d < max_expected_disparity ; d ++)
            {
                  "find the match value based on the chosen measure
                                          in a w1xw2 window"
            }

            disparity(i, j) = "d for which the match value was
                                                       the best";
      }
}
```

Algorithm 6.4 Structure of the point-oriented method

exit contains the disparity map, the expected maximum disparity range and finally dimensions of the rectangular search area. Their details are described in the function tag in Algorithm 6.5.

The matching process follows the scheme depicted in Figure 6.22. Size of a matching window is set by the two *match_area_cols* and *match_area_rows* parameters passed to the *ComputeDisparity_Local* method. The loops which realize iteration through all pixels of the left image are constrained inside the lines *L[44–74]* and *L[47–73]*, respectively. Around each pixel in the left image an area is created which is then matched with the *max_disp* number of areas in the right image. These iterations are organized by the innermost loop *L[53,69]* which traverses all possible disparity values. However, the actual matching of the areas is delegated in *L[57]* to the *ComputeAreaMatch* function.

The *ComputeAreaMatch* function has eight parameters, the details of which are described in Algorithm 6.6. The first seven of them are used to define the size and position of the areas to be matched in the two images. Let us observe that to define an area around a pixel at position (*col, row*) we pass the symmetrical offsets from this point to the left and to the right by *kColSwing* pixels, and to the top and to the bottom by *kRowSwing* number of pixels, respectively. In the lines *L[46,59]* of *ComputeAreaMatch* the border variables for the two loops are organized. Their purpose is also to ensure that all the accesses to pixels in the input images hit the valid positions in these images. Then the two loops are set around the lines *L[64,76]* which iterate

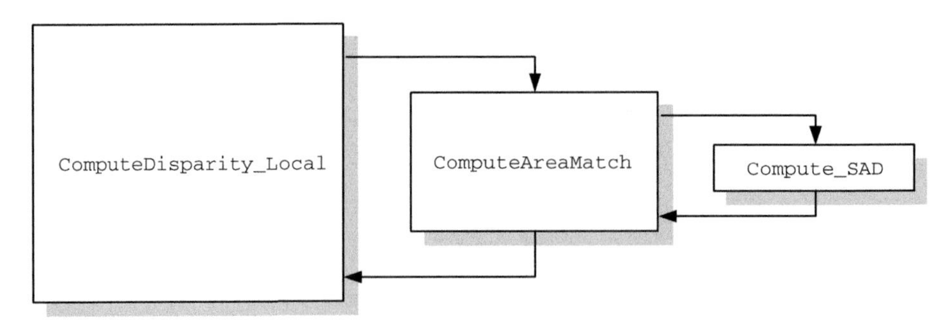

Figure 6.23 Mutual dependencies among functions for the point-oriented area matching

```
1   ///////////////////////////////////////////////////////////////
2   // This function computes a disparity map from the two
3   // monochrome input images. The point-oriented (local)
4   // algorithm is used. It is assumed that the input pair
5   // is already rectified, so only horizontal scanlines
6   // are searched for the matches.
7   ///////////////////////////////////////////////////////////////
8   //
9   // INPUT:
10  //          leftImage - a reference to the left image of a
11  //                  stereo-pair
12  //          rightImage - a reference to the right image of a
13  //                  stereo-pair
14  //          disparityMap - a reference to an image that
15  //                  upon return contains the disparity map;
16  //                  This image should be of the same size
17  //                  as are both images of a stereo-pair
18  //          match_area_cols - a horizontal size of the
19  //                  matching window
20  //          match_area_rows - a vertical size of the
21  //                  matching window
22  //          max_disp - expected maximum horizontal
23  //                  disparity value
24  //
25  // OUTPUT:
26  //          none
27  //
28  void ComputeDisparity_Local( const MonochromeImage & leftImage,
29                               const MonochromeImage & rightImage,
30                               MonochromeImage & disparityMap,
31                               const int match_area_cols,
32                               const int match_area_rows,
33                               const int max_disp )
34  {
35    const int kColSwing = match_area_cols / 2;
36    const int kRowSwing = match_area_rows / 2;
37
38    const int kTotalCols = leftImage.GetCol();
39    const int kTotalRows = leftImage.GetRow();
40
41    register int col, row, disp;
42
43    // Traverse each row
44    for( row = 0; row < kTotalRows; ++ row )
45    {
46          // Traverse each column
47          for( col = 0; col < kTotalCols; ++ col )
48          {
49                long prev_match = LONG_MAX;          // get a large value
50                int best_disp = -1;
51
52                // Traverse each possible disparity
53                 for( disp = 0; disp < max_disp; ++ disp )
54                {
55                      long area_match_value;
56                      // Compute match value for a given area and disparity
57                       if( ComputeAreaMatch( leftImage, rightImage, col, row,
58                                             disp,      kColSwing,      kRowSwing,
                                                             area_match_value )
59
                         == true )
60                       {
61                            // Compare this match to select the best one
62                             if( area_match_value < prev_match )
63                            {
64                                  // save the best values
65                                  prev_match = area_match_value;
66                                  best_disp = disp;
```

Algorithm 6.5 Listing of the *ComputeDisparity_Local* function which implements area-based matching with the point-oriented organization of the inner loops

```
67                              }
68                          }
69                      }
70
71                      // Store found the best disparity
72                      disparityMap.SetPixel( col, row, best_disp );
73                  }
74          }
75      }
```

Algorithm 6.5 (*Continued*)

```
1    ////////////////////////////////////////////////////////////
2    // This function computes an area match
3    // in a window which is set around a pixel at coordinates
4    // (col,row) in the left image.
5    ////////////////////////////////////////////////////////////
6    //
7    // INPUT:
8    //          leftImage - a reference to the left image of a
9    //                 stereo-pair
10   //          rightImage - a reference to the right image of a
11   //                 stereo-pair
12   //          col - a horizontal coordinate of a reference
13   //                 pixel in the left image
14   //          row - a vertical coordinate of a reference
15   //                 pixel in the left image
16   //          disp - a disparity value between two images
17   //          kColSwing - a horizontal swing from the
18   //                 col value; the search window is set
19   //                 around a reference point +/- kColSwing
20   //          kRowSwing - a vertical swing from the
21   //                 row value; the search window is set
22   //                 around a reference point +/- kRowSwing
23   //          match_value - a reference to the variable
24   //                 that upon successful return contains
25   //                 the computed match value between two
26   //                 matched areas
27   //
28   //
29   // OUTPUT:
30   //                       true - if success
31   //                       false - otherwise
32   //
33   bool ComputeAreaMatch(  const MonochromeImage & leftImage,
34                           const MonochromeImage & rightImage,
35                           const int col,
36                           const int row,
27                           const int disp,
38                           const int kColSwing,
39                           const int kRowSwing,
40                           long & match_value )
41   {
42     const int kTotalCols = leftImage.GetCol();
43     const int kTotalRows = leftImage.GetRow();
44
45     // Find border values for column index
46     int col_from = col - kColSwing;       // left
47     if( col_from < 0 )
48           col_from = 0;
49     int col_to = col + kColSwing; // right
50     if( col_to > kTotalCols )
51            col_to = kTotalCols;
52
53     // Find border values for row index
```

Algorithm 6.6 Listing of the *ComputeAreaMatch* function which returns a match measure for a single pair of the matched areas

```
54      int row_from = row - kRowSwing;        // top
55      if( row_from < 0 )
56              row_from = 0;
57      int row_to = row + kRowSwing; // bottom
58      if( row_to > kTotalRows )
59              row_to = kTotalRows;
60
61      match_value = 0.0;      // starting match value
62
63      register int r, c;
64      for( c = col_from; c < col_to; ++ c )
65      {
66              int right_c = c - disp;         // find column index
67              if( right_c < 0 )               // for the right image
68                      return false;           // exit - cannot compute this match
69
70              for( r = row_from; r < row_to; ++ r )
71              {
72                      // update match value for each pair of pixels
73                      match_value += Compute_SAD( leftImage.GetPixel( c, r ),
74                                                              rightImage.GetPixel(
                                                                        right_c, r ) );
75              }
76      }
77
78      return true;    // successful computation of a match value
79  }
```

Algorithm 6.6 (*Continued*)

through all the pixels in both matching areas. In *L[73]* the function *Compute_SAD* is called to compute a matching measure for a pair of pixels.

It is important to notice a scan order in the matched (right) image. A starting horizontal coordinate for that image is taken to be the same as the current horizontal coordinate in the reference image (left, in this case). Then, at each iteration step, from this value the current disparity value is subtracted, *L[66]*. This way, the horizontal coordinate in the matched (right) image progresses to the left from the corresponding horizontal position in the reference image. This is in accordance with the relative coordinate order for the *canonical* stereo setups, already discussed in section 3.4.2. If the order of images is changed, as for instance for the cross-checking procedure, the order of progressing the horizontal coordinate in the matched images also has to be changed. In our example in Algorithm 6.6, this can be achieved quite easily substituting subtraction for addition in *L[66]*.

The *Compute_SAD* in Algorithm 6.7 computes an absolute value of a difference of pixels. To change to another matching measure, for instance one of those presented in Table 6.1, it suffices to change only this call. However, some measures require computation of the mean value in the matched areas, which imposes additional iteration through the matched areas.

Although complete, Algorithms 6.5–6.7 present only an instructive implementation. They assume certain type of images and have a minimum number of controlling parameters. Also, they use only one comparison measure which is the sum of absolute differences D_{SAD}. The more advanced implementation would be realized with help of template classes which would allow different pixel types. It would also have overloaded methods for comparison measures which could be easily exchanged. This is left as an exercise.

```
1   ////////////////////////////////////////////////////////////
2   // This function computes an absolute value of a difference
3   // of two pixels.
4   ////////////////////////////////////////////////////////////
5   //
6   // INPUT:
7   //                      pixel_a - the first pixel
8   //                      pixel_b - the second pixel
9   // OUTPUT:
10  //                      | pixel_a - pixel_b |
11  //
12  long Compute_SAD(       const unsigned char pixel_a,
13                          const unsigned char pixel_b )
14  {
15    return abs( pixel_a - pixel_b );
16  }
```

Algorithm 6.7 Function for computation of the match value for a single pair of two pixels. It returns an absolute value of their difference

6.6.4 Disparity-oriented Implementation

Disparity-oriented implementation follows the flow chart presented in Algorithm 6.8. Its main difference from the point-oriented version, aside from the different loop organization, comes from the creation of the new data structure – a disparity space, depicted in Figure 6.24. Thus, this version of the algorithm necessitates much more memory than the point-oriented version. However, all match values for each compared pair of pixels are available when computing the disparity map. While the former can cause some problems for larger resolutions and expected disparity values, the latter feature of the algorithm opens qualitatively new possibilities. The main advantage is that the whole space can be used to infer disparity values based on the partial correlations among single pixels. Thus, the method can be used to perform a global search for a solution to the stereo problem.

One method of distinguishing the order of computations in the dense disparity maps (usually direct methods) is to analyse the way pixels are traversed on the two images to find a disparity map.

Figure 6.25 depicts the logical dependencies among functions and data structures for the presented method. The main function *ComputeDisparity_Global* (Algorithm 6.9) has the

```
1       for(int d = 0; d < max_expected_disparity; d ++)
2   {
3           for(int i = 0; i < max_image_rows; i ++)
4       {
5               for(int j = 0; j < max_image_cols; j ++)
6               {
7                   intermediate_value(i, j, d) =
8                                   comp_measure(I_L(i,j), I_P(i,j,d));
9               }
10          }
11
12      }
13
14      „find disparity based on the accumulated comp_measure (and other parameters, such as
        image contents, etc.)";
```

Algorithm 6.8 Structure of the disparity-oriented method

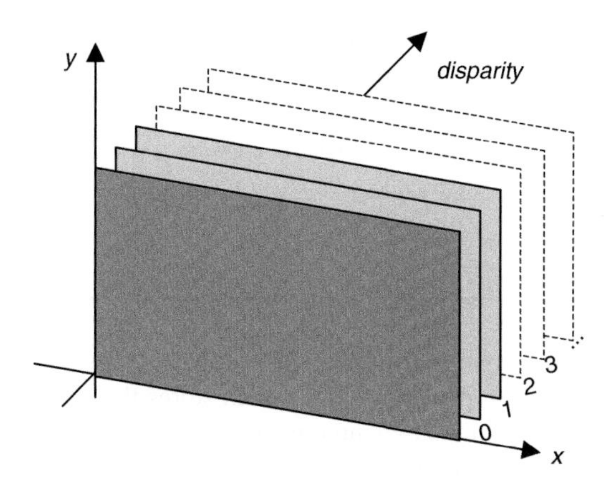

Figure 6.24 Disparity space built by disparity-oriented algorithms

same set of input parameters as the *ComputeDisparity_Local* method for the point-oriented algorithm.

In line *L[43]* of Algorithm 6.9 an array is created that stores pointers to each disparity plane in the disparity space. Each plane corresponds to a single disparity value. In this implementation we rely on the *vector<>* class which belongs to the STL library [231, 401]. However, a dynamically created array, with the help of the C++ *new* operator, can be used

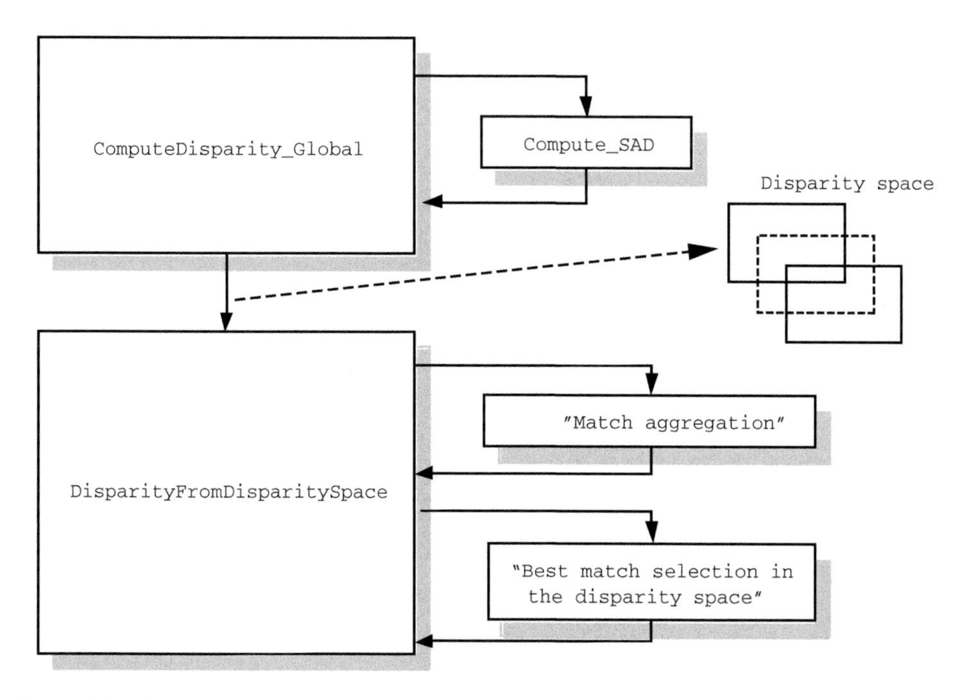

Figure 6.25 Logical dependencies among functions and data structures for disparity-oriented area matching

instead if STL is not available or not preferable. For instance this can be accomplished with the following code:

```
// Dynamically create an array of max_disp elements of
// type "LongImage *":
LongImage * * disparitySpace = new LongImage * [ max_disp ];

// ...

delete [] disparitySpace;          // delete the disparity array
```

The main disparity loop is organized in lines *L[46–69]*. The first action in each iteration consists of the creation of a new disparity plane which stores the chosen cost measures for each pair of compared pixels from the two input images.

The two loops in *L[53–68]* iterate trough all pixels in the input images and, for each pair of pixels from the two images, a match value *pixel_match_value* is computed in *L[62]*. Once again, this is done by invoking the *Compute_SAD* function (see Algorithm 6.7). By changing this function we can use different cost measures for single pixels. As alluded to previously in the case of the point-oriented implementation, the horizontal match position in the matched image is set at each iteration by subtracting a disparity value from the current horizontal position in the reference image – see *L[55]* in Algorithm 6.9. This has to be reversed (by changing subtraction into addition) if the order of input images is changed. It has to be remembered also that such a simple search works fine but only in the canonical stereo setups (section 3.4.2).

After the disparity space is completed, in *L[73]* of Algorithm 6.9 the *DisparityFromDisparitySpace* function is invoked. Its purpose is further processing of the disparity space for computation of the output disparity map (see Algorithm 6.10). *DisparityFromDisparitySpace* has two stages of computations: the first is match value aggregation; the second is finding the disparity map based on the best matches in the aggregated space.

```
1   ////////////////////////////////////////////////////////////
2   // This function computes a disparity map from the two
3   // monochrome input images. The disparity-oriented
4   // algorithm is used. It is assumed that the input pair
5   // is already rectified, so only horizontal scanlines
6   // are searched for the matches.
7   ////////////////////////////////////////////////////////////
8   //
9   // INPUT:
10  //         leftImage - a reference to the left image of a
11  //                 stereo-pair
12  //         rightImage - a reference to the right image of a
13  //                 stereo-pair
14  //         disparityMap - a reference to an image that
15  //                 upon return contains the disparity map;
16  //                 This image should be of the same size
17  //                 as are both images of a stereo-pair
18  //         match_area_cols - a horizontal size of the
19  //                 matching window
20  //         match_area_rows - a vertical size of the
21  //                 matching window
22  //         max_disp - expected maximum horizontal
23  //                 disparity value
```

Algorithm 6.9 Listing of the *ComputeDisparity_Global* function for a disparity-oriented computation of the disparity map

```
24    //
25    // OUTPUT:
26    //                          none
27    //
28    // REMARKS:
29    //
30    //
31    void ComputeDisparity_Global(  const MonochromeImage & leftImage,
32                                   const MonochromeImage & rightImage,
33                                   MonochromeImage & disparityMap,
34                                   const int match_area_cols,
35                                   const int match_area_rows,
36                                   const int max_disp )
37    {
38      const int kTotalCols = leftImage.GetCol();
39      const int kTotalRows = leftImage.GetRow();
40
41      register int col, row, disp;
42
43      vector< LongImage * > disparitySpace( max_disp );
44
45      // Traverse each possible disparity
46      for( disp = 0; disp < max_disp; ++ disp )
47      {
48              // Create a single disparity "plane" for disp disparity
49              disparitySpace[ disp ] =
50                                   new LongImage( kTotalCols, kTotalRows, LONG_MAX );
51
52              // Traverse each column
53              for( col = 0; col < kTotalCols; ++ col )
54              {
55                 int right_col = col - disp;
56                 if( right_col < 0 )
57                       continue;
58
59                      // Traverse each row
60                      for( row = 0; row < kTotalRows; ++ row )
61                      {
62                              long pixel_match_value = Compute_SAD(
63                                              leftImage.GetPixel( col, row ),
64                                              rightImage.GetPixel( right_col, row ) );
65                              disparitySpace[ disp ]->SetPixel( col, row,
66
                                        pixel_match_value );
67                      }
68              }
69      }
70
71      // At this point we have created the disparity space.
72      // In disparity space we can find a GLOBAL optimum
73      DisparityFromDisparitySpace(  disparityMap, disparitySpace,
74
        match_area_cols, match_area_rows );
75
76      // Finally, get rid of the disparity space
77     for( disp = 0; disp < max_disp; ++ disp )
78              delete disparitySpace[ disp ];
79    }
```

Algorithm 6.9 (*Continued*)

```
1    //////////////////////////////////////////////////////////////
2    // This function computes a disparity map from the disparity
3    // space.
4    //////////////////////////////////////////////////////////////
5    //
6    // INPUT:
7    //          disparityMap - a reference to an image that
8    //                  upon return contains the disparity map;
9    //          disparitySpace - a reference to the disparity
10   //                  space (a disparity vector); Number of
11   //                  elements in this vector is the same
12   //                  as the maximum horizontal disparity
13   //          match_area_cols - a horizontal size of the
14   //                  matching window
15   //          match_area_rows - a vertical size of the
16   //                  matching window
17   //
18   // OUTPUT:
19   //          none
20   //
21   void DisparityFromDisparitySpace(      MonochromeImage & disparityMap,
22                                          const vector< LongImage * > & disparitySpace,
23                                          const int match_area_cols,
24                                          const int match_area_rows )
25   {
26     const int kTotalCols = disparityMap.GetCol();
27     const int kTotalRows = disparityMap.GetRow();
28
29     register int col, row, disp;
30
31     const int max_disp = disparitySpace.size();
32
33     vector< long > theAggregMask_horz( match_area_cols, 1 );
34     vector< long > theAggregMask_vert( match_area_rows, 1 );
35
36     LongImage tmpImage( kTotalCols, kTotalRows );
37     // Smooth the disparity space
38     for( disp = 0; disp < max_disp; ++ disp )
39     {
40             Horz1DConvolve( * disparitySpace[ disp ],
41
       theAggregMask_horz, tmpImage );
42             Vert1DConvolve( tmpImage, theAggregMask_vert,
43                                                                   *
     disparitySpace[ disp ] );
44     }
45
46     // Traverse each row
47     for( row = 0; row < kTotalRows; ++ row )
48     {
49             // Traverse each column
50       for( col = 0; col < kTotalCols; ++ col )
51             {
52                     long prev_match = LONG_MAX;          // get a large value
53                     int best_disp = -1;
54
55                     // Traverse each possible disparity
56               for( disp = 0; disp < max_disp; ++ disp )
57                     {
58                             long area_match_value =
59                                     disparitySpace[ disp ]->GetPixel( col, row );
60                      if( area_match_value < prev_match )
61                             {
62                                     // save the best values
63                                     prev_match = area_match_value;
64                                     best_disp = disp;
65                             }
66                     }
67
68                     // Save found disparity
69                     disparityMap.SetPixel( col, row, best_disp );
70             }
71     }
72   }
```

Algorithm 6.10 Computation of the disparity map from the disparity space

In the implementation in Algorithm 6.10, the match aggregation in achieved simply by separated horizontal and vertical summations in the chosen window (section 6.5.1.1). For this purpose the two linear convolution methods *Horz1DConvolve* and *Vert1DConvolve* are employed. They use the linear masks with all values set to one. In effect, we simply sum up all elements in the *match_area_cols*×*match_area_rows* window. However, the size and values of this mask can be changed quite easily, for instance the Gaussian mask can be used instead, etc. We have to remember that in the presented solution integer arithmetic is utilized for computation speed. So, if a mask with fractional values were to be used, its elements should be scaled (multiplied) to take on integer values. This does not pose any problem since all local areas will be scaled in the same fashion which does not change their mutual relations. However, we should be aware of a limit set in the accumulation trait of the convolution operator (section 4.2.1).

The code in *L[47–71]* in Algorithm 6.10 contains three loops which once again traverse all pixel values in the disparity planes. In these iterations we select the best matches from the already aggregated values. This way we obtain the final disparity map.

In both implementations monochrome 8-bit input images were assumed. For the disparity planes images with integer values were chosen (*int* or *long* C++ built-in types). The computations are also done entirely on integer arithmetic. However, in other realizations this might need some modifications.

For speed improvement, the first place for inspection is the loops. Indeed, after profiling we notice that the random access methods *SetPixel* and *GetPixel*, which belong to the *TImageFor<>* class, require a significant number of arithmetic operations for computation of random positions of pixels. If a *sequential* access to *all* pixels in an image can be envisaged, then a much faster pointer-based implementation can be applied (section 3.7.1.2). The first candidates for such a change are the loops *L[53–68]* in the *ComputeDisparity_Global* function, as well as in *L[47–71]* of the *DisparityFromDisparitySpace* function.

To achieve subpixel accuracy of disparity values, discussed in section 6.4.2, it is sufficient to paste the code listed in Algorithm 6.11 into *L[67]* of Algorithm 6.10.

Computations of subpixel disparity values follow Equation (6.58). To stay in the integer domain, disparities are multiplied by a value of 100, in *L[3]* and *L[12]*. This is possible since disparities usually have a very limited dynamic range, so we can easily scale the values to

```
1    //////////////////////////////////////////////////////////////////////
2    // Sub-pixel estimation:
3    long sub_pix_disparity = 100 * best_disp;
4    if( best_disp > 0 && best_disp < max_disp - 1 )
5    {
6      long m_im1      = disparitySpace[ best_disp - 1 ]->GetPixel(col,row);
7      long m_i        = disparitySpace[ best_disp ]->GetPixel(col,row);
8      long m_ip1      = disparitySpace[ best_disp + 1 ]->GetPixel(col,row);
9
10     long denom      = m_im1 - m_i - m_i + m_ip1;
11     if( denom != 0.0 )
12                     sub_pix_disparity += ( 100 * ( m_im1 - m_ip1 ) ) /
13                                                      ( denom + denom );
14   }
15   //////////////////////////////////////////////////////////////////////
```

Algorithm 6.11 Disparity computation with subpixel accuracy

use integers and, at the same time, speed up computations compared to using floating point variables. Almost the same code can be used in point-oriented implementation as well.

6.6.5 Complexity of Area-based Matching

Analysing the point- and disparity-oriented algorithms for area-based matching we easily conclude that the computational complexity of these methods is of order

$$O(NMD_v D_h), \tag{6.92}$$

where the input images are of size $N \times M$ pixels and D_v and D_h are expected vertical and horizontal disparities. In the worst case $D_v \to N$ and $D_h \to M$. However, for the stereo setups the matching problem can be constrained to a 1D search, in which case $D_v = 1$ (the epipolar constraint; section 3.5). Further, in the canonical stereo setup, the positions of the epipolar lines coincide with the image scanlines which simplifies greatly the whole procedure. Although the computational complexity of area-based matching is given by (6.92), different versions of the algorithm will differ significantly in their execution time. This is caused mainly by two factors:

1. Organization of the iterations.
2. Additional time for match computations at each position of the output disparity map.

Regarding the organization of the iterations, we have already presented and discussed the point- or disparity-oriented realizations. However, their implications are much deeper than a mere setting of an order of iterations. In the point-based version we select the best match locally, i.e. for the current pixel position in the reference image. Thus, a match is computed at each step of the algorithm and then stored in the output disparity map.

In the disparity-oriented method, for each possible disparity value a separate data structure – a disparity plane – is created that stores the match value for a single pair of pixels, for each pixel position in the input images. Only after the whole disparity space, which consists of a number of disparity planes, is constructed is the output disparity map computed. Obviously this requires a significant amount of additional memory for storing of the intermediate results. However, having all values of pixel differences (or other single pair matching measures) creates new qualitative possibilities for best match selection, since the whole information is available at this stage. Interestingly, the memory complexity is of the same order as expressed in (6.92).

Moreover, in the disparity-oriented method we gain significantly in the execution time in the match aggregation stage (*L[39–44]* in Algorithm 6.10) if using a separable mask for summations (section 4.2.2). For the point-oriented version it is not easy to directly implement this technique at each step of an area correlation. Instead, some time improvement can be achieved when trying to reuse some of the already computed matching values or applying a smart match selection strategy (see section 6.3.6).

The additional time for match computation at each position depends mainly on the chosen matching measure and data dimensionality. These issues are discussed in section 6.3. However, some measures can be significantly slower, especially when they require prior computation of a mean value in each matched area or necessitate many multiplications, e.g. D_{CV} in (6.7). Some improvement can be achieved by means of the speed improving techniques discussed in section 6.3.6. However, this is at a cost of implementation complexity, which can be an issue for hardware realizations, etc.

Finally, there can also be some additional time necessary (which usually depends on a given image size) for an optional input pair transformation (stage (2) in Figure 6.21) and/or output disparity map postprocessing, such as cross-validation, filtering, etc.

6.6.6 Disparity Map Cross-checking

The cross-checking process has already been discussed in section 6.4.1. Algorithm 6.12 lists the complete function for computation of a cross-checked disparity map, based on the supplied left–right and right–left disparity maps. These are two versions of disparity maps, each with a different reference image, however.

```
1  ///////////////////////////////////////////////////////////////
2  // This function does cross checking of disparity maps.
3  ///////////////////////////////////////////////////////////////
4  //
5  // INPUT:
6  //                    d1 - left-right disparity map
7  //                    d2 - right-left disparity map
8  //                    d_out - outcome disparity map
9  //                    disparitySimilarityThresh - threshold value
10 //                            of allowable dissimilarity between
11 //                            the pairs of disparities
12 //                    kRejectPtMarker - in the rejected_points image
13 //                            this value is used to mark a rejected point
14 //                            (all other points are set to 0)
15 //                    rejected_points - map of rejected points
16 //
17 // OUTPUT:
18 //                    Number of rejected points
19 //
20 // REMARKS:
21 //
22 //
23 int DisparityMapCrossChecking(    const MonochromeImage & d1,
24                                   const MonochromeImage & d2,
25                                         MonochromeImage & d_out,
26                                   const int kDisparitySimilarityThresh = 0,
27                                   const int kRejectPtMarker = 1,
28                                         MonochromeImage * rejected_points = 0 )
30 {
30     int pt_counter = 0;
31
32     const int kCols = d1.GetCol();
33     const int kRows = d1.GetRow();
34
35     // Initialize d_out
36     d_out.SetAll( 0 );
```

Algorithm 6.12 Listing of the *DisparityMapCrossChecking* function for cross-checking of disparity maps

```
37
38     // Prepare rejected points map
39     if( rejected_points != 0 )
40             rejected_points->SetAll( 0 );
41
42     register int i, j;
43
44     unsigned char left_disp, right_disp;
45
46     for( i = 0; i < kRows; i ++ )
47     {
48             for( j = 0; j < kCols; j ++ )
49             {
50                     left_disp = d1.GetPixel( j, i );
51
52                     if( j + left_disp >= kCols || j + left_disp < 0 )
53                             continue;
54
55                     right_disp = d2.GetPixel( j + left_disp, i );
56
57                     if( abs(left_disp-right_disp)<=kDisparitySimilarityThresh )
58                     {
59                             d_out.SetPixel( j, i, left_disp );
60                     }
61                     else
62                     {
63                             ++ pt_counter;
64
65                             if( rejected_points != 0 )
66                                     rejected_points->SetPixel( j, i, kRejectPtMarker );
67                     }
68
69             }
70
71     }
72
73     return pt_counter;
74 }
```

Algorithm 6.12 (*Continued*)

The function *DisparityMapCrossChecking* accepts six parameters, from which three are obligatory. These are two disparity maps and one output map. The other optional parameters are a threshold for similarity of checked disparity values, as well as the image of rejected points and the value used to indicate such points in this image. The latter can be used for visual inspection of places in the matched images that have been rejected by the cross-checking process. The similarity threshold by default is set to 0. However, we can accept a disparity match if the values differ somehow, for example due to integer arithmetic. In such a case we set the threshold to a value greater than 0.

In *DisparityMapCrossChecking* we start with some organization statements and declarations. Then the two loops *L[46–71]* and *L[48–69]*, which traverse each pixel in the two input disparity maps, start. Then, in lines *L[50–55]*, the two disparity values are acquired, checking however the allowable range of indices. Observe that the index of the second disparity value in *L[55]* is computed from the first disparity value in *L[50]*. Finally, if the two disparities do not differ by more than a predefined threshold in *L[57]*, such disparity is set as valid. Otherwise, we indicate a position as rejected and increase the counter of the total number of rejected points.

Other details on cross-checking can be found in section 6.4.1 and Table 6.9.

6.6.7 Area-based Matching in Practice

First, let us analyse the areas of application of different versions of the area-based matching method.

Area-based matching is a very simple but still powerful matching technique that can be used to find correspondences among images whenever such comparison is necessary. This extends its application area not only to stereovision but also to multi-view matching, motion analysis and pattern detection.

Area-based matching can operate with a broad spectrum of input images with pixels which are transformed or not transformed, scalars, vectors, tensors, etc. The only interface that concerns the variety of pixel representations is the matching measure for a single pair of pixels. These were already discussed in section 6.3. So, if we can only compare values of pixels of a certain type, then we can easily use the area-based matching, overloading only the pixel comparison interface.

The situation becomes more complicated if information in the input signal (pixels) is used to control or modify in some way the behaviour of an algorithm. For instance, using tensor representation one can exclude from matching those areas which are characterized by small coherence value (section 4.6.2.1), i.e. which do not exhibit sufficient signal variations for reliable matching. Similarly, shape and size of the matching area can be adaptively adjusted to the image contents. In this way more powerful methods are created that can cope more easily with some problems inherent to the matching task.

Area-based matching produces a dense disparity map. However, its quality depends heavily on the contents of the input images and the chosen control parameters. Thus, area-based matching can constitute a prematching module of a more advanced matching scheme. This is a case of the hierarchical matching method, operating in scale-space (section 6.7.4). In this example, area-based matching is employed at each stage of the scale pyramid for an initial match. Then the disparity map found for a coarser level is refined at the next finer level, and so on until the final disparity map is built.

Regarding implementation issues, for software realizations the disparity-oriented approach can be recommended since it is much faster and provides global disparity space. However, for hardware implementation the point-oriented version seems a better solution since it does not require large memory blocks.

The following sections present area-based matching for different types of images as well as different settings of the method. All of the presented experimental results come from the software implementation, compiled with the Microsoft® Visual C++ 6.0 compiler. It was run on a PC computer with Intel® Pentium 4, operating with 3.4 GHz clock, 2 GB RAM and Windows XP Professional operating system.

6.6.7.1 Intensity Matching

Figure 6.26 presents results of execution of the two *ComputeDisparity_Local* and *ComputeDisparity_Global* procedures for two monochrome 8-bit versions of the test stereo-pairs 'Tsukuba' and 'Venus' (Table 3.4). Pixel matching is achieved with the very simple D_{SAD} measure, listed in Algorithm 6.7. The matching areas are squares 3×3, 5×5 and 11×11, from top to bottom in Figure 6.26. For small matching areas we notice many false matches, since no cross-checking was applied in this simple example. On the other hand, larger matching areas exhibit much smearing of disparity.

Figures 6.27 and 6.28 present execution times of the two matching procedures in relation to the size of the matching area. It is very interesting to observe how much faster is the disparity-oriented approach, implemented by the *ComputeDisparity_Global* function. For size of the matching window exceeding 11×11 it outperforms the simple point-oriented implementation by an order of magnitude!

Figure 6.29 presents a comparison of the disparity maps with and without subpixel accuracy. We easily notice that the subpixel computations allow much smoother maps since their disparity values are more adjusted to the matching measure.

Figures 6.30 and 6.31 present computation of the cross-checked disparity maps for the 'Tsukuba' and 'Sawtooth' images, respectively. First, the left–right and right–left disparity maps need to be computed (Figures 6.30(a, b) and 6.31(a, b)). Then the cross-checking is done which results in a cross-checked disparity map (Figures 6.30(c) and 6.31(c)) as well as a number of rejected points (white in Figures 6.30(d) and 6.31(d)).

We can observe in the cross-checked disparity maps in Figures 6.30(c) and 6.31(c) that most of the mismatched disparities occur at object boundaries.

6.6.7.2 Area-based Matching in Nonparametric Image Space

The matching examples presented in this section assume that the input images, prior to matching, are transformed from intensity signals into nonparametric space. Such organization follows the scheme depicted in Figure 6.21. Properties of the nonparametric transformation are discussed in section 6.3.7.

Figure 6.32 depicts results of the 'Tsukuba' image matching in the nonparametric 5×5 *Census* domain. Matching windows were 9×9 and 11×11 pixels, respectively. For comparison the Hamming D_{H} measure was used.

Figure 6.33 provides experimental results of the nonparametric matching of the 'Venus' stereo-pair in the 9×9 *Census* space. For matching the Tanimoto D_{T} measure was used. The disparity maps were cross-checked for validation. The disparity maps are obtained with 9×9 (Figure 6.33(a)) and 11×11 (Figure 6.33(c)) match areas. The ratio of cross-check rejected points is about 5.5% in Figure 6.33(b) and 5.2% in Figure 6.33(d).

Table 6.11 contains an assessment of accuracy and computation time for different nonparametric methods, test images and match settings; $n \times n$C stands for the size of the *Census* window W in (6.47), $k \times k$M denotes the size of the matching block; time is in seconds. Presented values concern the nonredundant Census coding, discussed in section 6.3.7.1.

Figure 6.26 Examples of area-based stereovision accomplished with *ComputeDisparity_Local* and *ComputeDisparity_Global*. Matching areas are squares 3 × 3, 5 × 5 and 11 × 11 (from top down). Pixel matching with the D_{SAD} measure; no cross-checking

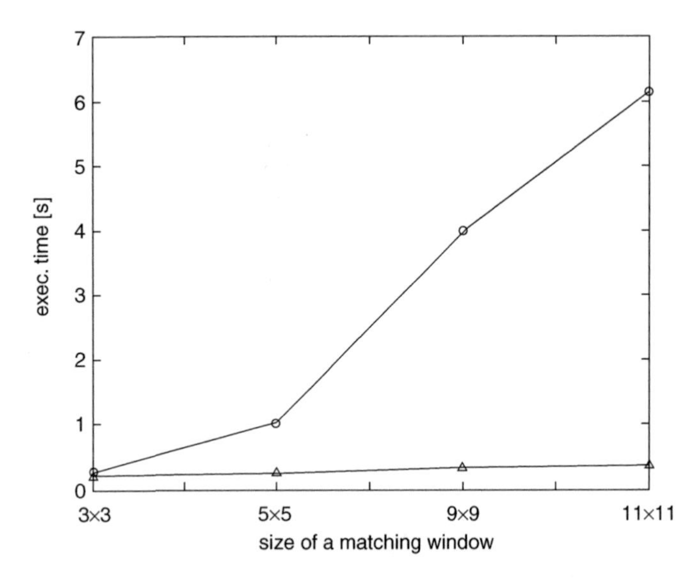

Figure 6.27 Execution times of the *ComputeDisparity_Local* (circles) versus *ComputeDisparity_Global* (triangles) for the 'Tsukuba' test pair

6.6.7.3 Area-based Matching with the Structural Tensor

As alluded to previously, the structural tensor provides valuable information on structure of local regions in the input image (section 4.6). However, in many applications it is desirable to use intensity *and* structural tensor together. For instance, Luis-García *et al.* propose such an extension for image segmentation [285]. Their idea consists of creating mixed products of

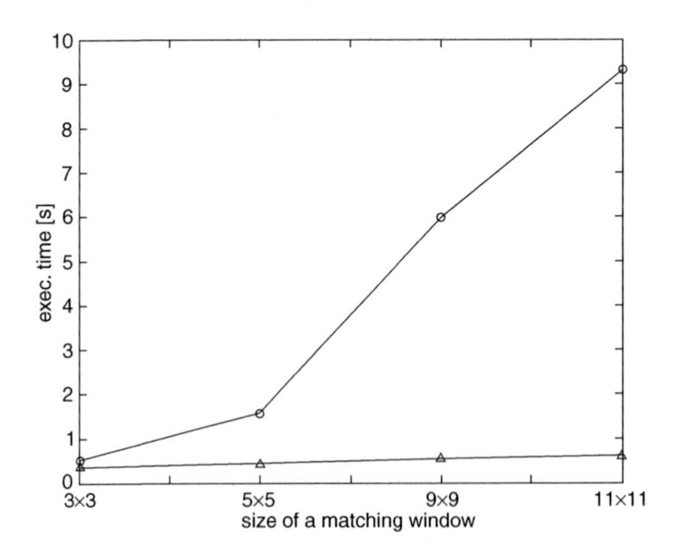

Figure 6.28 Execution times of the *ComputeDisparity_Local* (circles) versus *ComputeDisparity_Global* (triangles) for the 'Venus' test pair

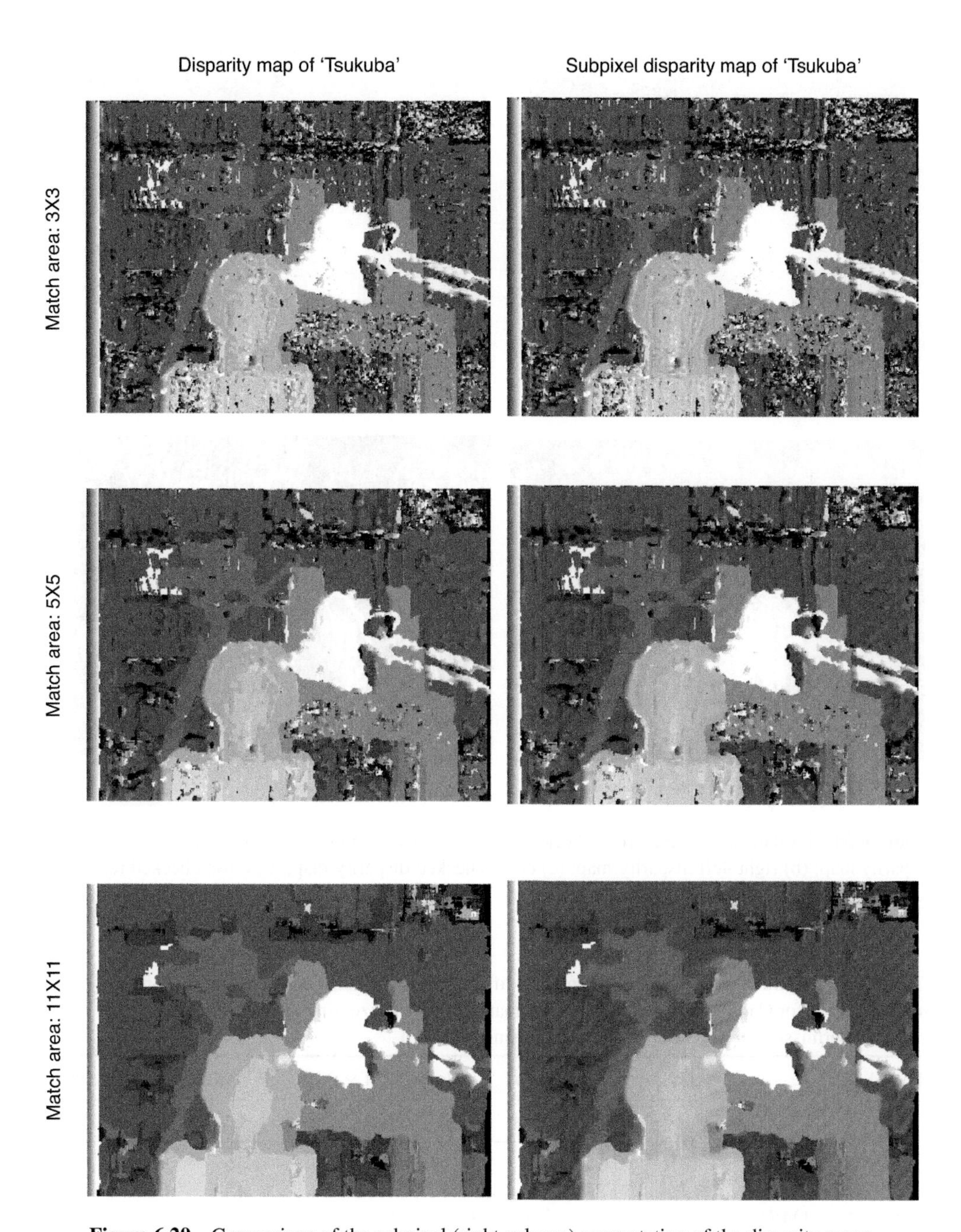

Figure 6.29 Comparison of the subpixel (right column) computation of the disparity maps

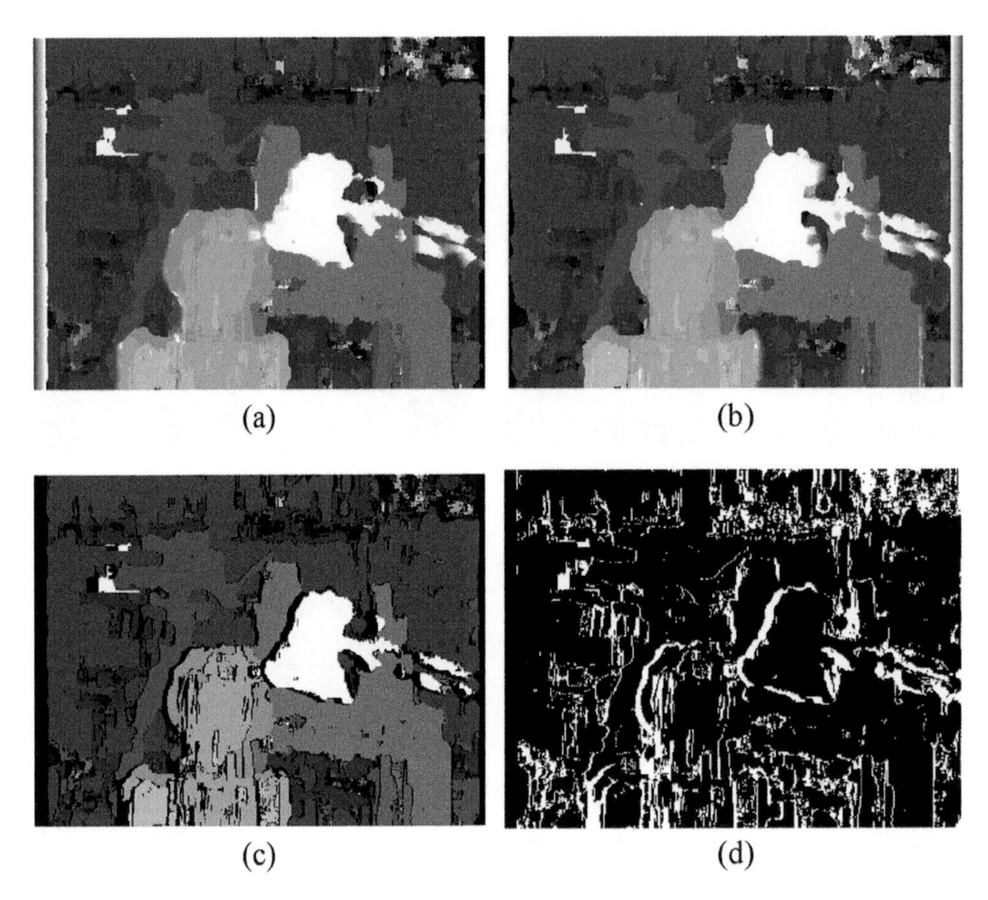

Figure 6.30 Explanation of the cross-checking process for the 'Tsukuba' stereo-pair: (a) left–right disparity map; (b) right–left disparity map; (c) cross-checked disparity map; (d) cross-checked rejected points

Table 6.11 Accuracy and computation time for different nonparametric methods, images and settings ($n \times n$C stands for Census size; $k \times k$M denotes matching block size; time in seconds). (From [90] with kind permission of Springer Science and Business Media)

Method	Tsukuba (384 × 288)			Venus (434 × 383)			Sawtooth (434 × 380)		
	Gr-T	Mis	Tme	Gr-T	Mis	Tme	Gr-T	Mis	Tme
5×5C, 5×5M D_H	19.4	0.21	0.7	18.2	0.21	1.1	16	0.18	1
5×5C, 7×7M D_H	17	0.17	1.2	17.4	0.13	1.8	15.8	0.19	1.8
5×5C, 11×11M D_H	13.3	0.12	2.4	16.9	0.2	3.8	13.1	0.1	3.8
11×11C, 3×3M D_{WT}	11.4	0.08	3.8	11.4	0.116	4.1	9.14	0.07	4
11×11C, 9×9M D_{WT}	10.2	0.05	4.7	10.2	0.055	6.6	7.2	0.06	6.2
11×11C, 13×13 D_{WT}	13.7	0.06	6.4	13.7	0.05	9.1	6.23	0.06	9.7

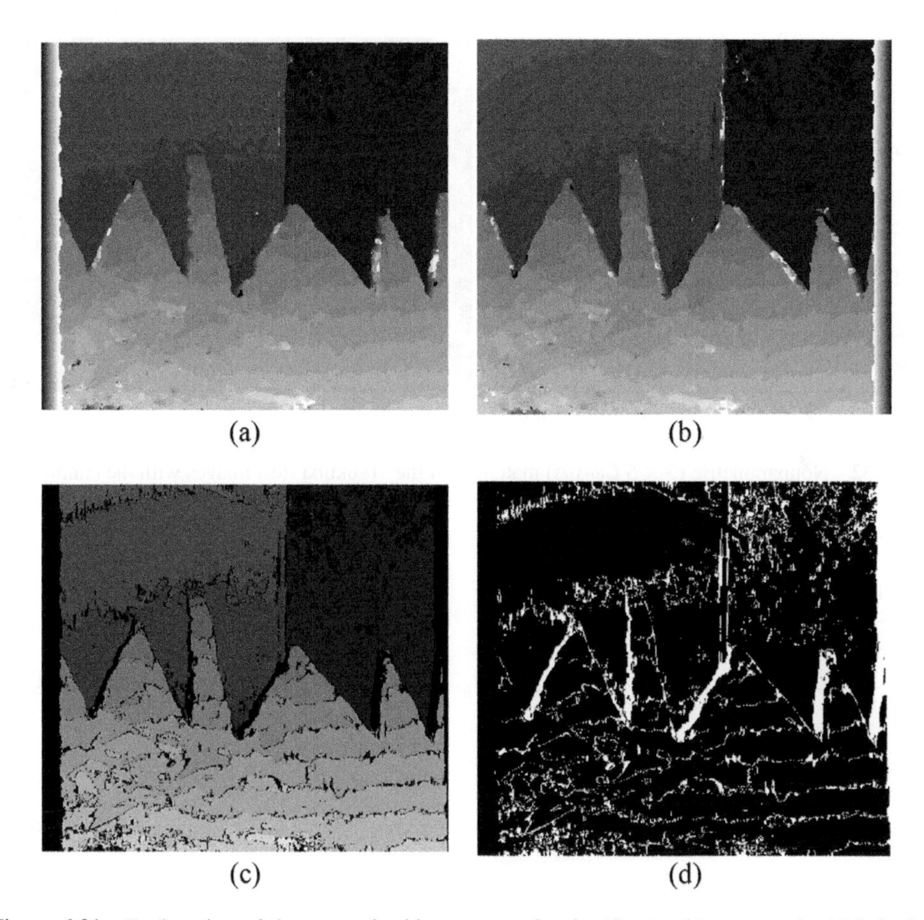

Figure 6.31 Explanation of the cross-checking process for the 'Sawtooth' stereo-pair: (a) left–right disparity map; (b) right–left disparity map; (c) cross-checked disparity map; (d) cross-checked rejected points

gradients and intensity signal. This way *the nonlinear extended structural tensor* is obtained, as follows:

$$\mathbf{T}^* = \begin{bmatrix} T_{xx} & T_{xy} & T_{xz} \\ T_{yx} & T_{yy} & T_{yz} \\ T_{zx} & T_{zy} & T_{zz} \end{bmatrix}, \tag{6.93}$$

$$\mathbf{T}^* = A\left(\begin{bmatrix} I_x \\ I_y \\ I \end{bmatrix}\begin{bmatrix} I_x & I_y & I \end{bmatrix}\right) = A\left(\mathbf{U}_s\mathbf{U}_s^T\right), \tag{6.94}$$

where $A(\ )$ denotes an averaging operator (section 4.3) and I_x and I_y denote x and y directional derivatives of scalar intensity signal I (section 4.3). For colour images, $\mathbf{U}_s$ in the above is

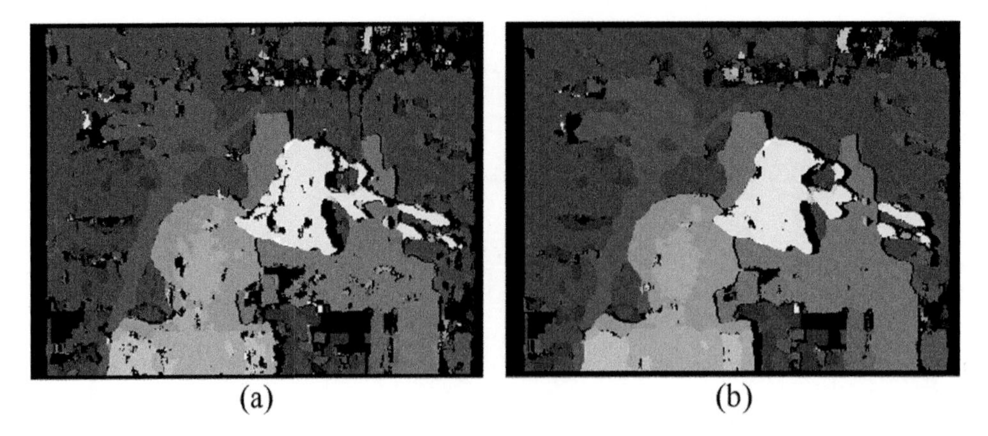

Figure 6.32 Nonparametric (5 × 5 *Census*) matching of the 'Tsukuba' test images with the Hamming D_{H} measure. Matching areas are (a) 9 × 9 and (b) 11 × 11. (From [90] with kind permission of Springer Science and Business Media)

further extended to $\mathbf{U}_c$, as follows:

$$\mathbf{U}_s = \left[I'_x \; I'_y \; I_R \; I_G \; I_B \right]^T , \tag{6.95}$$

where

$$I' = \frac{1}{3} \left(I_R + I_G + I_B \right). \tag{6.96}$$

A similar strategy was also proposed for stereo correlation [83, 85], the basic assumptions of which we describe in this section.

The idea here is very simple: instead of matching bare intensity signals, the input images are transformed into their structural tensor representation. It conveys much more intuitive information than the intensity signal alone. For instance, we have a direct knowledge of the type of structure in each of the local neighbourhoods of pixels. We can also quite easily find corners or straight lines in an image (section 4.6). This way we can tell textureless areas, which are not easy for matching, from the ones with well-developed structures. It was observed that an analysis of the coherence signal can help partitioning an image into areas which can be quite reliably matched with comparatively small matching windows. When observing histograms of some real test images, depicted in Figure 6.34, we see that we can define a certain level of 'structure' in a local neighbourhood which can lead to a reliable match. The other areas have to be treated with relatively larger matching windows. By this method false matches can be avoided.

Figure 6.35 depicts an architecture of the image correlation method that operates with the structural tensor representation computed from the input images. This is a version of a simple area-based matching (section 6.6), guided however by the coherence component of the structural tensor. Matching is also done for the augmented tensor signal, i.e. 4D data. However, instead of (6.93), the equivalent but more intuitive representation (4.130), augmented with the

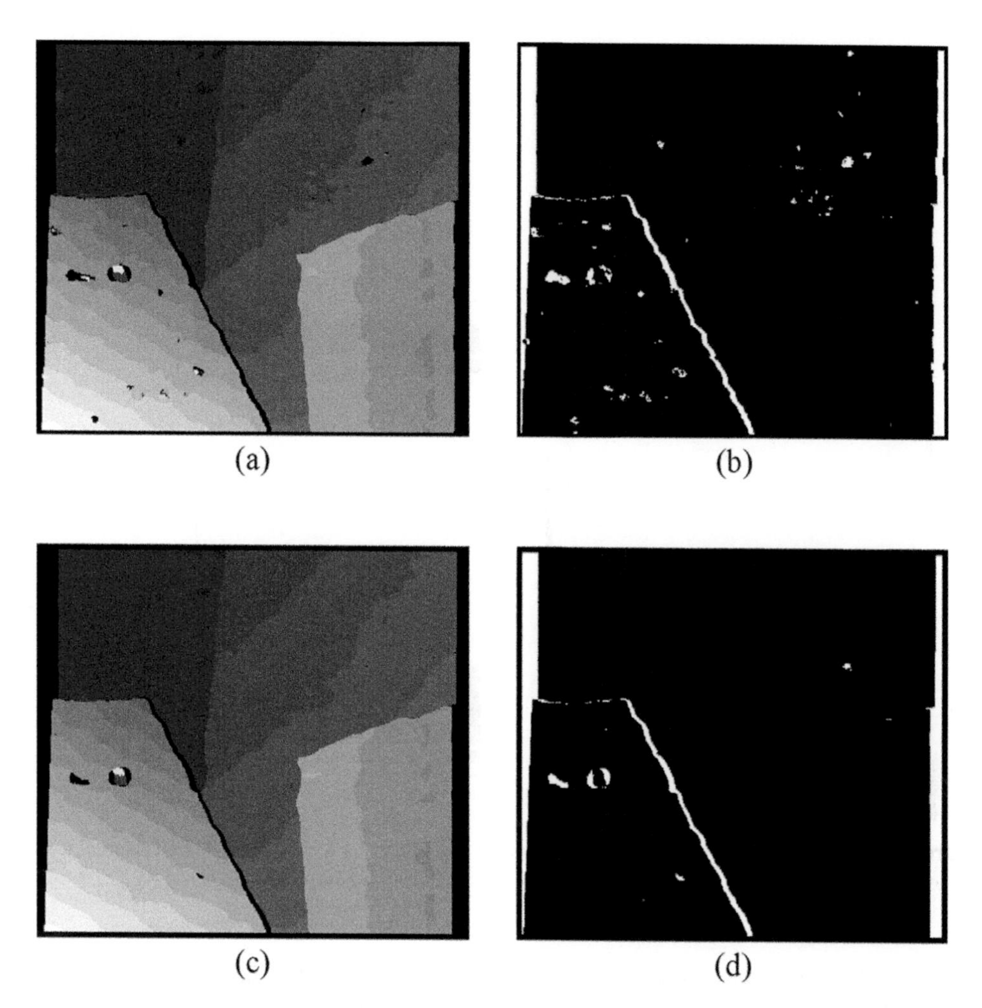

Figure 6.33 Nonparametric matching of the 'Venus' stereo-pair in the nonparametric 9×9 Census representation. For matching the Tanimoto D_{T} comparison measure is used and cross-checking for map validation. The disparity maps are obtained with (a) 9×9 and (c) 11×11 match areas. The cross-checking rejected points are (b) 5.5% and (d) 5.2%. (From [90] with kind permission of Springer Science and Business Media)

intensity signal, is used. It is given as

$$\hat{\mathbf{s}} = \begin{bmatrix} \hat{s}_1 \\ \hat{s}_2 \\ \hat{s}_3 \\ \hat{s}_4 \end{bmatrix} = \begin{bmatrix} T_{xx} + T_{yy} \\ \angle \mathbf{w} \\ c \\ I \end{bmatrix}, \tag{6.97}$$

where the meaning of the components is explained in section 4.5.4.

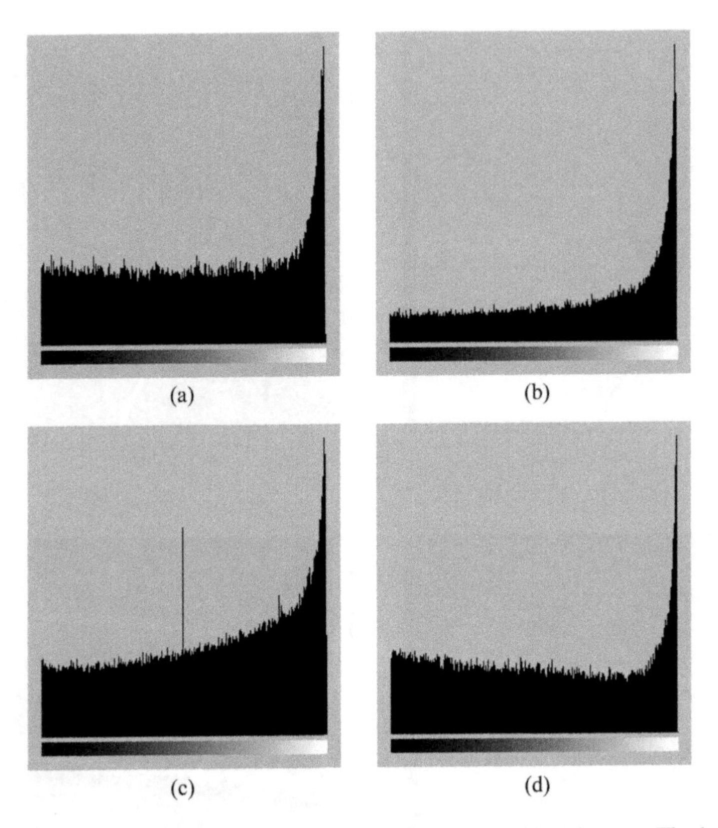

Figure 6.34 Histograms of the coherence component for some real test images. The horizontal axis denotes the coherence value; the vertical axis denotes frequency of occurrence

The matching process is twofold (module 4 in Figure 6.35). The first stage consists of image partitioning into regions with sufficient structure for reliable area-based matching. This is done by analysis of the coherence component c in (6.97) alone. The second stage is an area-based matching that uses all components of the augmented tensor $\hat{\mathbf{s}}$ given in (6.97). The main assumption here is that $\hat{\mathbf{s}}$ has more discriminative power than bare intensity.

The other stages of computation are analogous to the ones already presented for the area-based matching (section 6.6). Specifically, the two disparity maps are computed, one for each input image held as a reference, and used then for cross-checking (section 6.4.1). The final disparity map is filtered with the morphological filter to remove outliers. Since the method at one run provides only disparity values at structural places, then for a fully dense disparity map an additional run with much bigger matching window is necessary. Then the two disparity maps have to be merged. An alternative here is interpolation of missing disparity values based on existing ones. The linear interpolation method is described in Chapter 12. Other methods, such as bicubic (section 12.4) or spline based can be found in the literature (e.g. [352, 449]).

As alluded to previously, the method starts with image partitioning into regions with strong structure, which is given by the comparatively high coherence parameter c of $\hat{\mathbf{s}}$. Based on experiments it was found that binarization can be achieved with c thresholded around its

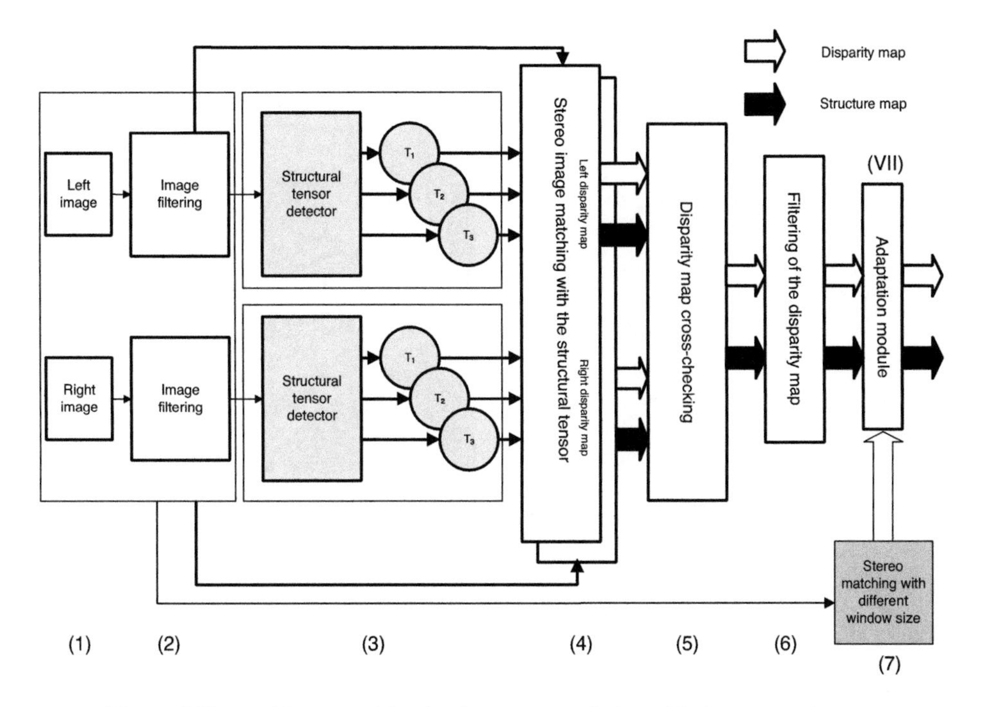

Figure 6.35 Architecture of the simple stereo correlation with the structural tensor

median value, as follows:

$$c_1(I) = B[c(I)], \tag{6.98}$$

where B denotes a thresholding operator around a median value of its argument. Next, $c_1(I)$ is morphologically dilated (section 4.8.2) to extract more concise structural areas. Once again we have to choose a structural element for dilation which is somehow arbitrary and can be based on experiments [351]. In many cases a simple square 3×3 or 5×5 for the structural element was a reasonable tradeoff. This way we obtain our map of structural places

$$c_2(I) = D_S[c_1(I)], \tag{6.99}$$

where D_S denotes a morphological dilation operator with the structural element S. Good results were also obtained when D_S was substituted by the median filter. Examples of this stage are presented in Figures 6.36 and 6.37 with some test images from Table 3.5.

The mask of structural places can be computed in the reference image only, or in the two images, and then joined together with the binary AND operation.

Now we are ready to design the evidence measure E for pixel matching, which will operate in the space of the structural tensor. It is proposed by the following formula:

$$E = \left[\sum_{i=1}^{4} w_i \left\| \hat{s}_i^{(L)} - \hat{s}_i^{(R)} \right\|^{\alpha} \right]^{1/\alpha}, \tag{6.100}$$

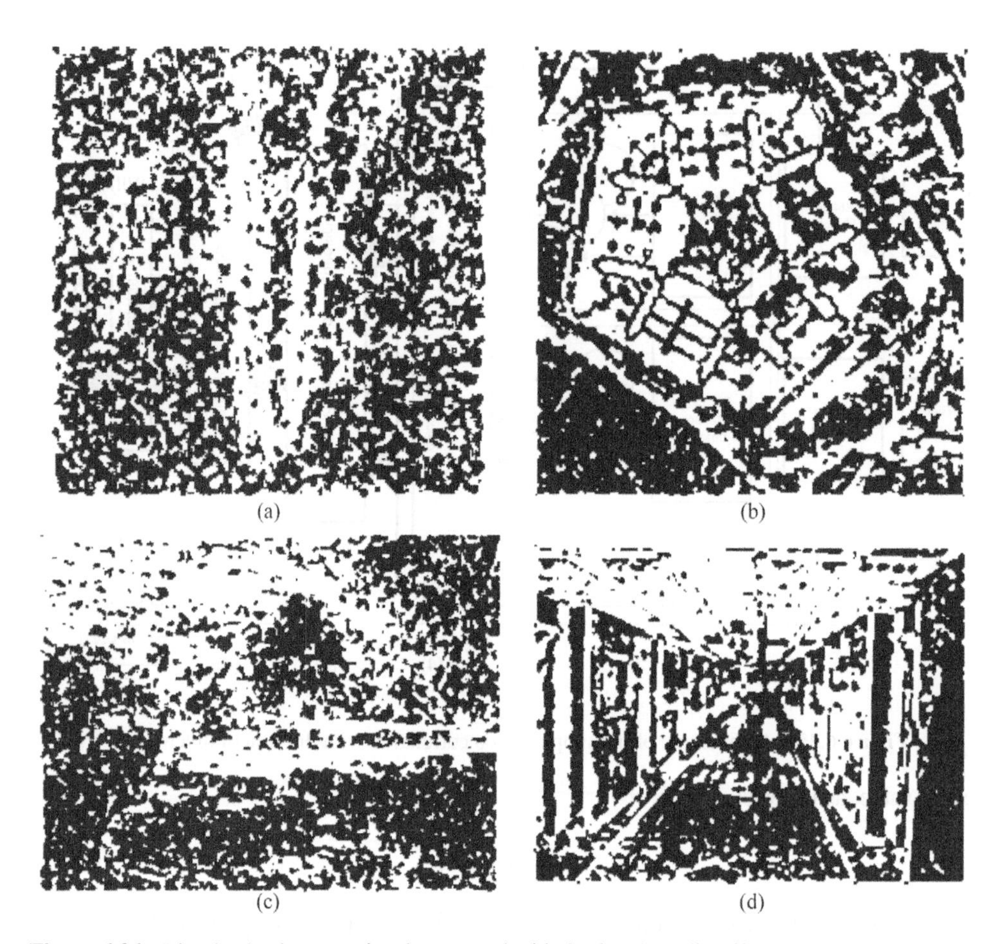

Figure 6.36 Binarized coherence signal processed with the 3 × 3 median filter: (a) 'Trees', (b) 'Pentagon', (c) 'Park', (d) 'AGH Corridor'

which is based on the Minkowsky measure (6.18) computed over the tensor representation $\hat{\mathbf{s}}$ (6.97); w_i are weights which control the influence of the tensor components on the evidence measure. Finally, the parameter α determines properties of the computed metric: for $\alpha = 1$ this is a SAD, for $\alpha = 2$ SSD, and so on, as discussed in section 6.3.3. For instance, in a particular case of $w_1 = w_2 = w_3 = 0$ and $w_4 = 1$, the above reduces to a simple scalar matching over the intensity values. Similarly, matching based on the pure phases of the local structures can be achieved setting $w_1 = 1$ with all other w_i simultaneously set to zero.

Assuming the canonical stereo setup, E at position (x, y) and disparity d can be written as follows:

$$E(x, y, d) = \left[\sum_{i=1}^{4} w_i \left\| \hat{s}_i^{(L)}(x, y) - \hat{s}_i^{(R)}(x + d, y) \right\|^{\alpha} \right]^{1/\alpha}, \qquad (6.101)$$

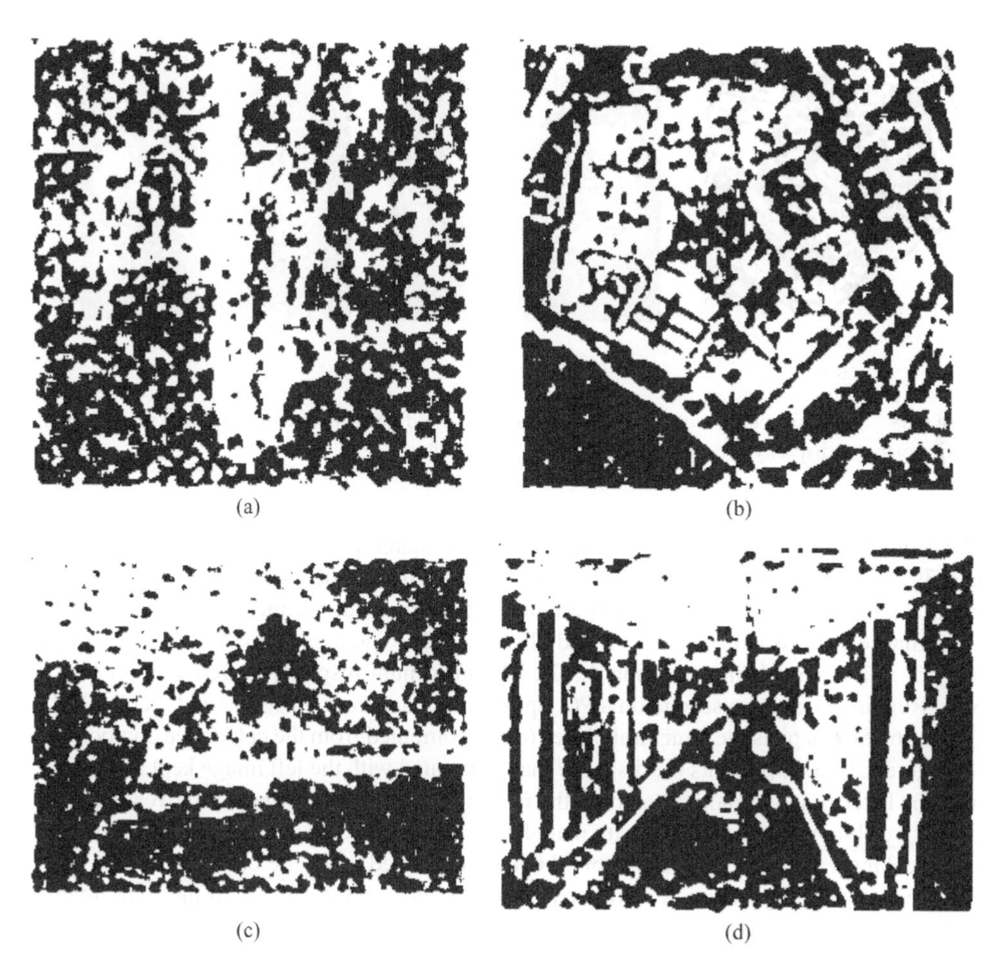

Figure 6.37 Binarized coherence signal processed with the 5 × 5 median filter: (a) 'Trees', (b) 'Pentagon', (c) 'Park', (d) 'AGH Corridor'

where $\hat{s}_i^{(L)}(x, y)$ denotes the i-th component of tensor $\hat{\mathbf{s}}$ at pixel position (x, y) in the left image L; a similar notation is used in respect of the right image.

In practice $\alpha = 1$ (SAD) or $\alpha = 2$ (SSD); the former requires less computational effort. Let us recall that the evidence measure E is computed only in the structural places, i.e. ones with $c_2(I) > 0$ (see (6.99)).

Computation of a disparity map can be done with the help of one of the already discussed methods. In the simplest case, the point-oriented algorithm (section 6.6.3) can be employed in which E is computed separately for each local window in the first window, then aggregated and compared with potentially corresponding windows in the second image. The best match determines a disparity value for a given position in the reference (first) image. In the disparity-oriented approach, all values of E will form a disparity space (section 6.6.4). Then the optimization process follows.

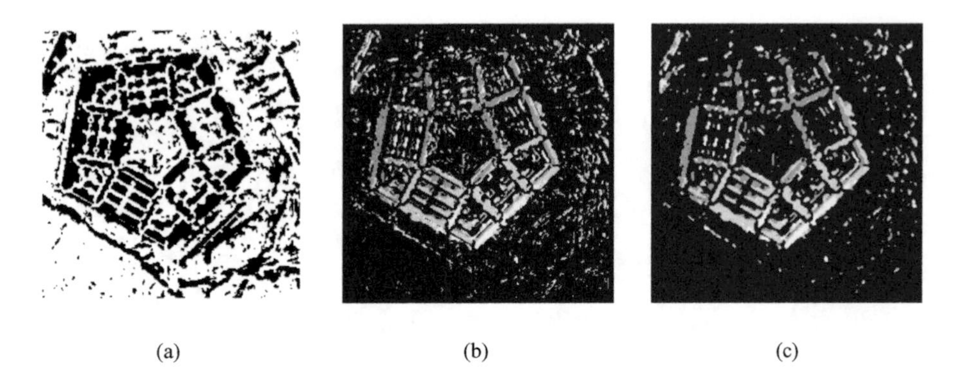

(a) (b) (c)

Figure 6.38 Stereo matching with the structural tensor method. (a) A map with structural places, (b) a disparity map, (c) disparity processed with a 3×3 median filter

An example of the method applied to the 'Pentagon' stereo pair is depicted in Figures 6.38 and 6.39. The structural places were computed as described, with the 3×3 square for the structural element in the morphological dilation filter. The evidence measure was computed in accordance with (6.101), for $\alpha = 2$ and all $w_i = 0.25$. The aggregation was done with a simple 7×7 binomial filter (section 4.3.2). The maximum disparity search was set to 11 pixels. Run time is about a second on a PC laptop with a duo core processor and 2 GB RAM.

Figure 6.38(a) depicts the structural places (dark) computed from the coherence component of $\hat{\mathbf{s}}$. Figure 6.38(a, b) presents the disparity map computed with the left image kept as a reference. The latter is additionally filtered by the nonlinear median filter which removes outliers. In effect we obtain a smoothed disparity map. However, this process is recommended after the cross-checking verification the results of which are presented in Figure 6.39(b). For this purpose a second disparity map was computed (Figure 6.39(a)) – this time with the right image being a reference, however. The final smoothed disparity map is depicted in Figure 6.39(c).

Figure 6.40 depicts a 3D visualization of the 'Pentagon' scene. Height values are directly provided from the disparity map in Figure 6.39(c).

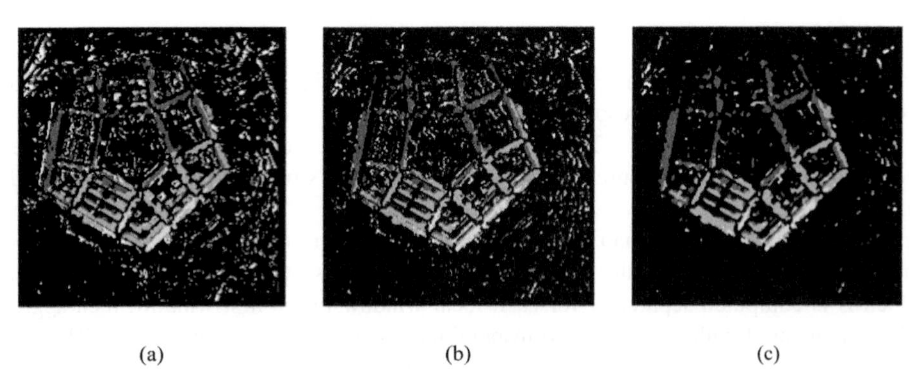

(a) (b) (c)

Figure 6.39 (a) Disparity map with a reference set to the right image of 'Pentagon'. (b) A cross-checked disparity map. (c) The same map after the 3×3 median filter

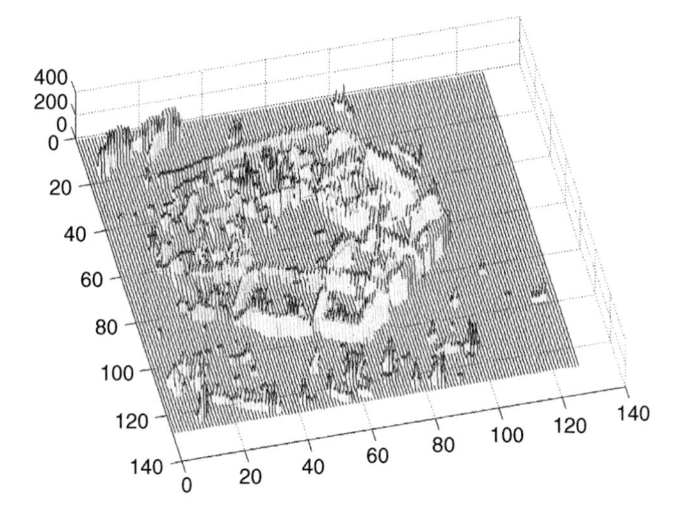

Figure 6.40 3D reconstruction of the test scene

A more advanced utilization of the structural tensor in stereo matching would be built with disparity computation based on an optimization process *guided* by the local values of the structural tensor. This way, the matching region is blocked on the sharp structure boundaries in the images.

6.7 Area-based Elastic Matching

We have now met the basic ingredients required to construct a practical area-based *elastic-warp* matching algorithm, including image metrics for comparing image patches and multi-resolution image data structures. This section illustrates how to combine the basic ingredients to produce a complete image-matching algorithm with the following properties.

- A wide dynamic range of pixel displacement search between images.
- Recovery of a dense disparity field.
- No requirement for prerectification of the input images, i.e. full 2D search as opposed to 1D scanline search.
- Ability to cope with perspective projection differences between the stereo-pair of images to be matched.
- Invariance to differences in gain and black-level in the stereo-pair of images to be matched.
- Subpixel disparity estimates.
- Immunity to false targets.
- Data confidence map.

6.7.1 Elastic Matching at a Single Scale

The stereo correspondence problem is essentially under constrained. In the context of an area-based matcher there are often a number of 'best matching' positions that could be found when searching for a patch in the test image that is most similar to a corresponding patch in the reference image. The reason for this is manyfold; image noise will ensure that projections of even identical parts of the scene on to the camera image plane will not necessarily produce a perfect correlation score when compared using area-based matching. Similarly occlusions and repeating patterns can confuse and subvert the matching process from the desired 'correct' solution. However, the stereo correspondence problem can be solved if suitable constraints can be applied to select plausible solutions. In this section we shall review the constraints required to complete the matching algorithm for matching at a single scale and review the range of disparities that such a process can recover.

We shall first deal with search range, feature stability and subpixel accuracy, and then consider matching as an elastic warping process and the implication for differences in perspective between the two views to be matched. Finally we shall investigate what is termed the match–warp–regularize cycle, an approach common to many matching schemes including surface manifold matching in three dimensions.

6.7.1.1 Disparity Match Range

In general when performing matching, it is assumed that the image patches to be matched will be filtered using a LoG filter prior to matching using the windowed correlation equation defined in (6.89), as outlined in section 6.6.2. Prefiltering the input images with a LoG function serves two purposes: the spatial scale of image structures that are compared in the matching process will be determined by the size of the Gaussian component of the LoG (and the overall band pass of the LoG in the spatial frequency domain); and the band pass nature of the LoG blocks the DC (mean) image component as described in section 4.5.3, thereby simplifying the correlation function itself. However, in order to simplify the analysis of certain matching relations, the LoG filtering step has been omitted as explained below.

A fundamental question that must be considered is what the relationship is between the spatial scale of the scale-defining Gaussian, σ_s, and the disparity search range limit D_{max}, i.e. Panum's fusional area. Marr and Poggio [299] estimated the D_{max}–σ_s relationship using statistical techniques and discovered that when $D_{max} = \sigma_s$ edge tokens based on detecting zero crossings could be detected with 95% reliability. However, to achieve this degree of reliability, a further constraint was imposed such that only those edge tokens whose respective orientations matched to within 30° were considered.

Jin [229] considers the scale-disparity range issue from a purely signal matching perspective, examining the 1D cross-correlation of a pair of Dirac impulses convolved with Gaussian envelopes and separated by disparity u. By making the simplification of not applying LoG filtering prior to matching in his analysis he shows that the (Gaussian spread) impulse autocorrelation function resembles a 1D LoG. It can be observed in Figure 6.41 that as we sweep the cross-correlation, i.e. as represented by shift d, from $-\infty$ to $+\infty$ that only the central portion of the cross-correlation function remains monotonically related to the spatial distance between the impulses being matched, and this range is thereby termed $[-D_{max} \ldots +D_{max}]$. The correlation score peak arises at $d - u = 0$ when the search shift equals the disparity present. Clearly, correlation scores within the region $[-D_{max} \ldots +D_{max}]$ are proportional to the spatial distance between the compared impulses and can therefore be used to guide the search process. The correlation scores returned outside of the range $[-D_{max} \ldots +D_{max}]$ are no longer

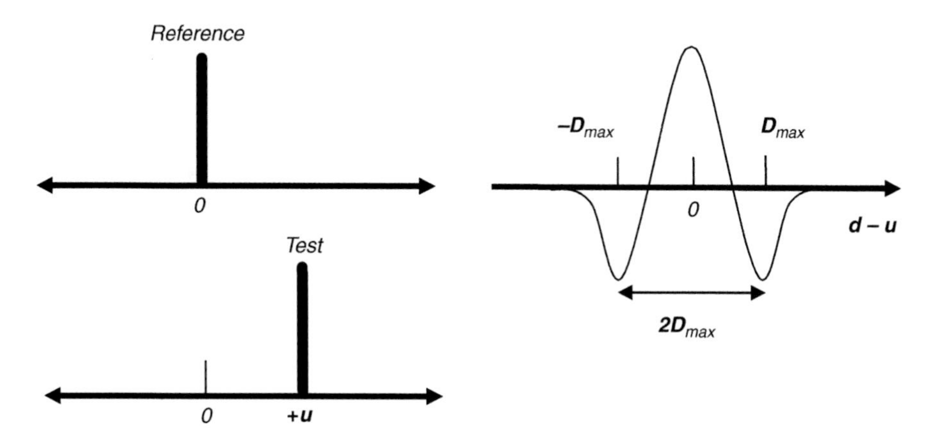

Figure 6.41 Cross-correlation function of a pair of impulses as a function of $d - u$ using Gaussian-weighted windows and statistical correlation: d is search shift and u is the disparity between the impulses. (Figure based on Jin [229])

monotonic with respect to the spatial distance between the compared impulses and therefore correlations in this region cannot be used unambiguously to determine the correct direction in which to search for potential match candidates.

Finally, as already presented, the correlation function itself is weighted by means of a Gaussian window of spatial size σ_w and the fusional range $[-D_{max} \ldots +D_{max}]$ is a function of both σ_s and σ_w. Jin computes the range $[-D_{max} \ldots +D_{max}]$ numerically for various ratios of σ_w/σ_s and concludes from this empirical investigation the following:

$$\begin{aligned} &\sigma_s \propto D_{max} \\ &D_{max} \to \infty, \frac{\sigma_w}{\sigma_s} \to \infty. \end{aligned} \tag{6.102}$$

To paraphrase Jin: Panum's fusional area is proportional to the scale factor of the spatial frequency channel of the LoG filter, while any given magnitude of disparity can be matched if the correlation window is sufficiently large. From Jin's empirical data, $D_{max} = 2.96$ when $\sigma_s = 1.0$ and $\sigma_w/\sigma_s = 1.0$.

6.7.1.2 Search and Subpixel Disparity Estimation

Given the preceding discussion on search range and spatial scale, we now have the basis on which to formulate search at a single scale. Knowing that we obtain a search range of

$$D_{max} = \lambda \cdot \sigma_s, \quad \text{when} \frac{\sigma_w}{\sigma_s} = 1.0, \quad \lambda = 2.96 \tag{6.103}$$

we can adjust σ_s to give any desired search range D_{max} under the constraint

$$\sigma_w \geq \sigma_s. \tag{6.104}$$

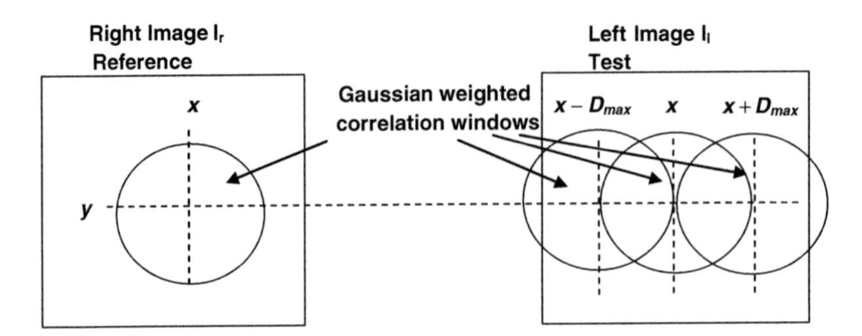

Figure 6.42 Basic 1D search by comparing a *reference* patch in the left image with three *test* patches extracted from the right image

In effect this relation gives rise to the well-known *continuity constraint*, whereby the continuity of the surface estimated via matching is guaranteed since the rate at which the match score can change is limited by the spatial Gaussian-weighted support of the correlation kernel.

Building on Jin's analysis we can now devise a single scale search algorithm that compares an image patch at (x, y) in the reference image at three locations in the test image, corresponding to $(x - D_{\max}, y)$, (x, y), $(x + D_{\max}, y)$, as in Figure 6.42.

The actual intermediate location of the 'true' correspondence can be estimated by means of interpolating between the correlation scores returned at the $x - D_{\max}$, x, $x + D_{\max}$ locations. Since we know that correlating pairs of impulses produces a correlation function comprising a single peak within the above search range, it is possible to approximate this function by means of a second-order polynomial as described in section 6.4.2. However, in our case we may have sampled the image at potentially *noninteger* locations $x - D_{\max}$ and $x + D_{\max}$ (achieved by shifting the centre of the Gaussian weight applied to the correlation window in the test image). However, for computational efficiency, it is clearly desirable to arrange for $D_{\max}$ to be an integer value.

Note that the integrity of this method hinges on the fact that the (Gaussian-weighted) test correlation windows *overlap*, and hence introduce continuity, i.e. spatial correlation, in their estimates which in turn results in the single turning point in the cross-correlation function and thereby leads to the possibility of interpolating the location of the correlation maximum by means of a second-order polynomial.

Section 6.4.2 introduces a generalized polynomial fitting scheme for subpixel disparity recovery. As described in section 6.4.2, we are going to shift to *local* coordinates centred on the current search position, i.e. anchored at $x = 0$, and in addition fix our interpolation function $c(x)$ in the domain $-D_{\max}$, 0, $+D_{\max}$ and now assume that the cross-correlation function takes the form:

$$c(x) = ax^2 + bx + c. \tag{6.105}$$

In order to find an estimate of the 'best' correlation score, d_x, we must locate the turning point at $2ax + b = 0$. Since we cannot guarantee that $D_{\max} = 1$, as opposed to using (6.57) here we solve the set of three equations directly as follows, substituting the sample values at

$x = -D_{\max}, 0, +D_{\max}$ into (6.105):

$$\begin{aligned} c(0) &= c \\ c(D_{\max}) &= aD_{\max}^2 + bD_{\max} + c \\ c(-D_{\max}) &= aD_{\max}^2 - bD_{\max} + c. \end{aligned} \quad (6.106)$$

Therefore from (6.106) we can deduce:

$$\begin{aligned} \alpha &= c(D_{\max}) - c(0) = aD_{\max}^2 + bD_{\max} \\ \beta &= c(-D_{\max}) - c(0) = aD_{\max}^2 - bD_{\max} \\ a &= \frac{\alpha + \beta}{2 \cdot D_{\max}^2} \\ b &= \frac{\alpha - \beta}{2 \cdot D_{\max}}. \end{aligned} \quad (6.107)$$

Solving for d_x yields the same result as before in (6.56):

$$d_x = \frac{-b}{2 \cdot a}. \quad (6.108)$$

We can now utilize the estimate of d_x and the direction of the curvature of the fitted parabola, given by the sign of the second derivative, i.e. the sign of a, within a search algorithm as follows.

1. If $a < 0$, and d_x is in the range $\pm 1.5D_{\max}$, then d_x serves as the disparity estimate for the current location.
2. If $a < 0$, and d_x is outside the range $\pm 1.5D_{\max}$, then shift the current location in the test image by $\pm 1.5D_{\max}$, and then continue to refine the search as in step 1.
3. If $a > 0$ implies that the search range contains a local minimum, then check which test patch correlation at $\pm D_{\max}$ has the greater value (i.e. is closest to the correct match location), and then shift the current location in the test image by $\pm 1.5D_{\max}$, and continue to refine the search as in step 1.
4. If $a = 0$, the interpolation solution is a flat surface and no refinement is possible; therefore return $d_x = 0$, i.e. record zero disparity at this reference location.

Back-substituting d_x into the polynomial gives $c(d_x)$ and this quantity can serve as a measure of the *confidence* of the retuned correlation score.

The above search and interpolation scheme can accommodate a 1D search within a stereo-pair of images that have been scanline registered. The matching process can be extended to a 2D search by making additional search comparisons at locations $y - D_{\max}$ and $y + D_{\max}$ as in Figure 6.43.

In order to estimate the subpixel x, y best match for the 2D search mentioned above, we can now simply fit 1D polynomials in the x and y directions independently and solve these as above to obtain correlation scores in x and y. A single confidence value C associated with each disparity map x, y location is computed by taking the average of the correlation score

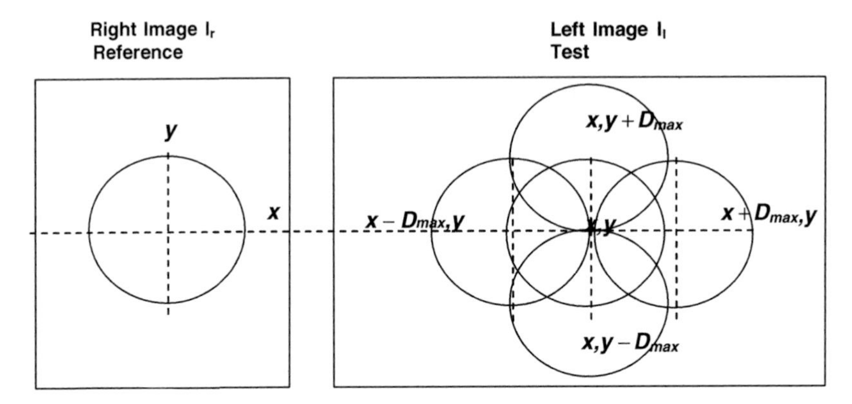

Figure 6.43 2D search by testing patches in both vertical and horizontal directions

maxima calculated independently in the x and y directions. The search process is applied at *every* pixel in the reference image to estimate the x and y disparity values that map to the corresponding locations in the test image. Hence x and y *disparity maps* D_x and D_y are so constructed along with the *confidence map* C that stores the value of the estimated correlation score at the interpolated disparity location calculated in the x and y directions. These three maps, D_x, D_y and C, are fundamental to developing a viable matching process as described in the following sections.

6.7.2 Elastic Matching Concept

Imagine matching one image to another by stretching the test image as though it were a rubber sheet such that it fits to (i.e. registers with) the reference image. Recall the fundamental property of a (dense) disparity field is to represent the degree of offset required to map (shift) each point on the reference image on to the corresponding point in the test image, with subpixel accuracy (6.77):

$$I_l(x', y') = I_r(D_x(x, y) + x,\ D_y(x, y) + y).$$

By resampling the test image using the disparity map (using the above equation) generated by matching with the reference image, we can warp the test image into the shape of the reference image (see Chapter 12 also). This assumes some form of local interpolation as the correspondences are specified with subpixel accuracy, using real-valued numbers. In this case a backwards warp is sufficient since the disparity map points to subpixel locations in the reference image that can be found using *bilinear* or *bicubic* interpolation and then mapped to integer locations which are *directly* comparable with the reference image. This warping process is the computational equivalent of stretching the test image into the same shape as the reference image.

Having stretched the test image into the same shape as the reference image by matching and warping, we can test how well these images now register. If the matching process were 'perfect' then we would expect that the root-mean-square difference ε between the reference

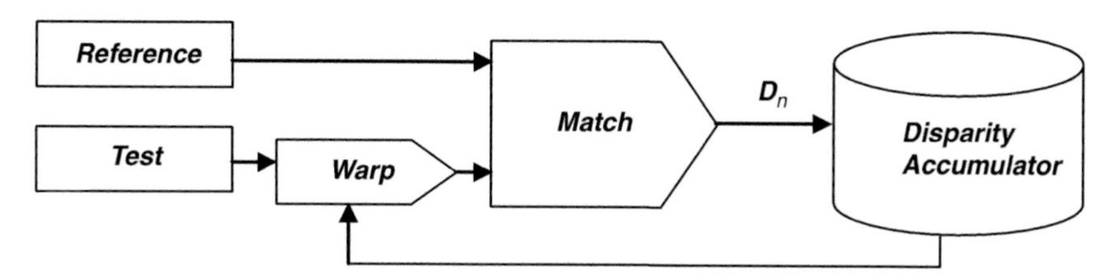

Figure 6.44 Match estimation via the basic match-warp process

(left) image and the warped test (right) would be zero:

$$\varepsilon = \sqrt{\frac{\sum_{x=1}^{X}\sum_{y=1}^{Y}(I_l(x,y) - (I_r(D_x(x,y)+x, D_y(x,y)+y))^2}{X \cdot Y}}. \tag{6.109}$$

The above measure can be used to evaluate the global quality of the match; it will never usually reach zero, but at least we can minimize ε. Indeed matching can be reapplied and the test image progressively warped into registration with the reference image, the process terminating when the global fitting error falls below some value ε. The loop shown in Figure 6.44 can therefore be established.

The matching process halts after the n-th iteration when the residual global disparity error, ε, falls below some preset threshold. This iterative rematching process forces the test image into correspondence with the reference image. This process is useful for reducing the effect of differing perspectives in each image, as by means of the current disparity field and the warping process, the test image is driven into the 'shape' of the reference image. Notice also that the backwards warping process ensures that the iterative refinement process takes place using kernels aligned to the image grid; this will be an important consideration when we investigate coarse-to-fine search in the next section.

The *smoothness constraint* described in section 6.7.1.2 allows us to make the assumption that we can use area-based operators to estimate local image (pixel) similarities and also expect that the disparity estimate we obtain will be locally continuous. This fundamental assumption is based on the observation that most of the world we see around us in 3D comprises continuous surfaces, as opposed to clouds of point-like particles, ribbon-like material or spiky barbed wire! While not entirely general, the smoothness constraint serves as a useful heuristic to allow the recovery of locally well-behaved surfaces, such as *human surface anatomy* or *navigable terrain*. Our matching loop can now be modified to include a filter to enforce smoothness upon the disparity surface (Figure 6.45).

A low-pass smoothing filter such as a Gaussian centre-weighted averaging filter can serve to smooth the disparity estimates. The smoothness constraint makes explicit the assumption that the simplest, i.e. smoothest, disparity surface that fits to within the desired error ε is likely to be the correct solution (the principle usually known as Occam's Razor [108]).

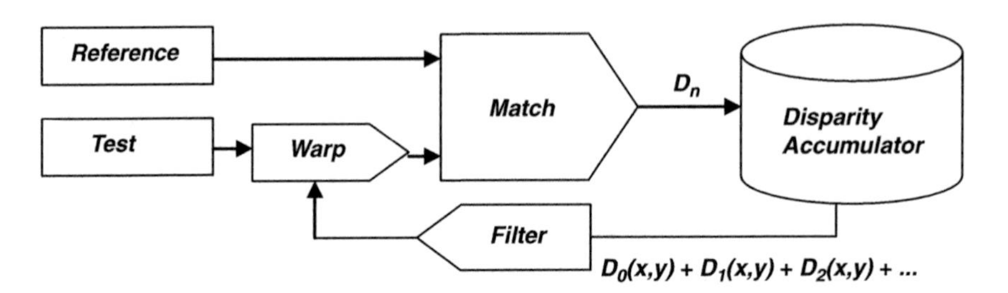

Figure 6.45 Match-warp estimation incorporating disparity filtering

The smoothness constraint is of course not valid at surface boundaries, i.e. depth edges. A mechanism is required to suppress smoothing at edges to avoid blurring over these. Approaches include explicitly detecting edges and gating the blurring filter. Or implicitly, measuring local image variance or gradient strength and weighting the filter accordingly:

$$\begin{aligned} D_{out}(x, y) &= D_{filtered}(x, y) \cdot \Omega(x, y) + D_{in}(x, y) \cdot (1 - \Omega(x, y)) \\ \Omega(x, y) &= \frac{\sigma_{\max}(x, y) - \sigma(x, y)}{\sigma_{\max}(x, y)} \end{aligned} \tag{6.110}$$

where $\sigma_{\max}(x, y)$ is the maximum local image standard deviation estimate found in the test image and $\sigma(x, y)$ is the current local image standard deviation estimate. There are potentially many such filtering schemes, the best known of which incorporates the concept of *anisotropic diffusion* [345] in order to preserve disparity surface edges.

6.7.3 Scale-based Search

The preceding sections describe how to match stereo-pairs of images by successive warping and filtering of the recovered disparity field. In order to be able to extend the range of disparities, $\pm D_{\max}$, that can be recovered beyond that of a single local Gaussian-weighted correlation window, while achieving the highest *stereo acuity* in terms of recoverable disparities, the single-scale matching approach must be extended within a multiscale framework. By constructing multi-resolution image pyramids, as described in Chapter 5, from the input test and target images, it becomes possible to implement a coarse-to-fine matching algorithm. Matching over a discretized multi-resolution scale-space confers a number of advantages.

- It is possible to match image structures at all scales using the same algorithmic machinery applied to each scale.
- If correctly structured, coarse-to-fine search can avoid the *false target* problem.
- As mentioned, the dynamic range of recoverable disparities can be greatly extended.
- The process can be made algorithmically and computationally efficient.

In order to analyse search over scale, Jin [229] considers the scenario depicted in Figure 6.46 where the task is to match two impulses in the reference image to two impulses in the test image.

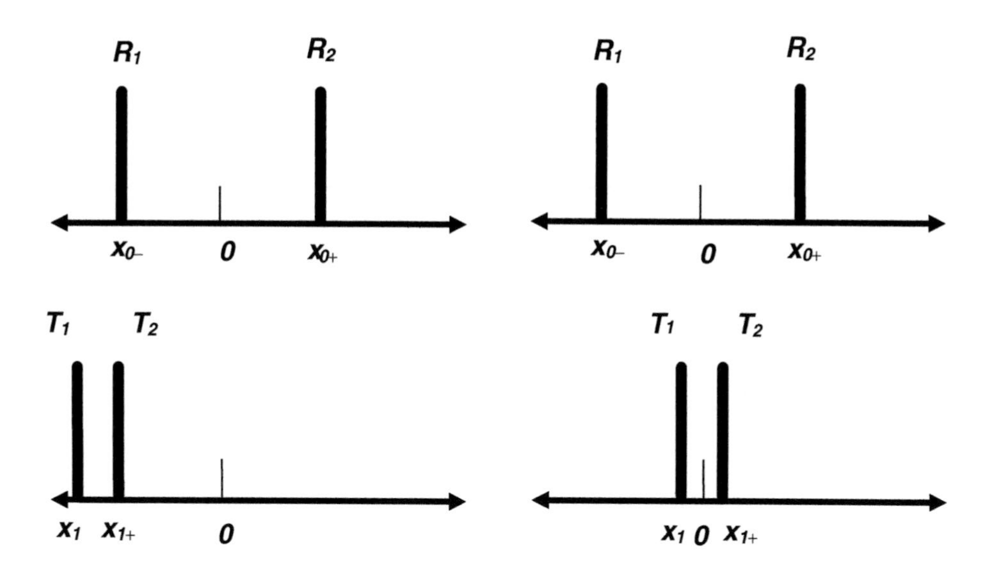

Figure 6.46 Matching a pair of impulses between reference and test images. Left: impulse position prior to gross alignment. Right: following gross alignment. (Figure based on Jin [229])

In the proposed multi-resolution matching scheme, impulses x_{1-} and x_{1+} must first be brought into the intermediate position shown in Figure 6.46 such that x_{1-} will be matched to x_{0-} and x_{1+} will be matched to x_{0+} by successive refinement over subsequent finer image scales. The key means to analysing the limits of the matching process is to consider when x_{1-} and x_{1+} are now so close that they can be considered to have merged into a single impulse (Figure 6.47). While successful matching cannot now be sustained, it does serve to illustrate the limiting condition for successfully centring the merged impulses between the attracting impulses. In a real situation, this limiting situation occurs at some arbitrarily coarse scale, when a pair of impulses would coalesce. At subsequent finer scales the impulses would resolve into individually identifiable locations when the separation distance of the impulses becomes significantly greater than the intrinsic blur of the current scale. This process is generally known as *scale-space tracing.*

The above discussion illustrates the effect of attempting to shift and correlate the double impulse pair with the single impulse for three different separations S of the impulse pair, namely large, small and intermediate critical separation distance. What the above discussion shows is that for a given spatial scale of matching, i.e. intrinsic level of blur, when S is sufficiently large, the intrinsic blur is not sufficient to coalesce these when making a correlation comparison to a single impulse, hence a double peak emerges on the correlation score graph. Consequently, it is not possible to register the impulse pair on the single impulse as the true location lies between the peaks as opposed to on either peak.

When S is sufficiently small, the blur present in the current image scale is sufficient to merge the impulses under the correlation operation, and hence a single peak in the correlation score emerges and this corresponds to the location to which the impulses should be shifted in order to register them with the single impulse.

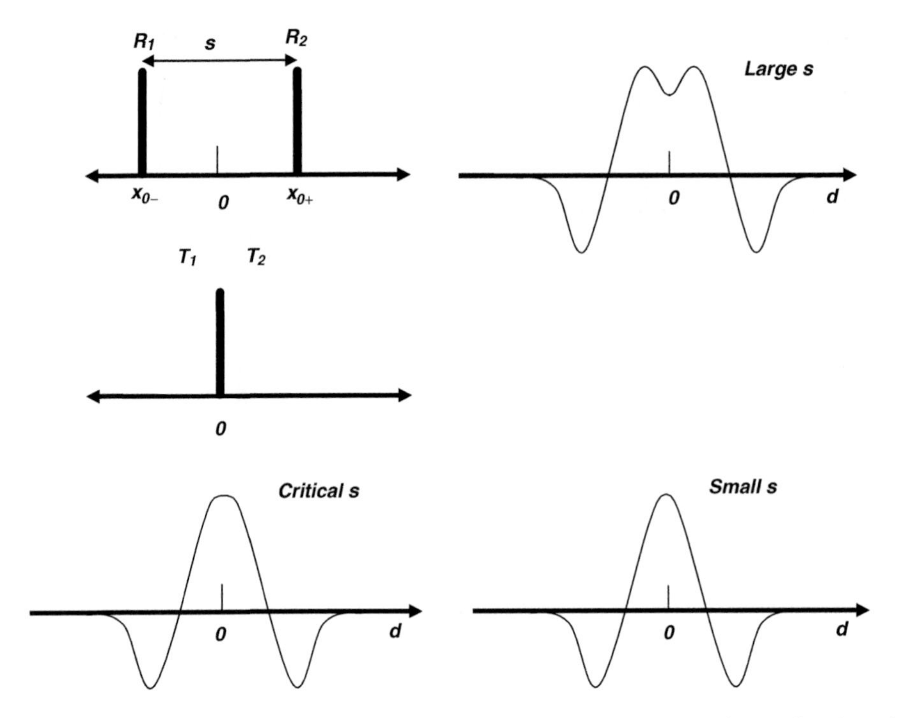

Figure 6.47 Matching competition between a single impulse and an impulse pair, as a function of the separation between the impulse pair. Figure based on Jin [229]

The middle graph in Figure 6.47 shows the critical situation when the pulses are just sufficiently close to merge under correlation and this maximum separation, $S_{\max}$, corresponds to the limit of scale-space tracing and can be used to determine the parameters relating to the separation between scales and fusional limit when searching over scale.

Jin discovered empirically the following relation between the intrinsic scale σ_s and the maximum separation distance $S_{\max}$, by plotting $S_{\max}$ against a range of values of σ_s:

$$\sigma_s \propto S_{\max}. \tag{6.111}$$

Jin also discovered that as the ratio of the correlation weighting Gaussian, σ_w, and the intrinsic image scale, σ_s, tends to infinity, $S_{\max}$ asymptotically approaches $\sqrt{2}$:

$$S_{\max} \to \sqrt{2}, \frac{\sigma_w}{\sigma_s} \to \infty. \tag{6.112}$$

Similarly, Jin also plotted the maximum recoverable disparity, $D_{\max}$, between the single impulse and the centre of the impulse pair for the critical case, i.e. the limits of scale-space tracing, and this was found to correspond to $\sigma_w/\sigma_s = 4.08$. Therefore as long as the above relation for $S_{\max}$ holds, the relation for $D_{\max}$ must also hold, as this is less stringent. Hence, in order to guarantee that correct match is made the following must hold:

$$\frac{\sigma_s}{D_{max}} \geq \sqrt{2}, \quad \text{where} \quad \sigma_w \geq \sigma_s. \tag{6.113}$$

In order to apply the above result under LoG filtering as opposed to Gaussian filtering, Jin applied the following corrections:

$$\text{when} \quad \sigma_{LoG} = 1.0, \quad S_{max} = 0.95, \quad \text{for} \quad \frac{\sigma_w}{\sigma_s} = 1.25. \tag{6.114}$$

We can infer the interscale sampling distance as follows: if we consider the critical situation where a single impulse (two fused impulses) has been brought into alignment, equidistantly between two competing impulses in the reference image, then the maximum distance between the competing impulse pair must be $S_{max} \approx \sigma_{LoG}$. At the next finer scale, where the previously fused impulses will now resolve into a pair of separate impulses, the distance between each impulse in the test image and its corresponding impulse in the reference image must be $<\sigma_{LoG}$. When the conditions in (6.114) hold and LoG filtering is applied, the maximum search range D_{max} will resolve corresponding to just over two pixels of disparity. Therefore a scale reduction factor of two, i.e. octave scale sampling, is (just) sufficient to continue to allow the impulses to be matched correctly.

In practice, it has been found that a more conservative interscale sampling factor of $\sqrt{2}$ gives significantly superior results to those obtained under octave scale separation. This is presumed to be due to a greater immunity to image noise when a smaller distance in scale-space must be spanned. Indeed, in current algorithm formulations arbitrarily finer interscale sampling is adopted as required to achieve successful image matching (section 6.7.5).

6.7.4 Coarse-to-fine Matching Over Scale

From the forgoing discussion on search over scale it is now possible to formulate a complete algorithm for coarse-to-fine image matching through scale-space, as in Algorithm 6.13.

It is important to note that the same search algorithm and scale parameters are applied to match each level of the image pyramids in Algorithm 6.13. Implicitly, the largest scale that can be matched corresponds to the basic range of $D_{max}\rho^k$, where k corresponds to the number of levels in the pyramid and ρ is the interlevel reduction factor. The matching resolution limit can be considered from different perspectives: the finest scale might simply be considered

```
1. Construct LoG or DOG image pyramids for the test and reference images
2. At the coarsest level of each pyramid:
a. Execute the single scale search algorithm in the x and y directions, for N match-
smooth-warp cycles.
3. While there are unmatched pyramid levels do:
a. Expand the x and y disparity maps and the match confidence by the same factor as the
pyramid reduction ratio, ρ, such that it is now equal in size to the images to be matched
at the next finer level in the test and reference pyramids.
b. Execute the single scale search algorithm in the x and y directions, for N match-
smooth-warp cycles.
```

Algorithm 6.13 Coarse-to-fine matching over scale

to be that at the highest resolution level of the pyramid, remembering that the pyramid can comprise two full-resolution levels, corresponding to the Laplacian filtered input image; in this case the intrinsic image blur defines the limiting scale. Alternatively, matching might be terminated at the highest resolution level following the application of the initial blur required to achieve a pyramid with the desired degree of blur at each level, as described in previous sections.

From a different perspective, the intrinsic image blur will affect the matching resolution limit. If a large degree of intrinsic blur is present, no extra information will be present in pyramid levels that sample more finely than that required to satisfy the Nyquist limit set by the intrinsic image blur. A further perspective is that of noise suppression: the degree of blurring provided by each level in the pyramid helps to suppress introducing noise into the disparity estimations. Therefore, matching may take place successfully until a scale is reached where the magnitude of image noise present begins to make a significant contribution to matching errors. However, match errors induced by image noise can also be mitigated by increasing the number of smoothing cycles at or within each matching iteration. Given the number of variables involved, certain parameters such as intrinsic image noise can be difficult to determine accurately, particularly when a sensor performance specification is not available. In practice, the usual recourse to obtaining acceptable results for any specific configuration is by experimental determination of parameter settings.

6.7.5 *Scale Subdivision*

Algorithm 6.13 can be extended to incorporate matching over subdivided scales between pyramid levels, to achieve arbitrary interscale sampling intervals. The process of scale subdivision in pyramids was covered in Chapter 5. When matching in the subdivided scales, the matching kernel parameters are no longer constant, as is the case when matching through a regular pyramid, and no subsampling or expansion of course takes place. However, the remainder of the matching process is essentially the same and the usual expansion process is applied prior to inter-resolution matching at the next finer pyramid level. Consider Figure 6.48. If we commence matching at the coarsest (undivided) pyramid, then upon completing a set of 'standard', i.e. as described previously, match–smooth–warp cycles at this level we must immediately expand to the next (subdivided) pyramid level.

We can calculate the required subdivision blurring factors as follows. Equation (5.20) gives the base scale blur factor in each level of a subdivided pyramid and this is essentially the N-th root of the pyramid division factor raised to the n-th power of ρ, the current subdivision level:

$$\sigma_n = \sigma_0 \left(\sqrt[N]{\rho} \right)^n, \quad n \in 1 \ldots N - 1. \tag{6.115}$$

Equation (5.23) gives the incremental increase in blur for each subdivided layer:

$$\sigma_{\text{sub}} = \sigma_0 \left[\sqrt[\frac{N}{2n+2}]{\rho} - \sqrt[\frac{N}{2n}]{\rho} \right]^{1/2}. \tag{6.116}$$

We see that upon expansion, the blur scale will increase by a factor of ρ and the size of the filter kernel support region must follow. Prior to matching the low-pass subdivided pyramid level can be straightforwardly band pass filtered by computing the Laplacian, i.e. summing the horizontal and vertical second-order partial derivatives (assuming that the intrinsic blur in the

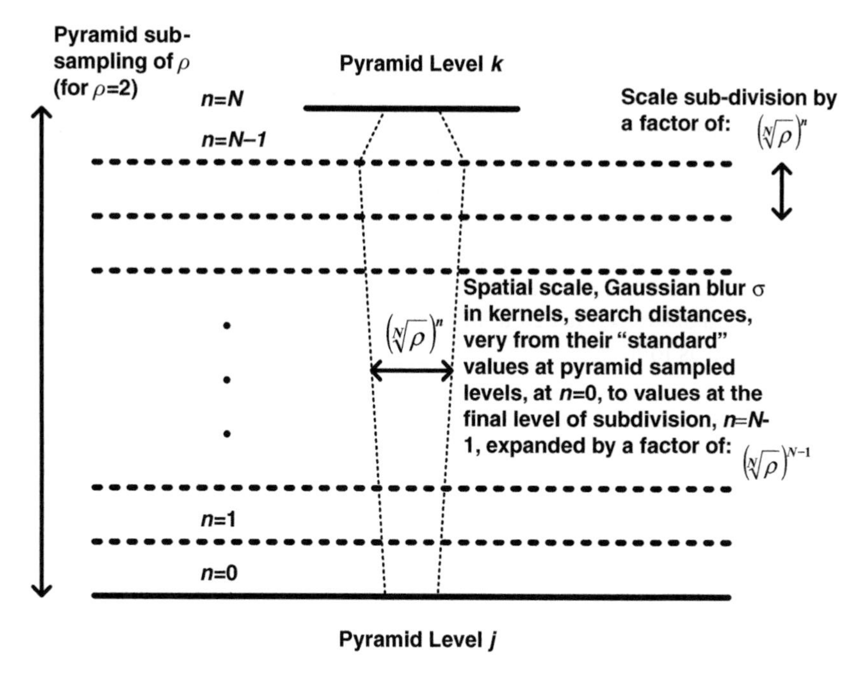

Figure 6.48 Scale subdivision by N within the layers of a regular pyramid subsampled at ρ pixels in each dimension

pyramid has been adjusted accordingly to achieve the desired disparity fusion limits). The blur in each subsequent subdivided layer will now *reduce* until the current base (nonsubdivided) level in the pyramid is reached and a 'standard' match cycle can then again be invoked.

Not only will the blur change (and band pass) in each subdivided level, but those search parameters that are directly linked to the scale blur, such as the search range $D_{\max}$, will also vary proportionally to the current subdivision blur as given in (6.20). Hence, the relative positions of the locations where the test image is sampled and compared by correlation to the reference image will change accordingly, i.e. the correlation kernels will be located in the test image at $\pm\rho D_{\max}$ (about the zero disparity position) in the first level of subdivision and this will reduce in factors of $\left(\sqrt[N]{\rho}\right)^n$ in each subsequent subdivision level n, until a factor of unity is reached at the next nonsubdivided level.

While the various subdivision parameters are easily computed as described above, the principal penalty of subdivision is the additional computation required to perform the additional matching steps in the subdivided layers with over-sampled convolution kernels. However, subdivision affords much greater matching continuity over scale and is often the only means to obtain correct matching between stereo-pairs containing large disparity ranges and severe disparity gradients/discontinuities.

6.7.6 *Confidence Over Scale*

Each matching iteration at each level in the pyramid generates a correlation score or *confidence value*. It is useful to propagate this confidence value to the finest resolution level in the pyramid at which matching takes place in order to provide a 'final' confidence map that indicates match quality of the computed disparity map on a per-pixel basis. A naive approach would be simply to output the final confidence map; however, this only contains match confidence values pertinent to the scale of structures matched at the finest level.

A satisfactory method for taking into account the match confidence at *all* levels of resolution is to expand and sum the final confidence values achieved at each level of the pyramid. Rather than store a pyramid comprising a confidence image for each resolution level, the confidence values $c_{n-1}(x, y)$ computed at the previous $(n - 1)$-th level are expanded to match the spatial resolution of the current pyramid level n at which matching is being carried out and the new confidence values c_{current} added to the prior values to generate an output confidence c_n via a weighted sum as follows:

$$c_n(x, y) = c_{\text{current}}(x, y)\alpha + c_{n-1}(x, y)(1 - \alpha), \tag{6.117}$$

Parameter α sets the fraction of confidence propagated between pyramid layers; a value of 0.7 has been found to work well by experiment [230].

6.7.7 *Final Multi-resolution Matcher*

The components described in the previous sections can now be assembled into a complete multi-resolution matcher that incorporates scale subdivision, illustrated in Figure 6.49. Following pyramidization of the input stereo-pair to be matched, the coarsest resolution level is matched and then expanded in preparation for subdivision matching, until the next level of resolution scale is reached. It is worth noting that initial disparity and confidence maps are present in this scheme. Where no additional information is available, these maps would normally be initialized with zero disparity and confidence values. However, there are occasions where initial disparity estimates are available from other processes, for example initial disparity estimates could be provided via a coarse, but reliable, boostrap stereo matcher, such as the *Census* algorithm described in section 6.3.7. Alternatively, it is possible to sum the entire match process for multiple iterations from start to finish, by spatially reducing the final disparity and confidence maps from an initial matching run and feeding these back into the matcher to reattempt matching based on reasonable starting disparities. While this is a very expensive approach computationally, it can yield satisfactory matches under difficult matching situations.

Figure 6.49 describes a matcher algorithm at the core of a stereo-photogrammetry system ultimately known as C3D that was originally implemented during the 1990s and continues to be developed at the time of writing. While the image matching techniques used within C3D were based primarily on Jin's doctoral thesis [229] (supervised by Dr Peter Mowforth, Turing Institute), Siebert [386], Urquhart [72, 433], van Hoff [198] and in particular Niblett [230] (the principal architect of the C3D system) also contributed to C3D's original development at the Turing Institute, as detailed in [230]. Further information and overviews of the C3D system are presented in [387, 389].

Figure 6.50 shows examples of a stereo-pair input to C3D, while Figure 6.51 shows the horizontal and vertical disparity maps and confidence maps generated. The bottom right image in Figure 6.51 also shows the *range* map that is generated from the disparity maps

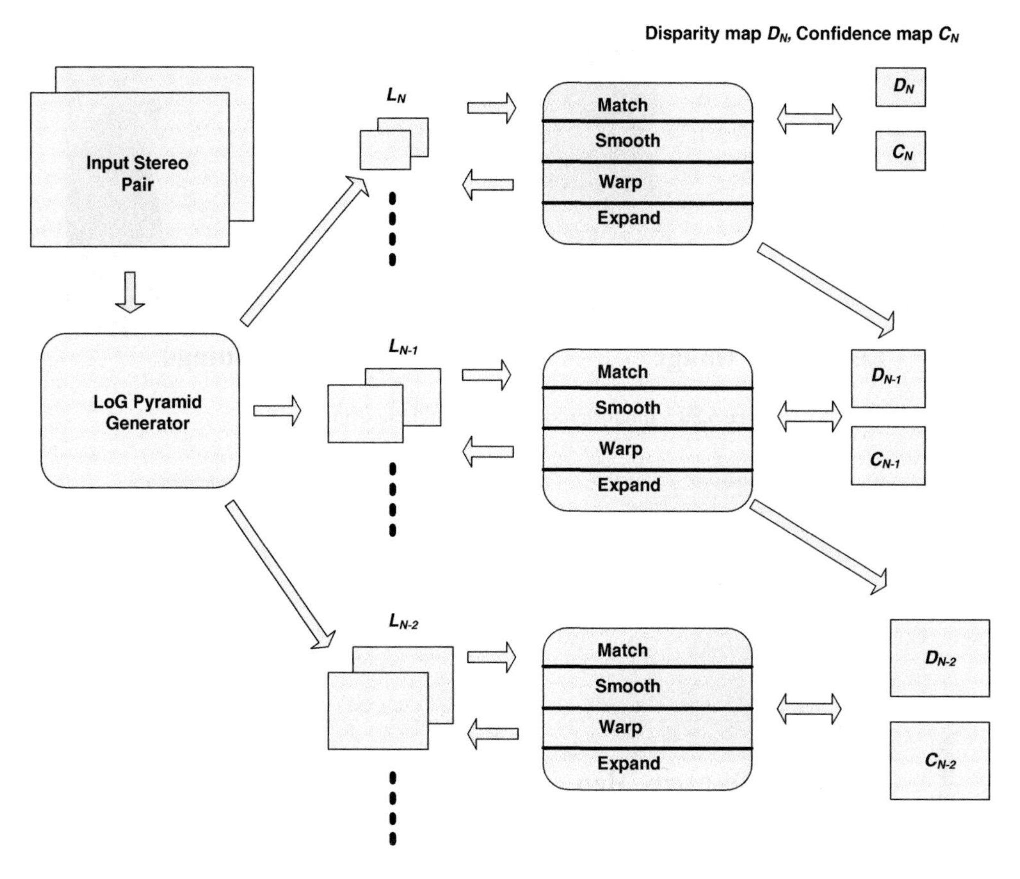

Figure 6.49 Overview of a complete multi-resolution scale-space tracing matcher

by means of photogrammetry. Note the striking resemblance between the horizontal disparity map and the reconstructed range map, principally due to the horizontal disparity map containing most of the displacement information in the horizontally aligned input stereo-pair. In the figures shown here the input images were 4504 × 3000 pixels in dimensions (but matched at approximately half resolution, 2250 × 1500 pixels), the image pyramids comprised 15 half-octave separated levels with 24 levels of interlevel subdivision and 5 smoothing cycles within each subdivided level. Figure 6.52 shows examples of the photorealistic 3D models generated in VRML format by the complete photogrammetry process and in the case of three of these models, photorealistically rendered by draping the left image of the stereo-pair on to the model surface.

Further examples of stereo-pairs matched using C3D and 3D surface models generated from these are presented in section 7.3, which discusses multiview integration of 2.5D surface models. Chapter 8 presents a number of case studies based on the use of C3D, in the context of face and body modelling in section 8.3, clinical veterinary applications in section 8.4 and archive/historical cine footage restoration in section 8.5.

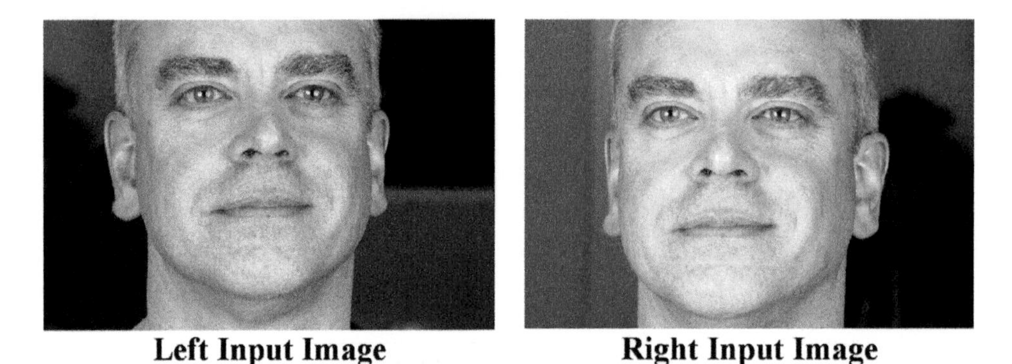

Figure 6.50 Input stereo-pair for matching using C3D

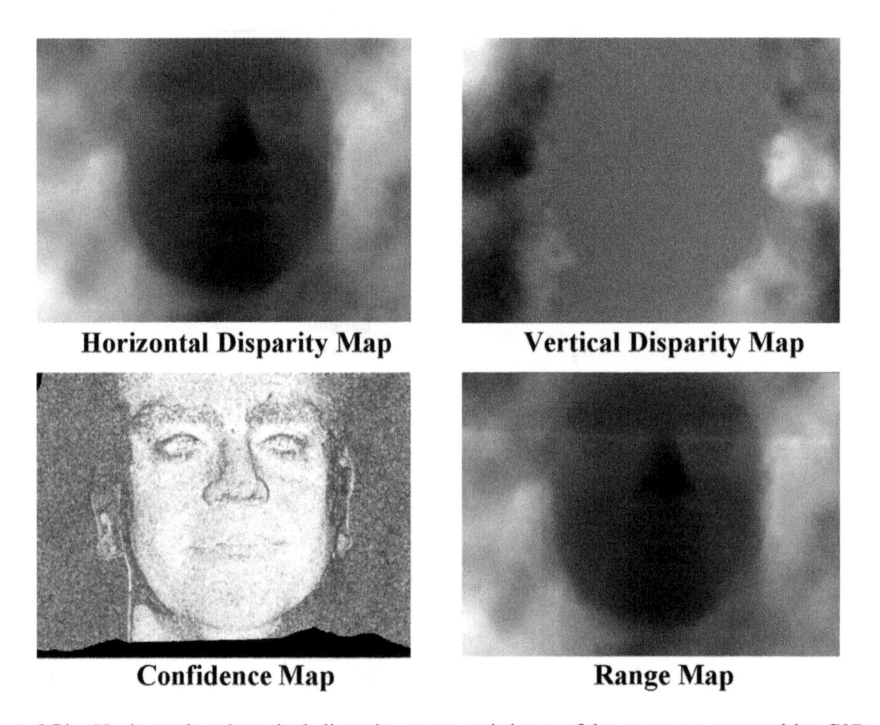

Figure 6.51 Horizontal and vertical disparity maps and the confidence map generated by C3D, followed by their photogrammetric conversion to a range map

6.8 Feature-based Image Matching

Features allow more reliable matching of images due to their discriminative properties. They are usually preserved after an image is subjected to geometrical transformations or its intensity signal is somehow modified, e.g. by noise. The most frequent features used in matching are lines or corners but also values of the structural tensor in highly coherent areas (section 4.6). The main drawback of feature-based image matching is a sparse disparity map. That is, disparity values are computed only for feature points. Thus, this group of methods is well

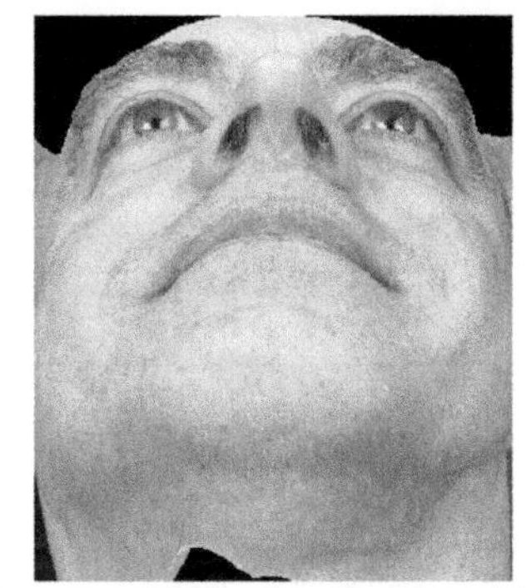

Figure 6.52 3D surface models generated in VRML format. The top left model has been rendered from the reconstructed surface using flat shading; the remaining models have the left image of the stereo-pair rendered on top

suited if only sparse point correspondences are required, as in the case of computation of the fundamental matrix or bifocal tensor.

In this section we discuss matching based on zero-crossings of the LoG operator and matching based on corners obtained from the structural tensor.

6.8.1 Zero-crossing Matching

It was shown by Mayhew and Frisby [302] and by Marr and Poggio [298, 299] that the human visual system (HVS) is endowed with the mechanism of edge detection which operates like the Laplacian of Gaussian (*LoG*) operator, discussed in (section 4.5.3). It was shown that the HVS uses this type of information in the perception of depth. That is, the stereo mechanism in the HVS can be modelled by five channels of *LoG*-like filters with different scale properties. The architecture of this matching scheme is presented in Figure 6.53.

The system consists of five matching channels, each operating at a different scale, i.e. with a different set of *LoG* filters. Results of coarser channels (low-pass filtered) propagate towards the finest channels. At each channel matching is done alongside epipolar lines [162]. The process is depicted in Figure 6.54(a). However, only the zero-crossed points are taken into

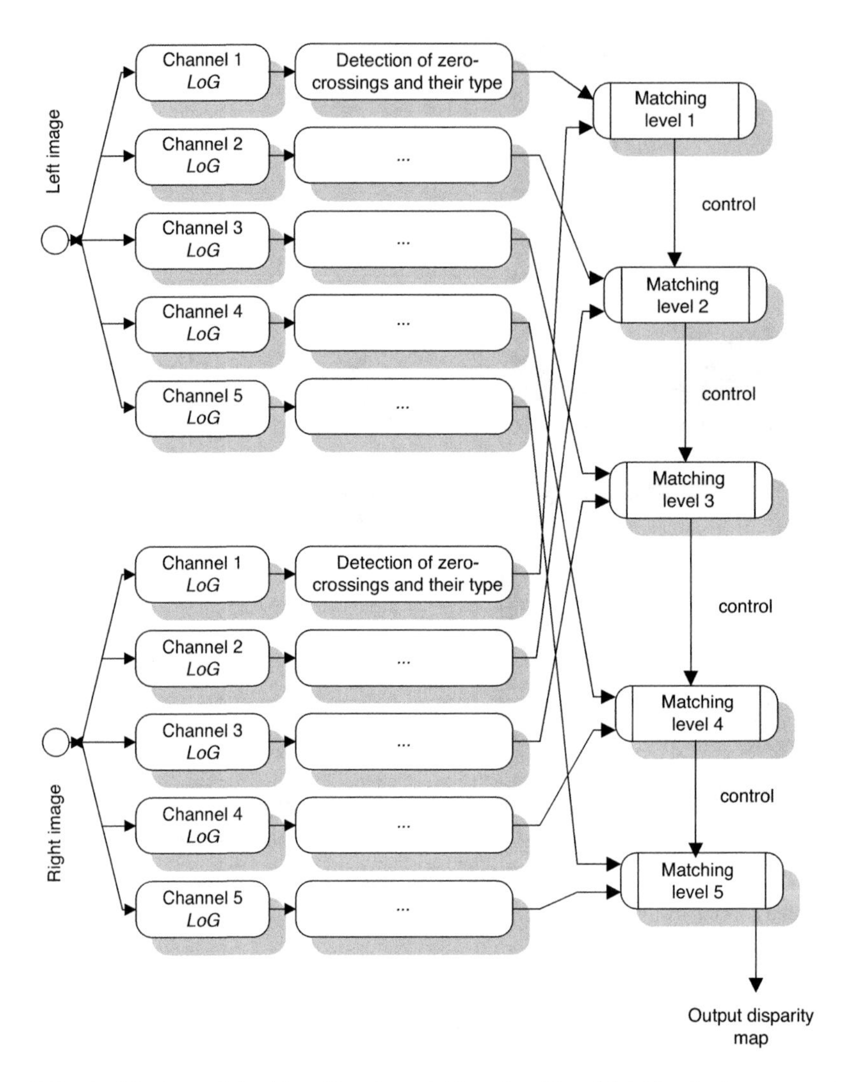

Figure 6.53 Architecture of the Marr–Poggio matching algorithm. Matching is done sequentially in five channels at different scales. In each channel *LoG* zero-crossings are detected and matched based on disparity values obtained in a previously processed channel

consideration. Additionally, the type of zero-crossing (i.e. positive-to-negative or vice versa) is used to clarify a match. In the original proposition of Marr and Poggio the masks of the *LoG* filter were 189, 105, 51, 27, 13, respectively.

For each matching candidate its disparity is checked to fulfil the validity conditions, as follows [162]:

$$d_{\min} = d_{\mathrm{av}} - w \leq d \leq d_{\mathrm{av}} + w = d_{\max}, \tag{6.118}$$

where d is checked disparity, d_{av} denotes an average disparity, known either from the previous coarsest channel or from some assumptions on the geometry of a scene, and w is the size of the *LoG* filter mask.

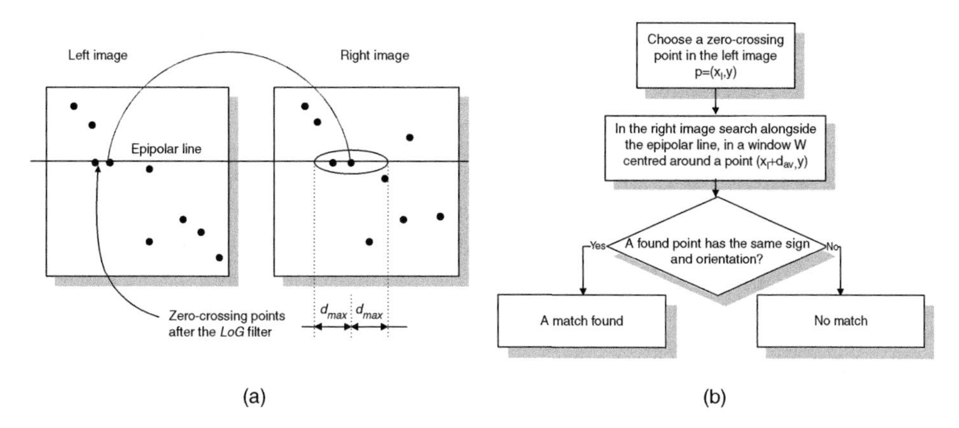

Figure 6.54 (a) Matching of zero-crossings alongside epipolar lines. (b) Flow chart of the algorithm for a single point match

At each stage of processing d_{av} has to be determined. This is done by the 'control' module in Figure 6.53. To clarify a match, the stereo matching constraints are employed. These are assumptions on disparity gradient, uniqueness and figural continuity constraints (section 3.5).

Figure 6.55 presents examples of two channels of matching based on the *LoG* zero-crossings. The 'Corridor' image was processed with 39 × 39 and 25 × 25 *LoG* filters, respectively. The plus-minus and minus-plus types of zero-crossings are denoted with different grey values (the middle column of Figure 6.55). Disparities found are visualized in the third column of Figure 6.55.

Figure 6.55 Two channels of matching of the 'Corridor' image, based on the *LoG* responses. Plus-minus and minus-plus zero-crossings are denoted with different grey values

6.8.2 Corner-based Matching

Corners convey very important information on characteristic points in images (section 4.7). Corner points with surrounding image patches can serve as very distinctive features in images which can be used for reliable point matching. This can be made stronger if before matching the patches are converted into nonparametric *Census* (section 6.3.7) or log-polar (section 6.3.8) representations. In this section we describe a method for matching of the corner points. These are detected by the structural tensor-based corner detector (section 4.7.2) and with arbitrarily set number of tiles. Then areas around corners are transformed into the log-polar domain which are finally matched in the extended log-polar search space (Figure 6.11) with the D_{CV} measure (6.7)).

Figure 6.56 presents stages of matching of corner points in the 'Venus' stereo-pair (see Table 3.4). Corners are detected with the tensor detector after dividing the left image into

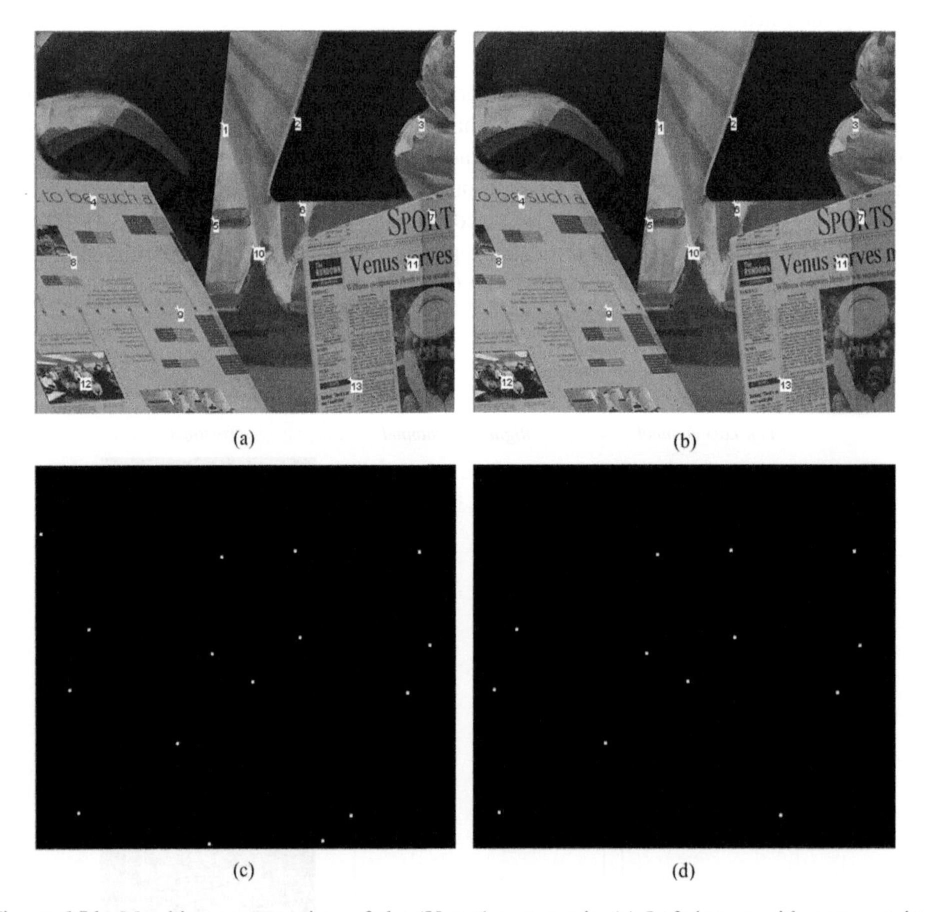

Figure 6.56 Matching corner points of the 'Venus' stereo-pair. (a) Left image with corner points. (b) Matched corner points in the right image. Corners are detected after dividing the left image into 4×4 tiles. Corner points in the (c) left and (d) right images

Table 6.12 Matched points and match values of 'Venus'. Best match denotes the best value of D_{CV}. Scale and rotation denote mutual change of scale and rotation of log-polar patches

No.	Left image corners	Right image corners	Best match value	Scale, rotation
1	(193, 91)	(188, 89)	0.994259	(0, 0)
2	(269, 85)	(263, 85)	0.998351	(0, 0)
3	(396, 86)	(389, 86)	0.998535	(0, 0)
4	(54, 163)	(44, 163)	0.994094	(0, 0)
5	(183, 187)	(177, 187)	0.997788	(0, 0)
6	(274, 171)	(267, 171)	0.994378	(0, 0)
7	(407, 179)	(395, 179)	0.934322	(0, 0)
8	(34, 223)	(21, 223)	0.974746	(0, 0)
9	(147, 276)	(135, 276)	0.988131	(0, 0)
10	(226, 215)	(219, 215)	0.982537	(0, 0)
11	(384, 226)	(372, 226)	0.95316	(0, 0)
12	(43, 345)	(26, 345)	0.95633	(0, 0)
13	(327, 348)	(314, 348)	0.95316	(0, 0)

4×4 tiles (Figure 6.56(c). After this, the log-polar representations of areas around corners in the left image are computed which are then matched with each point in the right image. The log-polar areas are 27×27 pixels. The best matches in the right image are depicted in Figure 6.56(b). Observe that some of the initial corner points were rejected before matching due to insufficient surrounding area (e.g. the leftmost corner in Figure 6.56(c)). We see that all points were matched correctly, although the process takes a few minutes on a standard PC machine.[4] This depends on the number of points and the size of the log-polar patches. However, the size of the chosen search space has a dominating influence on time consumption. The most general is a full 2D space, i.e. for each corner in the left image the whole space of the right image is searched. However, more often than not such wide space is barely justified. Thus knowledge about a camera setup can help greatly in search space reduction. We know that this can be done quite easily in the canonical stereo setup, since search space can be reduced to 1D search alongside the scanlines (section 3.5.1). Nevertheless, the search space can be reduced for other configurations as well, depending on the expected disparities. This in practice can be set to a rectangle around a point in the right image which corresponds to a zero disparity (i.e. has the same position as a test point from the left image). Depending on the camera configuration we usually expect prevailing horizontal or vertical disparity. Thus, the reduced search region should reflect this fact.

Quantitative results of the matching are contained in Table 6.12. Apart from coordinates of the corresponding points, the best match values of D_{CV}, as well as local change of scale and rotation, are included. The latter are obtained from the extended log-polar search space. It is interesting to observe that for nondistorted images (such as a stereo-pair) these two should be around zero from correct matches. Thus, if other values are encountered then the points should be checked for possible outliers.

[4] PC with Pentium Core Duo 2 GHz, 2 GB of RAM.

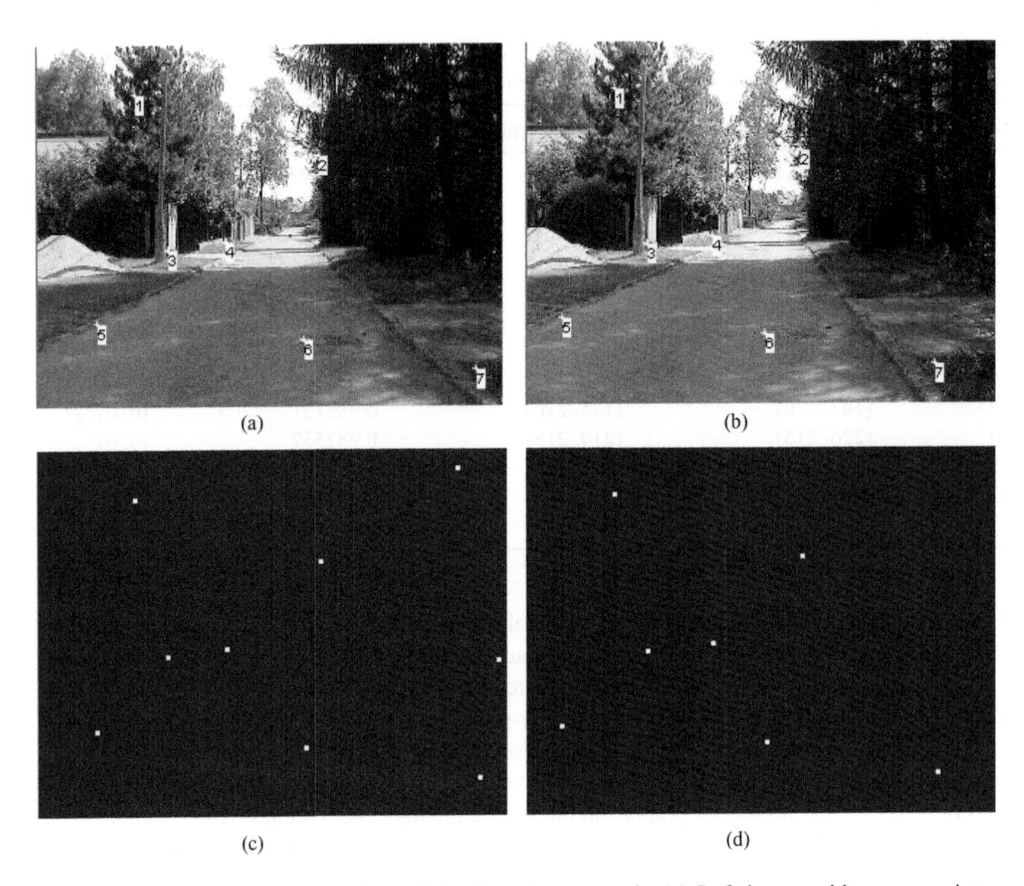

Figure 6.57 Matching corner points of the 'Street' stereo-pair. (a) Left image with corner points. (b) Matched corner points in the right image. Corners are detected after dividing the left image into 3×3 tiles. Corner points in the (c) left and (d) right images

Figure 6.57 shows matching of corner points in the 'Street' stereo-pair taken with a simple digital camera (see Table 3.5). This time the image is divided into 3×3 tiles in which corners are detected (see Figure 6.57(c)). The left and right images with matched points are depicted in Figure 6.57(a) and (b), respectively. The matching patches were chosen to 17×17 pixels. Table 6.13 contains coordinates of points found as well as match values and local scale and rotation parameters. Once again, the latter two are zero which indicates no internal change of scale or rotation in pairs of corresponding points.

Instead of D_{CV}, other matching measures can be tried as well. For instance similar results can be obtained with D_H (6.12) operating in the Census representation.

If matching images between which we expect a high degree of deformation (e.g. in registration of medical images) the SIFT method can perform better than a simple corner detector. SIFT is known to produce very discriminative features which are invariant to many even non-linear transformations, considering also a change of scale [283].

Table 6.13 Matched points and match values of 'Street'. Best match denotes the obtained value of D_{CV}. Scale and rotation denote mutual change of scale and rotation of log-polar patches

No.	Left image corners	Right image corners	Best match value	Scale, rotation
1	(66, 32)	(59, 30)	0.89668	(0, 0)
2	(193, 73)	(185, 71)	0.814084	(0, 0)
3	(89, 135)	(81, 133)	0.938967	(0, 0)
4	(129, 130)	(125, 128)	0.919967	(0, 0)
5	(40, 184)	(23, 182)	0.946919	(0, 0)
6	(183, 195)	(161, 193)	0.858052	(0, 0)
7	(301, 215)	(277, 213)	0.878956	(0, 0)

6.8.3 Edge-based Matching: The Shirai Method

An interesting combination of the feature- and area-based approaches was proposed by Shirai [246, 385]. The idea is to apply area matching but only of regions built around some edge points. These can be easily detected with the *LoG* operator (section 4.5.3) at a single scale (as opposed to the famous MPG algorithm; see section 6.8.1), or a Canny edge detector [60, 342, 381]. In the case of *LoG*, the sign of an edge is not taken into consideration. What is important in the Shirai method is that not only the edge points are matched. Instead, the regions *around* them are taken into correlation. For this purpose the $D_{SSD\text{-}N}$ (Table 6.1) measure was proposed. Moreover, the size of each region is *adapted* to the quality of the actual match.

The method starts with selection of the maximal search range R, and three threshold values, say t_1, t_2 and t_3. Then, a region of an initial size is selected in the reference image, around one of its edge points. This region is matched against possibly corresponding regions in the second image. For each match a decision is taken based on the following rules.

1. If the match is very good, say its value is below the first threshold t_1, then such a match is accepted.
2. Otherwise, if the match is very bad, say its value is above the second threshold t_2, or a maximal size of region has been reached, such a match is rejected.
3. Otherwise, the search area is reduced based on the third threshold t_3, with simultaneous increase of a size of a matching region. This corresponds to the case of a match with good indications, i.e. there are big chances that such a match will be properly classified with bigger matching region.

The pseudo-code of the above algorithm can be found in the book by Klette *et al.* [246]. Some examples of the Shirai method applied to the 'Corridor' stereo-pair are presented in Figure 6.58.

The method is not free from problems, however. The most important arises from the fact that more often than not matching regions around edge points correspond to occluded areas (section 6.4.1). Moreover, the output disparity map is sparse. The choice of so many threshold values is also troublesome.

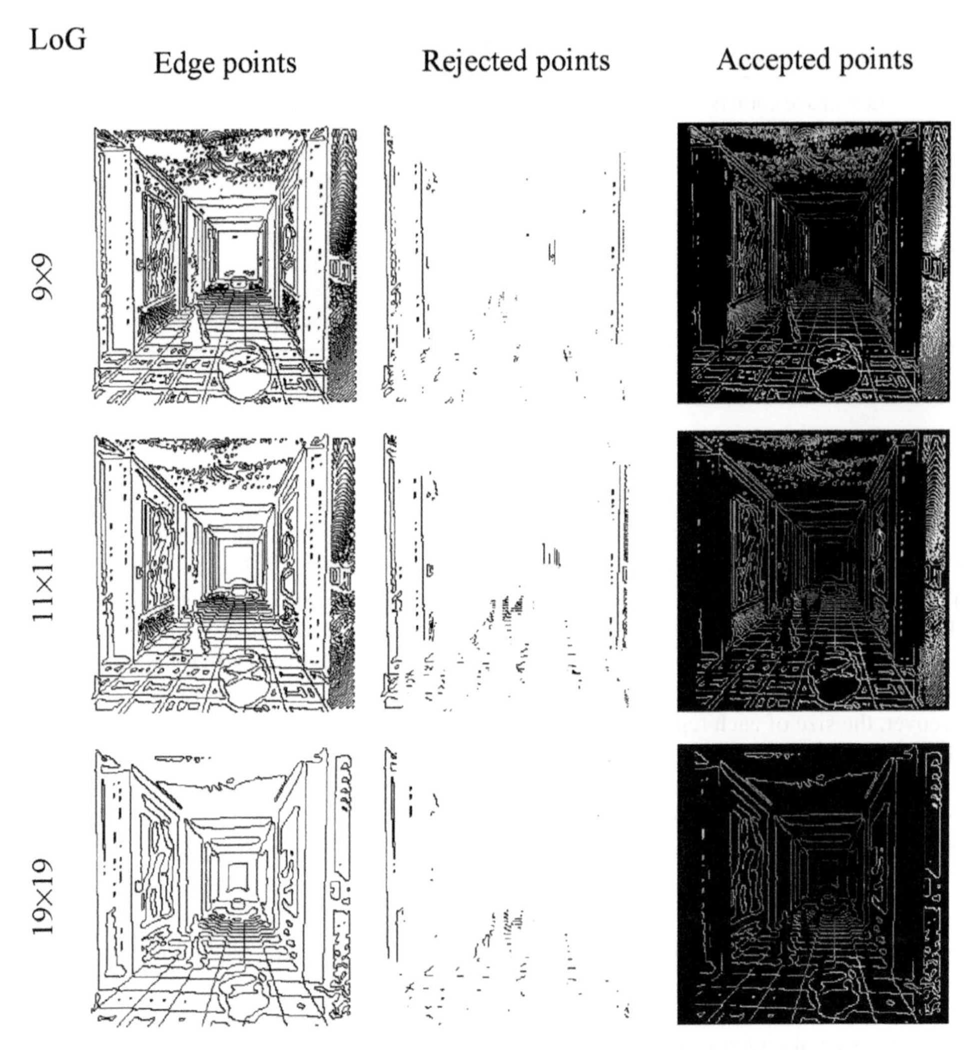

Figure 6.58 Results of the Shirai method for the 'Corridor' stereo-pair. The *LoG* edge detector is used with different mask size (rows)

6.9 Gradient-based Matching

Let us start once again from the image matching equations (6.65) and (6.66) for the standard stereo system, i.e. the one in which image scan and epipolar lines are collinear. For convenience we cite it once again:

$$I_1(x, y) = I_2(x + D(x, y), y)$$

In the gradient-based methods it is assumed that operator Ξ in (6.66) denotes a gradient, i.e. $\Xi(I) = \nabla(I)$. Such an approach was undertaken for instance by Wei *et al.* [444]. Differentiating[5] the above equation side by side with respect to the variables x and y, we obtain

$$\frac{\partial}{\partial x} I_1(x, y) = \left[1 + \frac{\partial}{\partial x} D(x, y)\right] \frac{\partial}{\partial \hat{x}} I_2(\hat{x}, y),$$

$$\frac{\partial}{\partial y} I_1(x, y) = \frac{\partial}{\partial y} D(x, y) \frac{\partial}{\partial \hat{x}} I_2(\hat{x}, y) + \frac{\partial}{\partial y} I_2(\hat{x}, y), \tag{6.119}$$

where

$$\hat{x} = x + D(x, y). \tag{6.120}$$

Observe that $I_2(\hat{x}, y)$ can be thought of as a second image deformed by an unknown disparity function $D(x, y)$. Based on the above equation the following conclusions can be drawn.

1. In the corresponding places of the original and the deformed second images, the phase and magnitude of their local gradient vectors are related by a linear equation.
2. There is a constraint on the allowable gradient of disparity.

A simple measure for gradient matching was proposed by Scharstein [369]. It is given as

$$E = \frac{1}{2} \left(\|\nabla I_1(x, y)\| + \|\nabla I_2(\hat{x}, y)\| \right) \\ -c \left\| \nabla I_1(x, y) - \nabla I_2(\hat{x}, y) \right\|, \tag{6.121}$$

where c is a constant (Scharstein suggests setting c to 1.0), and

$$\nabla I_i(x, y) = \left[\frac{\partial I_i(x, y)}{\partial x} \; \frac{\partial I_i(x, y)}{\partial y} \right]^{\mathrm{T}}.$$

The first term in (6.121) accounts for a match of gradient modulus whereas the second one concerns matching of gradient phase.

However, considering matching based solely on intensity gradients, a problem arises in places with no significant signal variation, i.e. in which gradient vanishes. The solution to the above was proposed for instance by Wei *et al.* [444]. They suggest splitting the image into regions with and without visible features. Based on this idea, their energy function for image

[5] This can be done if we assume that $I_{1,2}$ are differentiable functions. This can be assumed for discrete images, since they can be, for example, linearly interpolated to obtain their continuous representations.

matching based on gradients is given as

$$E = \sum_{(x,y)\in P} [I_1(x,y) - I_2(\hat{x},y)]^2 + \sum_{(x,y)\in F} \left\{ \left[\frac{\partial I_1(x,y)}{\partial x} - \left(1 + \frac{\partial D(x,y)}{\partial x}\right) \frac{\partial I_2(\hat{x},y)}{\partial \hat{x}} \right]^2 \right.$$
$$\left. + \left[\frac{\partial I_1(x,y)}{\partial y} - \frac{\partial D(x,y)}{\partial y} \frac{\partial I_2(\hat{x},y)}{\partial \hat{x}} - \frac{\partial I_2(\hat{x},y)}{\partial y} \right]^2 \right\}$$
$$+ c \sum_{(x,y)\in \bar{F}} \left[\left(\frac{\partial^2 D(x,y)}{\partial x^2} \right)^2 + 2 \left(\frac{\partial^2 D(x,y)}{\partial x \partial y} \right)^2 + \left(\frac{\partial^2 D(x,y)}{\partial y^2} \right)^2 \right]. \quad (6.122)$$

where P denotes a certain common region of the matched images, F is a subregion of P with visible features (gradient different from zero), $\bar{F}$ is the complement to F and c denotes a weight that controls smoothness of the solution. Observe that the last term involves second-order derivatives on disparity. Equation (6.122) follows the structure of the energy functions for image matching given by (6.67). The first term in (6.122) relates to the simple SSD-like matching of the intensity signal. The second denotes matching of the gradient signal and the last guarantees smoothness. When trying to solve (6.122) a problem arises due to image partitioning into feature and featureless regions. The solution proposed by Wei *et al.* consists of using neural networks with radial-based functions. More details can be found in [444].

Stereo matching with the gradient signal can be easily incorporated into the area-based matching frameworks (section 6.6.7). Some examples of matching with the measure (6.121) in the point-oriented fashion are presented in Figures 6.59 and 6.60. The bigger the matching windows, the more the chances are that there will be signal variations and in consequence nonzero gradient. Otherwise, the missing places are replaced by interpolated values.

6.10 Method of Dynamic Programming

Dynamic programming is an optimization method which is applicable to the multivariable problems in which not all variables are interrelated at the same time [33]. This method relies on problem decomposition into smaller ones and then assumes using partial results when trying to reach a global solution. By this strategy the computational complexity can be greatly reduced. For instance, the problem of matrix chain multiplication may be solved by dynamic programming [74]. The task is to find a product of a chain of matrices, for example

$$\mathbf{M} = \mathbf{M}_1 \mathbf{M}_2 \mathbf{M}_3, \quad (6.123)$$

assuming that the matrices in the chain have dimensions which allow their multiplication, i.e. if the dimensions of a matrix $\mathbf{M}_i$ are denoted as $r_i \times c_i$ (rows $\times$ columns), then for matrices $\mathbf{M}_i$ and $\mathbf{M}_j$ to be multiplied it must hold that $c_i = r_j$. The overall cost of such a computation, expressed in a number of multiplications, is of the order $r_i c_i c_j$.

In general the matrix product is not commutative but it is associative, i.e. the order of the multiplied matrices cannot be exchanged but the partial multiplications in a chain can be

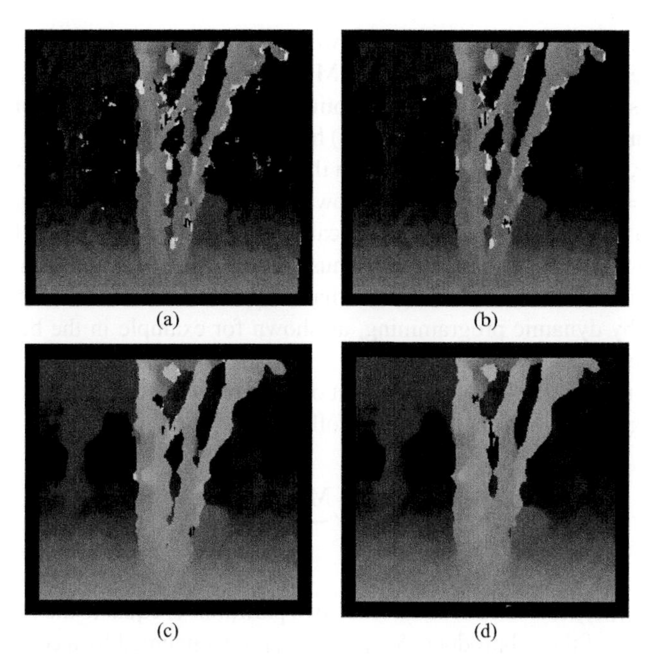

Figure 6.59 Results of the gradient matching of 'Trees'. Point-based implementation, match windows of size (a) 3×3, (b) 5×5, (c) 9×9 and (d) 11×11

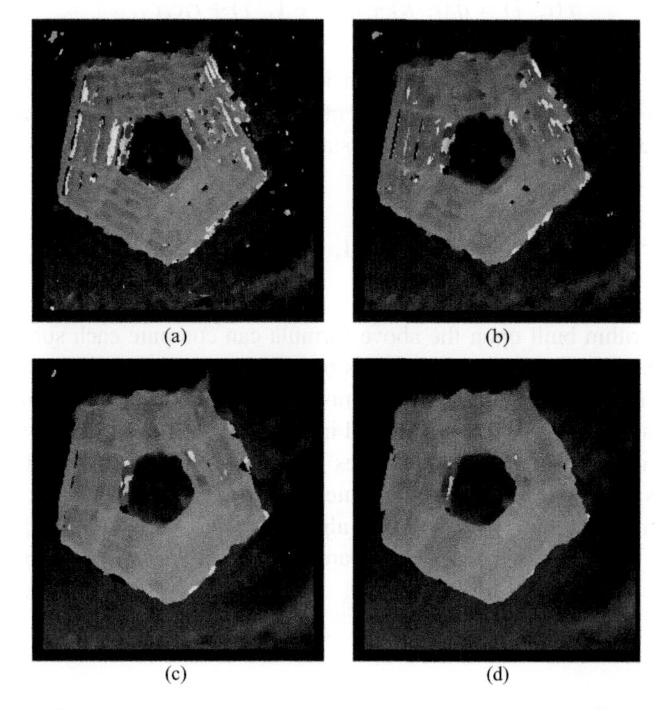

Figure 6.60 Gradient matching of the 'Pentagon' stereo-pair. Point-based implementation, match windows of size (a) 3×3, (b) 5×5, (c) 9×9 and (d) 11×11

done in any order. A first solution to (6.123) would be to multiply the matrices in a linear order, say from the left to the right, i.e. $\mathbf{M}_1$ and $\mathbf{M}_2$, then their product with $\mathbf{M}_3$. However, it is easy to show that such a strategy may not be optimal since the order of multiplication plays a role here. For example if the matrices in (6.123) have dimensions 10×10, 10×135 and 135×15, respectively, then the multiplication from the left to the right yields $(10 \times 10 \times 135) + (10 \times 135 \times 15) = 33\,750$ multiplications. However, if we first multiply $\mathbf{M}_{23} = \mathbf{M}_2\mathbf{M}_3$ and then $\mathbf{M}_1$ with $\mathbf{M}_{23}$ then the number of multiplications is greatly reduced to $(10 \times 135 \times 15) + (10 \times 10 \times 15) = 21\,750$ multiplications. Thus, determining the optimal parenthesizations allows multiplication of matrices with a minimal computational cost. This problem can be solved efficiently by dynamic programming, as shown for example in the book by Cormen *et al.* [74]. We briefly outline the method.

The first step consists of formulation of a cost of multiplication of the subproducts $S_{(i,k)}$ and $S_{(k+1,j)}$ partitioning the chain of n matrices, as follows:

$$\mathbf{M}_1 \ldots \underbrace{\mathbf{M}_i \mathbf{M}_{i+1} \ldots \mathbf{M}_k}_{S_{(i,k)}} \underbrace{\mathbf{M}_{k+1} \ldots \mathbf{M}_j}_{S_{(k+1,j)}} \ldots \mathbf{M}_n. \tag{6.124}$$

Then the total cost $q(i,j)$ of multiplication of the two partitions is equal to the minimum cost of partial computations of the subproducts $S_{(i,k)}$ and $S_{(k+1,j)}$, augmented by a cost of multiplying these products together, that is

$$q(i,\ j) = q(i,\ k) + q(k+1,\ j) + r_i c_k c_j, \tag{6.125}$$

since the partitions $S_{(i,k)}$ and $S_{(k+1,j)}$ are of dimensions $r_i \times c_k$ and $r_{k+1} \times c_j$, respectively, and it holds also that $c_k = r_{k+1}$. Certainly, if $i = j$ then $q(i,j) = 0$ since this is a single matrix. Hence, the total cost of the optimal partitioning can be stated as the following recursive formula [74]:

$$q(i,\ j) = \begin{cases} 0 & \text{if} \quad i = j \\ \min\limits_{i \le k < j} \left\{ q(i,\ k) + q(k+1,\ j) + r_i c_k c_j \right\} & \text{if} \quad i < j \end{cases}. \tag{6.126}$$

The recursive algorithm built upon the above formula can compute each subproblem many times. In such a case dynamic programming is of help. One of its paradigms is to compute *partial results* which are then used in a bottom-up fashion to find a solution to the whole problem. These partial computations are stored in a look-up table[6] which for the chain multiplication problem is of size $n \times n$ (n denotes the number of matrices in the chain). The algorithm requires additional storage of the same size for tracing whose index k achieved an optimal cost when computing $q(i, j)$. The full algorithm can be found in [74]. Its computational complexity is of the order $O(n^3)$, compared to the exponential time of a brute force approach.[7]

[6] Just the use of tables to store partial results was a reason why the method uses the word 'programming'.
[7] There are faster solutions which run in $O(n \log n)$ time.

Dynamic programming can help solve problems of minimization of the energy functionals E of many variables, which can be stated as follows [33, 276]:

$$\min_{\boldsymbol{\theta}} \{E(\boldsymbol{\theta})\} = \min_{\theta_1,\theta_2,\ldots,\theta_n} \{E(\theta_1, \theta_2, \ldots, \theta_n)\}. \quad (6.127)$$

If the energy functional in the above can be decomposed into a series of terms with independent variables

$$E(\boldsymbol{\theta}) = E(\theta_1, \theta_2, \ldots, \theta_n) = E_1(\theta_1, \theta_2) + E_2(\theta_2, \theta_3) + \ldots + E_{n-1}(\theta_{n-1}, \theta_n). \quad (6.128)$$

then the dynamic programming formulation leads to a series of functions with one variable:

$$\begin{aligned}
F_1(\theta_2) &= \min_{\theta_1} \{E_1(\theta_1, \theta_2)\}, \\
F_2(\theta_3) &= \min_{\theta_2} \{F_1(\theta_2) + E_2(\theta_2, \theta_3)\}, \\
&\vdots \\
F_i(\theta_{i+1}) &= \min_{\theta_i} \{F_{i-1}(\theta_i) + E_i(\theta_i, \theta_{i+1})\}, \\
&\vdots
\end{aligned} \quad (6.129)$$

Thus, solution to (6.127) with assumption (6.128) is

$$\min_{\boldsymbol{\theta}} \{E(\boldsymbol{\theta})\} = \min_{\theta_n} \{F_{n-1}(\theta_n)\}, \quad (6.130)$$

where $F_{n\text{-}1}(\theta_n)$ is given in (6.129). The formulation (6.130) leads to *global minimum* of E if E can be decomposed as in (6.128), i.e. into a series of terms with independent variables. Otherwise the solution can be suboptimal.

Dynamic programming has been employed in many tasks of computer vision, such as Markov random fields, curve detection, active contours and also in stereo correlation as will be shown in the next section [135, 276].

6.10.1 Dynamic Programming Formulation of the Stereo Problem

The stereo problem can be expressed in terms of dynamic programming when formulated as a problem of finding an optimal path through a set of nodes which represent possible matches between the scanlines in the left and right images respectively. The global cost of this optimal path represents a cumulative cost over the partial costs in each scanline. The local costs, in turn, can be determined for each pixel or for certain features, such as edges.

Dynamic programming formulation of the stereo problem in the two stages of the intra- and inter-scanline search was proposed by Ohta and Kanade [334]. This is a feature-based method since only points that belong to the edge intervals are matched. A goal of each intrascan search is to determine corresponding points alongside the scanlines which are also the epipolar lines since the canonical stereo setup is assumed (section 3.4.2). This can be treated as finding

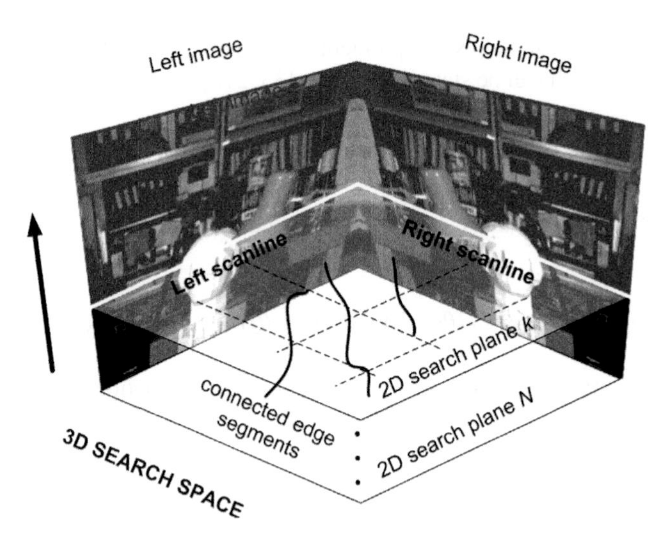

Figure 6.61 3D search space of the stereo methods employing dynamic programming

an optimal path in a 2D search space whose axes are the corresponding scanlines from the left and right images, respectively. However, taking each pair of scanlines separately does not provide information on figural continuity of the vertical edges which are expected in the observed scenes (section 3.5). Thus, to take advantage of this information Ohta and Kanade propose using the inter-scanline search in a 3D space composed of stacked 2D search planes already used in the intra-scanline process. The number of 2D planes equals vertical resolution of the stereo images. Thus, the correspondence problem is translated into finding an optimal matching surface in the 3D space. The cost of this matching equals the cost of the intra-scanline matches on the 2D planes, penalized however for those intra-scanline matches which violate the figural consistency assumption (details in [334]). Figure 6.61 illustrates a 3D search space composed of a series of 2D search planes.

Dynamic programming is employed in the two search stages which run simultaneously. The intra-scanline provides information on figural consistency, whereas the intra-scanline supplies the matching score. The latter is based on a similarity measure for edge intervals. However, application of dynamic programming requires strict ordering on computation of the partial results. That is, before computing a new result, all the previous partial results have already been processed. The second requirement is that computation of the current result does not depend on the history of previous computations. In terms of edge matching these translate to the requirement that if we are matching two edges from the left and right image respectively, all edges to the left of these in the two images must have been already processed. To fulfil this requirement edges are endowed with indexes in the left-to-right order on each scanline. In other words, application of dynamic programming requires fulfilment of the *uniqueness and ordering constraints* among the matches (section 3.5). A similar ordering constraint has to be superimposed on the intra-scanline search as well.

A dynamic programming solution to the stereo problem which optimizes a maximum likelihood cost function was proposed by Cox [76, 77]. This assumes that corresponding features in the matched images follow a Gaussian distribution with the mean being their true value.

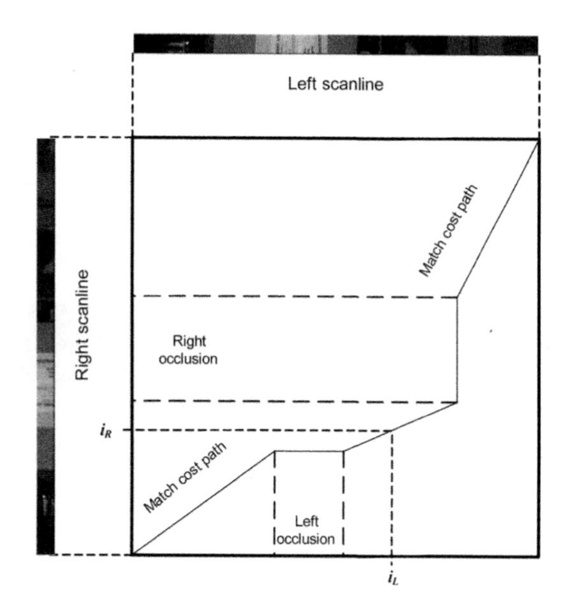

Figure 6.62 2D search space profile. (After [77])

If the features are matched then the cost function is represented as a weighted squared error term. Otherwise, if a feature is suspected to be occluded, the cost function is constant. However, instead of specific features, pixel intensities are used for matching. Using this method, one finds a dense disparity map, saving computations on feature extraction at the same time. Cox showed that if properly used the bare intensities can provide fair matching results.

Figure 6.62 depicts an exemplary match profile in a 2D search space for chosen right and left scanlines [77]. A point at index i_L in the left image is matched to a point at index i_R in the right image. However, these points have to fulfil the uniqueness and ordering constraints. Thus, it is only allowed that $j_L > i_L$ can be matched to a point at index $j_R > i_R$. The horizontal part of a profile represents left occlusion because many points in the left image are matched to the same point in the right image. Similarly the vertical profile denotes occlusion in the right image.

The cost function is formulated as the maximum likelihood (ML) problem which does not require knowledge of the prior probability density function, necessary when using the Bayes scheme [77, 237]. In this approach we are simply interested in direct maximization of the probability $p(z|\mathbf{X})$ of a likelihood of a measurement z if it originated in a point $\mathbf{X}$ in the scene. To clarify correspondences Cox proposed using N views instead of only two. From a selected pair of cameras the set of best correspondences is determined. From these the corresponding 3D points are reconstructed which are then back projected on to remaining $N - 2$ planes. The projections are used to verify the initial matches. The advantage of this approach is modelling of the occlusions.

Although dynamic programming helps in finding a global optimum with a polynomial complexity, stereo methods that rely on it are not free from problems, however. The most severe limitation is imposed by the requirements of the uniqueness and ordering constraints which are not always fulfilled in real scenes (see Figure 3.22). As a consequence errors can occur

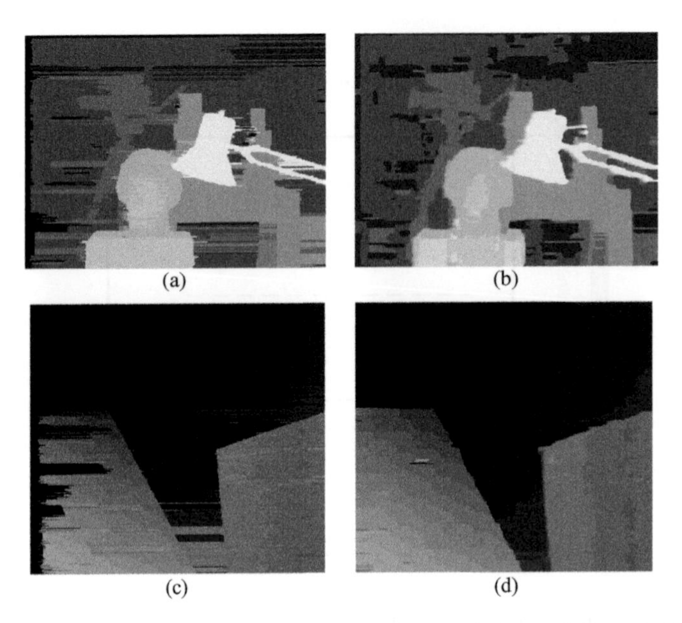

Figure 6.63 Stereo matching with dynamic programming: 'Tsukuba' and 'Venus'. (a, c) Method of Scharstein and Szeliski [371]; (b, d) method by Wang *et al.* [443]. (From [209])

which, if propagated along the scanlines, manifest with visible erroneous horizontal 'streaks' in disparity maps. This is noticeable in the results of the method by Scharstein and Szeliski in Figure 6.63(a, c).

Intille and Bobick propose a stereo method with dynamic programming that explicitly models occlusions and uses them to drive the matching [219]. The search for matches and occlusions is done with the help of a data structure called the disparity-space image. Matches with high confidence are used as ground control points to eliminate sensitivity to occlusion costs (details in [219]).

There are many other methods which employ dynamic programming for the stereo problem. For instance Meerbergen *et al.* propose a hierarchical stereo method that matches individual pixels with different cost functions [307]. Because of this approach the method offers low computational complexity which is independent of disparity range. The cost in each scanline is computed incrementally, i.e. the results are reused by adding new matches to a sequence. The optimal sequence is then found with dynamic programming.

Wang *et al.* [443] developed a stereo method that employs an adaptive aggregation step in a dynamic programming framework. First, a 3D cost space is built with the simple D_{SAD} measure computed over single pixel pairs and assumed range of disparities up to $d_{\max}$. The energy function for the stereo problem follows a general functional (6.67) with the smoothness term defined as

$$E_{\text{smooth}}(\theta) = \lambda \sum_{x} |d(x) - d(x+1)|, \tag{6.131}$$

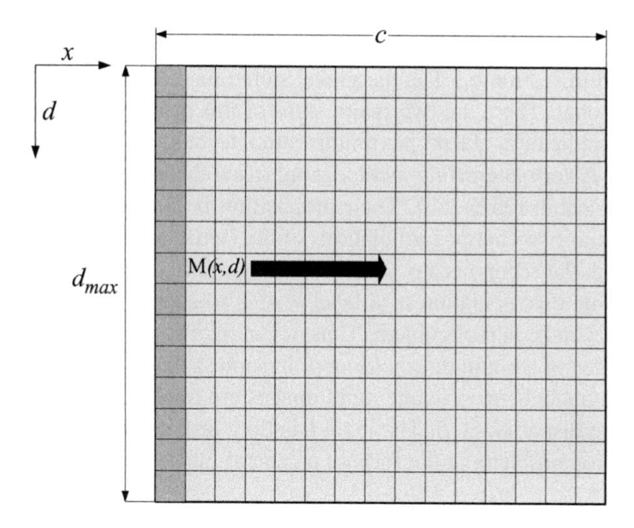

Figure 6.64 2D search space used in the method of Wang *et al.* [443]

where d denotes disparity (in the range $0...d_{\max}$) and λ is a parameter that penalizes depth discontinuities. Then, dynamic programming is employed: the 2D search space $\mathbf{M}$ is constructed which is a $c \times d_{\max}$ matrix, where c is the number of columns in the input images (similar to the method by Intille and Bobick [219] as depicted in Figure 6.64).

$\mathbf{M}$ is initialized from the computed D_{SAD} costs among pairs of pixels. Then the matrix is updated in the scanline direction, according to the formula [443]

$$\mathbf{M}(x,d) = \mathbf{M}(x,d) + \min\left\{\mathbf{M}(x-1,d-1)+\lambda, \mathbf{M}(x-1,d), \mathbf{M}(x,d+1)+\lambda\right\}. \quad (6.132)$$

The process ends when reaching the rightmost column, which corresponds to the last pixel in the scanline. The best path, which reflects the disparity for this scanline, is found by back tracking. This process is repeated for all scanlines, i.e. for the number of rows in the input images. The aggregation is computed as a weighted sum of per-pixel costs taking into account the colour and geometrical proximity. This feature allows effective processing on occlusion boundaries. However, the aggregation is done solely in the vertical direction. Because of this, the 'streaking' effect and also computation time are greatly reduced (see Figure 6.63(b, d)).

Dynamic programming on a tree for the stereo correspondence problem is proposed by Veksler [436]. A modification comes from Deng and Lin who propose a line segment-based stereo method that uses tree dynamic programming [97]. In their approach each epipolar line is segmented and then a tree is built with the obtained segments. Dynamic programming on this tree is used to find out correspondences of each line segment. Using line segments instead of pixels allows preservation of connections of neighbouring pixels and improves time performance of the method. For accurate labelling of occlusions Criminisi *et al.* propose dynamic programming based on a four state matching graph [79]. Their dense stereo matching is augmented with a view synthesis based on direct projection of the minimum cost surface. The method greatly eliminates the artefacts characteristic of many other stereo methods that rely on dynamic programming.

6.11 Graph Cut Approach

As already discussed in section 6.5.1.2, the global stereo methods rely on minimization of a certain energy functional. There are two major steps of this process: development of the energy functional and application of a proper minimization technique for the problem, as alluded to in section 6.5.1.2. A very interesting method comprising the above two steps was proposed by Kolmogorov and Zabih [252, 253]. Their proposition is formulated as a pixel labelling problem which leads to two energy formulations of the stereo problem with occlusions. The first is voxel labelling; the second is a pixel labelling algorithm.

Pixel labelling assumes association of a label $\theta_i \in L$ to each pixel $\mathbf{p}$ in an image $\mathbf{I}$. The meaning of a label depends on the problem. Thus, it can indicate an object in object detection tasks, it can be an index of a bin in thresholding or it can be a disparity value. However, in the latter case to associate labels in an image more images are used (e.g. two or more views are necessary to compute disparity, etc.). The pixel labelling task can be formulated in terms of an energy minimization problem, in the form of a general functional (6.67), as follows:

$$E(\theta) = \sum_{i \in \mathbf{I}} Q_i(\theta_i) + \sum_{\{i,j\} \in A} V(\theta_i, \theta_j), \tag{6.133}$$

where $\theta = (\theta_1, \theta_2, \ldots, \theta_{|\mathbf{I}|})$ is a labelling to be found, Q_i is a penalty term for assigning a label to a pixel i from image $\mathbf{I}$, V denotes a penalty term associated with a pair of labels to adjacent pixels (i.e. of splitting a local neighbourhood by different labels) and A is a set of adjacent (or, generally, interacting) pairs of pixels [51, 276]. In accordance with (6.67), the first term in (6.133) corresponds to data costs for the labelling θ, which actually makes θ fulfil the conditions of a modelled problem with input data. For the matching problem Q_i can be any matching measure discussed in section 6.3. However, due to computational properties usually SSD is a first choice, as follows:

$$Q_i(\theta_i) = [I_l(p_{i1}, p_{i2}) - I_r(p_{i1} + \theta_i, p_{i2})]^2, \tag{6.134}$$

where $\mathbf{p}_i = (p_{i1}, p_{i2})$ is the i-th point in an image and label θ_i denotes a disparity.

The second term in (6.133) forces the spatial smoothness. In matching tasks the penalty term V depends on scene geometry. If V has a strong influence then the solution will tend to oversmooth, the same effect as a large matching window in local area-based matching (section 6.6). The smoothness term is sometimes called a Potts model [51, 252, 253]. For the frontoparallel configurations it is usually proposed to define V as follows:

$$V(\theta_i, \theta_j) = \lambda T[\theta_i \neq \theta_j], \tag{6.135}$$

where λ is a penalty value and T is an indicator function which takes value 1 if its argument is fulfilled and 0 otherwise.

6.11.1 Graph Cut Algorithm

As alluded to previously, development of an energy functional for a problem is a first step to the solution of a problem. What is necessary now is an efficient minimization procedure that solves a minimization problem encoded in the energy functional [127, 331, 352]. Although many methods exist, such as the already mentioned neural networks (back propagation, Hopfield, RBF, etc.), genetic algorithms, simulated annealing, tabu search, dynamic programming and many more, the graph cut offers many advantages.

However, before trying to solve (6.140) or (6.145) with the graph cut, these constrained problems need to be converted into unconstrained ones. This can be accomplished by adding an additional term E_{valid} into the energy functionals [253].

The graph cut algorithm is inspired by the combinatorial optimization methods for maximum flow [74]. In all 'standard' optimization methods at each step of computing new energy value in accordance with (6.73) only a label of a single pixel can be changed. Contrary to this, in the graph cut approach larger moves are proposed. These are:

- the α–β swap;
- the α expansion.

These are explained in Figure 6.65. The initial labelling assumes the existence of three labels α, β and γ (Figure 6.65(a)). A standard move allows only change of a single pixel at a step. In Figure 6.65(b) this is a single pixel previously labelled β, exchanged into γ. In α–β swap

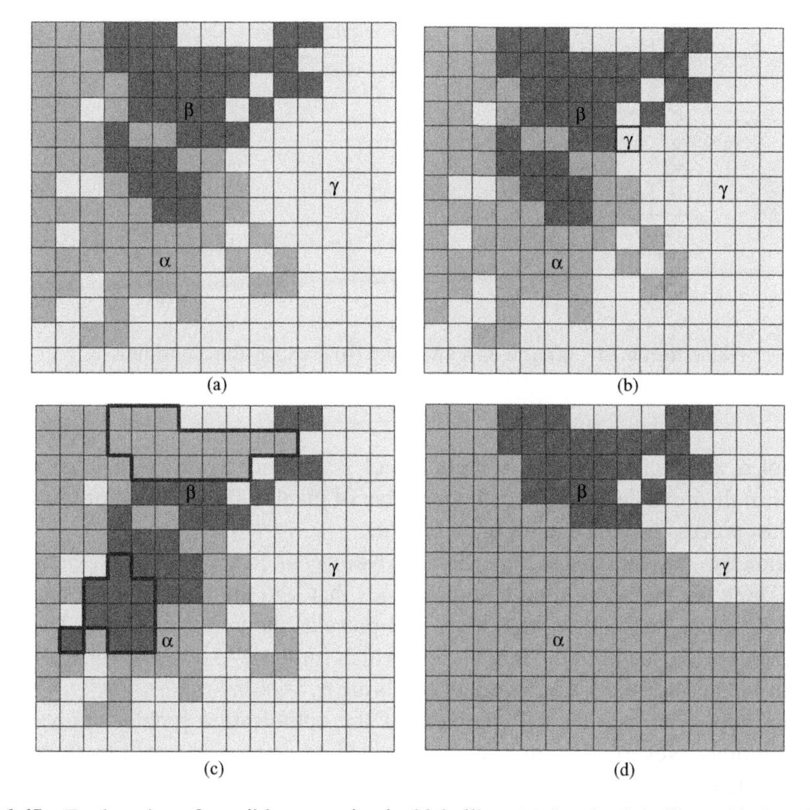

Figure 6.65 Explanation of possible moves in pixel labelling. (a) Starting labelling with three labels α, β, γ. (b) A standard move allows change of a single pixel at a step – a single pixel labelled β exchanged into γ. (c) In α–β swap some areas of α are exchanged with some areas of β and vice versa, with γ not changed. (d) In α expansion move a large number of pixels labelled β and γ is changed into α

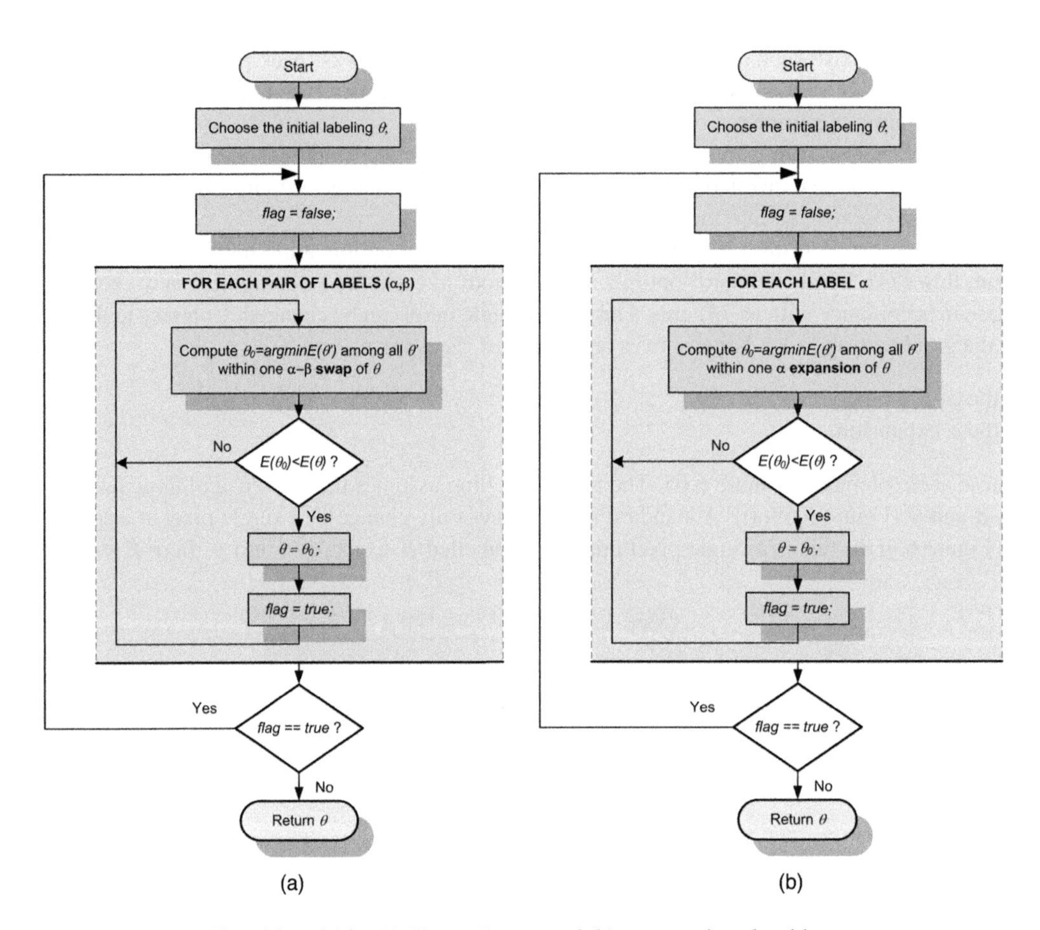

Algorithm 6.14 (a) The α–β swap and (b) α expansion algorithms

some areas of α are exchanged with some areas of β and vice versa, with γ not changed (Figure 6.65(c)). In α expansion move a large number of pixels labelled β and/or γ is changed into α. The last process is depicted in Figure 6.65(d).

Having defined α–β swaps and α expansions, Boykov *et al.* [51] propose two minimization algorithms. These are efficient graph-based methods that find the optimal α–β swap and α expansion for a given labelling θ. Algorithm 6.14 presents the two approaches. The two algorithms are identical except for the inner loops, which for the α–β swap traverse all pairs of (α, β) labels, whereas for the α expansion all labels are α.

Both algorithms overcome the NP-hardness of the optimization problem providing approximate solutions. Nevertheless in practice the method converges very rapidly [51], due to the large number of pixels changing their labels simultaneously in the α–β swaps or α expansions.

The algorithms are guaranteed to stop after the first unsuccessful run of the energy minimization loops (see Algorithm 6.14). In the worst case this can be reached after checking all (α, β) pairs for the α–β swap, or after α labels for the α expansion algorithm, respectively.

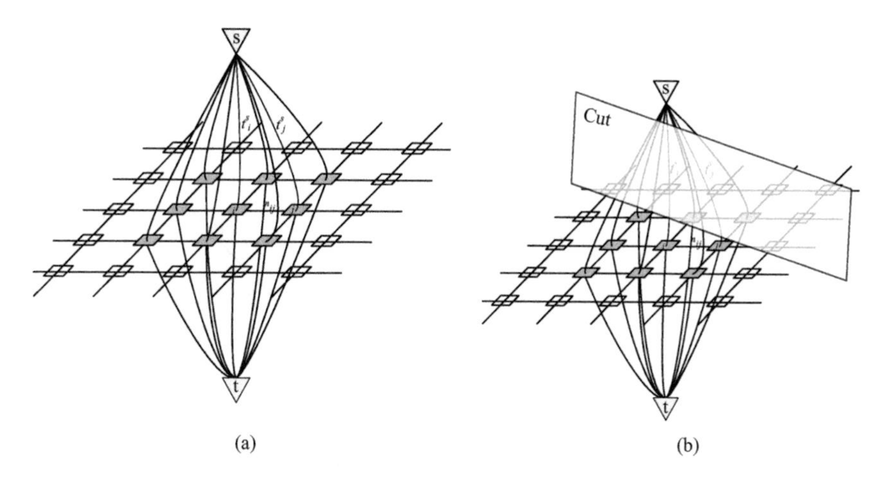

Figure 6.66 Graph representation used to model computer vision tasks. Pixels are connected by n-links. The terminal vertices are source 's' and sink 't'. (a) They denote pixel labels and are connected with pixels by the t-links. (b) A graph cut splits a graph into two subgraphs, each containing exactly one terminal

6.11.1.1 Graphs in Computer Vision

A graph G is defined as a pair $G = \{V, E\}$ where V denotes a set of vertices and E are edges among them. Additionally it is assumed that there are two specific vertices called terminals. These terminals are usually denoted as a *source* and a *sink*, and denoted by 's' and 't' respectively. With each edge $E(V_i, V_j)$ there is associated a weight $w_{ij} = W(E(V_i, V_j))$. Notice that in the general case it can hold that $w_{ij} \neq w_{ji}$, however.

In the graph representation of computer vision tasks it is common to represent each pixel as a separate vertex in the graph. The source and sink terminal vertices denote labels that can be assigned to pixels (a labelling problem). In this case we are usually concerned with two labels, since a problem with a higher number of labels can be divided into subproblems with two labels. In this case there are also two types of edges: the ones that connect pixel vertices and the ones that connect the terminals with pixels. These are called *n-links* and *t-links*, respectively. A cost of an *n-link* represents a penalty associated with discontinuity between adjacent pixels, i.e. it is denoted by the second term in (6.133). On the other hand, a *t-link* corresponds to cost of a label assigned to that pixel. This is the first term in (6.133).

A cut C in a graph G is a smallest set of edges ($C \subset E$) that when removed from G induce two subgraphs G_1 and G_2 such that each contains exactly one terminal node. A cost $|C|$ of a cut C is a sum of all weights of its edges, i.e. $|C| = \Sigma w_i$ for all $E_i \in C$. Figure 6.66(a) depicts a graph with two terminals. A cut in this graph is presented in Figure 6.66(b).

A minimum cut problem is to find a cheapest cut among all cuts that separate the terminal nodes.

As shown by Boykov *et al.* [51], finding a minimal energy $E(\theta')$ in the two algorithms in Algorithm 6.14 (first steps in shaded areas), is equivalent to solving the minimum cut problem on a graph with two terminals. This is a well-known problem in computer science which can be accomplished in polynomial time with the help of combinatorial algorithms [30, 74]. For instance, a minimum cut can be computed as a maximum flow between the terminal vertices

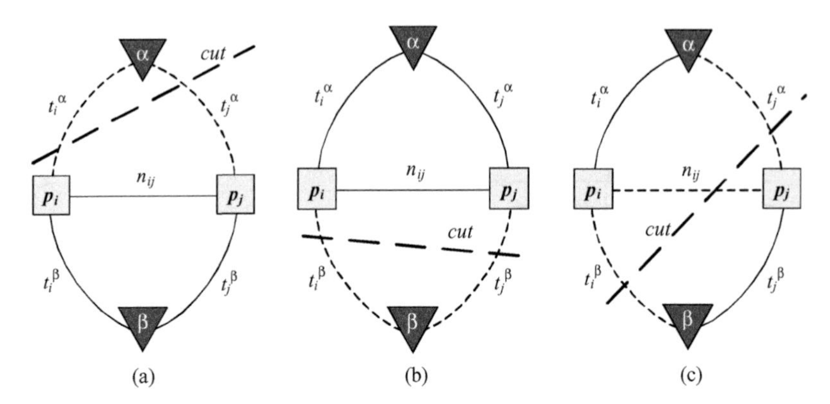

Figure 6.67 Possible swap cuts C on a graph $G_{\alpha\beta}$ for two pixels p_i and p_j. The pixels are connected to themselves by the n_{ij} link and to the terminals by *t-links*. Any allowable cut has to leave only one terminal connected to a pixel. Possible cuts are depicted from (a) to (c)

following the well-known method by Ford and Fulkerson [134]. An improvement to this was first proposed by Edmonds and Karp [112]. They noticed that choosing the shortest paths in each step of the flow increase reduces computational complexity to the polynomial, compared to the original formulation. Further improvement comes from Dinic who proposed to split the problem into separate stages which actually divide a graph into a layered network [104]. A suitable algorithm is also provided also in the work by Boykov and Kolmogorov [52].

6.11.1.2 Optimization on Graphs

Figure 6.67 presents possible cuts C for the swap moves on a graph G for the two neighbour pixels p_i and p_j, connected by the n_{ij} link and by t_i^α, t_i^β, t_j^α, t_j^β to the terminals, respectively.

Any cut in a graph has to sever exactly one *t-link* since otherwise there will be connection between terminals whereas it is assumed that any cut separates the terminals. As a consequence each graph cut leaves each pixel with exactly one *t-link*, which in turn defines a labelling θ^C corresponding to that cut C. In the light of (6.133) the weights associated with the edges are defined as follows [51]:

$$
\begin{aligned}
t_i^\alpha &: \quad Q_i(\alpha) + \sum_{j \in N(i), j \notin P_{\alpha\beta}} V(\alpha, \theta_j), \\
t_i^\beta &: \quad Q_i(\beta) + \sum_{j \in N(i), j \notin P_{\alpha\beta}} V(\beta, \theta_j), \\
n_{ij} &: \quad V(\alpha, \beta)\big|_{\{i,j\} \in N, (i,j) \in P_{\alpha\beta}},
\end{aligned}
$$

where $P_{\alpha\beta}$ denotes a union of pixels which are assigned the labels and $N(i)$ denotes a neighbourhood of pixels around a pixel indexed by i.

The key corollary stated in [51] says that the lowest energy labelling within a single α–β swap move from θ is θ^C, where C denotes a minimum cut on $\mathrm{G}_{\alpha\beta}$. The α–β swap allows V to be a semimetric. However, as shown by Boykov *et al.* [51] it does not guarantee the optimality

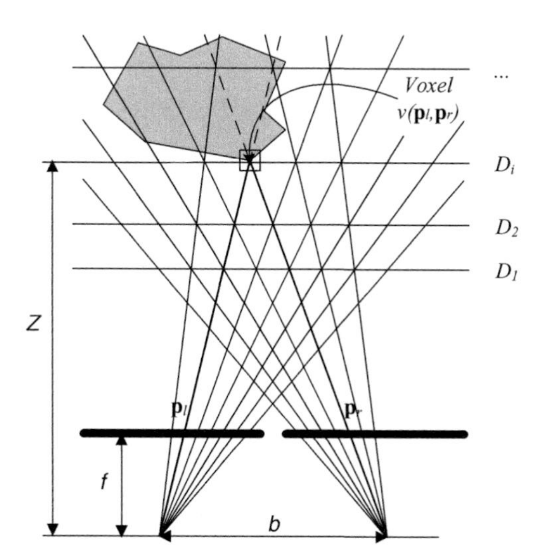

Figure 6.68 Stereo setup observing an object. A voxel $v = (\mathbf{p}_l, \mathbf{p}_r)$ is active since it corresponds to a real 3D point belonging to an observed object which is visible from the two cameras simultaneously

properties compared to the α expansion move. Description of the optimal expansion move as well as further details of the method can be found in [51, 52, 254].

6.11.2 Stereo as a Voxel Labelling Problem

Now let us return to the stereo problem formulation based on voxels. A voxel v is an unordered pair of pixels in the two images that correspond to a certain 3D point, i.e. $v = (\mathbf{p}_l, \mathbf{p}_r) = (\mathbf{p}_r, \mathbf{p}_l)$ (Figure 6.68).

In the canonical stereo setup it is characteristic of some disparity value $D_x(\mathbf{p}_l, \mathbf{p}_r) = p_{r1} - p_{l1}$ (3.39), with $p_{r2} = p_{l2}$ (3.42), as already discussed in section 3.4.2. In contrast to the traditional pixel-based formulation the voxel-based representation of stereo tries to explicitly model the whole 3D space of a scene and the cameras (each voxel belongs to the 3D space). The energy functional in the voxel formulation takes the following form [252, 253]:

$$E(g) = E_{\text{data}}(g) + E_{\text{occlusion}}(g) + E_{\text{smooth}}(g). \tag{6.136}$$

The first data term is defined as

$$E_{\text{data}}(g) = \sum_{v \in V} g(v)Q(v), \tag{6.137}$$

where $g: V \to \{0, 1\}$ is a labelling which for a given voxel $v = (\mathbf{p}_l, \mathbf{p}_r)$ assigns $g(v) = 1$ if that voxel contains pixels which correspond to each other, i.e. they are images of a real 3D point; $g(v) = 0$ for all other voxels. A voxel for which $g(v) = 1$ is called an active voxel, such as the one in Figure 6.68. $Q(v)$ is analogous to the formulation (6.134). The second term takes into account the existence of occlusions:

$$E_{\text{occlusion}}(g) = C_{\text{occ}} \left| P_{\text{occ}}(g) \right|, \tag{6.138}$$

where C_{occ} is a penalty value for an occlusion and $P_{\text{occ}}(g)$ denotes a set of occluded pixels in configuration g. These are such pixels as $\mathbf{p}_l$ which belong to such voxels v for which: $g(v) = 0$. Finally, the smoothness term requires that the neighbouring voxels have more or less the same disparity. This can be formulated as

$$E_{\text{smooth}}(g) = \sum_{\{v_i, v_j\} \in N_v} \lambda T \left[g(v_i) \neq g(v_j) \right], \tag{6.139}$$

where λ and T are defined in (6.135).

Now the stereo problem is computed as the following constrained optimization problem:

$$g_0 = \arg\min_{g \in C_{\text{valid}}} E(g), \tag{6.140}$$

where $E(g)$ is given by (6.136) and C_{valid} denotes a set of all valid configurations of voxels, i.e. the ones that fulfil the uniqueness constraint (section 3.5). In other words, these are such configurations g for which if there are two voxels v_1 and v_2 which contain the same first pixels $\mathbf{p}_{l1} = \mathbf{p}_{l2}$ and different second pixels $\mathbf{p}_{21} \neq \mathbf{p}_{22}$, then it holds that $g(v_1) = 0$ or $g(v_2) = 0$ or both.

6.11.3 *Stereo as a Pixel Labelling Problem*

In pixel labelling formulation of the stereo problem we directly follow the general functional (6.67):

$$E(h) = E_{\text{data}}(h) + E_{\text{smooth}}(h). \tag{6.141}$$

Its data term is defined as

$$E_{\text{data}}(h) = \sum_{v \in V} [h(\mathbf{p}_l) = h(\mathbf{p}_r) = D_x(v)] Q'(v), \tag{6.142}$$

where condition $h(\mathbf{p}_l) = h(\mathbf{p}_r) = D_x(v)$ means that a disparity $D_x(v)$ for a voxel $v = (\mathbf{p}_l, \mathbf{p}_r)$ is the same as a label $h(\mathbf{p}_l)$ and $h(\mathbf{p}_r)$; $Q'(v)$ is defined as

$$Q'(v) = \min\{0, Q(v) - C\}, \tag{6.143}$$

since for the graph cut method it should be nonpositive; C is a positive constant.

Now, the smoothing term is defined as

$$E_{\text{smooth}}(h) = \sum_{\{v_i, v_j\} \in N_v} \lambda T \left[h(v_i) \neq h(v_j) \right], \tag{6.144}$$

where again λ and T are defined in (6.135), assuming the Potts model.

As in the previous case, the constrained minimization problem

$$h_0 = \underset{h \in C_{\text{valid}}}{\arg\min}\, E(h) \tag{6.145}$$

has to be solved for $E(h)$ given by (6.141).

As pointed out in [253] the pixel labelling approach has some improvements over the voxel-based approach. The first property of pixel labelling is prohibition of holes in a scene, since it assumes that a layer with disparity 0 corresponding to the plane at infinity is totally filled. The second property is that the pixel labelling method allows models other than the Potts one. Nevertheless, both methods favour the frontoparallel surfaces in the observed scene.

Figure 6.69 presents results of stereo matching obtained by Kolmogorov and Zabih with their graph cut method for the test images with ground truth [209]. RGB colour images are used in the input. The Potts model is controlled by one parameter λ which also can depend on the pair of pixels. Such a strategy has the advantage of discouraging discontinuities between neighbouring pixels with very similar intensities. In other words, if it holds that $I(\mathbf{p}_l) \approx I(\mathbf{p}_r)$ then most likely the pixels $\mathbf{p}_l$ and $\mathbf{p}_r$ will have the same disparity as well (i.e. reversed situations are quite rare in real situations). This technique of adopting the contextual information is known as 'static cues' [51]. Thus, instead of independent λ we assume $\lambda(\mathbf{p}_i, \mathbf{p}_j)$ which depends on a relation of values of the pixels $\mathbf{p}_i$ and $\mathbf{p}_j$. In [253] it was proposed as

$$\lambda\left(\mathbf{p}_i, \mathbf{p}_j\right) = \begin{cases} 3K & \textit{for} \quad \left|I\left(\mathbf{p}_i\right) - I\left(\mathbf{p}_j\right)\right| < 5 \\ K & \textit{otherwise} \end{cases},$$

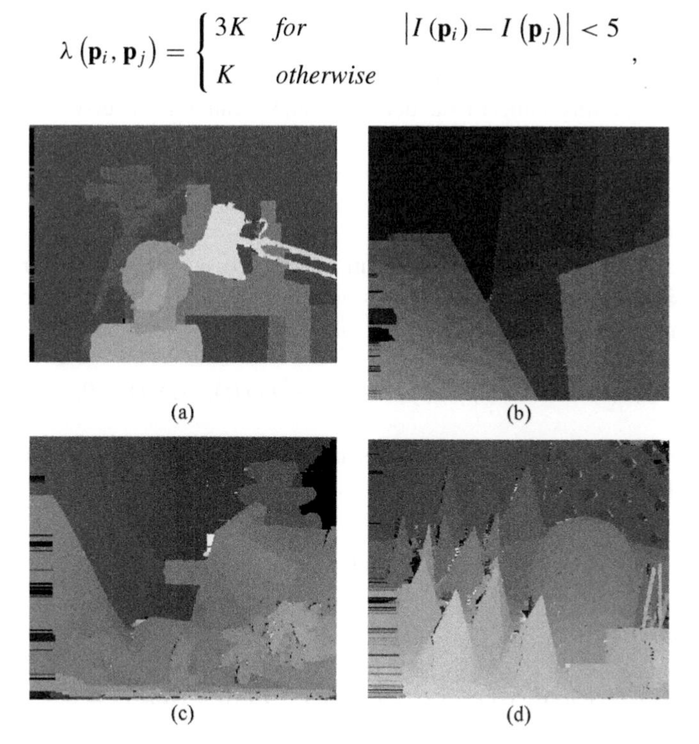

Figure 6.69 Stereo matching with graph cut method by Kolmogorov and Zabih [209]: (a) 'Tsukuba', (b) 'Venus', (c) 'Teddy' and (d) 'Cones'

for the pixel labelling (K is a parameter of the Potts model), and

$$\lambda\left(v_i, v_j\right) = \begin{cases} 3K & \textit{for} \qquad \max\left\{\left|I\left(\mathbf{p}_i\right) - I\left(\mathbf{p}_j\right)\right|, \left|I\left(\mathbf{q}_i\right) - I\left(\mathbf{q}_j\right)\right|\right\} < 8 \\ K & \textit{otherwise} \end{cases},$$

for the voxel labelling, where voxels are given as pairs $v_i = (\mathbf{p}_i, \mathbf{q}_i)$. $v_j = (\mathbf{p}_j, \mathbf{q}_j)$ and $\mathbf{p}_i$, $\mathbf{p}_j$ are pixels in the same image, and $\mathbf{q}_i$, $\mathbf{q}_j$ in the other one.

The qualitative parameters of the graph cuts and other stereo matching methods can be obtained from [209]. From Figure 6.69 we easily notice the sharp edges of the objects in the output disparity maps which result from the formulation of the energy functionals.

The maximum flow formulation of the stereo problem was also formulated by Ishikawa and Geiger [221]. Their proposed method computes a disparity map by solving a global optimization task that models occlusions and discontinuities.

6.12 Optical Flow

Optical flow refers to the problem of estimating a vector field of local displacements in a sequence of images (e.g. in a video stream). In the formulation of the optical flow problem we assume that a certain point (x_1, y_1) in an image acquired at instant t_1 will be matched by a point (x_2, y_2) in an image at instant t_2. The local displacement – which is assumed to be relatively not very distant, at least in terms of the image size – gives an answer on movements of objects observed in the subsequent images.

Thus, we see that the optical flow is an image matching problem, so we can start once again from (6.64). Now, assuming a linear local deformation, we can write a version of (6.64) aimed at solution of the optical flow problem [199]:

$$I_1(x, y) = I_2(x + \Delta x, y + \Delta y), \tag{6.146}$$

where I_1 and I_2 are two consecutive views from a sequence, acquired at time instants t_1 and t_2, respectively. Since a sequence of images is assumed, we can drop the indices of images adding a time stamp. Then, the above equation takes on the form

$$I(x(t) + \Delta x, y(t) + \Delta y, t + \Delta t) - I(x(t), y(t), t) = 0. \tag{6.147}$$

Relative displacements in the x and y directions are done with certain speeds, u and v, respectively. They are defined as

$$u \equiv \frac{\Delta x}{\Delta t} \quad \text{and} \quad v \equiv \frac{\Delta y}{\Delta t}. \tag{6.148}$$

With this notation we obtain

$$I(x(t) + u\Delta t, y(t) + v\Delta t,\ t + \Delta t) - I(x(t), y(t),\ t) = 0. \tag{6.149}$$

For the optical flow we assume small local displacements compared to image size, so it is justified to substitute the discrete displacements Δx, Δy and Δt with their infinitesimal counterparts dx, dy and dt. With this assumption the previous equation transforms to

$$I\left(x(t)+u\mathrm{d}t,\, y(t)+v\mathrm{d}t,\;\, t+\mathrm{d}t\right)-I\left(x(t),\, y(t),\;\, t\right)=0. \tag{6.150}$$

Now, assuming that I is a differentiable function, the first term in the above equation can be expanded around a point (x, y, t) using the Taylor[8] series. Taking only the first element of this series we obtain

$$\frac{\partial I}{\partial x}\frac{\mathrm{d}x}{\mathrm{d}t}+\frac{\partial I}{\partial y}\frac{\mathrm{d}y}{\mathrm{d}t}+\frac{\partial I}{\partial t}=0, \tag{6.151}$$

or in a more compact form

$$I_x u + I_y v + I_t = 0, \tag{6.152}$$

where I_x, I_y and I_t are partial derivatives of I with respect to x, y and t, respectively, while u and v express horizontal and vertical velocities in a sequence and are infinitesimal versions of (6.148). This equation is also called *a brightness constancy constraint* [199], since it expresses the idea of 'similar' brightness for the same objects observed in a sequence. When we fix our attention to a single point and measure velocities u and v flowing through that location then the problem is called the *optical flow*. Thus, the velocity vector $[u, v]^{\mathrm{T}}$ is called at a *single* location in an initial image. Otherwise, when we 'follow' with a given location and trace their position in consecutive images of a sequence, then the problem is called *feature tracking*.

As shown by Slesareva *et al.* [393], the optical flow approach can be embedded into the framework of stereo matching, so the two domains are also closely related.

Trying to solve a single equation, (6.152), for the two variables u and v denotes an underconstraint problem. Thus, the optical flow cannot be unambiguously determined – this is called an aperture problem. However, we can try to place a second constraint, such as the velocity vector is constant within a small neighbourhood Ω, placed around a certain point (x_0, y_0) in the input image – an idea originally proposed by Lucas and Kanade [284]. With this assumption, the optical flow problem can be approached by the following minimization task:

$$\underset{u,v}{\arg\min}\, E\left(u,\;\, v\right), \tag{6.153}$$

---

[8]Recall that the Taylor expansion for differentiable functions is given as follows:

$$\begin{aligned} f\left(x_1+dx_1, x_2+dx_2, \ldots, x_n+dx_n\right) &- f\left(x_1, x_2, \ldots, x_n\right) \\ &= df+\frac{1}{2!}d^2 f+\ldots+\frac{1}{(m-1)!}d^{m-1}f+\Re_m \end{aligned}$$

where $\Re_{\mathrm{m}}$ denotes a reminder and df is a total differential of an *n-dimensional* function f

$$df=\frac{\partial f}{\partial x_1}dx_1+\frac{\partial f}{\partial x_2}dx_2+\ldots+\frac{\partial f}{\partial x_n}dx_n$$

where the energy function is given as

$$E(u,\ v) = \int\limits_{\Omega(x_0, y_0)} \left(I_x u + I_y v + I_t\right)^2 \mathrm{d}x\mathrm{d}y. \tag{6.154}$$

A minimum of $E(u, v)$ has to meet the following conditions:

$$\frac{\partial E(u,\ v)}{\partial u} = 0 \quad \text{and} \quad \frac{\partial E(u,\ v)}{\partial v} = 0, \tag{6.155}$$

which leads to the linear system of equations

$$\begin{cases} \int\limits_{\Omega(x_0, y_0)} \left[2I_x^2 u + 2I_x I_y v + 2I_x I_t\right] \mathrm{d}x\mathrm{d}y = 0 \\ \int\limits_{\Omega(x_0, y_0)} \left[2I_y^2 v + 2I_x I_y u + 2I_y I_t\right] \mathrm{d}x\mathrm{d}y = 0 \end{cases}, \tag{6.156}$$

which can be expressed in a more compact matrix form as

$$\begin{bmatrix} \int\limits_{\Omega(x_0, y_0)} I_x^2 \mathrm{d}x\mathrm{d}y & \int\limits_{\Omega(x_0, y_0)} I_x I_y \mathrm{d}x\mathrm{d}y \\ \int\limits_{\Omega(x_0, y_0)} I_x I_y \mathrm{d}x\mathrm{d}y & \int\limits_{\Omega(x_0, y_0)} I_y^2 \mathrm{d}x\mathrm{d}y \end{bmatrix} \begin{bmatrix} u \\ v \end{bmatrix} = - \begin{bmatrix} \int\limits_{\Omega(x_0, y_0)} I_x I_t \mathrm{d}x\mathrm{d}y \\ \int\limits_{\Omega(x_0, y_0)} I_y I_t \mathrm{d}x\mathrm{d}y \end{bmatrix}. \tag{6.157}$$

However, let us compare now the above with the equations defining the structural tensor (section 4.6.2). Thus, we can express the close form solution to the optical flow problem with a local constancy constraint, in terms of the structural tensor [58]:

$$\begin{bmatrix} T_{xx} & T_{xy} \\ T_{yx} & T_{yy} \end{bmatrix} \begin{bmatrix} u \\ v \end{bmatrix} = - \begin{bmatrix} T_{xt} \\ T_{yt} \end{bmatrix}. \tag{6.158}$$

Thus, in places where the structural tensor is not singular, local velocities $[u, v]^{\mathrm{T}}$ at a point (x_0, y_0) can be found to be

$$u = \frac{T_{yt} T_{xy} - T_{xt} T_{yy}}{T_{xx} T_{yy} - T_{xy}^2} \quad \text{and} \quad v = \frac{T_{xt} T_{xy} - T_{yt} T_{xx}}{T_{xx} T_{yy} - T_{xy}^2}. \tag{6.159}$$

An even better approach is to endow the energy functional (6.154) with a regularization term Φ, as was already presented in the case of energy functions (6.67) designed for stereo correspondence (section 6.5.1.2). Now (6.154) takes the form

$$E(u, v) = \int_{\Omega(x_0, y_0)} \left[\left(I_x u + I_y v + I_t \right)^2 + c\Phi(\Delta I, \Delta u, \Delta v) \right] \mathrm{d}x\mathrm{d}y. \tag{6.160}$$

The regularization function Φ proposed by Nagel and Enkelmann [326] imposes smoothness everywhere except across the edges. The TV based regularization that allows discontinuity preserving smoothing was proposed by Cohen [71]. The high accuracy optic flow technique which is based on the theory of warping was proposed by Brox *et al.* [59].

To facilitate comparison of different optical flow methods, very useful are common test sequences with ground truth. In this respect a very good collection has been prepared by Baker *et al.* [22] from Middlebury University [210].

Figure 6.70(a, b) presents two frames from the 'Yosemite' test sequence, originally created by Lynn Quam (a version with clouds), and available from [203]. The vector fields of the ground-truth [203] and computed velocities $[u, v]^T$ are depicted in Figure 6.70(c) and (d), respectively. Estimated motion in this sequence starts from about 2 pixels per frame in the upper right area up to about 5 pixels per frame in the lower left corner.

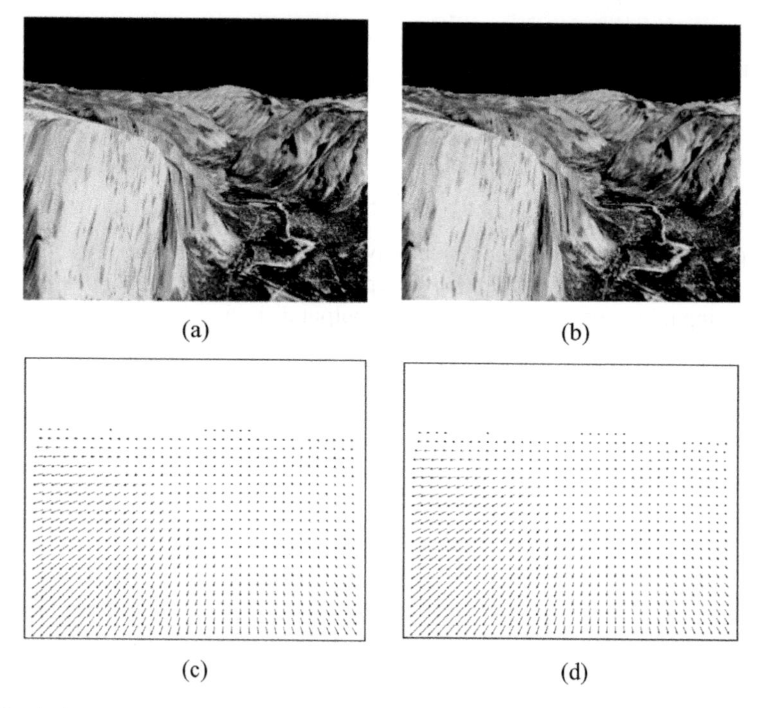

Figure 6.70 (a, b) Two frames from the 'Yosemite' test sequence. (c) Ground-truth velocities $[u, v]^T$ vector field, (d) computed vector field. (From [210])

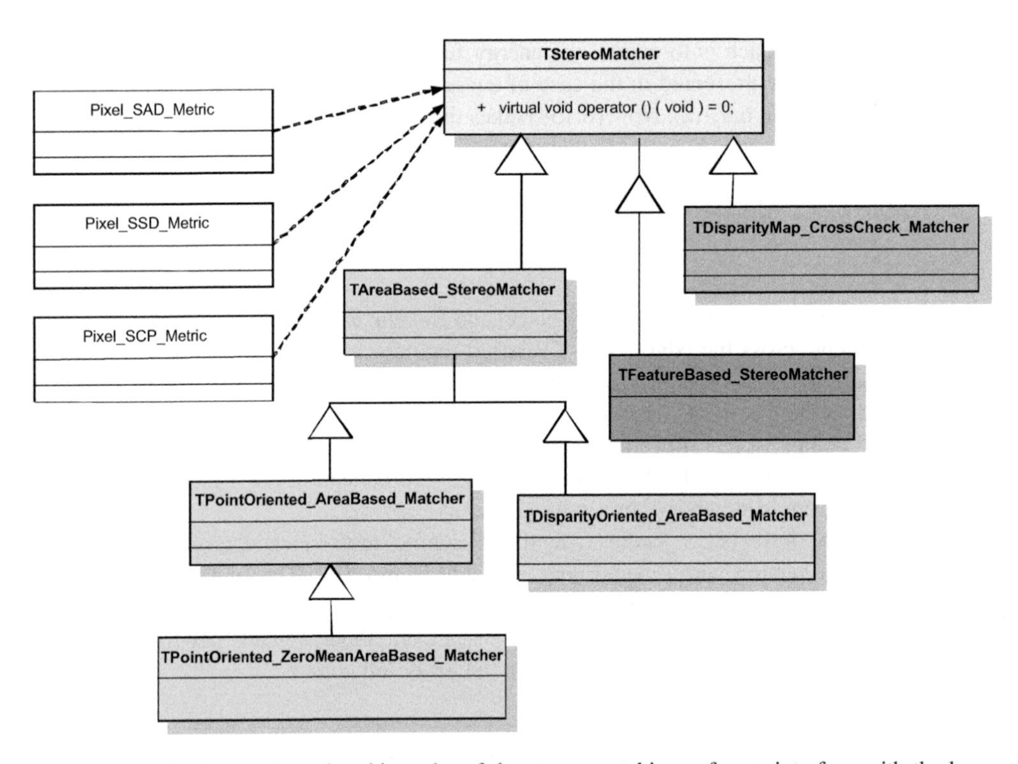

Figure 6.71 Basic template class hierarchy of the stereo matching software interface with the base class *TStereoMatcher*

6.13 Practical Examples

6.13.1 Stereo Matching Hierarchy in C++

Figure 6.71 presents the template class hierarchy of the image matching interface. The pure virtual base template class *TStereoMatcher* accepts two template parameters which are types of pixels in the input images, as well as a type of output disparity values.

From the *TStereoMatcher* are derived the following classes.

1. *TAreaBased_StereoMatcher* – this is the base class for all area-based stereo matching methods.
2. *TDisparityMap_CrossCheck_Matcher* – the auxiliary class for disparity map cross-checking.
3. *TFeatureBased_StereoMatcher* – a branch of feature-based matching methods.

The base class *TAreaBased_StereoMatcher* defines a common framework for image matching. This is achieved by the overloaded function operator which is then defined in derived classes.

The basic matching measures, such as SAD or SSD, have been designed as auxiliary classes: *Pixel_SAD_Metric*, *Pixel_SSD_Metric* and *Pixel_SCP_Metric*. The main member is the virtual functional operator. Other matching measures can be added in the same way. These classes are used then as template parameters for the matching classes.

The *TAreaBased_Matcher* template class is a base for the group of point-oriented area-based matching methods, such as *TPointOriented_AreaBased_Matcher* and *TDisparityOriented_AreaBased_Matcher*.

Finally, the *TFeatureBased_StereoMatcher* starts a branch of stereo matching methods that rely on feature matching. These can be the ones discussed in section 6.8, but also new classes can be easily added.

6.13.2 Log-polar Transformation

The log-polar transformation can be computed with the image warping modules described in Chapter 12. The bilinear interpolation is usually sufficiently accurate. The only thing to do is definition of the log-polar warp engines whose role is to convert coordinates from the Cartesian system into the log-polar space. Examples of the two are listed in Algorithms 6.15 and 6.16.

```
//
// The log-polar coordinate transformation engine.
//
class TLogPolar_TransformEngine : public TNonLinearTransformEngine
{
	protected:

		double fLogBase; // actually this is not a base but log( base )
		Real_2D_Point fCenterPoint;

	public:

		// =====================================================
		TLogPolar_TransformEngine( const Real_2D_Point & centerPoint,
									double
								logBase = 10.0 )
		{
			REQUIRE( logBase != 0.0 && logBase != 1.0 );
			fLogBase = log( logBase );
		}

		// class virtual destructor
		virtual ~TLogPolar_TransformEngine() {}
		// =====================================================


		///////////////////////////////////////////////////////////
		// This function converts a given point based to its
		// log-polar representation.
		///////////////////////////////////////////////////////////
		//
		// INPUT:
		//				in - the input point
		//
		// OUTPUT:
		//				the output (converted) point
		//
		// REMARKS:
		//
		//
```

Algorithm 6.15 Listing of the warp engine for the log-polar transformation of coordinates. (Reproduced with permission of Pandora Int. Inc., London)

```
        virtual Real_2D_Point operator () ( const Real_2D_Point & in )
        {
                double dx = in.x - fCenterPoint.x;
                double dy = in.y - fCenterPoint.y;

                double r = sqrt( dx * dx + dy * dy );
                r = r == 0.0 ? - DBL_MAX : log( r ) / fLogBase;
                double phi = dx == 0.0 ? kPiHalf : atan( dy / dx );

                return Real_2D_Point( r, phi );
        }
};
```

Algorithm 6.15 (*Continued*)

```
//
// The inverse log-polar coordinate transformation engine.
//
class TInvLogPolar_TransformEngine : public TLogPolar_TransformEngine
{
    public:

        // ===================================================
        TInvLogPolar_TransformEngine( const Real_2D_Point & centerPoint,

                                                  double logBase = 10.0 )
        {
                REQUIRE( logBase != 0.0 && logBase != 1.0 ); //wrong input
        // we need to copy it again since the base constructor
                // have already changed it
                fLogBase = logBase;
        }

        // class virtual destructor
        virtual ~TInvLogPolar_TransformEngine() {}
        // ===================================================


        ///////////////////////////////////////////////////////////
        // This function converts a given point based to its
        // inverse log-polar representation.
        ///////////////////////////////////////////////////////////
        //
        // INPUT:
        //                      in - the input point
        //
        // OUTPUT:
        //                      the output (converted) point
        //
        // REMARKS:
        //
        //
        virtual Real_2D_Point operator () ( const Real_2D_Point & in )
        {
                // The input "x" is "r", whereas "y" is "theta":
                double _power = pow( fLogBase, in.x );
                return Real_2D_Point( _power * cos( in.y ) + fCenterPoint.x,
                                                                  _power
                                          * sin( in.y ) + fCenterPoint.y );
        }
};
```

Algorithm 6.16 Listing of the warp engine for the inverse log-polar transformation of coordinates. (Reproduced with permission of Pandora Int. Inc., London)

Algorithm 6.15 presents a definition of the simple *TLogPolar_TransformEngine* class, derived from the *TNonLinearTransformEngine* framework (see Figure 12.5). Its main method is overloaded function operator whose role is to convert the input Cartesian coordinate into its log-polar representation. There are two parameters, the base of the logarithm and the central point. These are discussed in section 6.3.8.

Algorithm 6.16 lists the definition of the inverse log-polar transformation class *TInvLogPolar_TransformEngine*. This is derived from the *TLogPolar_TransformEngine*, presented in Algorithm 6.15. It operates in accordance with (6.52). Since the backward warping scheme is usually preferred, this class fits into the inverse transformation scheme required by this type of warping.

The two classes can be made more optimal since if operating in an image patch the same values are computed many times, they can be stored and reused to save on computations. For instance, the value of B^r can be processed this way. This nicely fits into the *TGenericTransformEngine*, presented in Figure 12.5, which builds a transformation look-up table. This data structure allows much faster processing than on-line computation of each value. However, it is at the cost of memory consumption.

6.14 Closure

In this chapter the basic matching methods and techniques are discussed. These attract much attention from the vision research community since they constitute the basic mechanisms of depth perception, motion analysis and object detection in digital images. Therefore we try to give an overview of the classic methods in this field, with special stress on the ones which find direct practical applications.

We start with an outline of the most common groups of comparison measures, for image regions, for bit streams, statistical, as well based on theory of information. Most of them find applications in all computer tasks which necessitate comparisons of different types of data. One such task is stereovision in which comparison measures are used in the search for point correspondences. The computational aspects of stereo processing are discussed next. Among many, these are problems of occlusions and subpixel depth estimation.

The rest of this chapter is devoted to provide basic information on the diversity of stereo matching methods. We start with overall classification of the methods and describe the main processing steps. Then the major groups of stereo methods are discussed, such as area-based matching, area-based elastic matching and the feature-based and gradient-based methods. The chapter ends with an introduction to the dynamic programming, graph cut and optical flow methods.

Some C++ implementations of the basic methods are also discussed; their full implementations are available from the accompanying web page.

6.14.1 Further Reading

Additional information on matching measures can be found in many texts on image processing and computer vision, such as the books by Gonzalez and Woods [157] or Pratt [351]. One of the best textbooks on the theory of information is that by Cover and Thomas [75].

There is a relatively large number of publications on stereo matching, though scattered in many different scientific publications. An excellent source of up-to-date information on stereo matching methods is the paper by Scharstein and Szeliski [370], as well as the web page of Middlebury University devoted to comparison of stereo methods [209]. It is also an ample source of further references and synthetic test data, which are also used in this book. A good overview can also be found in the paper by Brown *et al.* [57]. Other sources of information on some stereo matching techniques can be accessed in the books by Faugeras and Luong [119], Hartley and Zissermann [180], as well as in Faugeras [122], Scharstein [369] and Klette *et al.* [246].

A good overview of visual labelling, as well as local and global optimization methods in the context of computer vision is provided in the book by Li [276]. Another source of information on this subject is the already mentioned paper by Boykov *et al.* [51], which also provides ample references to other works in this area. For linear programming and network flows, very recommended is the book by Bazaraa *et al.* [30].

6.14.2 Problems and Exercises

1. Using the simple matching model for two images, design and implement an algorithm for matching histograms computed in local regions of two images. As a histogram matching measure assume the Kullback–Leibler measure D_K given in (6.43). What can we tell about this method? What matching measures other than D_K can be used?
2. Prove Equation (6.10).
3. Starting from the code for area-based matching (see Algorithm 6.6), implement the Shirai method (section 6.8.3).

7

Space Reconstruction and Multiview Integration

7.1 Abstract

Space reconstruction relates to the techniques of recovering information about the structure of a 3D space based on direct measurements or depth computation from stereo matching. This gives positions and dimensions of the sensed object surfaces and this information can, for instance, be used for robot navigation or to guide surgery procedures.

In this chapter we deal with the basics of space reconstruction and multiview integration. Depending on the available parameters of the acquisition system(s) different parameters of the space can be determined. A basic triangulation gives rise to the so-called 2.5D depth reconstruction. However, if full 3D surface manifold information is required, multiple view integration techniques come into play in order to achieve volumetric integration of recovered 2.5D surfaces. Hence, in this chapter we also discuss 3D surface construction methods based on implicit surfaces and marching cubes, as well as direct mesh integration.

7.2 General 3D Reconstruction

The essence of multiple view processing is to acquire some information about the 3D structure of the observed scene. However, it need not always be in the form of absolute Euclidean coordinates of visible objects in a predefined coordinate system attached to that scene. For many applications either it is not necessary or it is not even possible to get such coordinates, for example due to missing camera calibration data [122, 164, 188, 369]. It is an interesting observation to recall here that the human visual system does not perform tedious camera calibrations and numerical 3D reconstruction, and yet we are able to move and orient quite easily, e.g. driving a car [302, 442]. Thus, the soft computing methods that mimic biological behaviour in many aspects can come into play.

The accuracy of 3D reconstruction depends on availability and accuracy of data of the camera setup. A detailed analysis of the 3D reconstruction with respect to the accuracy of the camera calibration parameters was presented by Grimson [164]. He showed that the reconstruction process based on available disparities extracted from stereo-pair images has

---

An Introduction to 3D Computer Vision Techniques and Algorithms Bogusław Cyganek and J. Paul Siebert

Table 7.1 Breakdown of 3D data reconstruction in respect of the available calibration parameters

	Available calibration data	Possible 3D space reconstruction
I	The extrinsic and intrinsic parameters of the camera setup	The 3D Euclidean coordinates. (Precise reconstruction, called also triangulation)
II	Only the intrinsic parameters available	Reconstruction up to a certain scaling factor
III	Extrinsic and intrinsic data not available	Reconstruction up to a certain projective transformation

a critical and nonlinear dependency on the accuracy of the camera calibration parameters. Especially important is the precise computation of the camera central points (section 3.3.2, Figure 3.60), as well as the deviation angle of the camera optical axes.

An important role in the task of object recognition is played by so-called image invariants, especially those that do not require a precise 3D reconstruction. Pattern matching with the help of image invariants can be made much simpler because they convey important information on encountered image objects regardless of their scale, position, luminance, etc. One of the most common invariants is the cross ratio (section 9.7) [63, 180, 322, 380].

As alluded to previously, depending on availability and accuracy of calibration data associated with the camera setup used, there are different possible degrees of 3D reconstruction. Generally three characteristic cases can be distinguished here [2, 122, 428, 430, 459].

1. Full reconstruction of the Euclidean 3D space.
2. Reconstruction up to a certain scaling factor.
3. Reconstruction up to a certain projective transformation.

Table 7.1 compares the three characteristic reconstruction possibilities given the available calibration data. We shall present foundations of each in the next sections.

7.2.1 Triangulation

Triangulation is a process of finding coordinates of a 3D point (Figure 3.7) based on its corresponding image points $\mathbf{p}_r$ and $\mathbf{p}_l$, lying on the camera planes, as well as knowledge of calibration data (section 3.4). In this sense it seems to be a straightforward technique since it is assumed that the calibration was already done and thus the calibration parameters are assumed to be known beforehand (section 3.6). In practice, however, due to discrete space and limited accuracy of found positions of the points $\mathbf{p}_r$ and $\mathbf{p}_l$, the two rays from these points through camera centres $\mathbf{O}_l$ and $\mathbf{O}_r$ do *not* intersect in a single point [171]. To overcome this problem we can try to find an approximating crossing point $\mathbf{P}_E$, such that it lies a minimal distance from the two rays simultaneously [2, 180, 430]. This situation is illustrated in Figure 7.1.

Our task is to determine a position of the approximating 3D point $\mathbf{P}_E$. This can be found from the linear equations (7.1). Then a middle distance, on the segment parallel to the vector $\mathbf{J}$, and connecting the two rays crossing through the points $\mathbf{p}_l$ and $\mathbf{p}_r$, has to be found. This is given by the following equation [430]:

$$a\mathbf{p}_l + c\mathbf{J} = \mathbf{T} + b\mathbf{R}^T\mathbf{p}_r, \tag{7.1}$$

where coordinates $a, b, c \in \Re$; $a\mathbf{p}_l$ is an equation of the ray crossing through the central point $\mathbf{O}_l$ (for $a = 0$), as well as $\mathbf{p}_l$ (for $a = 1$); and $\mathbf{T} + b\mathbf{R}^T\mathbf{p}_r$ is an equation of the ray crossing

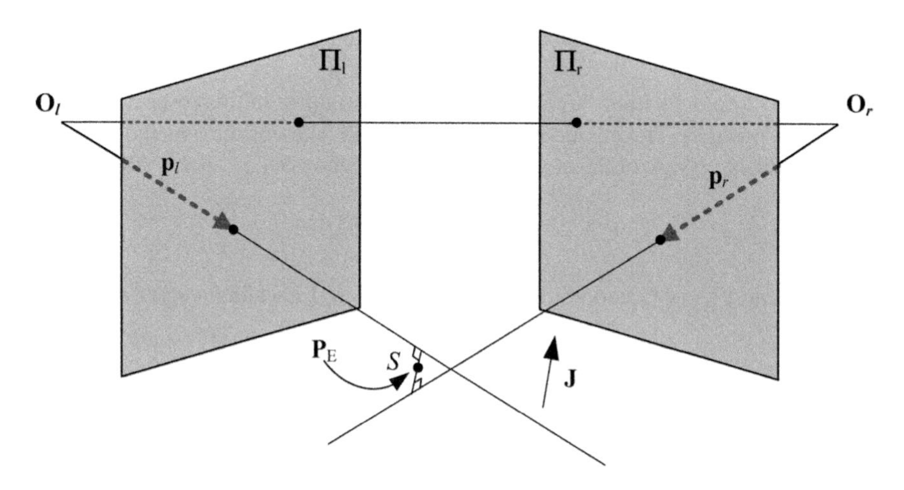

Figure 7.1 Triangulation with an approximated ray crossing point

through the central point $\mathbf{O}_r$ (for $b = 0$) and $\mathbf{p}_r$ (for $b = 1$), in respect of the coordinate system of the left camera. $\mathbf{J}$ has to be a vector parallel to the two rays. Thus, it has to fulfil the following

$$\mathbf{J} = \mathbf{p}_l \times \mathbf{R}^T \mathbf{p}_r, \tag{7.2}$$

where $\mathbf{R}$ denotes the rotation matrix of the stereo setup (Equation (3.111)). Combining the two equations we obtain

$$a\mathbf{p}_l + c\left(\mathbf{p}_l \times \mathbf{R}^T \mathbf{p}_r\right) = \mathbf{T} + b\mathbf{R}^T \mathbf{p}_r. \tag{7.3}$$

The precise triangulation algorithm working with the presented scheme was given by Ahuja [2] and by Trucco and Verri [430]. The input parameters are the corresponding pairs of points $(\mathbf{p}_{li}, \mathbf{p}_{ri})$, as well as the matrices $\mathbf{R}$ and $\mathbf{T}$. Further, the method consists of solving (7.3) for the end points of the segment S joining the two rays and parallel to $\mathbf{J}$. The ends of the segment are $a\mathbf{p}_{li}$ and $\mathbf{T} + b\mathbf{R}^T \mathbf{p}_{ri}$, respectively. Point $\mathbf{P}_E$ is simply the midpoint of the segment S.

Notice that in the canonical stereo system (section 3.4.2) the procedure simplifies since $\mathbf{R} = \mathbf{I}$, and $\mathbf{T}$ has only one nonzero value.

Rothwell *et al.* provided an analysis of the ray convergence entirely in the projective space [363], which without any additional assumptions does not belong to the group of metric spaces. In consequence it is not easy there to state any minimization task, such as the problem just discussed of finding the smallest distance between two rays.

7.2.2 Reconstruction up to a Scale

In the case when only the intrinsic camera parameters are known then reconstruction is possible only up to a certain scaling factor. This also complies with intuition, since if the external calibration parameters are not known then the position of cameras with respect to an external 'world' coordinate system is also not known – and thus can have arbitrary values. It is now evident that the reconstructed coordinates of 3D points cannot be unique since the positions of cameras are not given.

Many reconstruction algorithms are proposed in the literature. In this respect we incorporate the approach given by Trucco and Verri [430]. The first step of their method consists of building a normalized version of the essential matrix **E** (section 3.4.5), in the following form:

$$\mathbf{E}^{\mathrm{T}}\mathbf{E} = (\mathbf{RA})^{\mathrm{T}}(\mathbf{RA}) = \mathbf{A}^{\mathrm{T}}\mathbf{A}. \tag{7.4}$$

Taking into account **A** from Equation (3.18) with normalized translation vector **T** (3.4):

$$\hat{\mathbf{T}} = \mathbf{T}/\|\mathbf{T}\| = \left[\hat{T}_1, \hat{T}_2, \hat{T}_3\right]^{\mathrm{T}}, \tag{7.5}$$

after some multiplications we obtain

$$\hat{\mathbf{E}}^{\mathrm{T}}\hat{\mathbf{E}} = \begin{bmatrix} 1-\hat{T}_1^2 & -\hat{T}_1\hat{T}_2 & -\hat{T}_1\hat{T}_3 \\ -\hat{T}_1\hat{T}_2 & 1-\hat{T}_2^2 & -\hat{T}_2\hat{T}_3 \\ -\hat{T}_1\hat{T}_3 & -\hat{T}_2\hat{T}_3 & 1-\hat{T}_3^2 \end{bmatrix}, \tag{7.6}$$

that is $\hat{\mathbf{E}}$ is a normalized version of the essential matrix **E** (3.22). Assuming now that we know a sufficient number of point correspondences, as well as the intrinsic parameters, then the essential matrix can also be determined. Then, parameters of the translation vector are obtained from (7.6).

Coefficients of the matrix **R** can be obtained from (3.22) and (7.4):

$$\mathbf{R}_i = \mathbf{w}_i + \mathbf{w}_j \times \mathbf{w}_k, \tag{7.7}$$

where

$$\mathbf{w}_i = \hat{\mathbf{E}}_i \times \hat{\mathbf{T}}. \tag{7.8}$$

From the above and based on Equation (3.2) we can determine the parameter Z_{l} denoting the depth of the point $\mathbf{P}_{\mathrm{l}}$ [430]:

$$Z_{\mathrm{l}} = f_{\mathrm{l}} \frac{(f_{\mathrm{r}}\mathbf{R}_1 - x_{\mathrm{r}}\mathbf{R}_3)^{\mathrm{T}}\hat{\mathbf{T}}}{(f_{\mathrm{r}}\mathbf{R}_1 - x_{\mathrm{r}}\mathbf{R}_3)^{\mathrm{T}}\mathbf{p}_{\mathrm{l}}}, \tag{7.9}$$

where Z_{l} is a third coordinate (denoting depth) of the point $\mathbf{P}_{\mathrm{l}}$, f_{l} and f_{r} are focus lengths of left and right cameras, respectively, $\mathbf{R}_1$ and $\mathbf{R}_3$ are first and third rows of the matrix **R**, x_{r} is the first coordinate of the point $\mathbf{p}_{\mathrm{r}}$ and $\hat{\mathbf{T}}$ is a normalized translation vector.

One of the consequences of such an approach is ambiguity associated with the sign of coordinates of the vector $\hat{\mathbf{T}}$, determined from (7.4), since these coordinates are in second-degree polynomials. However, there is only one solution that gives positive values of Z_{l} for all matched points.

The unknown scaling factor can be determined quite simply, e.g. finding the image of an object of known 3D dimensions [286].

7.2.3 Reconstruction up to a Projective Transformation

When intrinsic and extrinsic parameters are not known then it is still possible to perform a certain kind of reconstruction; however, only up to a projective transformation that usually is not known [99, 358, 362, 430]. Only if this projective transformation is given explicitly is the full reconstruction possible.

It was already shown that having at least eight pairs of matched points it is possible to determine the fundamental matrix **F** as well as position of the epipoles, which are null spaces of the transformation associated with this fundamental matrix. Then due to the properties of the projective spaces it is possible to perform transformation of five points $\mathbf{P}_1, \ldots, \mathbf{P}_5$ of the 3D space on to the standard 3D projective space $\wp^3$, in such a way that none of their triples is collinear and none of their quadruples is coplanar. Based on this feature it is possible to determine projective matrices of cameras. In consequence, knowing the camera projective matrices allows the determination of the 3D point positions based on their images and by means of triangulation.

As alluded to previously, the five points are transformed on to the standard 3D projective space $\wp^3$ because of a projective transformation V (section 9.6.1):

$$\mathbf{P}_1 = \begin{bmatrix} 1 \\ 0 \\ 0 \\ 0 \end{bmatrix}, \quad \mathbf{P}_2 = \begin{bmatrix} 0 \\ 1 \\ 0 \\ 0 \end{bmatrix}, \quad \mathbf{P}_3 = \begin{bmatrix} 0 \\ 0 \\ 1 \\ 0 \end{bmatrix}, \quad \mathbf{P}_4 = \begin{bmatrix} 0 \\ 0 \\ 0 \\ 1 \end{bmatrix}, \quad \mathbf{P}_5 = \begin{bmatrix} 1 \\ 1 \\ 1 \\ 1 \end{bmatrix}. \tag{7.10}$$

Now using Equation (3.7) for the left camera we obtain

$$\mathbf{M}_l \mathbf{P}_k = a_k \mathbf{p}_{lk}, \tag{7.11}$$

where $\mathbf{M}_l$ is the sought matrix of the projective transformation in respect of the coordinates of the left camera, defined up to a certain multiplicative coefficient $a_k \neq 0$, $\mathbf{P}_{lk}$ is one of the points defined in (7.10), and $\mathbf{p}_{lk}$ is its image on the left camera plane.

To simplify further considerations, also the image points p_{lk} are transformed by means of a certain projective transformation U, this time, however, on to the standard 2D projective space $\wp^2$:

$$\mathbf{p}_{l1} = \begin{bmatrix} 1 \\ 0 \\ 0 \end{bmatrix}, \quad \mathbf{p}_{l2} = \begin{bmatrix} 0 \\ 1 \\ 0 \end{bmatrix}, \quad \mathbf{p}_{l3} = \begin{bmatrix} 0 \\ 0 \\ 1 \end{bmatrix}, \quad \mathbf{p}_{l4} = \begin{bmatrix} 1 \\ 1 \\ 1 \end{bmatrix}. \tag{7.12}$$

This way we define the transformation U that is then used to transform *all* other image points.

Applying $\mathbf{P}_1, \ldots, \mathbf{P}_4$ and $\mathbf{p}_{l1}, \ldots, \mathbf{p}_{l4}$ into (7.11) we obtain an expression for the matrix $\mathbf{M}_l$:

$$\mathbf{M}_l = \begin{bmatrix} a_1 & 0 & 0 & a_4 \\ 0 & a_2 & 0 & a_4 \\ 0 & 0 & a_3 & a_4 \end{bmatrix}. \tag{7.13}$$

Taking now a certain point $\mathbf{p}_{l5}$, expressed in standard coordinates of the $\wp^2$ base, based on (7.10) and removing a_1, a_2, a_3 from the set of equations (7.11), the formula (7.13) can be put in the following form:

$$\mathbf{M}_l = \begin{bmatrix} ap_{l51}-1 & 0 & 0 & 1 \\ 0 & ap_{l52}-1 & 0 & 1 \\ 0 & 0 & ap_{l53}-1 & 1 \end{bmatrix}, \quad \text{where} \quad a = \frac{a_5}{a_4}. \tag{7.14}$$

The central point $\mathbf{O}_l$ constitutes a centre of the projective transformation, described by the matrix $\mathbf{M}_l$. Thus

$$\mathbf{M}_l\mathbf{O}_l = 0. \tag{7.15}$$

The matrix $\mathbf{M}_l$ has rank three and therefore there exists a nontrivial solution of (7.15) in the form

$$\mathbf{O}_l = \begin{bmatrix} \frac{1}{1-ap_{l51}} & \frac{1}{1-ap_{l52}} & \frac{1}{1-ap_{l53}} & 1 \end{bmatrix}^{\mathrm{T}}. \tag{7.16}$$

Similar considerations for the right camera lead to the analogous relation, as follows:

$$\mathbf{O}_r = \begin{bmatrix} \frac{1}{1-bp_{r51}} & \frac{1}{1-bp_{r52}} & \frac{1}{1-bp_{r53}} & 1 \end{bmatrix}^{\mathrm{T}}. \tag{7.17}$$

To determine parameters a, b in (7.16) and (7.17), we use the relation that the central point $\mathbf{O}_r$ of the *right* camera is transformed by the camera matrix into the *left* epipole $\mathbf{e}_l$. Analogous conditions hold for the left camera (Figure 3.7), and therefore

$$\mathbf{M}_l\mathbf{O}_r = s_l\mathbf{e}_l, \quad \mathbf{M}_r\mathbf{O}_l = s_r\mathbf{e}_r, \tag{7.18}$$

where s_l and s_r are certain multiplicative constants, pertaining to the characteristics of the projective transformations (section 9.6).

Taking (7.16)–(7.18) it is possible to determine parameters a and b in the following form [430]:

$$\begin{aligned} a &= \frac{\mathbf{e}_r^{\mathrm{T}}(\mathbf{p}_{l5}\times\mathbf{p}_{r5})}{\mathbf{v}_r^{\mathrm{T}}(\mathbf{p}_{l5}\times\mathbf{p}_{r5})}, \quad \mathbf{v}_r = \begin{bmatrix} p_{l51}e_{r1} & p_{l52}e_{r2} & p_{l53}e_{r3} \end{bmatrix} \\ b &= \frac{\mathbf{e}_l^{\mathrm{T}}(\mathbf{p}_{l5}\times\mathbf{p}_{r5})}{\mathbf{v}_l^{\mathrm{T}}(\mathbf{p}_{l5}\times\mathbf{p}_{r5})}, \quad \mathbf{v}_l = \begin{bmatrix} p_{r51}e_{l1} & p_{r52}e_{l2} & p_{r53}e_{l3} \end{bmatrix} \end{aligned} \tag{7.19}$$

Thus to find the matrix $\mathbf{M}_l$ (and also $\mathbf{M}_r$) it is necessary to know projective coordinates of the points $\mathbf{p}_{l5}$ and $\mathbf{p}_{r5}$, as well as the epipoles $\mathbf{e}_l$ and $\mathbf{e}_r$. However, it is important to remember the assumptions that have been put forward.

1. The transformation V that transforms the points $\mathbf{P}_1, \ldots, \mathbf{P}_5$ on to the standard projective base $\wp^3$; however, these points have been already transformed from the 3D Euclidean into the projective space. Thus, only if their exact coordinates in the 3D Euclidean space are known beforehand can the transformation V be unambiguously determined.

2. The transformation U (in fact, there are two transformations, U_l and U_r, for the left and right camera, respectively), that transforms from the projective spaces associated with the camera planes on to the standard projective basis $\wp^2$. This transformation can be determined relatively more simply than V since in this case only local point positions in the camera planes are necessary.

Applying (7.14) and (7.19), after determination of the matrices $\mathbf{M}_l$ and $\mathbf{M}_r$ it is possible to perform a reconstruction of any point from the $\wp^3$ space based solely on its corresponding pair of matched points in the left and right camera planes. However, further reconstruction to the Euclidean space is possible only after determination of the transformation V. Ray equations through the points $\mathbf{O}_l$ and $\mathbf{p}_l$ as well as $\mathbf{O}_r$ and $\mathbf{p}_r$ are given by the equations of the projective space [196, 314, 380]:

$$\begin{aligned} r_l &= g_l\mathbf{O}_l + h_l\mathbf{p}_l \\ r_r &= g_r\mathbf{O}_r + h_r\mathbf{p}_r \end{aligned}, \tag{7.20}$$

where coordinates $g_l, h_l, g_r, h_r \in \Re$, while r_l and r_r are left and right rays, respectively.

To find conditions on $r_l = r_r$, the following set of equations has to be solved in respect of unknown parameters a, b, c and d:

$$\begin{bmatrix} O_{l1} & p_{l1} & -O_{r1} & -p_{r1} \\ O_{l2} & p_{l2} & -O_{r2} & -p_{r2} \\ O_{l3} & p_{l3} & -O_{r3} & -p_{r3} \\ 1 & 0 & -1 & 0 \end{bmatrix} \begin{bmatrix} g_l \\ h_l \\ g_r \\ h_r \end{bmatrix} = 0. \tag{7.21}$$

This can be accomplished by the SVD decomposition (section 4.8.3) by taking a column of the matrix D corresponding to the lowest singular value from the matrix V.

7.3 Multiview Integration

The techniques described in this book for recovering depth from matched stereo-pairs of images result in a set of points, the range map, being acquired that can be triangulated to describe a single 2.5D manifold in 3-space. In many practical applications of 3D imaging, there is a requirement to generate a complete and closed 3D surface manifold by fusing together, i.e. integrating, multiple 2.5D range maps corresponding to multiple views captured of a single object. In this case sufficient views of the object must be captured to ensure that a closed 3D mesh can be formed.

Curless and Levoy [82] stipulate that, ideally, the above process should make use of all the range data collected, take into account the quality or certainty of each local range measurement, generate the same 3D mesh irrespective of the order in which range maps are processed and allow incremental addition of range maps. In addition, the integration process should also undertake steps to ameliorate defects in the captured data, for example by detecting and removing range map outliers or filling holes in the constructed 3D mesh (to thereby construct a *watertight* mesh). Finally, the process should not be restricted to objects of any specific topological configuration and should also be computationally efficient, as tens or even hundreds of range maps might have to be integrated in real applications.

A very large body of work on range surface integration has been reported in the literature; however, two techniques are most commonly adopted, volumetric integration and direct mesh integration, as described in the following sections.

7.3.1 *Implicit Surfaces and Marching Cubes*

The basic idea behind volumetric integration, as described by Curless and Levoy [82], involves decimating the 3D space in which the captured range surfaces lie by means of a *voxel* data structure. Representation and integration of multiple range surfaces within individual voxel elements rely on the concept of a *signed distance function* $D(x)$ that records the distance in space from an imaged surface to the centre of each voxel. This distance is recorded from each voxel centre to the range surface, following the path of the sightline from the imaging sensor, for positive distances. Negative distances occur where the sensor sightline penetrates the range surface prior to reaching a specific voxel. Figure 7.2 illustrates this geometric configuration.

Accordingly, the zero-surface, i.e. when $D(x) = 0$, represents each range surface and $D(x)$ is therefore termed an *implicit function*. For each range surface to be integrated the corresponding signed distance functions, $d_i(x)$, are constructed and *accumulated* in voxel space to provide a new zero-surface that in effect averages their relative displacements. This

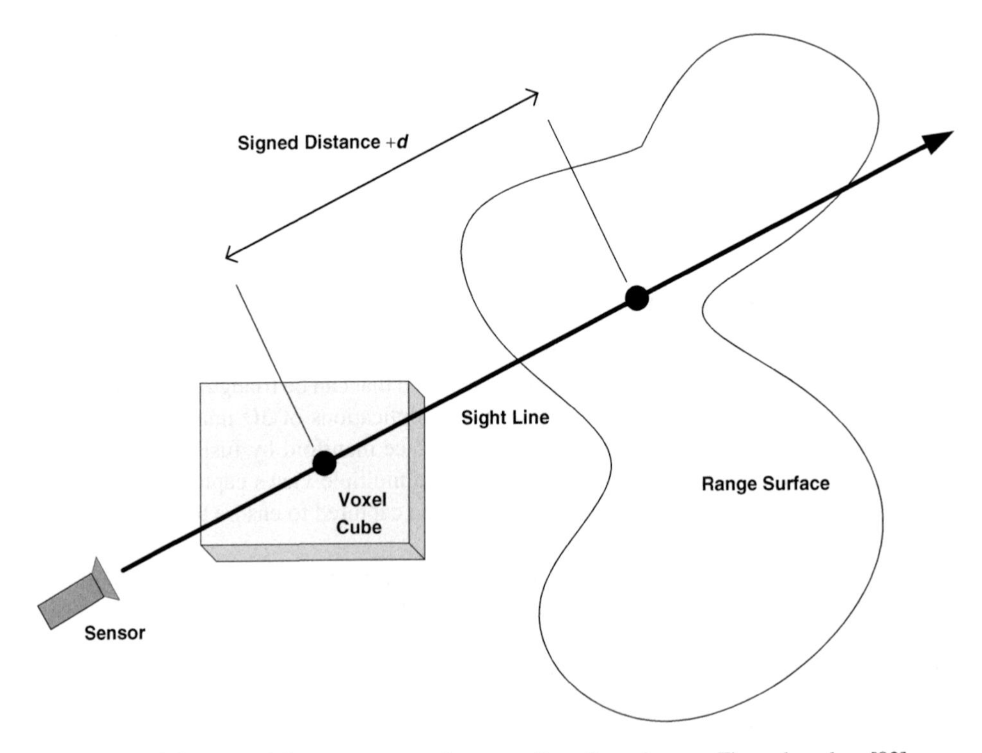

Figure 7.2 Signed distance range surface encoding of voxel space. Figure based on [82]

accumulated zero-surface can then be extracted and triangulated using a technique such as marching cubes [280], described briefly in section 7.3.1.4.

A further complication is that the quality of the physical sensing mechanism is also taken into account when forming the implicit function for each range surface. When viewing a surface from the reference camera (say) of a stereo-pair, assuming frontoparallel geometry for simplicity, then the optimum surface reconstruction will take place when the observed surface normal is collinear to this camera's sightline. A cosine angle weighting function $w_i(x)$ can therefore be formed by taking the dot product of the (dominant) camera sightline unit vector and the observed surface normal. This function is used to weight each implicit function sampled. It is then possible to sum the sampled implicit functions within each voxel (x) from each range i map by taking the first-order moments as follows:

$$D(x) = \frac{\sum w_i(x) d_i(x)}{\sum w_i(x)}. \tag{7.22}$$

Furthermore, the above sum can be formed incrementally as follows to give the signed distance function $D_i(x)$ and weight function $W_i(x)$ accumulated for the i-th range map:

$$\begin{aligned} D_{i+1}(x) &= \frac{W_i(x) D_i(x) + w_{i+1}(x) d_{i+1}(x)}{W_i(x) + w_{i+1}(x)}, \\ W_{i+1}(x) &= W_i(x) + w_{i+1}(x). \end{aligned} \tag{7.23}$$

The distance x in voxel space over which signed distances are formed in front of and behind the range map has to be restricted to avoid the surfaces of self-occluding manifolds from interfering with each other. This places a limit on the minimum thickness of closed manifold that can be constructed by this method. In the implementation of Curless and Levoy the implicit function is formed for half the maximum uncertainty interval in the range measurements in front of (and behind) the range map surface.

7.3.1.1 Range Map Pre-segmentation

Due to the characteristics of range maps captured by stereo-photogrammetry a number of preprocessing steps are usually required to assist segmenting a cleanly closed volume from multiple captured range maps of an object. Another reason for segmenting out only surfaces of interest is due to the n^3 memory cost of constructing a voxel space, where n is the sample size of each dimension of the imaged volume, it is vital to reduce the imaged volume size to encompass only valid data.

It is not uncommon to adopt 'blue screen' colour segmentation to isolate the object surface of interest in the intensity images of the stereo-pairs and then use these as segmentation masks to isolate the corresponding relevant area in the range maps for integration. Clearly this imposes the limitation of capturing the desired object in front of an appropriately coloured backdrop, typically blue, green or orange, that does not correspond to the colour of the surface to be reconstructed. As this segmentation process may result in fragmentation of the range map, usually a number of morphological operations are applied to smooth the resultant binary segmentation mask, select the largest contiguous blob (assumed to correspond to the surface of interest) and then fill any holes remaining in this blob.

When colour segmentation is not possible or inappropriate (monochrome stereo-pairs might only be available or it might not be possible to constrain the capture conditions) then it is possible to threshold the confidence map produced by the stereo matching algorithm in order to identify contiguous regions comprising viably matched surfaces.

7.3.1.2 Volumetric Integration Algorithm Overview

The principal basic algorithmic steps set out by Curless and Levoy [82] are as follows.

1. Initialize the voxel space with zeros.
2. Construct triangles on the nearest neighbour elements of the range maps such that triangles are not formed over steep discontinuities by detecting triangles whose side lengths, when taken as ratios of each other, exceed a threshold limit. In this manner the surface normal for each triangle can be extracted and its dot product formed with the (dominant) camera line of sight to produce a weight value for each observed element of each range surface.
3. The signed distance for each voxel for each range map is computed by casting a ray from the principal point in the dominant camera through each element in the range map, and the distance noted to each voxel within a distance of $\pm D_{\max}$ voxels as determined by the range measurement uncertainty.
4. Each voxel element is updated by accumulating the weighted signed distances using Equation (7.24).
5. Isolate a new isosurface for $D(x) = 0$.

7.3.1.3 Hole Filling

While the above algorithm will generate a good approximation to an isosurface, unseen areas can result in holes when attempting to triangulate the isosurface. Since there are often situations where the presence of holes in a surface is unacceptable, e.g. when measuring volume change due to surface displacement, a common clinical requirement when assessing the outcome of certain therapies or surgery procedures on the body or face, a means of filling holes is required to produce a watertight surface reconstruction.

The standard approach is to label the voxels according to one of the following states: *unseen*, *empty* or *surface* (within $\pm D_{\max}$ of the accumulated surface). Figure 7.3 illustrates this approach, and it can be seen that surface holes arise at the boundary of empty and unseen voxels. Therefore, placing surfaces at these boundaries provides a simple means of generating a watertight surface based on the minimum of assumptions.

The previously described algorithm is now modified as follows.

1. Assign all voxels to an initial *unseen* state.
2. Compute the weighted signed distance accumulation on those voxels within $\pm D_{\max}$ of the accumulated surfaces as before, labelling these voxels as *surface*.
3. Reset all the voxels between those labelled as surface, following the dominant camera viewing direction, back to the boundary of the voxel space closest to this camera with the state *empty*.
4. Once more extract the isosurface for $D(x) = 0$ and also extract a surface at the interface between those regions labelled unseen and those regions labelled empty.

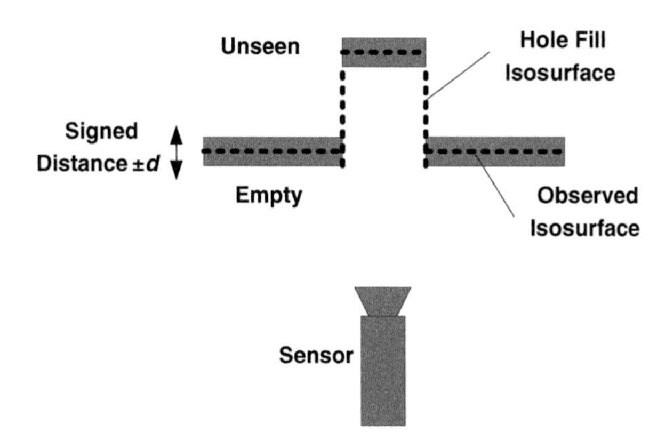

Figure 7.3 Geometric configuration for hole filling. Figure based on [82]

Where hole filling is applied then surface discontinuities are likely to arise, thereby generating artefacts. This obvious effect of such hole filling can be reduced by applying local filtering (averaging nearest neighbouring vertices) only to the hole regions themselves. In this way it is possible to avoid blurring the remainder of the mesh while suppressing artefacts. The support region of the filter weights is allowed to taper between the hole-filled region and the observed data in order to smooth their transition [82].

7.3.1.4 Marching Cubes

The final step of the above volumetric integration method requires that the isosurface for $D(x) = 0$ is constructed from the voxels containing the accumulated signed distance functions. An algorithm called *marching cubes,* first reported by Lorensen and Cline [280], provides a standard method for constructing a watertight polygon mesh from a volumetrically sampled space. Marching cubes was originally developed to allow isodensity surfaces to be extracted from CT (Computed Tomography) or MRI (Magnetic Resonance Imaging) voxel data; in the case described here we are concerned only with extracting the zero isosurface.

Marching cubes comprises two principal steps: the intersection of the surface is detected within a local cube generated by eight voxel samples; thereafter the surface within the cube is represented by triangles and their vertex locations computed. Following these steps, we *march* to the next cube and repeat the process until the entire voxel space has been triangulated.

The essential beauty and simplicity of the algorithm stems for the observation that there are a finite number of possible intersections of a surface with a cube (as depicted in Figure 7.4). Therefore, if we can determine which particular type of local surface intersection is taking place, we can triangulate the cube accordingly, the particular form of local mesh being known in advance as shown in Figure 7.4.

In order to determine the local intersection configuration, we must first determine how each vertex of the cube is positioned with respect to the surface we are attempting to extract. Each vertex can be in one of two states, *inside* the surface (including being *on* the surface) or *outside* the surface. We assign either a one or a zero to each vertex depending on whether it is in the inside or outside state respectively. Clearly, in order for the surface to intersect with a

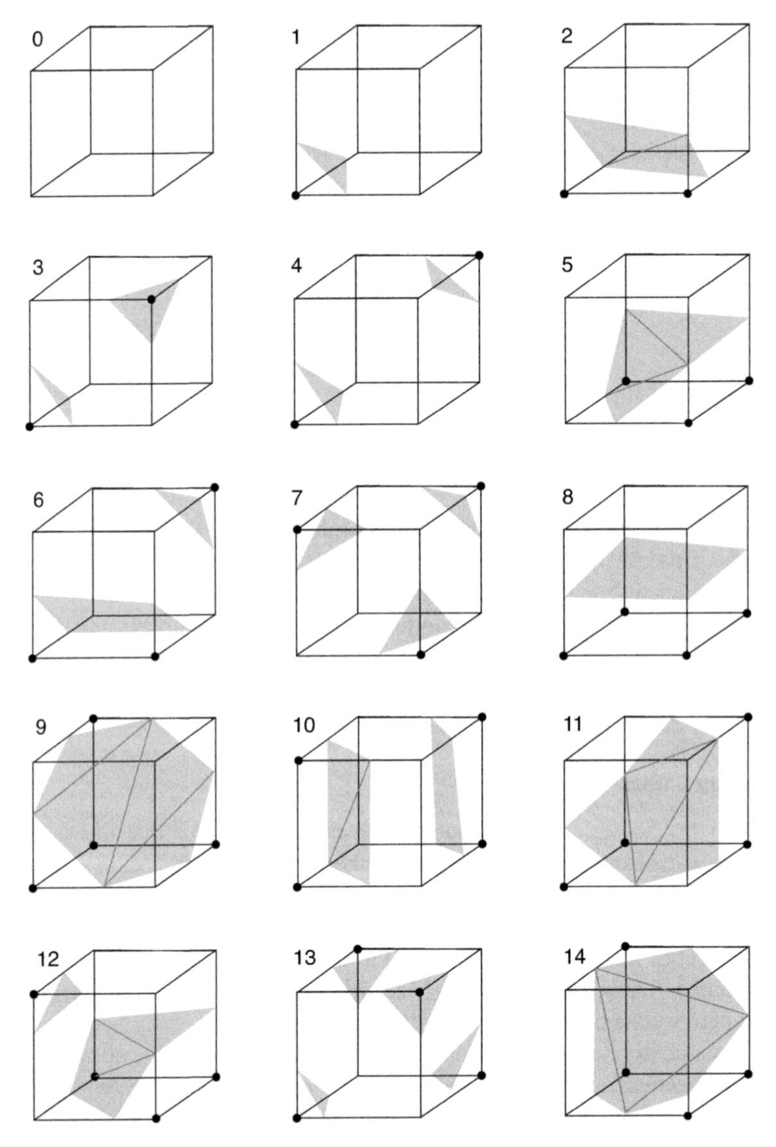

Figure 7.4 The 15 unique possible intersections of a triangulated surface and a voxel cube. (Reproduced from [69])

cube, the cube must straddle the surface and therefore the inside–outside state of at least one vertex must be different from the remainder of those neighbours in the cube being tested. By considering the inside–outside states of the vertices of the cube, we can determine, i.e. index, which local form of intersection is taking place. As Lorensen and Cline point out, each cube has eight vertices and each vertex can be in one of two states, therefore there can only be 2^8 (256) possible intersections. When symmetries are taken into account, only 15 unique states (including the empty state) remain.

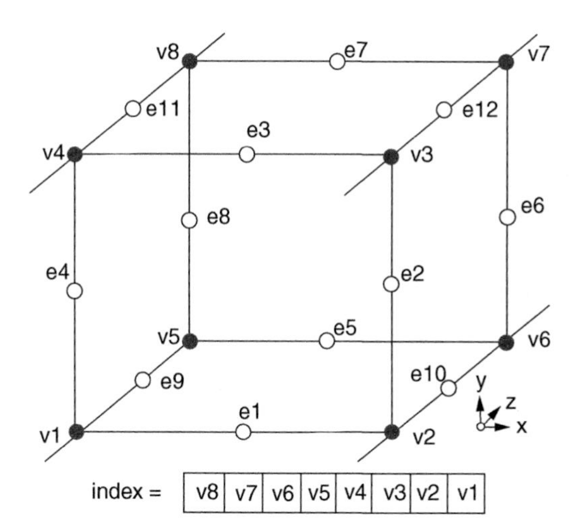

Figure 7.5 Cube numbering. (Reproduced from [280] (copyright ACM, Association for Computing Machinery))

Having determined the local surface intersection, we must now determine exactly where each vertex of the local triangulation intersects on the appropriate edges of our cube. Cube edge numbering is defined in Figure 7.5 and given a particular intersection vector we can estimate the (zero) position of each triangle vertex by simple first-order interpolation. For example, given that two adjacent cube vertices, located at positions i and j, sample signed distance values of v_i and v_j respectively, then the location of an interpolated edge e_{ij} offset from j will be

$$e_{ij} = \frac{(1 + v_j)\, j - v_j i}{v_i - v_j}. \tag{7.24}$$

Figures 7.6–7.9 show the results of an algorithm originally developed by the Turing Institute, Glasgow, UK, based on the above techniques and implemented within the C3D stereo-photogrammetry package [389]. In Figure 7.6 two views of a human head have likewise been integrated.

In a museum artefact scanner application, two stereo-pairs of cameras have been configured to view objects set on a turntable. One stereo-pair views the object side-on, while the second views the object from a raised perspective looking obliquely down. Figure 7.7 shows the dominant camera view of each of the stereo-pairs of images have been captured, in this case a total of eight stereo-pairs. This set of stereo-pair views has been matched to produce eight depth maps which were then integrated to form a complete 3D model, by means of a version of the volumetric integration techniques described in conjunction with marching cubes, in Figure 7.8.

The integration process itself can be somewhat unpredictable as to surface selection during merging as can be observed in Figure 7.9, showing the merged contributions from different views; notice the 'islands' of range surface that can appear and the jagged intersection boundary of the merged surfaces.

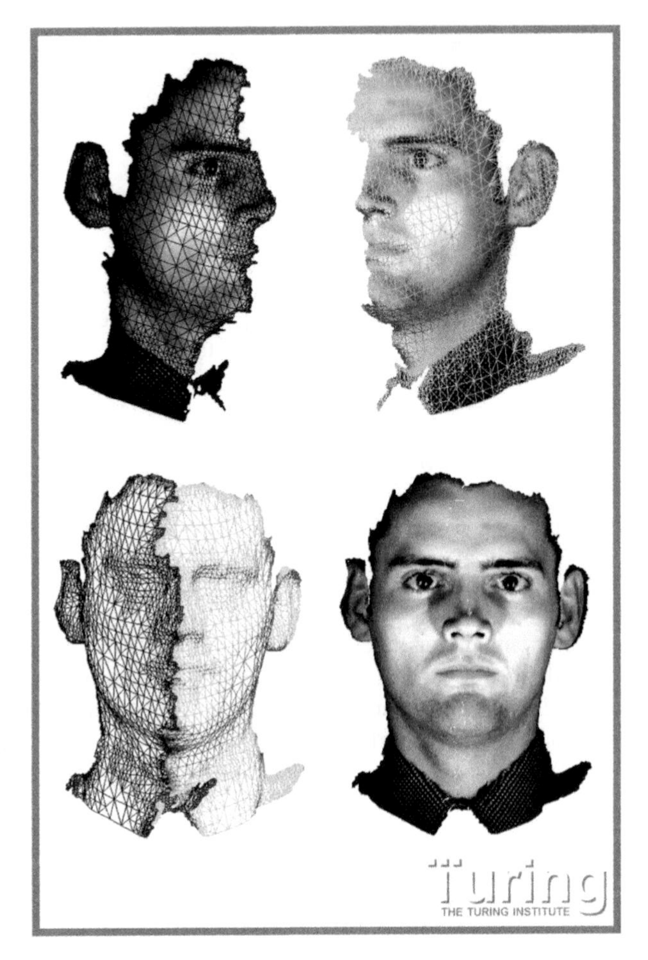

Figure 7.6 Integration of two range surfaces based on marching cubes. (Copyright University of Glasgow)

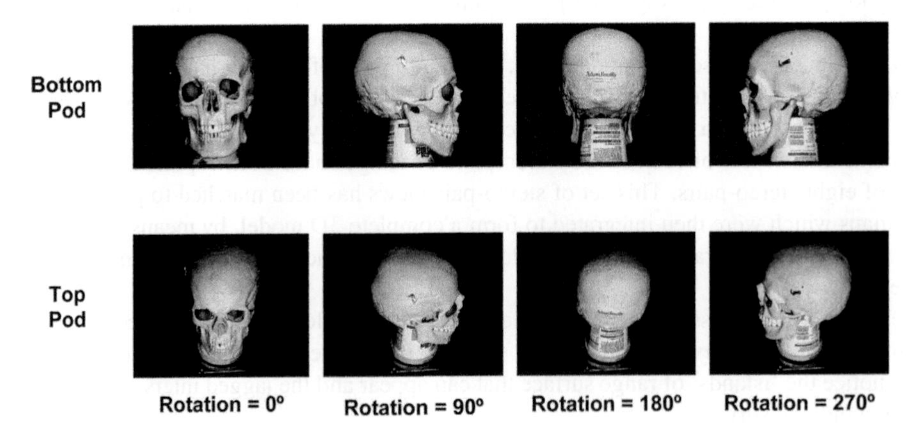

Figure 7.7 Eight dominant camera views of a skull (Plate 8)

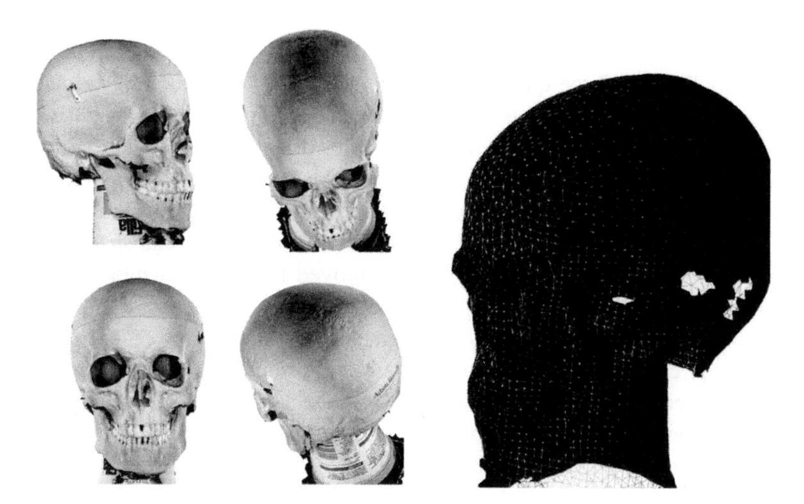

Figure 7.8 Five views (four of these have been texture-pasted) of a single complete 3D skull model computed by marching cubes integration of eight range surfaces (Plate 9)

It should be mentioned that the volumetric integration approach based upon marching cubes has a number of serious limitations beyond those mentioned at the start of section 7.3.1. The dimensions of the voxel set the effective sampling density of the final model, therefore an overly coarse voxel tessellation will produce aliasing artefacts in the form of stepping contours on the reconstructed model surface. At the same time, fine surface detail captured in the original range images and evident prior to integration may be lost. Typically corners and thin surfaces can be destroyed. However, the severe memory requirements set by sampling the range data using a voxel space usually dictates that the voxel size is considerably larger than the underlying range sampling interval and consequent artefacts and loss of detail become inevitable.

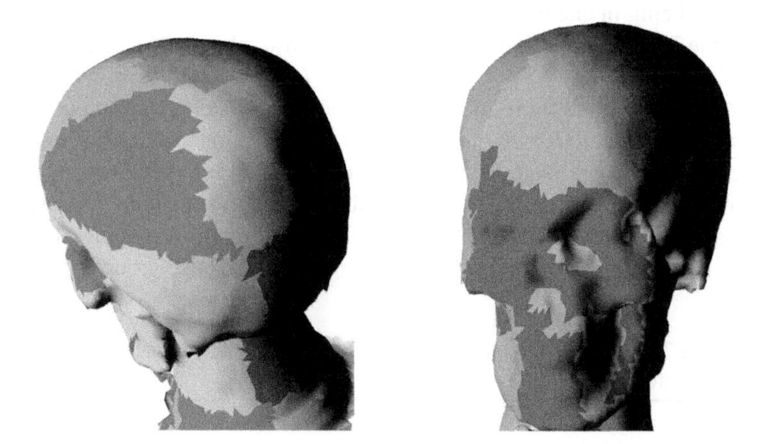

Figure 7.9 Two views of the integrated skull model showing the colour-coded contributions from different range maps (Plate 10)

7.3.1.5 Implementation Considerations

Due to the potentially very large quantities of data and large volume of space that must be voxelized with potentially high precision, many mechanisms have been proposed to make the computational cost of this approach more tractable. It is possible to improve greatly the speed of updating the voxel space with a new range map by resampling this map such that its scanlines align with the voxel grid when traversed [82]. The enormous memory cost of voxelization can be mitigated by means of run-length encoding this data structure [82]. An oct-tree decimation of voxel space can also be employed to reduce memory requirements. This approach is particularly efficient as the signed distance voxels representing the input manifolds usually occupy a comparatively small fraction of the total voxel space.

A large number of citations in the literature report extensions and improvements to the basic algorithm described above, some of the more significant publications including resolving topological ambiguities inherent in the original marching cubes formulation [69, 193] and also adaptive generation of surface meshes [324].

7.3.2 Direct Mesh Integration

As an alternative to volumetric approaches to range surface integration, direct mesh integration offers the possibility of retaining more of the original detail contained in the range maps at greatly reduced memory requirements. The principal difficulty encountered when attempting to merge range maps directly is the very large number of potential intersection cases between the triangles representing the merge boundary between surfaces.

A recently proposed direct mesh integration approach [232, 233] circumvents mesh intersection issues by ensuring that the meshes to be integrated do not overlap. Each range map is assumed, as before, to be referenced to a common coordinate system, established though prior multiview camera calibration. As before, it is also assumed that it is possible to segment the target surface from the dominant images of the matched stereo-pair from each view (the left image in this case). In addition, this approach relies on the availability of the match confidence maps associated with each range map. The basic idea of the approach is to label each pixel of each range map as being in one of four states, *visible*, *occluded*, *overlapping* or *unprocessed*, as defined in Table 7.2. Figure 7.10 illustrates the geometric relationships in a two range map example.

Since the range maps A and B are in the same coordinate system, we can determine those range pixels that are common to *both* maps and their classification state as in Table 7.2.

Having labelled each range map, each map is then grouped into patches comprising pixels of the same label. Furthermore, to resolve the ambiguity inherent in assigning visible or occluded labels, a *confidence competition* is run on entire groupings based on the masked confidence images associated with each range image. Those groupings that lose the

Table 7.2 Criteria for labelling each range map as being in one of the four states listed

visible	*if* $r_{\mathrm{B}}(m,n) > O_{\mathrm{B}}P_{\mathrm{A}} + \varepsilon$
occluded	*if* $r_{\mathrm{B}}(m,n) < O_{\mathrm{B}}P_{\mathrm{A}} - \varepsilon$
overlapping	*if* $\mid r_{\mathrm{B}}(m,n) - O_{\mathrm{B}}P_{\mathrm{A}} \mid \leq \varepsilon$
unprocessed	*if* $p_{\mathrm{B}}(m,n) \notin \mathrm{B}$

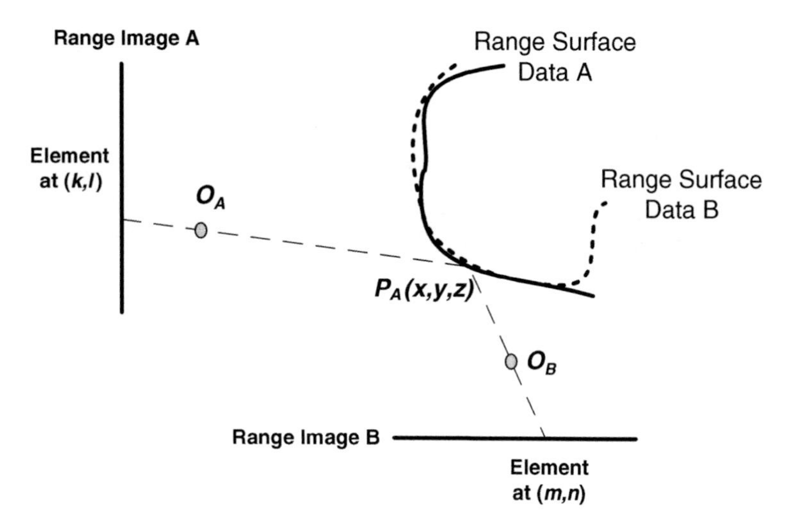

Figure 7.10 Geometric configuration of range maps to be merged. (Adapted from [233], © IEEE Computer Society Press)

competition are removed by assigning their corresponding segmentation mask areas to zero (i.e. deselected).

The confidence competition determines if the range pixels associated with a contiguous grouping are on average of greater confidence than the corresponding range pixels of all other range maps being integrated. Consider the correlation score C_A at element $p_A(k, l)$ in range image A and the score C_B of the corresponding element $p_B(m, n)$ in range image B:

$$\begin{cases} w\,(k.l) = 1 & \text{if } C_A \geq C_B \\ 0 & \text{otherwise} \end{cases}. \tag{7.25}$$

We can then determine the winning patch by finding the average winning confidence contribution:

$$w(k, l) = \sum_{(k,l)\in S} \frac{w(k, l)}{N}, \tag{7.26}$$

where N is the number of elements in the grouped patch S. If $W > 0.5$ it is retained, otherwise it is removed. Following patch deletion for all of the range images, the unmasked region of each range image is then triangulated. Where meshes generated from different range maps overlap, the overlapped region on each mesh is eroded until each region becomes disjointed from any other region. The set of nonoverlapping meshes are then joined by triangulation, which also fills any gaps that appear between meshes to produce a single continuous mesh. In a further refinement, a cosine surface normal map is employed to cull all range pixels pointing away from the reference camera by more than a preset angle, as is used in volumetric merging. In this case an angle of approximately 80° to the viewing angle proved to be viable for removing surface regions that point steeply away from the camera and are therefore likely to be unreliable.

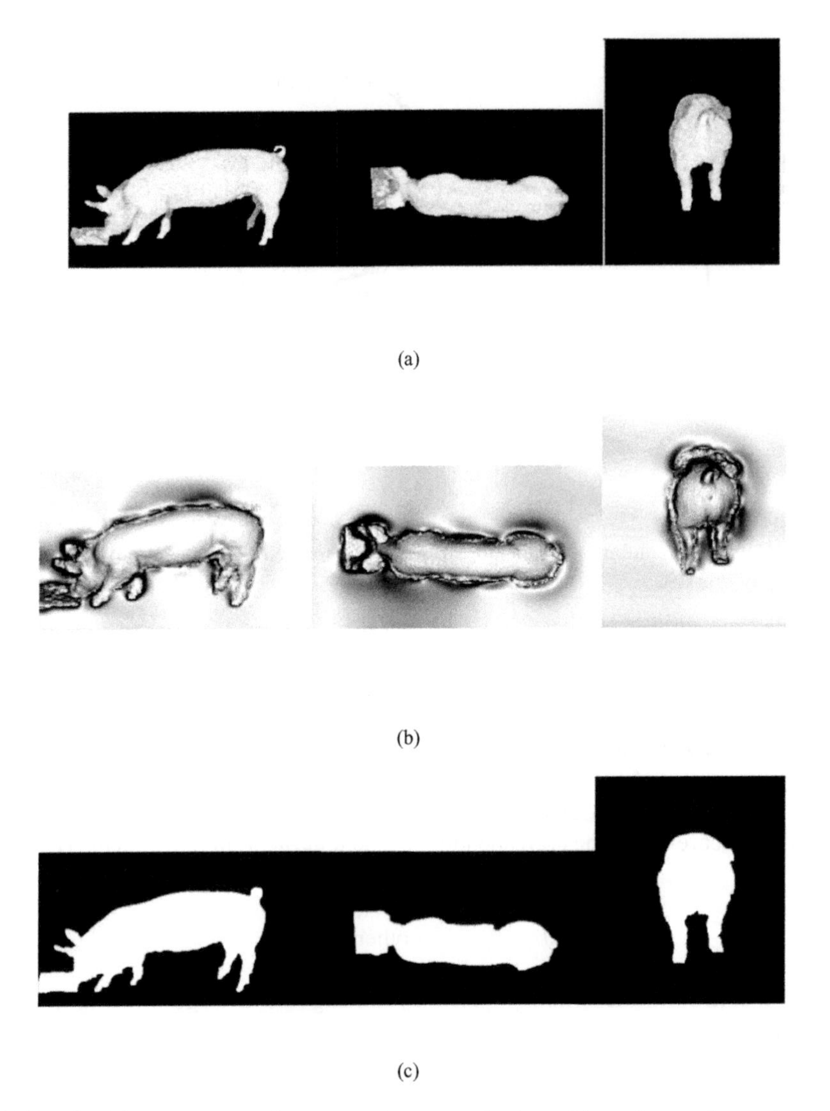

(a)

(b)

(c)

Figure 7.11 Confidence, range and segmentation maps for top, side and rear views of a live pig: (a) match confidence map; (b) range surface maps; (c) blue-screen segmentation masks. (Reproduced from [234], Copyright (2007) Chinese Society of Theoretical and Applied Mechanics)

In Figure 7.11 example range images of a pig (Figure 7.11(a)), along with their associated match confidence images (Figure 7.11(b)) and foreground segmentation masks (Figure 7.11(c)) are shown. The results of direct mesh integration as described above are illustrated in Figure 7.12 (right) and the eroded surfaces' preintegration are depicted in Figure 7.12 (left). Finally, examples of surface shaded pig models integrated using direct mesh merging and marching cubes are presented in Figures 7.13 and 7.14, respectively. Notice the artefacts present under marching cubes integration which do not appear on the direct mesh integrated model.

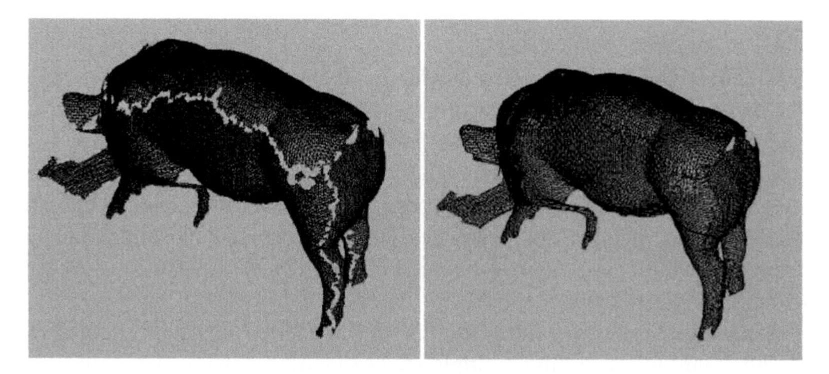

Figure 7.12 Left: surface meshes after patch deletions and boundary erosion. Right: integrated mesh with triangle insertions (i.e. surface join). (Reproduced from [234], copyright (2007) Chinese Society of Theoretical and Applied Mechanics)

Figure 7.13 Range surface integration using direct mesh merging. (Reproduced from [232], © IEEE Computer Society Press)

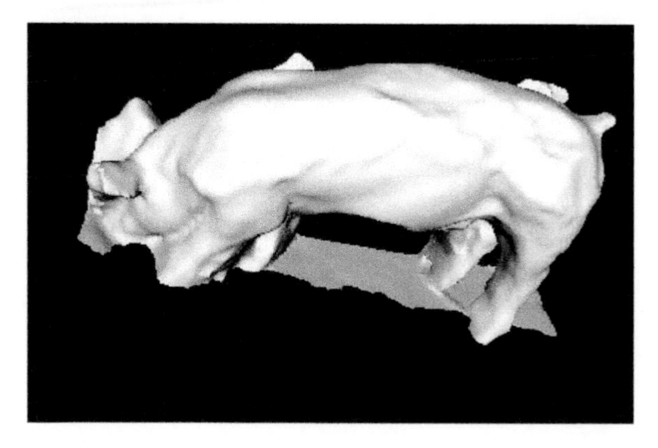

Figure 7.14 Range surface integration using marching cubes. (Reproduced from [232], © IEEE Computer Society Press)

7.4 Closure

In this chapter an overview of the space reconstruction methods is outlined. Depending on the available camera parameters there are basically three types of reconstruction.

We have presented multiview integration methods that cater for the situation where the relative orientation relationship between captured 2.5D surface manifolds is known (i.e. 6 degrees of freedom transformation between manifold coordinate frames), to illustrate how integrated and complete 3D reconstructions can be achieved. In the context of close-range photogrammetry, it is comparatively easy to obtain the required multiview coordinate transformations via standard calibration protocols.

It should be noted that multiview integration is very much the subject of ongoing research, particularly with regard to integrating multiple views of surfaces that self-occlude or contain complex topologies, often in the context of capturing the human form (section 8.3), with minimal pose restrictions. Other issues revolve around how best to combine fragments of the same surface captured from different views and how to deal with 'difficult' objects comprising semitransparent surfaces or filaments (e.g. hair). Human hands are perhaps a good example of a form of surface that represents a particularly difficult 3D imaging and integration challenge.

7.4.1 Further Reading

The three types of reconstruction presented are based on the methods described in the books by Ahuja [2] and Trucco and Verri [430].

Further information on reconstruction can be found in the books by Hartley and Zisserman [180] and by Faugeras and Luong [119]. The proceedings of the International Conference on Computer Vision and also the European Conference on Computer Vision are both good sources of publications relevant to the discussion presented in this chapter and the other sections of this book dealing with image processing and computer vision.

8

Case Examples

8.1 Abstract

This chapter provides several applications which make use of the techniques and methods presented in previous chapters. These are rather short descriptions of top-level ideas rather than detailed descriptions of implementations. However, they can be influential for further developments in these and related areas. We begin with a description of a 3D system which serves as a video aid for visually impaired persons and then present examples of face and body modelling based on the data collected by 3D vision systems. This is followed by clinical and veterinary applications. Finally, an application of image matching techniques used to synthesise missing frames in archive cine footage is presented.

8.2 3D System for Vision-Impaired Persons

It appears that the techniques presented for inferring 3D information based on images provided by digital cameras could be of help to people with limited vision abilities. 'Artificial eyes' may offer a practical means of guiding a person to avoid obstacles on his or her way. There have been many attempts to build such systems – for an overview, refer to the work by Molton [316]. A system that employs stereo processing for detection of obstacles was also presented by Molton *et al.* [315]. Their method relies on a comparison of recovered disparity values with the expected position of the ground. A similar idea was proposed by Se and Brady [375]. The same authors also developed systems for detecting zebra crossings [377] and staircases [376] for blind people.

In this section we present a simple system that processes a stereo-pair of images to construct a relative depth map (i.e. a disparity map) which is then transformed into some form of sound sensations that can be perceived by a user. We will focus on the first task, i.e. 3D recovery in quasi-real time and in a real, rather than laboratory, environment. However, development of a proper sound coding scheme that can be learned by a blind person for navigation also poses a challenge. This problem touches upon psychophysiology at least as regards the limits of human perception to discriminate between potentially chaotic sound mixtures, as well as construction of a code, or alphabet, to describe accoustic building blocks that represent 3D structures in the environment.

An Introduction to 3D Computer Vision Techniques and Algorithms Bogusław Cyganek and J. Paul Siebert

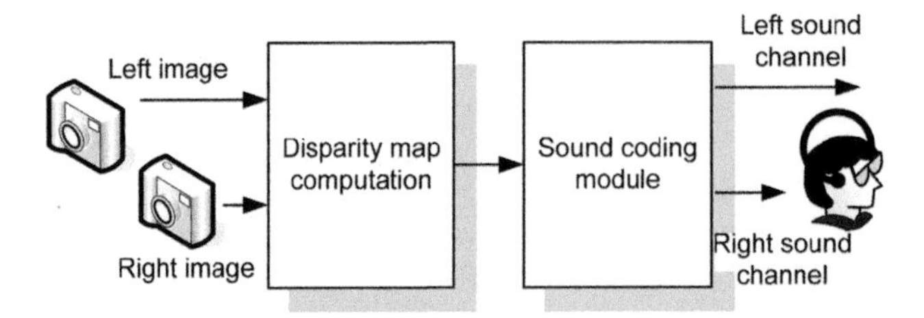

Figure 8.1 Architecture of the vision-aid system for conversion of depth information into sound sensations for blind people

The first block of the system (Figure 8.1) computes a dense disparity map from the stereo-pair but only from subareas which correspond to the direction of the head. This limitation in the input space speeds up the computations. Since we need only relative depth information, a disparity map is sufficient and no calibration or scene reconstruction is necessary. A simple block matching algorithm was used which operates on the *Census* transformed images (section 6.3.7).

At each time step, the acquisition window is chosen at the virtual centres in the two input images (i.e. a kind of a cyclopean eye [201, 235]) that correspond to the direction of sight of a person. The cameras are configured in fronto-parallel stereo rig and head-mounted on spectacles. Thus, disparity is computed only for the central window, as depicted in Figure 8.2. Further, this window is divided into vertical stripes and only one value is selected that represents the global disparity in each stripe. The best results were obtained when the selected value was the median of the values in a stripe, since this method rejected outlying disparities. Then, the selected values were divided into two equal partitions that were finally encoded into sound and fed into two channels R and L of the headphones (Figure 8.1). The encoding transformation consists of uniform noise modulated with disparity values, as follows:

$$S[i,k] = D[k]N[i], \tag{8.1}$$

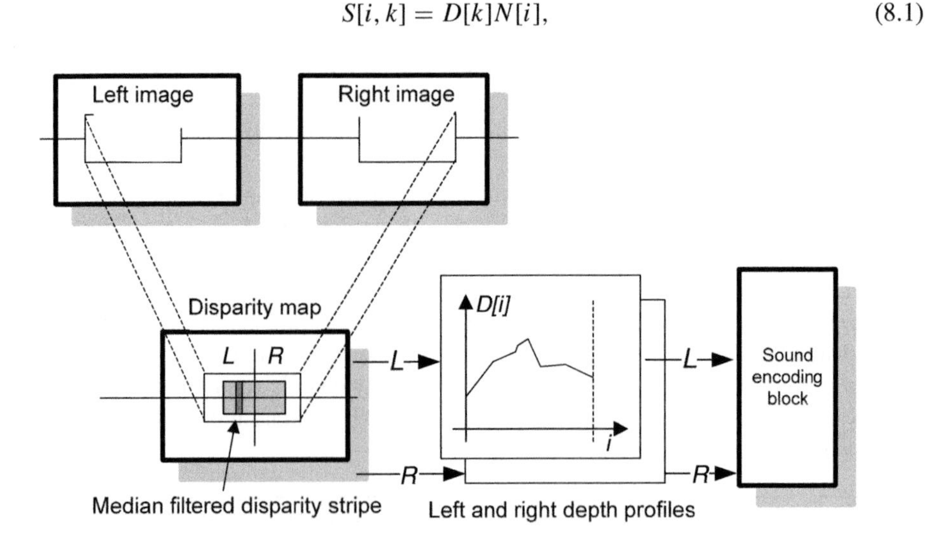

Figure 8.2 Sound coding scheme from the profiles drawn from the disparity map

where $D[k]$ denotes disparity at a point k of a scanning profile and $N[i]$ denotes a synthesized noise at a point i. This means that for far objects a person hears a quiet noise, whereas for close objects the noise increases and depth gradient can then be distinguished by the stereo sound effect.

The size of the acquisition window was chosen to be 10–50 pixels vertically by 30–180 horizontally, depending on scene type.

8.3 Face and Body Modelling

8.3.1 Development of Face and Body Capture Systems

3D capture systems based on processing stereo-pairs of images started to appear in the early 1980s. One of the earliest systems developed was the capture system by Nishihara [330] which used the Marr–Poggio matcher [298, 299] to recover surfaces. By the mid-1980s, in the UK a collaboration between University College London and Thorn EMI Ltd. produced a close-range photogrammetry system that was marketed by the British Technology Group and eventually served as the basis for the system developed and marketed by Tricorder Ltd. When Tricorder was liquidated, the technology was relaunched by 3DMD Inc. [202] which currently markets devices based on the Tricorder technology in the USA. At the same time, a stereo-photogrammetry system was developed in Scotland in a collaboration between Bolt Beranek and Newman Ltd (BBN; Edinburgh) and the Turing Institute (Glasgow). Figure 8.3 shows a single-pod (stereo-pair) capture system marketed by the Turing Institute in 1997 [230]. When BBN closed in Edinburgh in the early 1990s, the technology was further

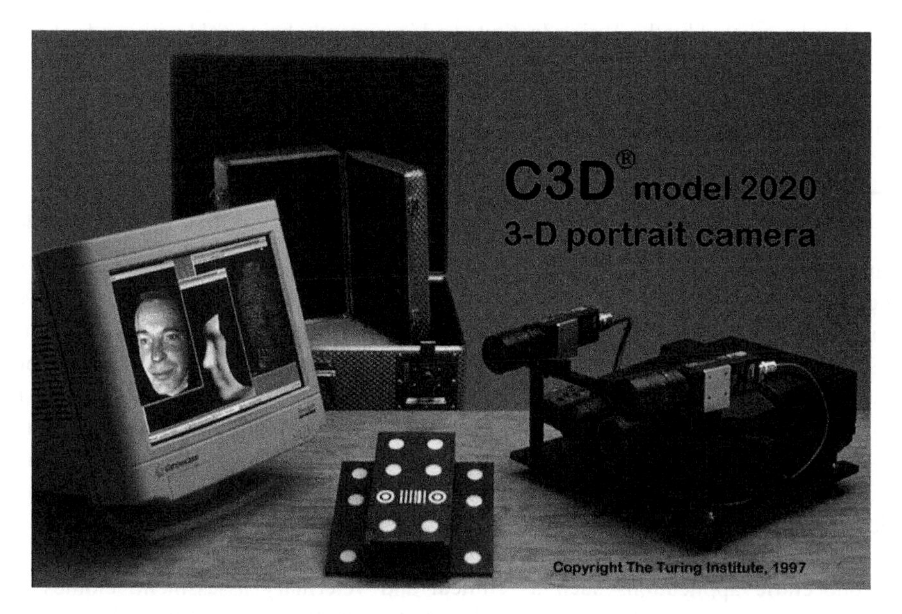

Figure 8.3 Example of a complete commercial stereo-based 3D imaging system marketed in 1997 by the Turing Institute, Glasgow. Note the stereo-pair of TV cameras mounted over a computer-controlled slide projector that illuminates the scene with a speckle textured light [386, 389]. (Copyright University of Glasgow)

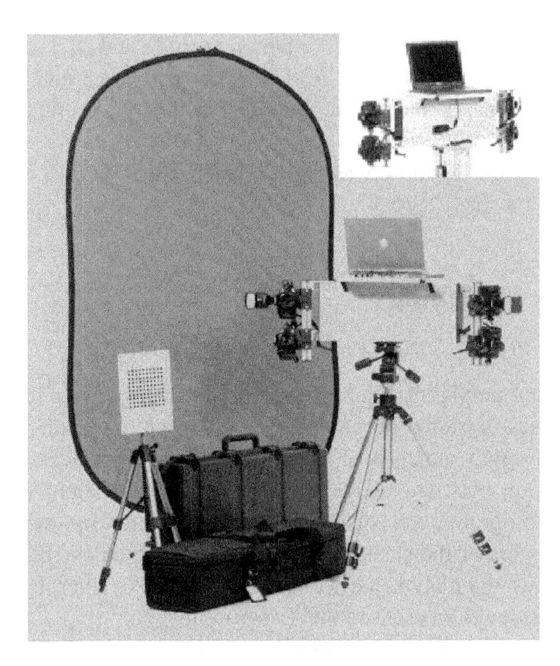

Figure 8.4 Two-pod (four-camera) high-resolution 3D imaging system manufactured by Dimensional Imaging Ltd complete with calibration target and blue-screen background. (The inset shows the rear pod view.) (Copyright Dimensional Imaging Ltd.)

developed and marketed by the Turing Institute and is currently being sold within a foot scanning device marketed by Precision 3D Ltd [211] (Figure 8.9). In a follow-up development, a system using similar technology is now currently being developed and sold by Dimensional Imaging Ltd [207] (Figure 8.4).

8.3.2 Imaging Resolution, 3D Resolution and Implications for Applications

A common feature of early video camera-based 3D photogrammetry systems of the 1980s was the limited available imaging resolution (575 × 786 pixels in Europe and 640 × 480 pixels in North America and Japan) of the then current TV camera technology. Under such image sensor resolution constraints, depth imaging resolutions of the order of 0.5 mm could be achieved for working distances of the order of a few metres and working volumes of the order of 220 × 280 × 150 mm. At the same time a reasonably small inter-camera separation (stereo-baseline) could be maintained (of the order of 300 mm) while maintaining similar resolutions in all three spatial measurement axes (X, Y and Z). While this level of resolution is inadequate for most traditional metrology applications, such as surface inspection and parts dimensioning, it is sufficient for measuring human face and body surfaces.

Anthropocentric applications such as clinical and veterinary assessment, clothes and footwear fitting and virtual actor avatars for the creative media and computer gaming industry then followed this 3D measurement capability. Applications involving human simulation such as crash test dummies and human factors analysis (e.g. for scenarios such as operating

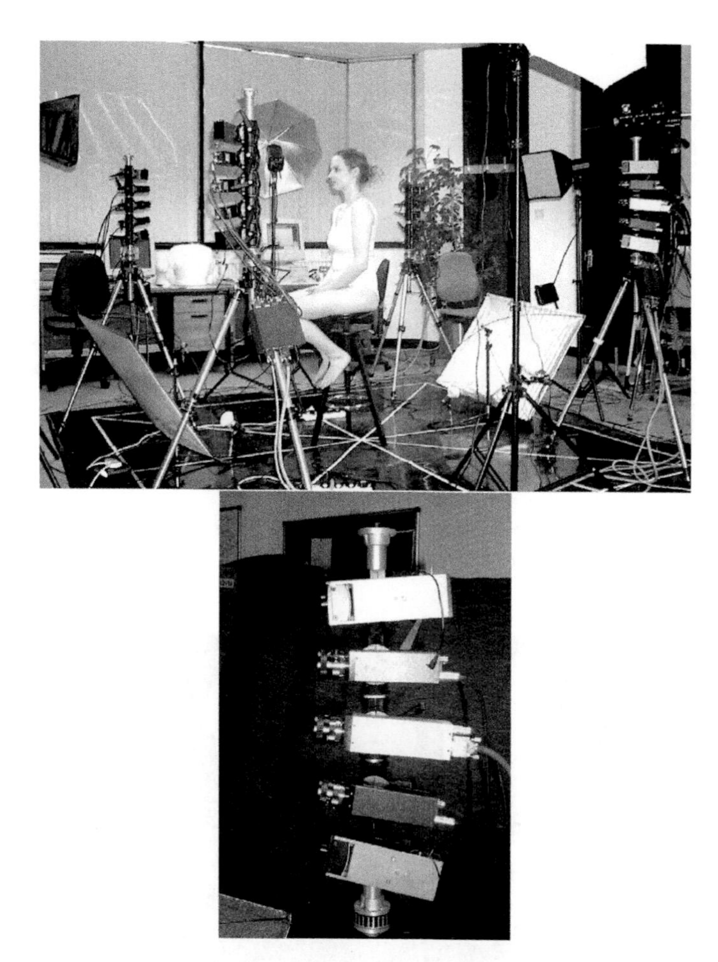

Figure 8.5 Top: an experimental multiview all-round head capture system. Bottom: an imaging pod comprising a stereo-pair of monochrome cameras (in black), a colour camera centre, and texture illumination flash units top and bottom

machinery or vehicles, or evacuating buildings) could now be supplied with virtual humans based on real-world data via photogrammetry-based 3D imaging.

Faces tended to be the first human body surfaces to be imaged using single stereo-pairs of cameras [18] and systems with greater surface coverage soon followed using multiple stereo-pairs [19]. Thereafter, multiview systems were developed to image the complete body [389] or specific parts such as the foot [211], breast [389] and back (scoliosis assessment) [418]. Figure 8.5 shows a prototype whole head scanner (this system could also be configured to image the front or back of the body; Figure 8.8) developed by the Turing Institute for the CREATEC unit at Ealing Studios, London (results from this scanner are shown in Figure 8.7). Figure 8.6 shows a prototype whole body scanner developed at Glasgow University based on the Turing technology and Figure 8.8 gives examples of frontal and rear 3D body scans. Figure 8.9 shows a multiview foot scanner developed by Precision 3D Ltd based on Turing's C3D technology.

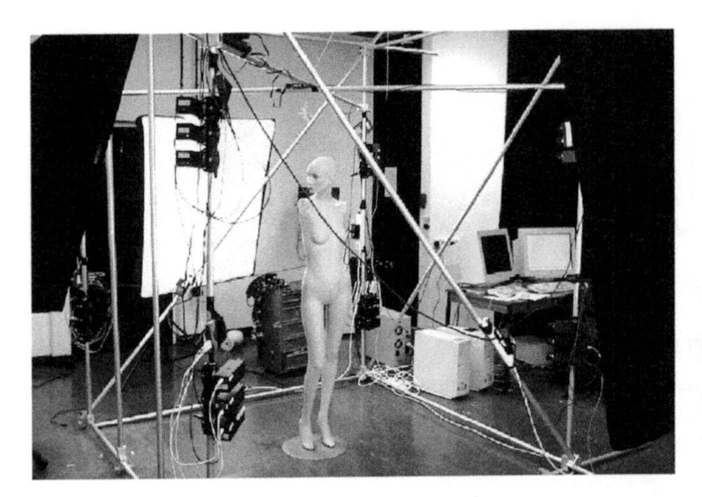

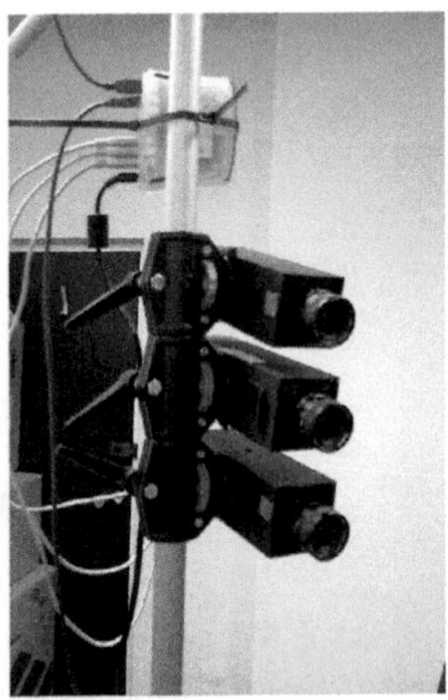

Figure 8.6 Left: prototype whole body scanner. Right: an imaging pod comprising a stereo-pair of monochrome cameras (top and bottom) and a colour camera (centre)

Figure 8.7 Four rendered views of a 3D model captured by an experimental five-pod head scanner (Plate 11). (Subject: His Excellency The Honourable Richard Alston, Australian High Commissioner to the United Kingdom, 2005–2008)

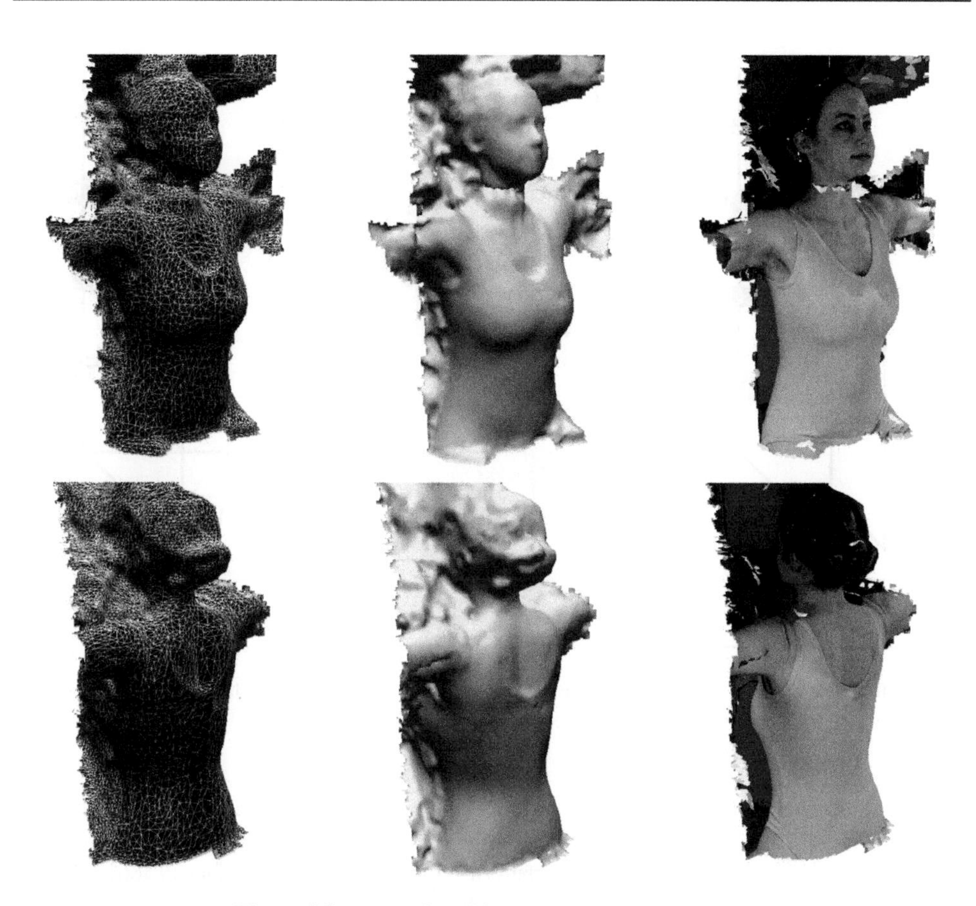

Figure 8.8 Examples of front and rear 3D body scans

Figure 8.9 An example of a commercial 3D scanner that captures all-round the foot, including the sole (Images copyright precision 3D Ltd.)

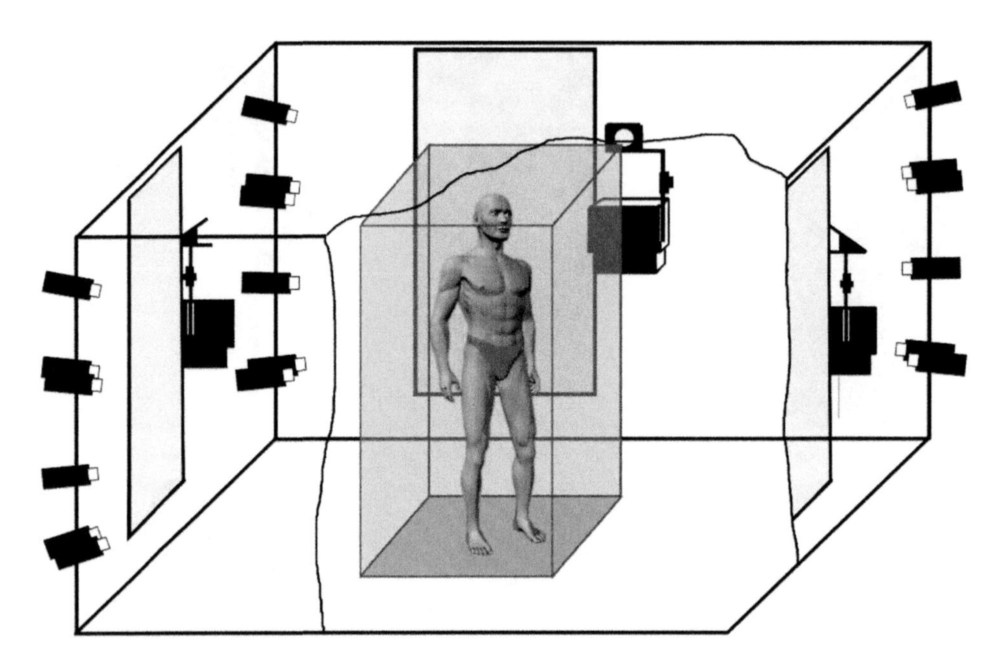

Figure 8.10 Schematic of prototype real-time immersive 3D scanner. (Reproduced from [327] with permission of the Institute of Electrical Engineers)

The systems described above share a common characteristic, namely in order to ensure that good 3D models can be recovered, adequate surface texture (required for stereo-pair matching) is guaranteed by projecting a speckle pattern on to imaged body surfaces [386]. A key development in face and body capture was the realization of 'texture projection free' systems that comprised only stereo-pairs of cameras and standard studio flash illumination. These latter systems adopted high-resolution digital still cameras (comprising imaging sensors of the order of 6M pixels or greater, typically double this figure for human face imaging) that were capable of resolving the indigenous surface texture, such as pores, present on human skin.

A further recent development in 3D human surface measurement resulting from the availability of high-resolution digital video cameras is the development of real-time 3D surface capture, i.e. 3D cine. Figure 8.10 shows a prototype system developed at Glasgow University [327, 466, 467] that was capable of capturing eight stereo-pairs of surface data at a rate of 25 frames per second. However, this system used low-resolution 640×480 pixel cameras (Figure 8.11) and utilized texture projection in combination with strobe lighting. A more recent offering from Dimensional Imaging is based on HDTV resolution cameras and does not require texture projection [207].

8.3.3 3D Capture and Analysis Pipeline for Constructing Virtual Humans

The basic pipeline required to construct virtual humans starts with 3D model capture: multiview stereo-pair capture, image matching, space intersection (to recover depth values) and integration of multiple depth maps to form a single polygonized 3D model. At this stage, the interpretation of this data typically relies upon manual landmarking to delineate known

Figure 8.11 Real-time capture pod comprising a stereo-pair of black-and-white cameras and a colour camera mounted above. (Reproduced from [467])

anchor points on the model surface, from which it is possible to align a pre-annotated generic 3D model to the captured 3D data using nonrigid registration. Given that the semantics of the generic model have been assigned in advance, it is then possible to automate subsequent processes such as inserting a skeleton within the model and then animating this model using motion capture data (usually derived directly from human actors undertaking prescribed (choreographed) actions such as dancing or any sequence of animation required by a script). However, automatic means for generating human animation via kinematic modelling have also been the subject of research activity in the field. Prior to animation, other cosmetic changes can be made to the virtual human such as the application of virtual makeup, the attachment of virtual hair (and hairstyle) and of course attachment of virtual clothing.

The above pipeline from 3D cloning to animation is illustrated in Figures 8.12 and 8.13. In Figure 8.12 we see an example of an individual (Dr Gegang Tao) who has been scanned by

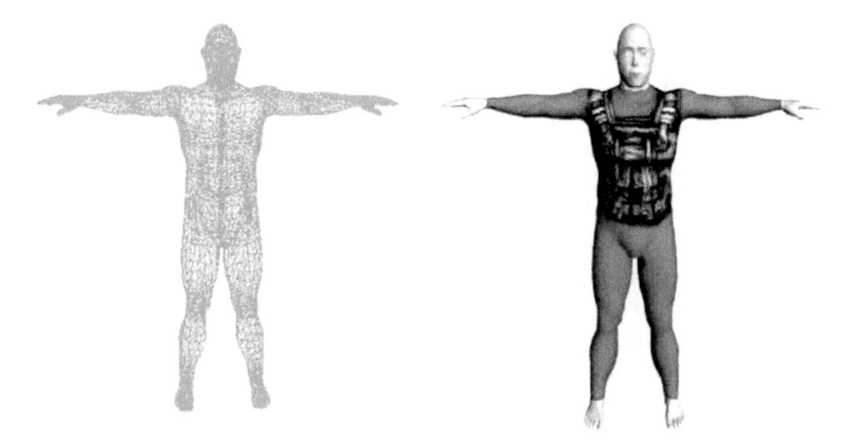

Figure 8.12 Left: a generic body model conformed to a 3D whole body scan. Right: a photorealistically rendered version of the generic body model. (Frames from a 3D animation sequence generated by Dr J.C. Nebel and composited by George Barbour)

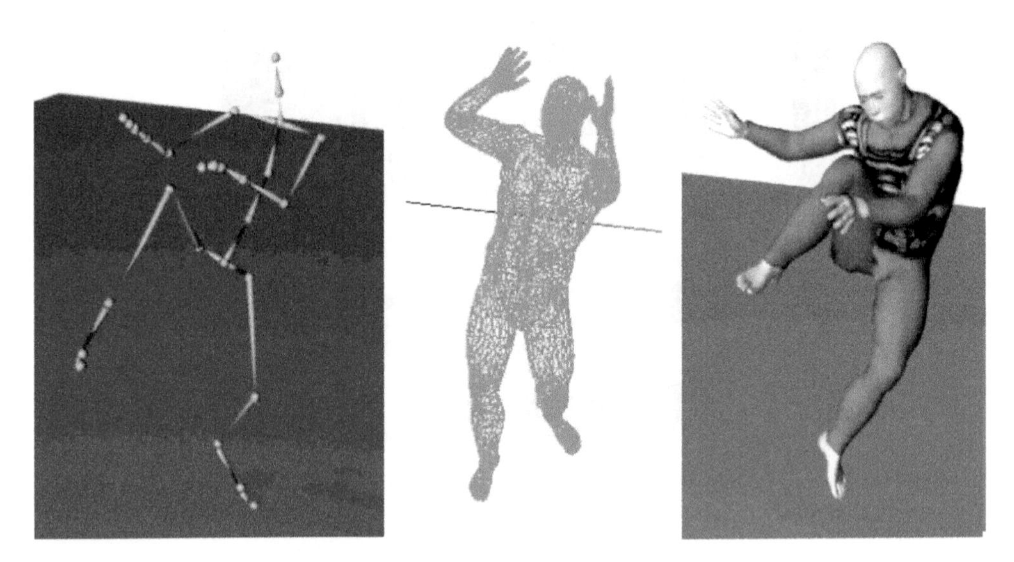

Figure 8.13 Left: a kinematic skeleton instantiated by means of motion capture data. Middle: the generic body mesh with a skeleton inserted. Right: a complete frame from a cine sequence showing texture-pasted conformed and animated using the skeleton and motion capture data. (Frames from a 3D animation sequence generated by Dr J.C. Nebel and composited by George Barbour)

means of a Wicks & Wilson [217] commercial whole body scanner (this particular scanner utilizes Moiré fringing to achieve depth estimation). The same figure also shows Dr Tao's conformed generic model 'skinned' with virtual clothing. Figure 8.13 shows a skeleton instantiated with motion capture data that is inserted into the generic model, and the result of animating the unskinned and skinned body models is shown also.

8.4 Clinical and Veterinary Applications

8.4.1 Development of 3D Clinical Photography

As explained in section 8.3.2, the spatial resolution in x, y and z afforded by stereo-photogrammetric capture using commercially available cameras lends itself to clinical applications involving 3D capture of human (and animal) surface anatomy. The clinical motivation for 3D surface anatomy capture flows from the need to make quantitative and objective measurements of the surface of the body before and after surgery or some other clinical intervention. Such measurements may be required to be made longitudinally to assess a patient's longer term postintervention progression. Simple 2D photographs require a considerable degree of subjectivity to make an assessment or comparison longitudinally or between patients or control subjects. The current trend in *evidence-based* medicine is that treatment and outcome evaluation should be based on good-quality assessment protocols, in turn implying that more quantitative and objective assessment techniques are required than can be afforded by conventional photographs alone. Therefore, there has been considerable research effort expended in extending clinical photography into the third dimension and also in developing appropriate tools to allow clinicians to make appropriate body surface measurements.

Clinical applications for 3D facial assessment include cleft lip and palate repair, reduction/enlargement of the jaws, trauma surgery, cranial remodelling in babies and congenital defect repairs. In addition, the link between development and facial appearance has recently also been explored in order to detect the potential of certain syndromes and also schizophrenia.

8.4.2 Clinical Requirements for 3D Imaging

The first requirements for clinical photography concern capture speed and 3D accuracy. Normally better than 0.5 mm measurement error is required for maxillofacial applications and better than 2.0 mm error is required for general body surface anatomy measurement. Since the appearance of the skin is critical to clinical interpretation, a life-like 3D photorealistic model of the surface anatomy under investigation is required; hence the natural photographic appearance of the skin must be rendered on top of the constructed surface. Photorealistic surface rendering provides the clinician with the means to locate critical landmarks and evaluate skin appearance in addition to 3D anatomy shape. Capture time is also critical for many applications, particularly those involving children or even babies, and this is only really satisfied when using studio flash illumination (of the order of 1.5 ms duration). Hence the need to be able to process simple stereo-pairs of flash illuminated images to generate range surface from which surface anatomy models can be computed and assessed. Finally, sufficient coverage of the anatomy under investigation must be achieved to provide a useful 3D model. In the case of the face, a maxillofacial surgeon requires the largest plausible face to be captured, occupying a volume of approximately 220 mm wide by 280 mm in length and 150 mm in depth. It is important that the entire face is captured from the hairline to the hyoid bone and also from the tip of the nose to the back of each pinna (the flap of the ear that extends outside of the head).

8.4.3 Clinical Assessment Based on 3D Surface Anatomy

The types of question that a clinician wants to answer include: How ‘abnormal’ is this patient’s face? To what degree has the appearance of this face improved (or deteriorated) following correction by surgery? Which surgery protocol works best for a given type of defect and under what specific considerations regarding the patent, for example, which method of breast reconstruction works best for a given size and particular shape of breast? Increasingly, questions are asked such as: Is there any statistically significant difference between the shape of this reconstructed/repaired face and the general face shape of the population to which this patient belongs? What level of reconstruction quality is a particular surgeon, or surgery unit, capable of achieving? Therefore, potentially both clinical and medicolegal issues are at stake when undertaking surface anatomy assessment. To answer the above questions, clinicians are interested in establishing statistical models of shape, growth, longitudinal change and symmetry.

The overall approach to clinical assessment of surface anatomy, termed *anthropometry*, involves the collection of surface measurements of a population of individuals (control group) in order to establish *population norms*, i.e. the normal shape and shape variability [107] of a particular area of surface anatomy must be determined in order to model this area in the healthy individual. Where this area of surface anatomy has been compromised, either through disease, birth defect or trauma, the measurements modelling this area of surface anatomy in a healthy population can serve as a standard against which to compare measurements of a specific individual or individuals exhibiting the same pathological condition. Thereby it becomes possible to evaluate the degree to which the individual or individuals deviate in the shape of their pathological surface anatomy compared to a given control population. It also becomes possible to evaluate if the shape variation of the pathological group is statistically significant, i.e. is within acceptable variation given the population or could be expected within the normal shape variation found within the healthy population.

Therefore the above approach aims to evaluate whether the shape of an area of anatomy has been restored through surgical intervention such that no statistical difference between an evaluated individual (or group of individuals) postsurgery and individuals within a control population can be determined to within a specific degree of confidence (i.e. *confidence interval*). The measurement of surface anatomy shape and shape change over time serves the role of determining outcome success in adults. In addition to shape measures, symmetry measures are also important, especially for evaluating the face. In children there is the additional dimension of growth and growth variation over time that needs to be captured and deconvolved from change variation in order to establish and monitor postoperative outcome success objectively. By undertaking longitudinal 3D assessment of surface anatomy, there is also the potential to discover information that would otherwise be difficult to capture, such as the trajectories of growth centres and the evolution of surface shape over time, which may provide insights into developmental processes.

8.4.4 Extraction of Basic 3D Anatomic Measurements

In order to evaluate 3D surface anatomy shape surgeons require basic tools to allow 3D surface measurements to be collected and compared. The most fundamental traditional anthropometric measurements are based on Euclidean distances between landmarks defined on the surface anatomy. However, as outlined below, the availability of a metrically accurate 3D surface anatomy model opens the possibility for a far wider range of measurements that can be afforded by the traditional calliper-based measurements of the anthropometrist. In order to capture these measurements, the clinician requires a basic tool that allows him or her to display and interact with captured 3D surface anatomy models on a computer workstation, such that these can be landmarked by point placement on the displayed surface. There are several instances of such tools having been developed and an example of the Facial Analysis Tool (FAT) [293] is given in Figure 8.14 that provides facilities for facial surface anatomy display and landmarking.

Based upon specific sets of landmarks, standard *surface area* measurements can be captured and compared longitudinally, while the enclosed volumetric differences between compared surfaces can also be measured over time. In this latter case it is necessary to register the compared 3D manifolds accurately and this is usually a three-stage process. Firstly, landmarks are placed at corresponding anatomically defined locations on each of the surfaces to be aligned. In order to obtain accurate results, two sets of landmarks will be required, one set defining corresponding regions of the compared manifolds which are not expected to have changed, e.g. the forehead. A second set of landmarks is then used to define the region of interest which is expected to have changed by displacement, for example the jaw following an ostiotomy procedure (jawbone shift) to extend or retract the jaw as required. Secondly, the (unchanging) corresponding landmarks serve to anchor an initial alignment of the surfaces by means of the Procrusthese algorithm [11], which finds the rigid body transformation that

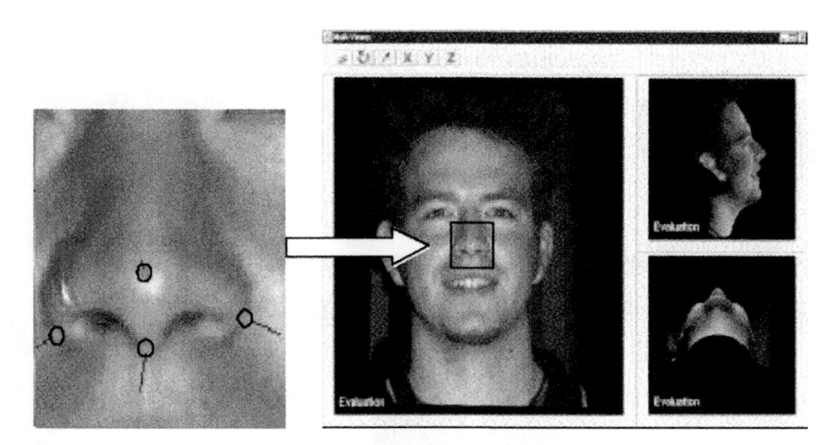

Figure 8.14 Left: anatomic landmarks placed on a 3D face model. Right: a screenshot of the Facial Analysis Tool. Note the three orthographic views of the 3D model are displayed simultaneously to assist clinicians place landmarks accurately and consistently. (Reproduced from [293])

minimizes the RMS error between these landmarks. However, this rigid body alignment over the corresponding *areas* defined by these landmarks is further refined by means of a modified version of the iterated closest points (ICP) algorithm [38], termed HICP, developed by Mao *et al.* [293]. In basic form, ICP measures the distance between each vertex on one surface and the closest triangle surface on the other. ICP attempts to find the 3D rigid body transformation that minimizes the average of these distances. Since 3D scanned data may contain artefacts such as holes, missing data and outliers, the HICP algorithm minimizes the median of the intersurface distances (as opposed to the average intersurface distance) and thereby improves the stability and robustness of the registration procedure under real-world 3D imaging conditions. Finally, the two 3D surface patches (defined by the landmarks to contain the region over which volume change is to be measured; Figure 8.15) are joined together to form a watertight volume. It is then possible to measure this closed volume by projecting the mesh triangles as extruded volumes on to an arbitrary plane for each surface and subtracting the overall volumes to find the volume difference. In this approach longitudinal differences on the face of 0.7 cm^3 have been measured reliably [168], while considerably lower volume differences (of the order of 0.1 cm^3) have been measured on average over populations [106]. Figure 8.15 illustrates area measurement and landmark-based registration of compared surfaces and their areal and volumetric differences within the FAT. Figure 8.16 shows the effect of registering 3D data collected from the same face pre-/postoperatively where an ostiotomy procedure has been conducted to displace the jaw back into the face by removing sections of bone [468].

In addition to Euclidean distances, surface areas and volume change over time, geodesic distances are also extracted in order to characterize surface shape. Surface geodesics are usually defined between standard landmarks and have the advantage of sampling the shape of the surface to a greater degree than the landmarks alone. Rather than compute the *true* geodesic distances between pairs of landmarks, often a *pseudo* geodesic is sufficient to capture the underlying surface shape consistently. The pseudo geodesic developed by Mao and co-workers [294, 295] is computed by finding the shortest path intersection of a plane that can rotate about a line terminated by the two anchoring landmarks defining the start and end points of

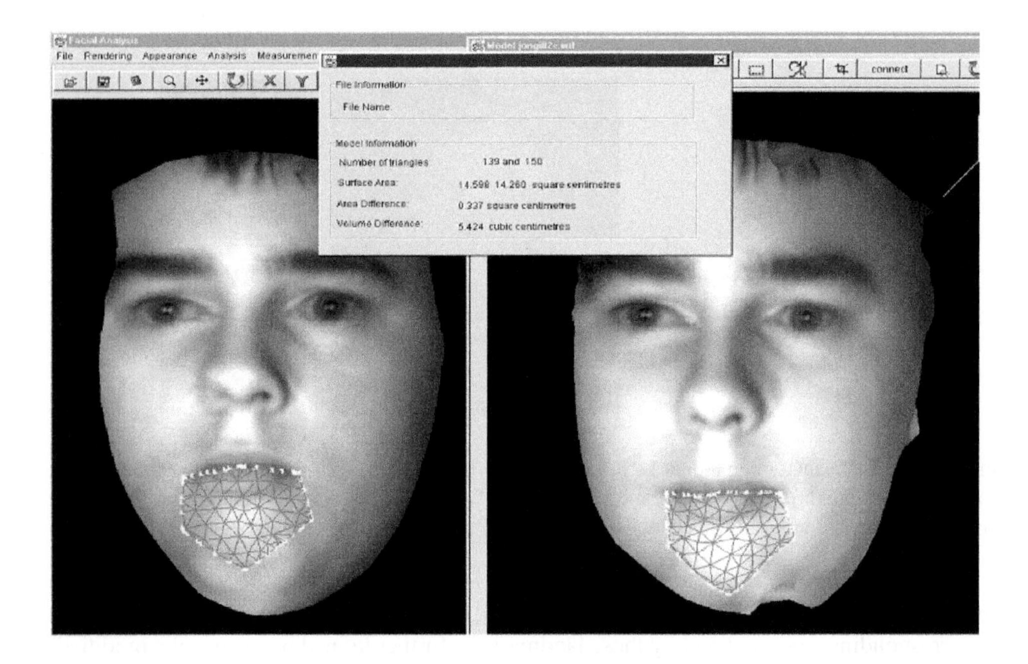

Figure 8.15 Triangulated region, defined by 6 landmarks, is shown on each face and the area of each region is measured. The difference in area and volumetric difference between the triangulated regions (following HICP registration) are also computed within the FAT. (Reproduced from [293])

the extracted trajectory. Figure 8.17 shows the initial plane anchored by the landmark pair, the series of planes generated to sample possible surface intersection paths and the appearance of the paths on the surface of a sampled model. While this pseudo geodesic path clearly will not recover the shortest surface distance between the two landmarks, for many situations it will provide a good approximation and may also provide a measure that behaves more

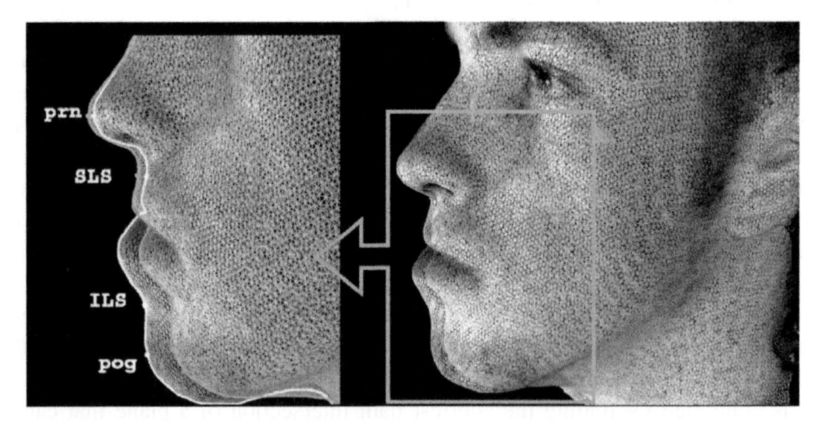

Figure 8.16 Example of HICP registration on patient 3D face data collected before and after an ostiotomy procedure to displace the jaw back into the face. (Reproduced from [468])

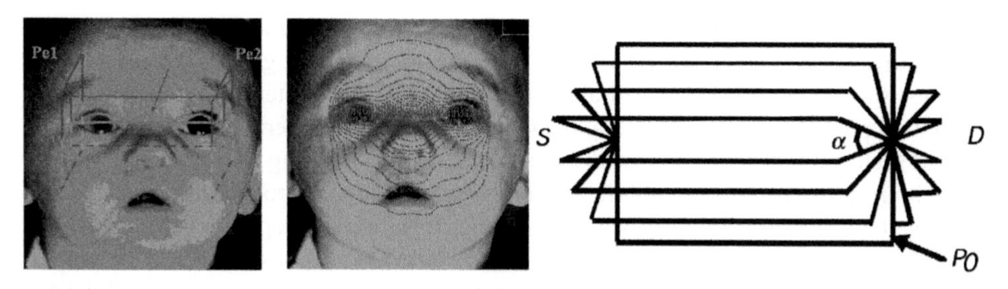

Figure 8.17 Cutting planes. Left: the initial plane P_0. Middle: the series of surface sampling planes $\{P_i\}$. Right: the intersection of the triangulated surface and the series of planes $\{P_i\}$. (Reproduced from [295])

intuitively to the clinician (Figure 8.17). Appropriate situations for the use of pseudo geodesic measurement are given in [333].

8.4.5 Vector Field Surface Analysis by Means of Dense Correspondences

While geodesic measurements capture more 3D information from the manifold, they do not fully exploit the potential of representing and comparing the entire 3D anatomy surface geometry. To this end techniques based on *dense correspondences* [233, 295, 296] have been developed that allow anatomic surfaces to be represented using a consistent set of measurements that can then be compared between different individuals or the same individual longitudinally. The basic idea requires the conformation of a generic 3D surface model that represents the gross shape of the area of anatomy under investigation (as mentioned in the human modelling section above), to a 3D surface model captured from an individual undergoing assessment. Figure 8.18 illustrates this process showing a generic facial model prior to conformation, 3D data captured from a real face and the final conformation of the generic mesh into the shape of the real face (using Mao's basic shape-similarity method [295].

Conformation for dense correspondence extraction comprises an iterated process [233]. Firstly the generic 3D mesh is brought into approximate rigid-body correspondence with the captured 3D surface data by means of pairs of corresponding landmarks, one landmark placed on each surface. These landmarks are then forced into correspondence and the mesh

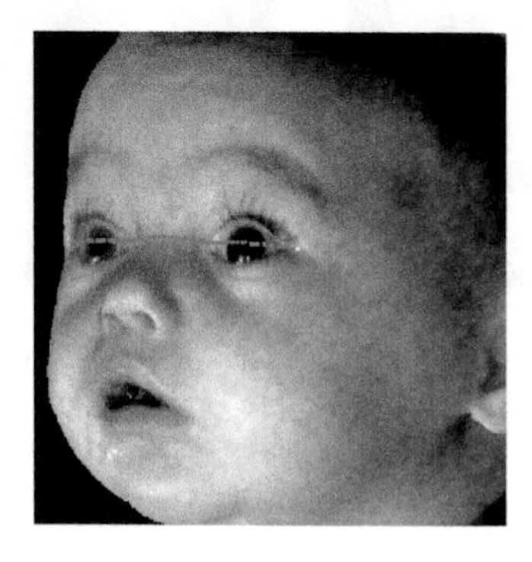

Figure 8.18 Estimated shortest path using the pseudo geodesic. (Reproduced from [295])

neighbouring each landmark is displaced according to the weight of a radial basis function (RBF) centred at each landmark. Thereafter, the distance from each generic mesh vertex to the closest surface point is calculated along the direction of the vertex average normal. An optimization process is established to allow the positions of the generic mesh vertices to be displaced such that the closest surface distance is reduced, by bringing the spring forces generated within mesh edges into equilibrium with the force required to produce the displacement for each vertex. Therefore the mesh behaves as though it has elastic properties that tend to prevent it becoming 'crumpled' if it were simply clipped on to the closest surface of the captured data [233]. This match-smooth vertex displacement cycle is applied many times until the algorithm converges on a stable solution exhibiting a low global surface *geometric* registration error.

In order to improve the *topological* registration error between the generic mesh and the captured data Mao [295, 296] developed a surface compatibility function that takes into consideration not only the local distance between the surfaces, but also the relative angle between their respective surface normals and the difference in their respective principal surface curvatures. Therefore, this method drives locally corresponding surfaces into registration using shape information as well as closest distances. In a further improvement, the topological compatibility of the local destination and target surfaces is established by computing the local Gaussian and mean curvatures which allows the *type* of local surface (peak, pit, ridge, valley, saddle ridge, saddle valley, minimal, flat) to be classified and cross-checked.

If the generic mesh is sufficiently detailed to match the spatial resolution of the captured data, it is possible to 'clone' the shape reasonably faithfully as can be observed in Figure 8.19. As a consequence, the shapes of captured anatomy can be compared directly since these

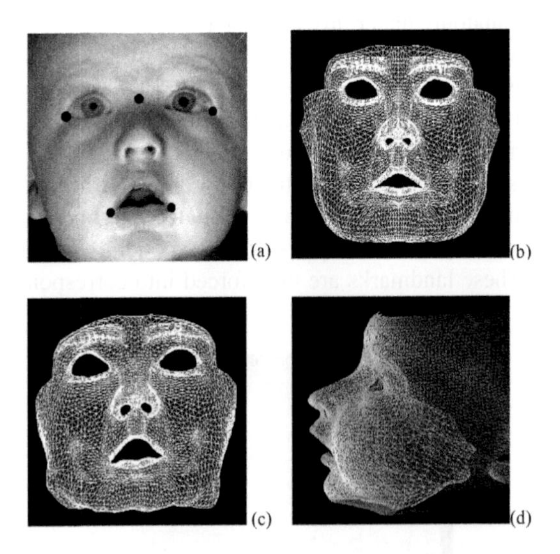

Figure 8.19 Result of the conformation process, using Mao's basic method, reproduced from [296]: (a) the scanned model with 5 landmarks placed for the global mapping; (b) the generic model; (c) the conformed generic model; from [295]: (d) the scanned model aligned to the conformed generic model. The (smaller) lighter mesh is the conformed generic model, the darker mesh (representing the whole face) is the scanned model (Plate 13)

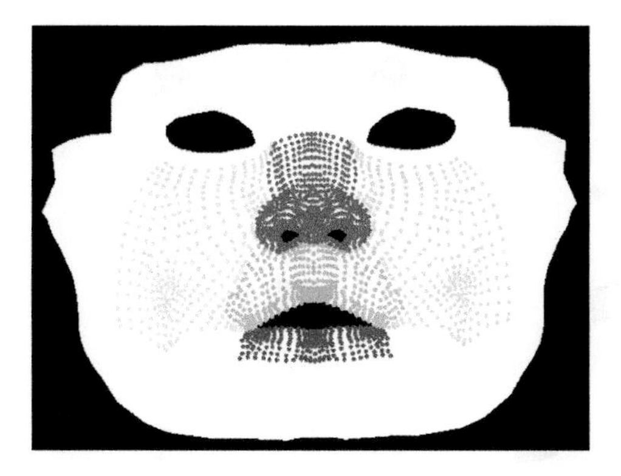
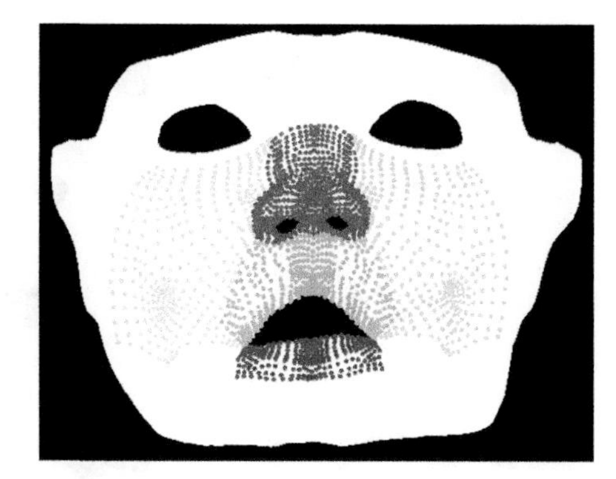

Figure 8.20 Left: a generic mesh colour coded to label different anatomic regions of the face. Right: the generic mesh conformed into the shape of a captured 3D face mesh (Plate 12). (Reproduced from [295])

have been resampled on a topologically consistent basis. Following rigid body registration as described above, the distance and direction between corresponding vertices of compared conformed generic meshes can be computed directly. Accordingly the difference between the compared anatomy samples forms a *vector field* that represents the residual displacements between corresponding vertices. The magnitude of each vector represents the direction of the local displacement between the compared surfaces and the angle of each vector gives the direction of this displacement. While it is possible to utilize the vector field directly for comparing surface shape and symmetry [295], a more powerful analysis in terms of statistical shape variation can be performed as described in the next section.

A further consequence of being able to clone the shape of a surface onto a predefined generic mesh is that semantic labelling attached to the generic mesh can be transferred to the cloned anatomy. Figure 8.20 shows a generic mesh labelled in terms of the principal parts of the face and this mesh and the associated labelling cloned to the 3D model of a real face.

8.4.6 Eigenspace Methods

Since the conformation of a generic mesh to captured data provides a consistently sampled representation for specific regions of captured 3D surface anatomy, it becomes possible to compute the modes of variation of the anatomy when sampled over a population [295, 296]. In her doctoral thesis [295], Mao shows the result of comparing the mean shape of a control population of baby faces with a surgically managed group (postfacial cleft repair) (Figure 8.21). Figure 8.22 illustrates the first principal mode of variation of the control group using principal components analysis (PCA) and Figure 8.23 shows the corresponding variation for the surgically managed group [295]. Notice the distortion about the nostrils in the surgically managed group.

Mao applies the same technique to the investigation of facial shape change over time by PCA modelling the vector displacement fields captured between individuals longitudinally [295] (Figure 8.24). In addition, Mao characterises facial asymmetry by computing the displacement vector field between a conformed generic mesh and a bilaterally reflected version of *itself*. Figure 8.25 shows corresponding points reflected about the facial midplane on a

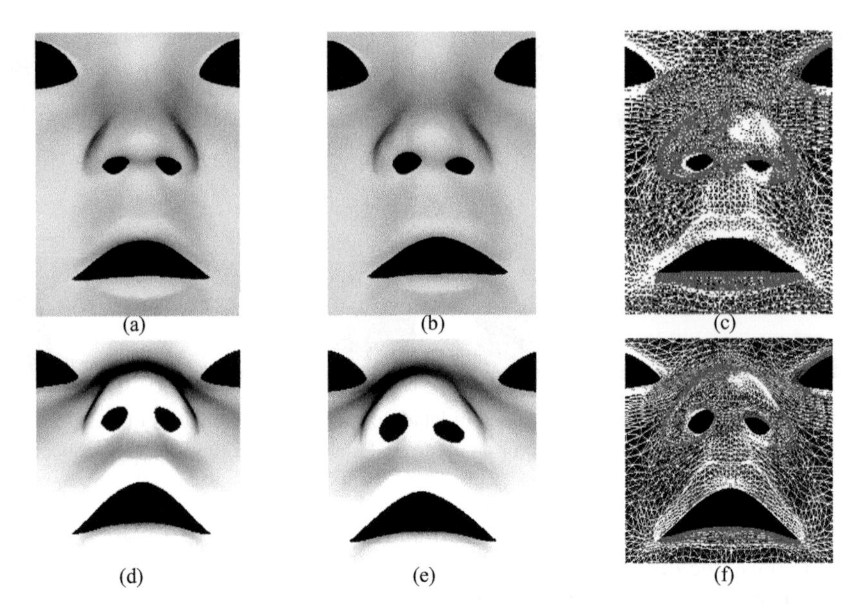

Figure 8.21 Mean shape of the control and surgically managed group: (a, b) the mean shape of the control group; (c, d) the mean shape of the surgically managed group; (e, f) the mean shape of the control group (white mesh) aligned to the mean shape of the surgically managed group (red mesh). The top row shows the front view, the bottom row shows the bottom view. (Reproduced from [295])

generic face mesh [295]. The vector field then represents the displacements required to bring the left-hand side of the mesh into correspondence with the right-hand side (assuming in this case that a bilateral axis of symmetry is exhibited by the anatomy under assessment). Figure 8.26 shows the complete process required to produce a vector field representing displacements between the left- and right-hand sides of the face. Again, it is possible to characterize the normal modes of facial symmetry fluctuation within control (Figure 8.27) and surgically managed (Figure 8.28) populations by means of PCA [295].

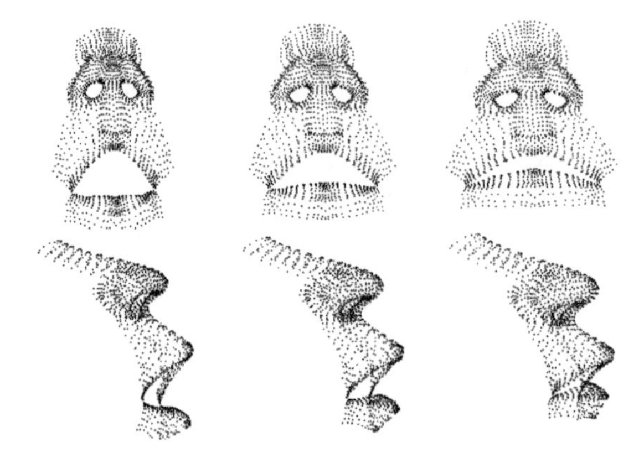

Figure 8.22 First principal component of the control group, between -3 (the first column) and $+3$ (the third column) standard deviations. (Reproduced from [295])

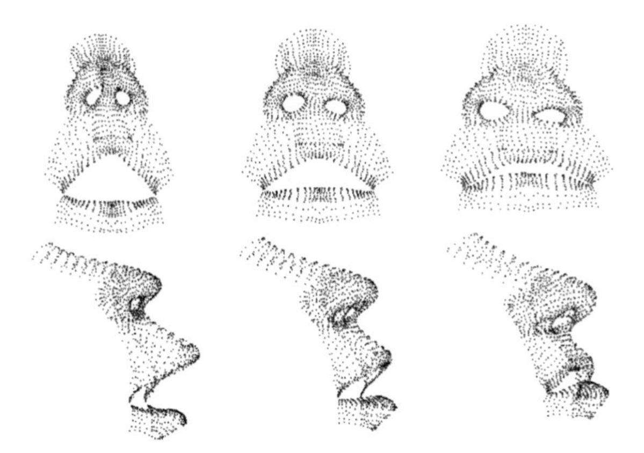

Figure 8.23 First principal component of the surgically managed group, between −3 (the first column) and +3 (the third column) standard deviations. (Reproduced from [295])

Figure 8.24 Comparison of corresponding vertices between the mean shapes for 3D face models of 1- and 2-year-old children in a surgically managed group (unilateral facial cleft): light points indicate no statistically significant difference, while the dark points indicate a significant difference between the models captured at the two different ages (0.05 significance) (Plate 14). (Reproduced from [295])

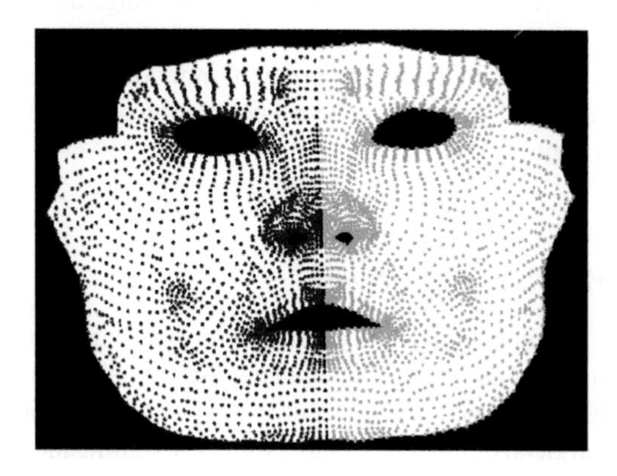

Figure 8.25 Corresponding points reflected about the facial midplane on a generic face mesh. (Reproduced from [295])

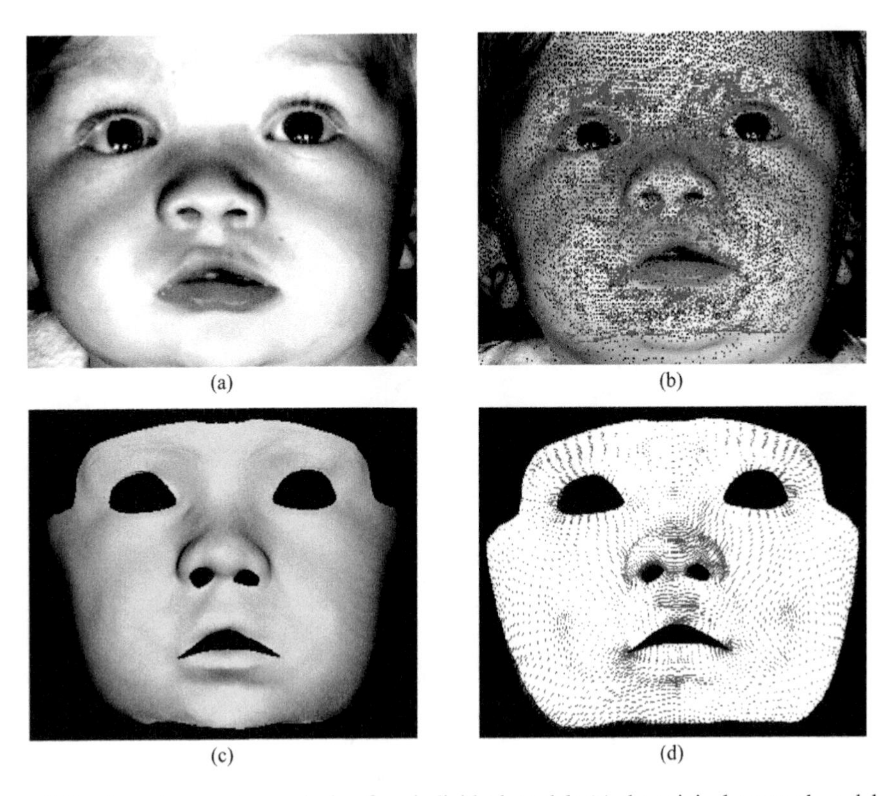

Figure 8.26 Facial symmetry analysis of an individual model: (a) the original scanned model; (b) the corresponding conformed model; (c) the original scanned model (the light mesh) aligned to the conformed model (the dark mesh); (d) the calculated symmetry vector field (Plate 15). (Reproduced from [295])

Having constructed PCA models of facial shape, growth/change and asymmetry for a specific control population, this can be used as a reference by which to test if the shape, facial change or facial asymmetry of an individual or group of individuals when projected into PCA space falls within the range exhibited by the control population or is statistically significant in its deviation [295, 296] (Figure 8.29).

8.4.7 Clinical and Veterinary Examples

In this section several examples of clinical and veterinary applications are presented. Figure 8.23 shows the variation of facial shape of cleft cases in babies that have been repaired surgically. Comparing this result with that of Figure 8.22 for the corresponding control group it can be seen that the surgically repaired group exhibits a significant residual deformation about the area of the nostrils [295]. A similar result for facial asymmetry can be observed by comparing the symmetry variation within the control and surgically managed groups shown in Figures 8.27 and 8.28, respectively.

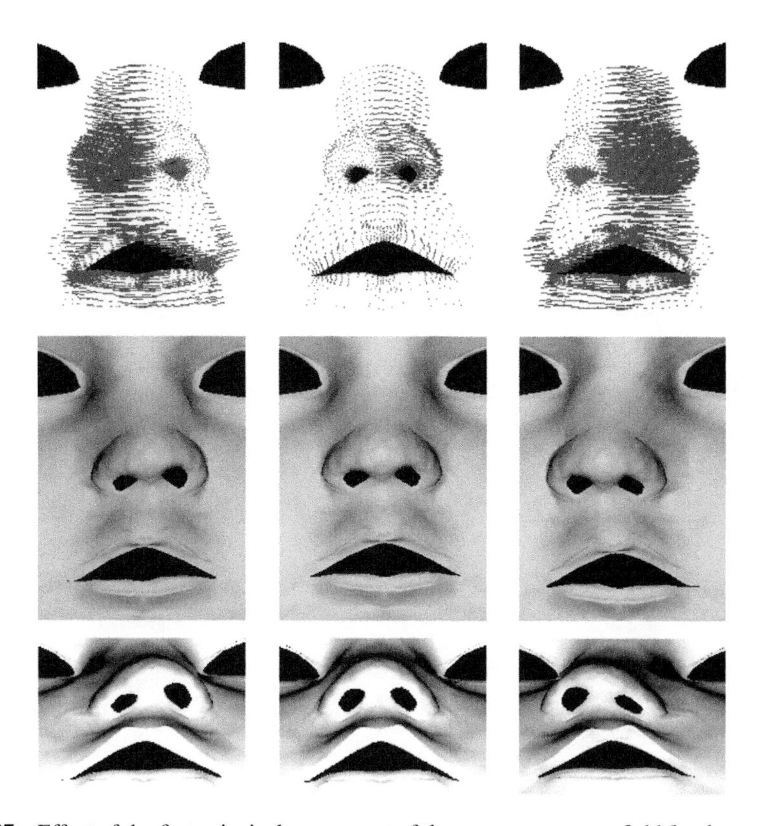

Figure 8.27 Effect of the first principal component of the symmetry vector field for the control group between −3 (the first column) and +3 (the last column) standard deviations: the top row shows the constructed symmetry vector field; the middle and the bottom row show the front view and the bottom view of the generic model modulated according to the symmetry vector field at the top row within the same column. (Reproduced from [295])

The ability to align facial models over time has been used to evaluate the persistence over time of collagen injections in facial creases. Although only a fraction of a cubic centimetre is injected (typically 0.3 cm^3), it is possible to detect this difference and track its persistence over a number of months, when averaging over the entire group of samples taken at each time step (in this case of the order of 15 subjects) [106].

More challenging assessment situations require capture facilities comprising three or four stereo-pairs in order to capture complex areas of anatomy such as the breasts. Figure 8.30 shows a four stereo-pair breast capture rig (developed in a collaboration between the Canniesburn Plastic Surgery Unit, Glasgow Royal Infirmary, UK, and the University of Glasgow, UK). A Breast Analysis Tool (BAT) has been implemented [333] to allow the immediate area of the breast to be defined within an area bounded by four manually placed landmarks. These landmarks define the corners of a Coons patch [128] which is used to model the chest wall. The intersection of the Coons patch with the breast forms a closed volume comprising a segmented breast whose volume and skin area can be measured in the BAT in

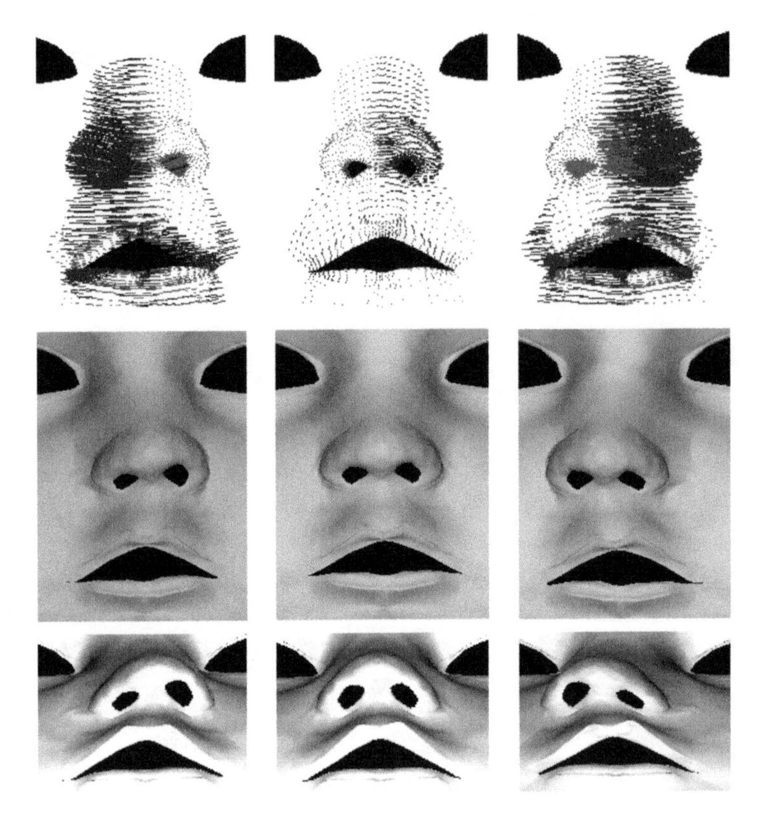

Figure 8.28 Effect of the first principal component of the symmetry vector field for the surgically managed group between −3 (the first column) and +3 (the last column) standard deviations. Layout as per Figure 8.27. (Reproduced from [295])

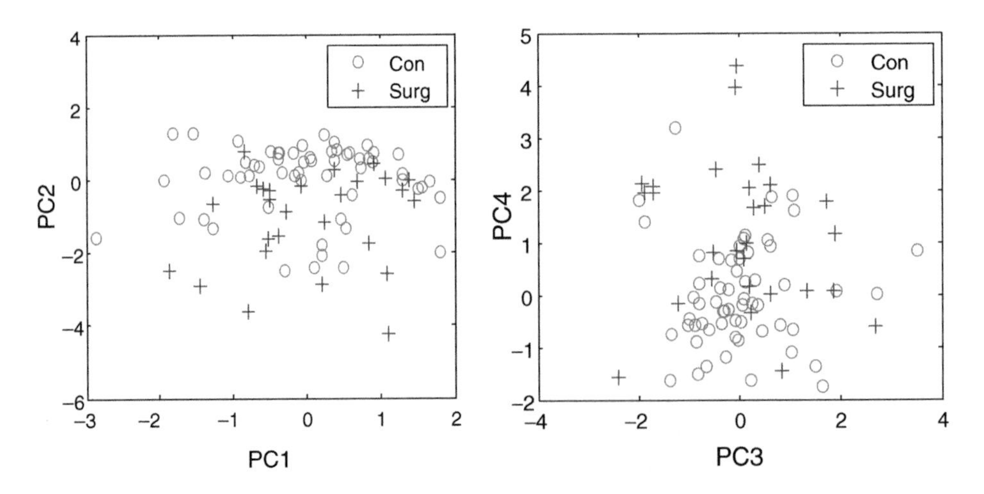

Figure 8.29 Projection of the models in the surgically managed group and in the control group, on to the face shape space formed by the principal components of the control group: the second versus first and fourth versus third principal components. (Reproduced from [296])

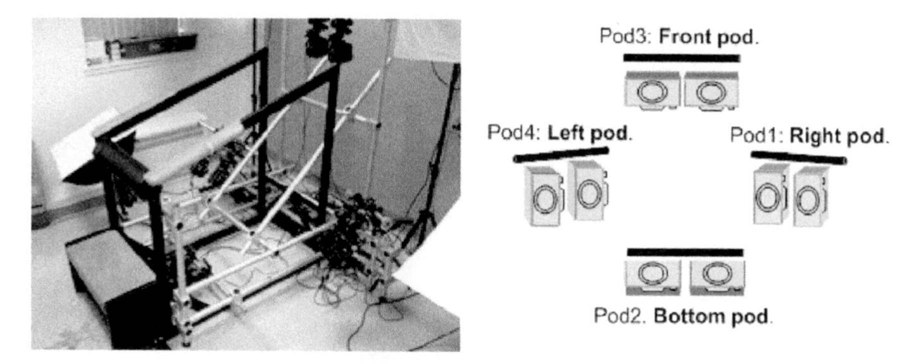

Figure 8.30 Left: an experimental four imaging pod breast capture system. Right: the imaging pod schematic configuration of this breast capture system. (Reproduced from [333])

order to inform a surgeon of the required tissue volume and skin area required to effect a postmastectomy reconstruction (Figure 8.31).

In addition to human anatomy assessment tasks, there are many potential applications for 3D imaging in the field of veterinary medicine and food production. A three stereo-pair rig has been developed (in a collaboration between the Silsoe Research Institute, UK, and Glasgow University, UK) (Figure 8.32) [451] to allow live pigs to be 3D captured from side, top and rump views. Having merged this data [233] (Figure 8.33), it was possible to correlate the shape of the animal with final weight gain at the end of the growth period. In this case a 14-week trial was conducted on two cohorts of 16 pigs to construct a statistical relationship between weight gain, feed composition (high or low lysine) and 3D body shape. In this case body shape was characterized by the cross-sectional shape of the pig as illustrated in Figure 8.34. Curvature analysis based on differential geometry applied to the surface of the pig provided curvature

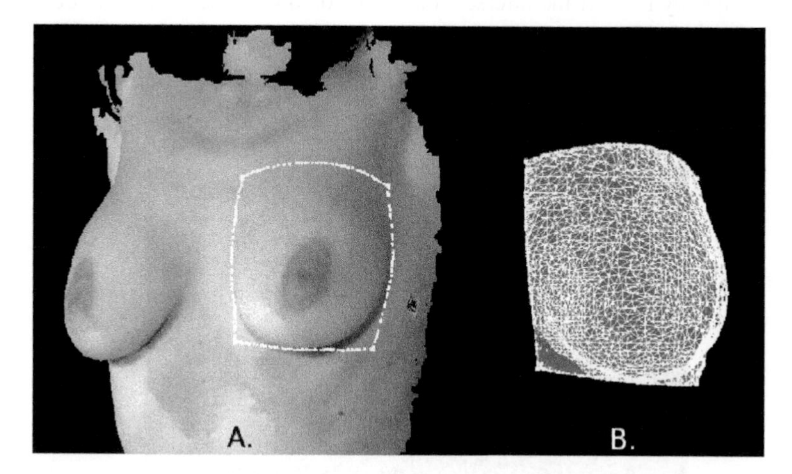

Figure 8.31 (A) 3D breast model with a segmentation region defined manually by means of four landmarks. (B) The result of segmenting the breast by means of the Coons patch defined by the landmarks to form a closed volume that can be measured by the BAT. (Reproduced from [333])

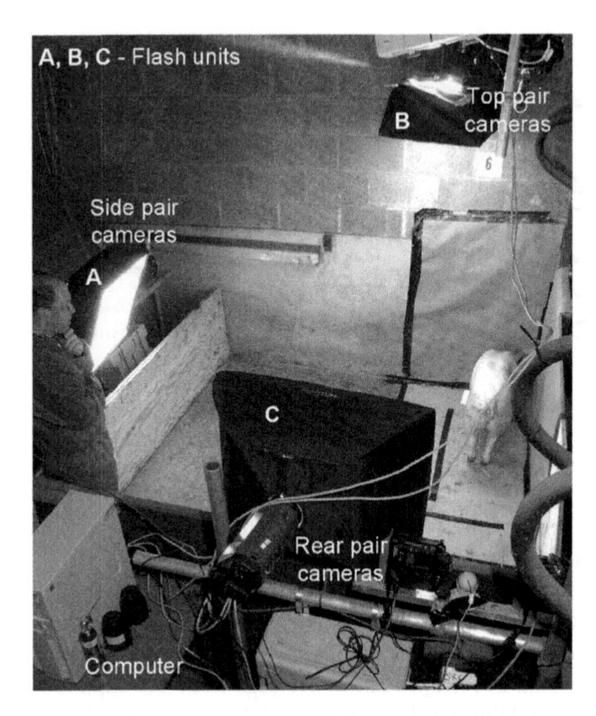

Figure 8.32 Three-imaging pod rig for 3D capture of live pigs. (Reproduced from [417])

features on the otherwise bland surface of the animal. It was possible to extract the spine as the line of minimum curvature along the back of the animal via *parabolic* point analysis. Manually placed anatomic landmarks were used to define a plane that served to cross-section the animal and thereby recover the intersection of the torso with the plane. Full details of this analysis are given in [417].

Figure 8.33 Integrated 3D surface model [233] of a live pig captured using the rig shown in Figure 8.32. (Reproduced from [233])

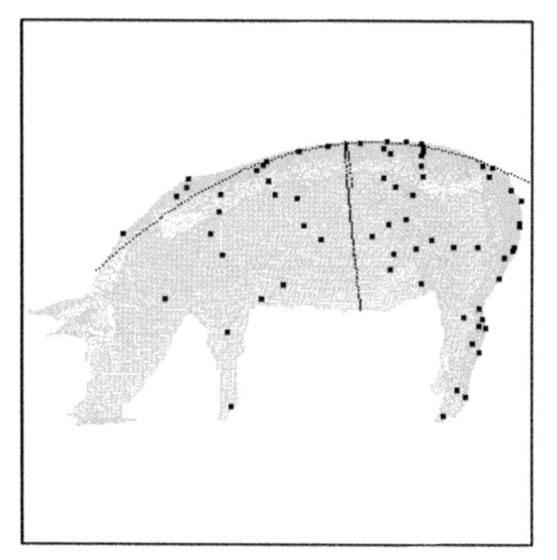

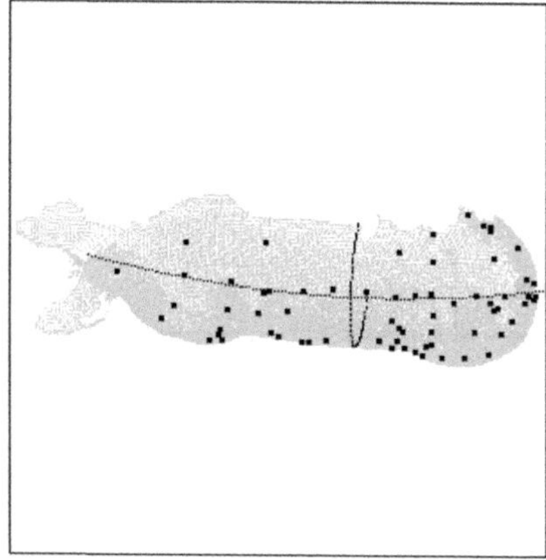

Figure 8.34 Two views of recovered pig spine and measurement of cross-sectional plane. (Reproduced from [417])

8.4.8 Multimodal 3D Imaging

This section completes the overview of 3D imaging applications by presenting examples of 3D models constructed by combining different imaging modalities. In a medical context, it is frequently difficult for both the clinician and patients to interpret the rather bland and featureless skin surface generated by segmenting the air–soft tissue boundary created when segmenting the 3D *voxel* data set produced by MRI or CT imaging, e.g. of the human head. What is really required is a photorealistically rendered surface 3D anatomy model combined with the underlying 3D voxel data. Such a model would allow clinicians and patients to visualize how their internal organs relate in position to both their surface features, and also well-defined anatomic landmarks used to guide surgery procedures.

Figure 8.35 illustrates such a model: we see three views of a skull voxel model combined with a photogrammetrically generated 3D surface anatomy model. The skull, comprising 400 slices, was imaged using a Marconi spiral CT scanner, MX8000, Kv 120, MaS 200 (with bone filter). The settings for this instrument comprised: 0.75 seconds rotation, 0.625 pitch

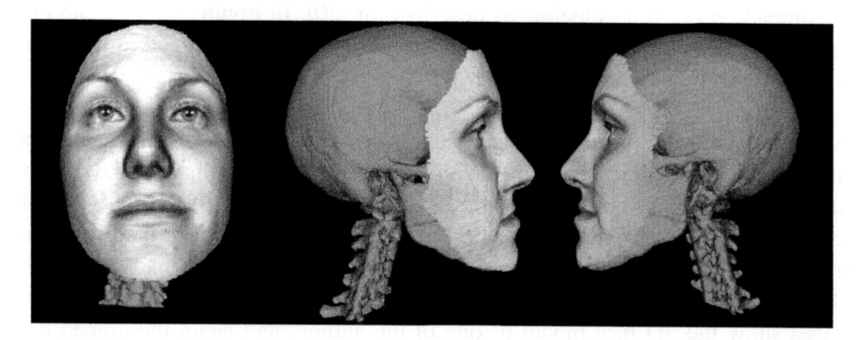

Figure 8.35 Three views of a 3D face surface model combined with a segmented and polygonised CT voxel model. (Reproduced from [20] with permission of Elsevier Science)

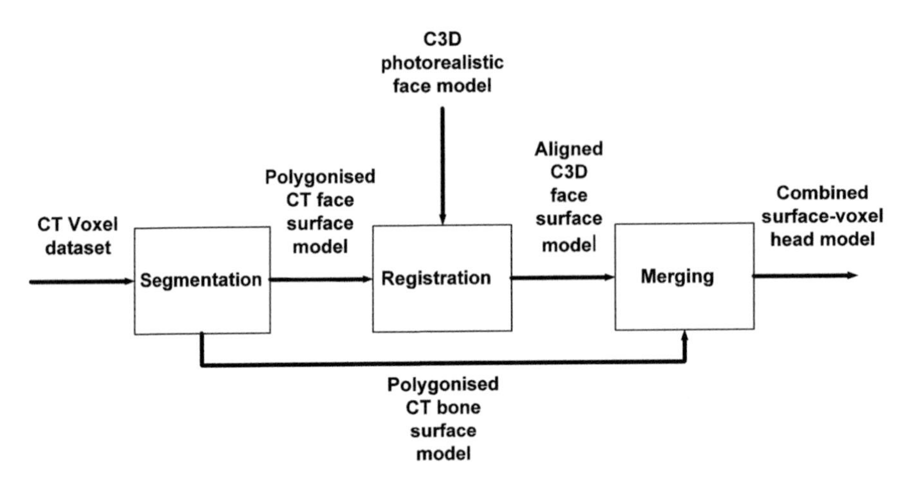

Figure 8.36 Pipeline for combining 3D surface and 3D voxel models. (Reproduced from [20] with permission of Elsevier Science)

angle, to produce a slice thickness of 1.3 mm. A DI3D [207] stereo-photogrammetry system was used to capture the 3D surface model of the face.

Following multimodal data acquisition, the key issue was how to combine the data generated by the photogrammetric and CT imaging modalities, since these have radically different representations, namely 3D surface manifolds and 3D voxels. Integration was achieved by adopting a common polygon representation. The skull slice images in DICOM format were segmented in a commercial clinical package (Amera®, TGS Europe) to reveal the air–soft tissue boundary and then polygonized. Likewise it was possible to segment the soft tissue–bone boundary to produce an internal polygon model of the bony structures, also in the same coordinate system as the segmented facial surface model. At this point the segmented polygon models were in VRML format that could be read into the Facial Analysis Tool mentioned in section 8.4.4 along with the 3D facial surface anatomy model. The face surface model generated by photogrammetry was initially registered to the polygonized segmented voxel face model using Procrusthese-based alignment of corresponding landmarks on each surface, followed by vernier registration using the HCIP algorithm. Figure 8.36 illustrates the complete processing pipeline as described in [20]. In Figure 8.37 we can observe how the photorealistic facial model can be rendered with semitransparency to reveal underlying structures and their relative alignment with respect to features as can be seen on the surface of the face.

In the second example of multimodal 3D imaging, we combine a 3D surface model captured using conventional stereo-photogrammetry with a thermal image, by aligning and draping the thermal image over the underlying surface model [5, 469]. Figure 8.38 shows two examples of combined 3D thermal imaging produced at the University of Glasgow: in the upper images, the thermal images draped on the human arm surface models clearly depict the underlying areas of hot (light) and cold (dark) to reveal sub-skin structures such as blood vessels. The lower images show the 3D face model of one of the authors and again the draped thermal images reveal underlying areas of hot and cold to expose blood vessels and colder internal cavities.

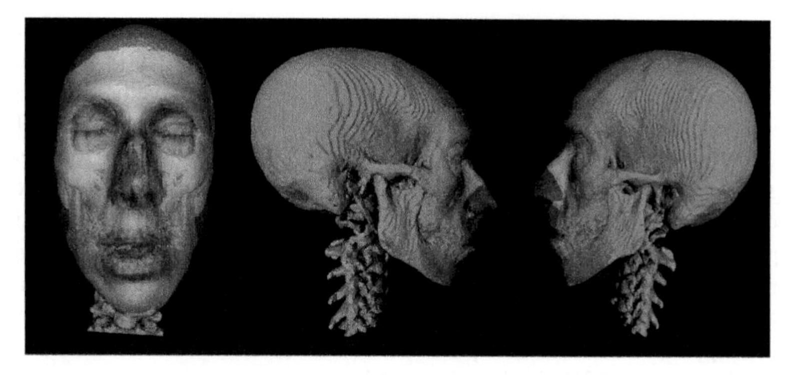

Figure 8.37 Combined surface–voxel models rendered with partial transparency of the surface model. (Reproduced from [20] with permission of Elsevier Science)

The thermal image was captured using a Merlin® Indigo thermal camera having a Stirling cooled InSb sensor operating in the 3–5 μm band (with cold filter) and better than 25 mK thermal sensitivity, producing images of 320 × 256 pixels. It is a comparatively simple procedure to produce a photorealistic 3D surface model (e.g. as depicted in Figure 8.31) by draping an image whose camera geometry is known with respect to the stereo-pair used to capture the 3D

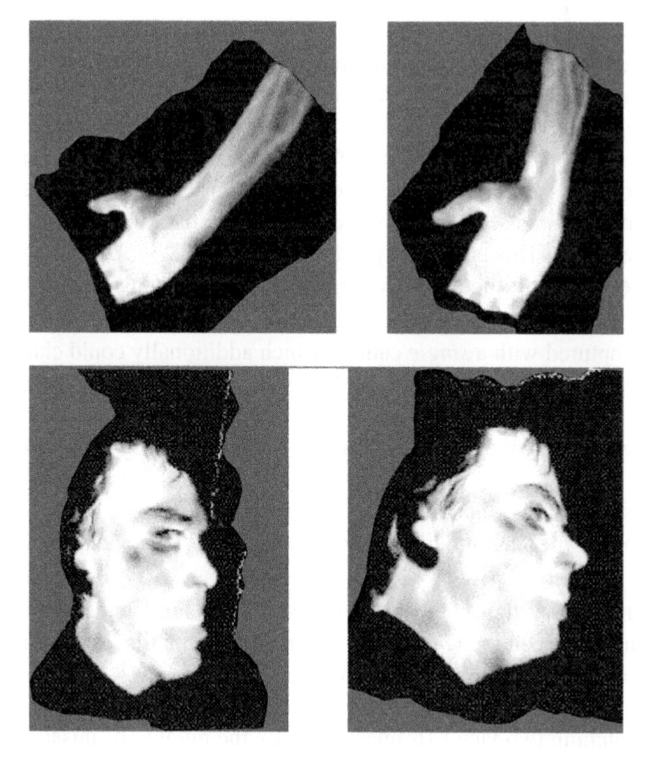

Figure 8.38 3D surface models draped with 3–5 μm band thermal images

surface model. Typically one of the stereo-pairs themselves is used as the photorealistic drape. When the stereo-pair of acquisition cameras are monochrome, a third colour camera is used to capture the image that serves as the photorealistic surface drape and calibration images are captured by this camera at the same time as calibrating the acquisition stereo-pair. Thereby the intrinsic and extrinsic parameters become known for all three cameras. However, when substituting a thermal camera for a colour camera, the problem arises that the calibration target must be observable in *both* the visible and thermal wavebands. Producing a calibration target comprising patterns of both visible and thermal contrast is not always easy to arrange, requiring special materials to generate the required thermal and visible contrasts and a means of heating/thermally illuminating the target [5, 469]. However, although not without practical challenges, the approach does have the potential to reveal otherwise invisible structures and relate these in the context of 3D physical shape and position.

The preceding sections have listed only a few practical examples of human face and body 3D imaging. Every part of the human body surface can potentially be imaged, modelled and interpreted in 3D for creative or clinical applications. Likewise animal forms may be similarly captured and analysed, and arguably, in this case, an even greater range of potential applications can be envisaged. However, it should be remembered by researchers and developers that when human 3D data is being captured, manipulated and potentially distributed, then ethical, privacy and security issues become paramount considerations.

8.5 Movie Restoration

For over a hundred years the film industry has created thousands of movies. The majority of them were stored on celluloid film which has a tendency to deteriorate over time. Thus, film archives around the world contain great movies whose quality degrades each year. We have to save that heritage, and one of the ways is to digitize and restore what is left today. It happens sometimes that in the processed movie some frames are in such a bad a condition that they cannot be recovered. However, under some assumptions the remaining frames either side of the missing frame can be used to restore, i.e. synthesize, the missing data without visible deterioration of the movie. This process is presented in Figure 8.39. Unfortunately the task of film repair can be extremely difficult, especially due to the following factors.

1. The scene was captured with a *single* camera which additionally could change its position and optical parameters during acquisition. More often than not these parameters are not known.
2. There can be many different objects in the scene, possibly moving with different speed and directions.
3. Signal quality is poor (noise, material scratches, spurious artefacts, etc.).
4. The exact number of missing or damaged frames may not be known.

Nevertheless, assuming that the scenes to be recovered are sufficiently static we can try to build a disparity map between the two frames around a gap in a film and then synthesize the missing frames to achieve visual continuity over the repaired gap.

Frame recovery is performed by means of view synthesis guided by disparity maps obtained by image matching two views (frames). Usually the number of missing frames is also unknown and has to be empirically chosen by the operator.

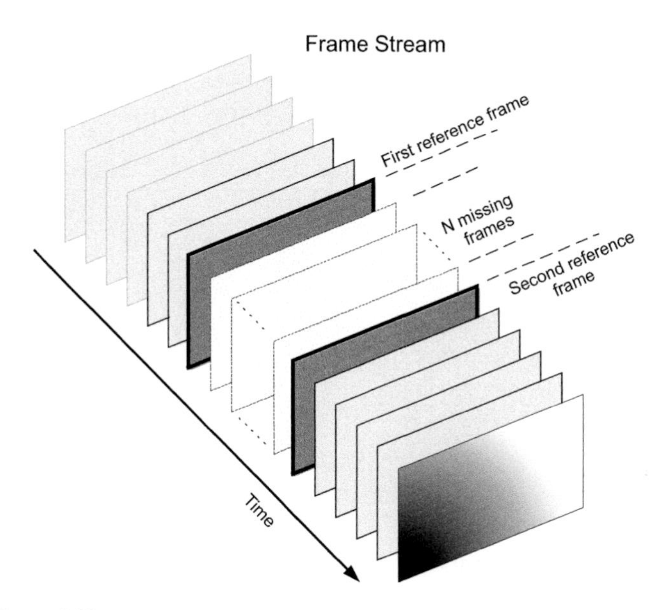

Figure 8.39 Stream of frames with selected reference and missing frames

Figure 8.40 contains two abutting frames, nos. 201 and 202 used for initial matching. Figure 8.41 depicts horizontal disparity map obtained from these frames. These disparities can be used as initial values for iterative matching and should also be scaled in accordance with expected number of missing frames in the gap. To some extent this has to be guided by an operator.

The disparity maps are computed with the hierarchical matching algorithm presented in section 6.7. Actually, two disparity maps were created: one with the left image being a reference, then in the reverse order, i.e. with the right image in a role of a reference. The purpose of this strategy is twofold. First, it was possible to use the left–right checking procedure to remove occluded points (section 6.4.1). The second reason comes from the requirements of the view synthesis algorithm. View synthesis was implemented by warping the existing frames

(a) (b)

Figure 8.40 Reference frames (colour in original version) from the input data stream: (a) frame no. 201; (b) frame no. 202. (Reproduced by permission of Arnold & Richter Cine Technik GmbH & Co. Betriebs kg)

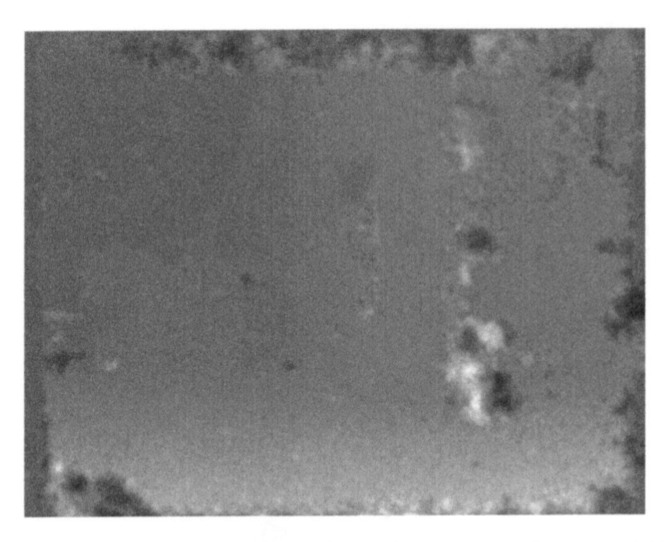

Figure 8.41 A horizontal disparity map computed from the consecutive frames no. 201 and 202 shown in Figure 8.40

guided by disparity values recovered at each pixel. Note that the hierarchical matching algorithm provides 2D disparities (i.e. horizontal and vertical) at each run.

The idea behind view synthesis is very simple: recall that if a disparity map D_{LR} (left is taken as reference) is given between two images I_L and I_R then for corresponding points in the matched images it holds that

$$I_R(x, y) = I_L\left(x + d_x, y + d_y\right), \tag{8.2}$$

where the 2D disparity for a point with local coordinates (x, y) in the left image is denoted as $D_{LR}(x, y) = (d_x, d_y)$. A similar equation can be written when the right image is the reference. Now, since we assume that the scene is sufficiently static and that the base frames were taken in approximately the same conditions with only slight camera shift, we can approximate the missing frames using a linear combination of (8.2). More specifically, half of the missing frames are synthesized from the left base image, whereas the other half are synthesized from the right base image. However, the disparity map has been already cross-checked to remove occluded areas.

Synthesis is a twofold process. First we find new coordinates of a frame, as follows:

$$I_i(x, y) = I_r\left(x + \alpha_i d_x, y + \beta_i d_y\right), \tag{8.3}$$

where $0 < \alpha_i < 1$, $0 < \beta_i < 1$ are scaling parameters for the horizontal and vertical disparities, respectively. I_r denotes one of the base images, I_L or I_R. For instance, if we assume that ten frames are missing and have to be synthesized then as a first attempt we can assume $\alpha_i = \beta_i = 0.1i$ and half of the frames are synthesized from the left base image. Note that in general the new coordinates in the i-th frame computed from (4.3) do not align with the integer pixel coordinate grid of that image. Thus, the second step in the synthesis consists of pixel value interpolation. Usually this is done by bilinear or cubic interpolation (section 12.4). Note also that the synthesized views are in colour, therefore after finding new point position (which is the same for all channels) the interpolation is applied in three colour channels independently.

Usually we do not know how many frames are missing. Moreover, if the scenes are dynamic the problem is more complicated and recovery of the scene geometry and camera parameters requires factorization methods (see section 3.6.3).

Figure 8.42 depicts six reconstructed frames based on the two reference frames in Figure 8.40 and computed disparity maps (see Figure 8.41). The first half length of the reconstructed stream (Figure 8.42(a–c)) is based on the left reference image and left-to-right (i.e. forward)

Figure 8.42 Six reconstructed frames. (a–c) Three frames synthesized from the left reference frame; (d–f) the next frames synthesized from the right reference image

disparity maps, while the second half (Figure 8.42(d–f)) was created from the right image and the right-to-left (i.e. reverse) disparity maps. The visible distortions in the reconstructed views are a direct consequence of drawbacks encountered in the disparity maps. However, the general quality of the reconstructed frames is sufficient for film restoration. Small locations with distortions (mostly occlusions or dynamically changing areas) can be repaired with external graphical tools normally used for visual special effects in the final production.

As alluded to previously, the process of film reconstruction is semiautomatic and should be guided by an operator in accordance with image contents and artistic expectations (usually we strive for a visually plausible effect). Image contents have to be assessed by an operator and then the proper choice of control parameters can be made.

8.6 Closure

In this chapter we present examples of systems that utilize techniques discussed in previous chapters of the book. Stereo matching systems find broad applications in many areas of industry and science. We show their usage in a system designed to assist the visually impaired to navigate, in digital cinema for frame synthesis and in systems for face and body modelling and analysis.

Stereo-photogrammetry is rapidly becoming the preferred means of digitizing human surface anatomy in three dimensions. In turn, this development underpins the future of clinical 3D photography and the raft of objective assessment, diagnosis and planning applications that are now in the early production/ongoing development stage. The entertainment industry is also benefiting from the widening availability of human-form digitization and we have shown here how it is now possible to construct animated 3D graphics models of complete individuals from their face and body scans to serve games and cine production.

Although the examples presented in this chapter do not touch on other more traditional areas of photogrammetry activity, such as quality control, parts inspection, DEM (digital elevation model) construction from aerial/satellite images, surveying and active binocular robot vision, these are all burgeoning areas of research and development underpinned by stereo matching and digital photogrammetry. Perhaps the application with the greatest potential impact has yet to be realized, namely high-definition *immersive* 3D TV based on multiview 3D capture, compression, transmission and 3D reconstruction of a studio or outdoor space in real time.

8.6.1 Further Reading

Many developments in computer vision were driven by specific problems. These are, for instance, food inspection, robot navigation, document analysis, traffic management, face recognition and medical imaging, to name a few. Information on applications of 3D image processing techniques can be found in large numbers of conference and journal publications. These include *Transactions on Pattern Analysis and Machine Intelligence*, *Transactions on Image Processing* and *Transactions on Medical Imaging* by IEEE, *International Journal of Computer Vision*, *Machine Vision & Application* by Springer and *Computer Vision & Image Understanding* by Elsevier Science, the SIGGRAPH proceedings and also the 3D Modelling symposia held annually in Paris.

Part III

9

Basics of the Projective Geometry

9.1 Abstract

The algebraic projective geometry constitutes a convenient mathematical tool for the description of such geometrical objects as a point, a line or a plane under a group of transformations. Thus, knowledge of the basic concepts of the algebraic projective geometry is very helpful in understanding 3D machine vision.

The intention of this chapter is to give the reader a very basic, very short and by no means complete, but intuitive, overview of the most important concepts of the algebraic projective geometry.

9.2 Homogeneous Coordinates

Let us analyse a projective transformation of a 3D point $\mathbf{P} = (X, Y, Z)$ on to a plane in the Euclidean coordinate system. Let us assume also that the centre of this projection is in the centre of the coordinate system, i.e. at a point $\mathbf{O} = (0, 0, 0)$ and that the projective plane is at distance f from $\mathbf{O}$. Under this assumption, an image of the point $\mathbf{P}$ lies on the projective plane at a point[1] $\mathbf{p} = (x, y)$ (see Figure 9.1). The coordinates of the two points are constrained by the simple relations which follow from the triangle equation:[2]

$$x = f\frac{X}{Z}, \quad y = f\frac{Y}{Z}. \tag{9.1}$$

Let us observe that (9.1) describes a mapping of the Euclidean space $\Re^3$ on to Euclidean space $\Re^2$, into which a camera plane is embedded. Unfortunately this equation is nonlinear in respect of the Z coordinate. However, we can make it linear at the cost of additional coordinates (i.e.

[1] The third coordinate of this point is constant.

[2] These formulas have already been derived: see Equation (3.2).

An Introduction to 3D Computer Vision Techniques and Algorithms Bogusław Cyganek and J. Paul Siebert

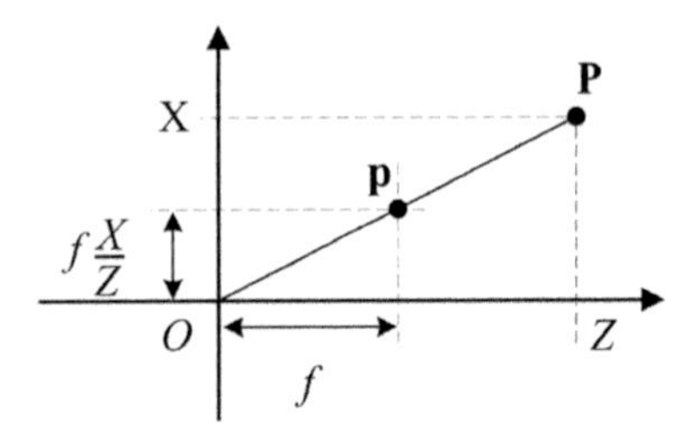

Figure 9.1 Image of a point **P** in a projection with a centre **O**

extending dimensions of the space), as follows:

$$\begin{bmatrix} x \\ y \end{bmatrix} \Leftrightarrow \begin{bmatrix} \tilde{x} \\ \tilde{y} \\ \tilde{z} \end{bmatrix} = \begin{bmatrix} f & 0 & 0 & 0 \\ 0 & f & 0 & 0 \\ 0 & 0 & 1 & 0 \end{bmatrix} \begin{bmatrix} X \\ Y \\ Z \\ 1 \end{bmatrix}, \tag{9.2}$$

where it is now assumed that

$$x = \frac{\tilde{x}}{\tilde{z}}, \quad y = \frac{\tilde{y}}{\tilde{z}}; \quad \tilde{z} \neq 0. \tag{9.3}$$

Coordinates $\tilde{x}, \tilde{y}, \tilde{z}$ in (9.2) and (9.3) are called *homogeneous coordinates* of a point.

Definition 9.1. An affine space $\Re^n$ is transformed to a projective space $\wp^n$ by the following mapping:

$$(x_1, x_2, \ldots, x_n)^{\mathrm{T}} \mapsto (\tilde{x}_1, \tilde{x}_2, \ldots, \tilde{x}_n, \tilde{x}_{n+1})^{\mathrm{T}} = (x_1, x_2, \ldots, x_n, 1)^{\mathrm{T}}. \tag{9.4}$$

The inverse mapping, from the projective space $\wp^n$ to the affine space $\Re^n$, is given as

$$(\tilde{x}_1, \tilde{x}_2, \ldots, \tilde{x}_n, \tilde{x}_{n+1})^{\mathrm{T}} \mapsto (x_1, x_2, \ldots, x_n)^{\mathrm{T}} = \left(\frac{\tilde{x}_1}{\tilde{x}_{n+1}}, \frac{\tilde{x}_2}{\tilde{x}_{n+1}}, \ldots, \frac{\tilde{x}_n}{\tilde{x}_{n+1}} \right)^{\mathrm{T}}, \tag{9.5}$$

where in (9.5) we assume that $\tilde{x}_{n+1} \neq 0$.

The special case constitutes a set of points of the projective space for which $\tilde{x}_{n+1} = 0$. These points are called *points in infinity* or *ideal points*, and their set is called *a set of ideal points* P_∞.

We can easily observe (Figure 9.2) that the mapping from $\Re^n$ to $\wp^n$ is an injection but it is not surjective (i.e. an image of $\Re^n$ does not cover the whole space $\wp^n$). We notice also that for a space $\wp^n$ induced from $\Re^n$ its points are described with $n + 1$ coordinates (9.4).

Another important observation follows directly from this: we see that a point with homogeneous coordinates $(a, b, c)^{\mathrm{T}}$ as well as a point with coordinates $s(a, b, c)^{\mathrm{T}} = (sa, sb, sc)^{\mathrm{T}}$, where $s \neq 0$ is a scaling coefficient, describe *the same point* in the space $\Re^n$. Thus, all such points belong to the one equivalence class. This leads to the conclusion that all

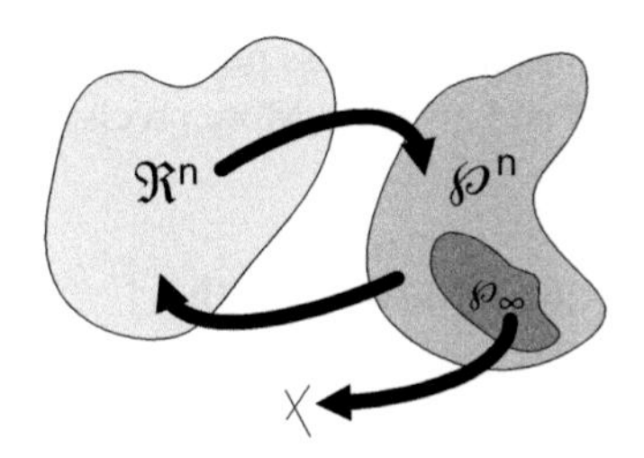

Figure 9.2 Mappings of the $\Re^n$ and $\wp^n$ spaces

projective transformations are defined up to a certain scaling factor. For example, the point **p** is an image of the point **P** but also of all other points lying on the line **OP** (Figure 9.1).

9.3 Point, Line and the Rule of Duality

A point $\mathbf{x} = (x, y)^T$ in the 2D Euclidean space belongs to a line $\mathbf{l}$, represented by three values a, b, c, so $\mathbf{l} = (a, b, c)^T$ if the following condition holds:

$$ax + by + c = 0. \tag{9.6}$$

Going from the Euclidean into the projective space we change coordinates of a point $\boldsymbol{x}$ into the homogeneous representation, in accordance with (9.4). Thus (9.6) can be written as a scalar product:

$$(x, y, 1)(a, b, c)^T = \mathbf{x} \cdot \mathbf{l} = \mathbf{x}^T\mathbf{l} = 0. \tag{9.7}$$

Observe that the above can be written as

$$\mathbf{x}^T\mathbf{l} = \mathbf{l}^T\mathbf{x} = 0, \tag{9.8}$$

which comes directly from the definition of the scalar product. This means that because of the homogeneous transformation of coordinates we obtained a nice symmetry in the formulas for a point and for a line. Thus, it is possible to exchange the roles of **x** and **l** in (9.8), and keep the whole formula untouched at the same time. This is the so-called *duality principle*.

We find a common point of two lines $\mathbf{l}_1 = (a_1, b_1, c_1)^T$ and $\mathbf{l}_2 = (a_2, b_2, c_2)^T$ by taking their cross product:

$$\mathbf{x} = \mathbf{l}_1 \times \mathbf{l}_2. \tag{9.9}$$

It is easy to verify that such a found point **x** lies on both lines, since we have

$$\mathbf{x} \cdot \mathbf{l}_1 = \mathbf{x} \cdot \mathbf{l}_2 = (\mathbf{l}_1 \times \mathbf{l}_2) \cdot \mathbf{l}_1 = (\mathbf{l}_1 \times \mathbf{l}_2) \cdot \mathbf{l}_2 = 0, \tag{9.10}$$

which after (9.7) shows that **x** belongs to $\mathbf{l}_1$ and simultaneously to $\mathbf{l}_2$. The interesting property of the projective space is that even the parallel lines have a common point – a point at infinity.

Sometimes it is convenient to represent the cross product of two vectors as a multiplication of a certain skew-symmetric matrix and a vector. More precisely for two vectors $\mathbf{a} = (a_1, a_2, a_3)^T$ and $\mathbf{b} = (b_1, b_2, b_3)^T$, we have

$$\mathbf{a} \times \mathbf{b} = [\mathbf{a}]_\times \mathbf{b} = \left(\mathbf{a}^T [\mathbf{b}]_\times\right)^T, \tag{9.11}$$

where $[a]_\times$ is given as

$$[\mathbf{a}]_\times = \begin{bmatrix} 0 & -a_3 & a_2 \\ a_3 & 0 & -a_1 \\ -a_2 & a_1 & 0 \end{bmatrix}. \tag{9.12}$$

Based on the above we can rewrite (9.9) as

$$\mathbf{x} = \mathbf{l}_1 \times \mathbf{l}_2 = \begin{bmatrix} 0 & -c_1 & b_1 \\ c_1 & 0 & -a_1 \\ -b_1 & a_1 & 0 \end{bmatrix} \begin{bmatrix} a_2 \\ b_2 \\ c_2 \end{bmatrix}. \tag{9.13}$$

Using the duality principle we obtain a dual to (9.9) of a line crossing two points $\mathbf{x}_1$ and $\mathbf{x}_2$, as follows:

$$\mathbf{l} = \mathbf{x}_1 \times \mathbf{x}_2. \tag{9.14}$$

Finally, it is interesting to notice that having two different points $\mathbf{x}_1$ and $\mathbf{x}_2$ in the projective space, the set $\mathbf{Q}$ of all points lying on the line passing through these points can be expressed as

$$\mathbf{Q} = \alpha\mathbf{x}_1 + \beta\mathbf{x}_2, \tag{9.15}$$

where α and β are certain scalar values. The above can be easily verified, since, taking (9.7) and (9.14), for any point $\mathbf{Q}$ we obtain

$$\mathbf{Q} \cdot \mathbf{l} = \mathbf{Q} \cdot (\mathbf{x}_1 \times \mathbf{x}_2) = \alpha\mathbf{x}_1 \cdot (\mathbf{x}_1 \times \mathbf{x}_2) + \beta\mathbf{x}_2 \cdot (\mathbf{x}_1 \times \mathbf{x}_2) = 0. \tag{9.16}$$

In other words, a condition for a point $\mathbf{x}_i$ to lie on the line joining two points $\mathbf{x}_1$ and $\mathbf{x}_2$ can be stated as a condition of a zero mixed product of these points:

$$\left| \mathbf{x}_1 \ \mathbf{x}_2 \ \mathbf{x}_i \right| = 0. \tag{9.17}$$

9.4 Point and Line at Infinity

In (9.3) and (9.5) we assumed that the last coordinate is different from zero. However, such points, which can be represented as $(x_1, x_2, \ldots, 0)^T$, exist in the projective space and are called points at infinity or ideal points.

It is interesting to observe that all ideal points in the $\wp^2$ space lie on the line $\mathbf{l} = (0, 0, l_3)^T = \mathbf{l}_\infty$, since $(x_1, x_2, 0)(0, 0, l_3)^T = 0$, for all values of x_1, x_2 and l_3. From this observation and

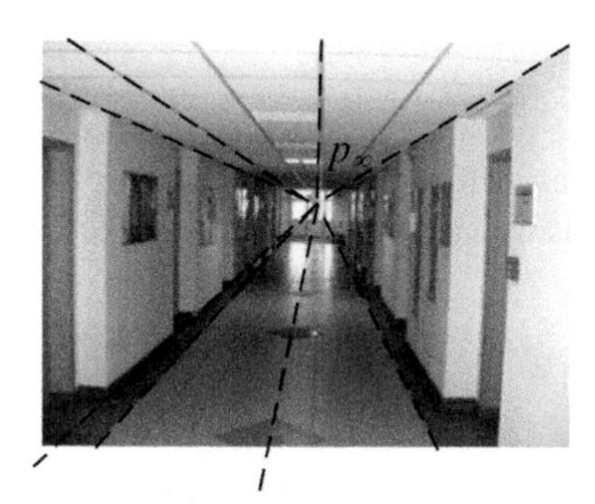

Figure 9.3 Finding an ideal point $\mathbf{p}_\infty$ from a crossing of lines which are parallel in the 3D Euclidean space

from (9.9) we draw an important conclusion that in the projective space even parallel lines cross and their crossing point is a point at infinity. This can be easily shown starting from (9.13) which gives us coordinates of a crossing point for two lines

$$\mathbf{x} = \begin{bmatrix} -c_1 b_2 + b_1 c_2 \\ c_1 a_2 - a_1 c_2 \\ -b_1 a_2 + b_2 a_1 \end{bmatrix}. \tag{9.18}$$

If the two lines are parallel then from (9.6) it holds that

$$\frac{a_1}{b_1} = \frac{a_2}{b_2}, \quad \text{where } b_1 \neq 0,\ b_2 \neq 0, \tag{9.19}$$

but in such case the last coordinate of a point $\mathbf{x}$ in (9.18) is 0, which means that this is a point at infinity.

Figure 9.3 shows how to find an ideal point in an image from a known set of lines which we know are parallel in the Euclidean 3D space. After the projective transformation of the camera's optical system the lines on the camera's image plane are no longer parallel. Instead they have a common crossing point which is an ideal point $\mathbf{p}_\infty$ (i.e. a point at infinity).

Figure 9.4 shows how to find an ideal line $\mathbf{l}_\infty$ which constitutes a set of ideal points.

Having two different sets of lines which are parallel in the Euclidean 3D space we find their ideal points. Two such points are sufficient to determine an ideal line $\mathbf{l}_\infty$.

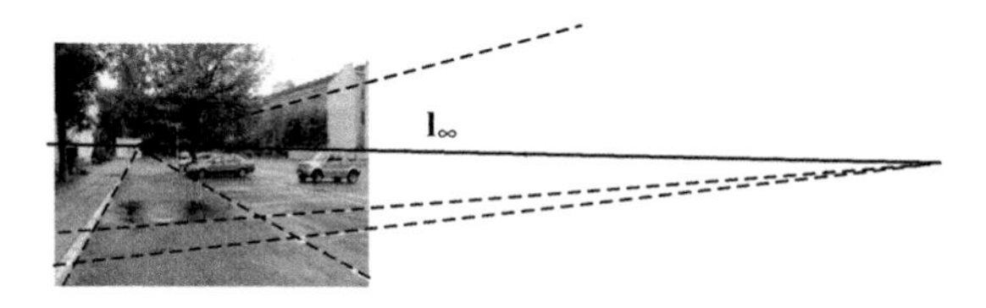

Figure 9.4 Line at infinity $\mathbf{l}_\infty$ can be found from a set of lines which are parallel in the Euclidean space

9.5 Basics on Conics

Conics are geometric objects the parameters of which allow determination of important features of the spaces to which they belong. In this section we briefly outline the notion of a conic in the $\wp^2$ and $\wp^3$ spaces, as well as the notion of a dual conic, circular points and finally absolute and dual conics.

9.5.1 Conics in $\wp^2$

The conic on a plane is a curve described by an equation of second degree. In the Euclidean space it can be given as

$$ax_1^2 + bx_1x_2 + cx_2^2 + dx_1 + ex_2 + f = 0, \tag{9.20}$$

where a, b, c, d, e, f are scalars from $\Re$.

Changing the coordinates in the above to the homogeneous space, in accordance with (9.5), we obtain

$$a\left(\frac{\tilde{x}_1}{\tilde{x}_3}\right)^2 + b\frac{\tilde{x}_1}{\tilde{x}_3}\frac{\tilde{x}_2}{\tilde{x}_3} + c\left(\frac{\tilde{x}_2}{\tilde{x}_3}\right)^2 + d\frac{\tilde{x}_1}{\tilde{x}_3} + e\frac{\tilde{x}_2}{\tilde{x}_3} + f = 0, \tag{9.21}$$

which after some arrangements and skipping the tilde in the names of the homogeneous coordinates simplifies to

$$ax_1^2 + bx_1x_2 + cx_2^2 + dx_1x_3 + ex_2x_3 + fx_3^2 = 0. \tag{9.22}$$

This in turn can be expressed simply as

$$\mathbf{x}^{\mathrm{T}}\mathbf{C}\mathbf{x} = 0, \tag{9.23}$$

where the symmetrical matrix $\mathbf{C}$ is given as

$$\mathbf{C} = \begin{bmatrix} a & \frac{b}{2} & \frac{d}{2} \\ \frac{b}{2} & c & \frac{e}{2} \\ \frac{d}{2} & \frac{e}{2} & f \end{bmatrix}. \tag{9.24}$$

It can be shown that five points on a plane determine (up to a scale) a conic passing through those points [180].

Finally, a line $\mathbf{l}$ tangent at a point $\mathbf{x}$ to a conic described by a matrix $\mathbf{C}$ is given as

$$\mathbf{l} = \mathbf{C}\mathbf{x}. \tag{9.25}$$

This can be verified by comparing (9.8) and (9.23).

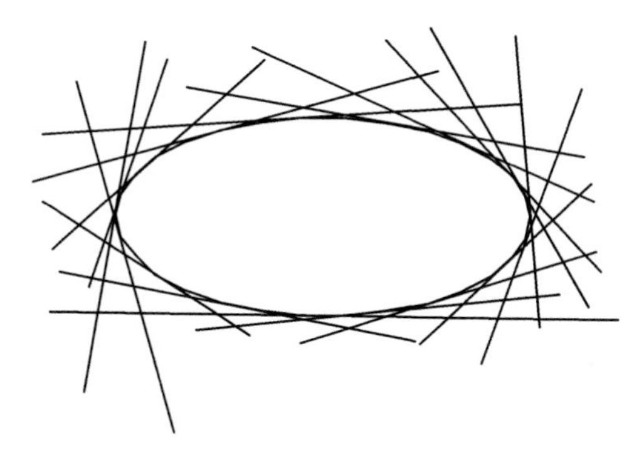

Figure 9.5 Dual conic obtained from the tangent lines

9.5.1.1 The Dual Conic

Equation (9.23) holds for points that belong to a conic given by a matrix $\mathbf{C}$. Therefore such representation of a conic is called a *point conic*. However, once again we can call on the duality rule and express a conic in terms of lines which are tangent to that conic (Figure 9.5). This way we obtain a line conic, called also a dual conic, and denoted by a dual matrix $\mathbf{C}^*$.

To find a relation between a conic and its dual, let us assume that the matrix $\mathbf{C}$ is invertible, so from (9.25) we obtain that

$$\mathbf{x} = \mathbf{C}^{-1}\mathbf{l}.$$

Taking the above to (9.23) leads us to the following expression:

$$\left(\mathbf{C}^{-1}\mathbf{l}\right)^{\mathrm{T}} \mathbf{C} \left(\mathbf{C}^{-1}\mathbf{l}\right) = 0.$$

Because $\mathbf{C}$ is symmetric, then $\mathbf{C}^{-\mathrm{T}} = \mathbf{C}^{-1}$, and the above equation transforms to

$$\mathbf{l}^{\mathrm{T}}\mathbf{C}^{-1}\mathbf{l} = 0. \tag{9.26}$$

This is the dual conic representation. Finally, we notice that

$$\mathbf{C}^* = \mathbf{C}^{-1}. \tag{9.27}$$

9.5.1.2 Circular Points

Setting $a = c$ and $b = 0$ in (9.22), the conic is reduced to a circle, which is given as follows (in the homogeneous coordinates):

$$ax_1^2 + ax_2^2 + dx_1x_3 + ex_2x_3 + fx_3^2 = 0. \tag{9.28}$$

Let us now find the cross points of the above and the line at infinity $\mathbf{l}_\infty$ (section 9.4). Setting $x_3 = 0$ in (9.28) we obtain

$$x_1^2 + x_2^2 = 0, \tag{9.29}$$

the solution of which is given by the two *circular points* **I** and **J** with coordinates as follows:

$$\mathbf{I} = \begin{pmatrix} 1 \\ i \\ 0 \end{pmatrix}, \quad \mathbf{J} = \begin{pmatrix} 1 \\ -i \\ 0 \end{pmatrix}. \tag{9.30}$$

Since $\mathbf{I} \cdot \mathbf{J} = 0$, then algebraically the circular points give two orthogonal directions in the Euclidean space, given, however, by the conjugate complex numbers of the form $(1, 0, 0)^{\mathrm{T}} \pm i(0, 1, 0)^{\mathrm{T}}$ [180]. Because of this property, knowledge of the circular points gives us some metric properties of the space, such as orthogonality or an angle between lines.

9.5.2 Conics in $\wp^3$

Positions of points lying on a plane at infinity π_∞ allow determination of some metric properties of that space – for instance, the geometric objects are parallel if their intersection points belong to π_∞. The other properties can be deduced from the position of the so-called absolute conic and its dual conic. Their parameters allow determination of angles in the observed space. In this section we discuss these concepts.

9.5.2.1 The Absolute Conic

Definition 9.2 The absolute conic Ω_∞ is a conic on the plane at infinity π_∞ and such that the following conditions hold:

$$\begin{cases} x_1^2 + x_2^2 + x_3^2 = 0 \\ x_4 = 0 \end{cases}, \tag{9.31}$$

where x_i are coordinates of a point in a projective space $\wp^3$ and belonging to Ω_∞.

Ω_∞ has a very interesting properties – for instance it is invariant with respect to the figural congruity. Moreover, all circles intersect with Ω_∞ in two points, whereas intersection of all spheres with π_∞ is just Ω_∞ [180]. However, probably most important is that knowledge of a position of Ω_∞ in the $\wp^3$ space allows determination of the metric properties of that space, such as angles between lines or distances between points.

It is well known that having two lines, determined by their directional vectors $\mathbf{k}_1$ and $\mathbf{k}_2$, an angle α between them in the Euclidean space can be found from their dot product:

$$\cos\alpha = \frac{\mathbf{k}_1^{\mathrm{T}}\mathbf{k}_2}{\sqrt{\left(\mathbf{k}_1^{\mathrm{T}}\mathbf{k}_1\right)\left(\mathbf{k}_2^{\mathrm{T}}\mathbf{k}_2\right)}}. \tag{9.32}$$

In a projective space, however, we have the following property [180]:

$$\cos\alpha = \frac{\mathbf{k}_1^{\mathrm{T}}\Omega_\infty\mathbf{k}_2}{\sqrt{\left(\mathbf{k}_1^{\mathrm{T}}\Omega_\infty\mathbf{k}_1\right)\left(\mathbf{k}_2^{\mathrm{T}}\Omega_\infty\mathbf{k}_2\right)}}, \tag{9.33}$$

where $\mathbf{k}_1$ and $\mathbf{k}_2$ are respectively coordinates of points of intersection of the two lines with the plane at infinity $\boldsymbol{\pi}_\infty$, containing Ω_∞. In the Euclidean space (9.33) reduces to (9.32), since $\Omega_\infty = \mathbf{I}$.

From the last equation if follows simply that two lines are orthogonal if the following holds:

$$\mathbf{k}_1^{\mathrm{T}} \Omega_\infty \mathbf{k}_2 = 0. \tag{9.34}$$

9.5.2.2 The Dual Absolute Conic

The absolute conic is given by the system of two equations (9.31). However, an easier representation is given by *the dual conic* Ω_∞^*, which can be interpreted as a set of contact points of the tangent planes with Ω_∞.

The dual conic Ω_∞^* can be defined as a uniform 4×4 matrix of rank three, which in a metric space takes on the form

$$\Omega_\infty^* = \begin{bmatrix} 1 & & & \\ & 1 & & \\ & & 1 & \\ & & & 0 \end{bmatrix}. \tag{9.35}$$

Finally knowing Ω_∞^*, the angle α between two planes $\boldsymbol{\pi}_1$ and $\boldsymbol{\pi}_2$ is given as [180]

$$\cos\alpha = \frac{\pi_1^{\mathrm{T}} \Omega_\infty^* \pi_2}{\sqrt{\left(\pi_1^{\mathrm{T}} \Omega_\infty^* \pi_1\right)\left(\pi_2^{\mathrm{T}} \Omega_\infty^* \pi_2\right)}}. \tag{9.36}$$

9.6 Group of Projective Transformations

In this section we present a brief introduction to a group of projective transformations. We start with a definition of a canonical projective base, then present hyperplanes which are generalizations of concepts of points and lines, and finally we focus on the projective transformations.

9.6.1 Projective Base

A base of the projective space $\wp^n$ constitutes $n+2$ points of that space, from which none of the $n+1$ points belongs to a hyperplane. This is equivalent to the statement that a $(n+1) \times (n+1)$ size matrix of those points is not singular [314].

It is easy to verify that the set of points $\mathbf{P}_i$

$$\mathbf{P}_1 = \begin{bmatrix} 1 \\ 0 \\ 0 \\ \cdots \\ 0 \end{bmatrix}_{(n+1)\times 1}, \mathbf{P}_2 = \begin{bmatrix} 0 \\ 1 \\ 0 \\ \cdots \\ 0 \end{bmatrix}, \ldots, \mathbf{P}_n = \begin{bmatrix} 0 \\ \cdots \\ 1 \\ \cdots \\ 0 \end{bmatrix}, \mathbf{P}_{n+1} = \begin{bmatrix} 0 \\ 0 \\ 0 \\ \cdots \\ 1 \end{bmatrix}, \mathbf{P}_{n+2} = \begin{bmatrix} 1 \\ 1 \\ 1 \\ \cdots \\ 1 \end{bmatrix} \tag{9.37}$$

is a base of the $\wp^n$ space, which is called *a canonical projective base*. It is composed of n ideal points (i.e. points at infinity), corresponding to the n axes of a coordinate system, as well as of its middle point $\boldsymbol{P}_{n+1}$ and a unit point $\boldsymbol{P}_{n+2}$.

9.6.2 Hyperplanes

A hyperplane is a natural extension of a notion of a point and line in a projective space. In the Euclidean space it is defined by the following equation:

$$a_1x_1 + \cdots + a_nx_n + a_{n+1} = 0, \quad \mathbf{x} \in \Re^n. \tag{9.38}$$

After change to the homogeneous coordinates in accordance with (9.4) the above equation transforms to (again, the tilde symbol is omitted)

$$a_1x_1 + \cdots + a_nx_n + a_{n+1}x_{n+1} = \mathbf{a} \cdot \mathbf{x} = 0, \quad \mathbf{x} \in \wp^n. \tag{9.39}$$

It is interesting to observe that also in this case the duality rule holds. Specifically, we obtain immediately that

$$\mathbf{x}^{\mathrm{T}}\boldsymbol{\pi} = \boldsymbol{\pi}^{\mathrm{T}}\mathbf{x} = 0, \tag{9.40}$$

where $\boldsymbol{\pi} = (\pi_1, \pi_2, \pi_3, \pi_4)^{\mathrm{T}}$ represents a plane in $\wp^3$, while $\mathbf{x}$ is a point belonging to that plane.

A plane given by the four coefficients $(0, 0, 0, 1)^{\mathrm{T}}$ is called a plane at infinity and is denoted as $\boldsymbol{\pi}_\infty$. Based on the concept of a plane at infinity, two conclusions can be drawn.

1. Two planes are parallel if their common line belongs to $\boldsymbol{\pi}_\infty$.
2. Similarly, a line is parallel to another line or to a plane if their crossing point belongs to $\boldsymbol{\pi}_\infty$.

9.6.3 Projective Homographies

Definition 9.3. For a projective space $\wp^n$ a projective homography is defined as a nonsingular matrix $H_{(n+1)\times(n+1)}$ with elements belonging to $\Re$, and defined up to a certain scalar value, called a scaling coefficient. A point $\mathbf{x}$ is projectively transformed to $\mathbf{x}'$ as follows:

$$\mathbf{x}' = \mathbf{Hx}, \quad \mathbf{x}, \mathbf{x}' \in \wp^n, \tag{9.41}$$

where the matrix $\mathbf{H}$ denotes a projective transformation.

It can be shown that projective homographies constitute a group, since an inverse transformation as well as a composition of such transformations are also projective homographies.

In the special case of a projective space $\wp^2$ a point $\mathbf{x_i}$ is transformed into a point $\mathbf{x_i}'$, in accordance with (9.41). Two different points $\mathbf{x_1}$ and $\mathbf{x_2}$ unambiguously define a line $\mathbf{l}$ that passes through these points, i.e. it holds that

$$\mathbf{x}_1^{\mathrm{T}}\mathbf{l} = \mathbf{x}_2^{\mathrm{T}}\mathbf{l} = 0,$$
$$\left(\mathbf{x}_1^{\mathrm{T}} - \mathbf{x}_2^{\mathrm{T}}\right)\mathbf{l} = 0.$$

The points $\mathbf{x_1}$ and $\mathbf{x_2}$ are transformed by $\mathbf{H}$ into $\mathbf{x_1}'$ and $\mathbf{x_2}'$ respectively, which both lie on a line $\mathbf{l}'$, so it holds also that

$$\left(\mathbf{x}'^{\mathrm{T}}_1 - \mathbf{x}'^{\mathrm{T}}_2\right)\mathbf{l}' = 0.$$

Combining the two last equations we obtain that

$$\left(\mathbf{x}'^{\mathrm{T}}_1 - \mathbf{x}'^{\mathrm{T}}_2\right)\mathbf{l}' = \left(\mathbf{x}_1^{\mathrm{T}} - \mathbf{x}_2^{\mathrm{T}}\right)\mathbf{l}. \tag{9.42}$$

Now, inserting (9.41) into (9.42) we get that

$$\left((\mathbf{H}\mathbf{x}_1)^{\mathrm{T}} - (\mathbf{H}\mathbf{x}_2)^{\mathrm{T}}\right)\mathbf{l}' = \left(\mathbf{x}_1^{\mathrm{T}} - \mathbf{x}_2^{\mathrm{T}}\right)\mathbf{l},$$
$$\left(\mathbf{x}_1^{\mathrm{T}} - \mathbf{x}_2^{\mathrm{T}}\right)\mathbf{H}^{\mathrm{T}}\mathbf{l}' = \left(\mathbf{x}_1^{\mathrm{T}} - \mathbf{x}_2^{\mathrm{T}}\right)\mathbf{l}.$$

Now assuming that $\mathbf{x_1} \neq \mathbf{x_2}$ and considering the above equation, we see that the lines transform as

$$\mathbf{l}' = \mathbf{H}^{-\mathrm{T}}\mathbf{l}, \tag{9.43}$$

where $\mathbf{l}$ and $\mathbf{l}'$ are lines in $\wp^2$, and under an assumption that $\mathbf{H}$ is invertible. Because of the difference between (9.41) and (9.43) we say that points follow *a contravariant*, whereas lines follow *a covariant* transformation.

9.7 Projective Invariants

Recognition of the relationships among points and lines that are unchanged regardless of a projective transformation are called projective invariants. These are very useful in the object recognition process [108]. More on such invariants and their applications can be found in a classic text by Mundy and Zisserman [322].

Figure 9.6 depicts a projection with a centre $\mathbf{O}$ of four coplanar points $\mathbf{X}_1$, $\mathbf{X}_2$, $\mathbf{X}_3$ and $\mathbf{X}_4$, on to a line $\mathbf{l}$. The relation of the distances of their image points $\mathbf{x}_1$, $\mathbf{x}_2$, $\mathbf{x}_3$, $\mathbf{x}_4$ appears to be invariant in respect to a position of the line $\mathbf{l}$. This is called a cross-ratio which is defined as follows.

Definition 9.4. Given four collinear points $\mathbf{x}_1, \mathbf{x}_2, \mathbf{x}_3, \mathbf{x}_4$, a cross-ratio $d(\mathbf{x}_1, \mathbf{x}_2, \mathbf{x}_3, \mathbf{x}_4)$ is given by the relation

$$d\left(\mathbf{x}_1, \mathbf{x}_2, \mathbf{x}_3, \mathbf{x}_4\right) = \frac{|\mathbf{x}_1\mathbf{x}_2|\,|\mathbf{x}_3\mathbf{x}_4|}{|\mathbf{x}_1\mathbf{x}_3|\,|\mathbf{x}_2\mathbf{x}_4|}. \tag{9.44}$$

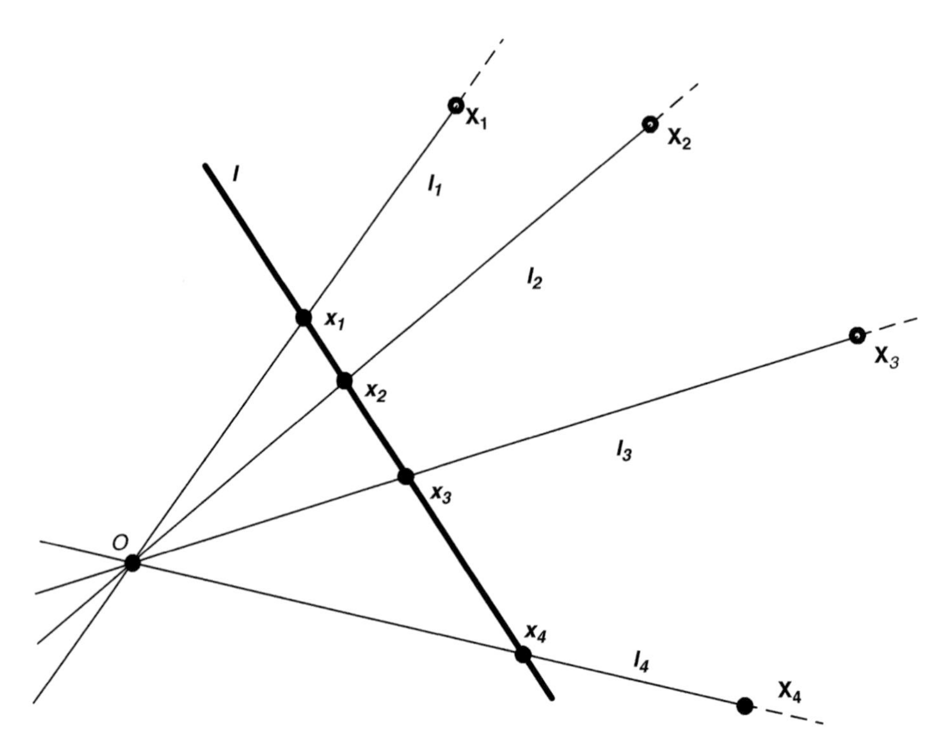

Figure 9.6 Four convergent lines define a cross-ratio on a plane which is invariant in respect to the projective transformations

where $|\mathbf{x_p}, \mathbf{x_q}|$ denotes a distance between the points $\mathbf{x}_p$, $\mathbf{x}_q$:

$$\left|\mathbf{x}_p\mathbf{x}_q\right| = \det\begin{bmatrix} x_{p1} & x_{q1} \\ x_{p2} & x_{q2} \end{bmatrix}. \tag{9.45}$$

As already mentioned, $d(\mathbf{x}_1, \mathbf{x}_2, \mathbf{x}_3, \mathbf{x}_4)$ is invariant with respect to the projective transformation of a line, i.e. if $\mathbf{x}_i' = \mathbf{H}_{2\times 2}\mathbf{x}_i$, where $\mathbf{H}_{2\times 2}$ is a matrix of such a transformation, then it holds that $d(\mathbf{x}_1', \mathbf{x}_2', \mathbf{x}_3', \mathbf{x}_4') = d(\mathbf{x}_1, \mathbf{x}_2, \mathbf{x}_3, \mathbf{x}_4)$.

Equation (9.44) holds even if one of the points is an ideal point, i.e. a point at infinity.

9.8 Closure

In this chapter we present a short introduction to the algebraic projective geometry. It constitutes a very useful mathematical tool since it allows description of geometric objects under perspective transformation in the language of algebra.

9.8.1 Further Reading

Much more complete introductions to the area of projective geometry can be found, for example, in the classic book by Semple and Kneebone [380] or in a recent book by Casse [64]. Projective geometry in the light of computer vision is provided in the seminal works by Hartley and Zisserman [180] or Faugeras and Luong [119]. There are also some short overviews on the subject: the appendix in the book by Mundy and Zisserman [322] or in the paper by Mohr and Triggs [314].

10

Basics of Tensor Calculus for Image Processing

10.1 Abstract

The tensor calculus has found many applications in computer vision and image processing. It offers a more compact representation of many quantities, for instance relations among corresponding points in many views of the same scene. Tensors are mostly used in physics, especially in mechanics and theory of relativity. In this section we give a very brief introduction to this area, trying to explain this mathematical tool.

10.2 Basic Concepts

A characteristic feature of the tensor notation is lower and upper indices which in this case do not mean a power of a variable. Number and position of indices indicate a type of tensor, as will be discussed later on. The other specifics is the summation convention which simply assumes that if in one equation different variables have the same index, then this means *summation* in respect to that index. Simultaneously the summation sign $\sum$ in this expression is usually *omitted*. This is known also as the Einstein summation rule. For instance, instead of $\sum_{1=1}^{n} a_i x_i$, we simply write $a^i x_i$. Notice that it is customary that the summation indices (i in our example) are repeated, however at different positions – in the first factor it is an upper index (contravariant), and lower (covariant) in the second, or vice versa. It means also that we know the summation range (n in the above example).

Let us now consider a vector $\mathbf{x}$ in an n-dimensional vector space with given base $\mathbf{b}_i$. Such a vector can be thought of as an ordered set of n real values, or a directed straight line connecting two points, say point $\mathbf{O}$, the centre of the coordinate system, and a point $\mathbf{P}$. It can be represented with respect to the base $\mathbf{b}_i$ as follows:

$$\mathbf{x} = \sum_{i=1}^{n} x^i \mathbf{b}_i, \tag{10.1}$$

An Introduction to 3D Computer Vision Techniques and Algorithms Bogusław Cyganek and J. Paul Siebert

or in the Einstein notation:

$$\mathbf{x} = x^i \mathbf{b}_i. \tag{10.2}$$

Thus, knowing the base, we can write $\mathbf{x} = (x^1, x^2, \ldots, x^n)$.

There are two basic mathematical concepts which are important prerequisites to understand tensors. These are:

- linear operators;
- change of coordinate systems – Jacobians.

We will discuss them in the next sections.

10.2.1 Linear Operators

For a given vector $\mathbf{x}$ let us assign a vector $\mathbf{y}$. This way we have defined a vector function f as follows:

$$\mathbf{y} = f(\mathbf{x}). \tag{10.3}$$

We say that f is linear if for any scalar values r, s and any vectors $\mathbf{x}$, $\mathbf{y}$ the following is fulfilled:

$$f(r\mathbf{x} + s\mathbf{y}) = rf(\mathbf{x}) + sf(\mathbf{y}). \tag{10.4}$$

Now, taking (10.1) into (10.4) we obtain

$$f(\mathbf{x}) = f\left(\sum_{i=1}^{n} x^i \mathbf{b}_i\right) = \sum_{i=1}^{n} x^i f(\mathbf{b}_i), \tag{10.5}$$

or in a short form

$$f(\mathbf{x}) = x^i f(\mathbf{b}_i). \tag{10.6}$$

Thus, to find a value of a linear function f of *any* vector $\mathbf{x}$ it is sufficient to know only n values of this function on the base vectors $\mathbf{b}_i$ (which we can actually precompute once and then use for any new vector $\mathbf{x}$). This is all that we need to completely describe f. However, from (10.3) we see that $f(\mathbf{b}_i)$ is also a vector, and therefore it can be expressed in accordance with (10.1), which holds for all vectors:

$$f(\mathbf{b}_i) = T_i^{\,1}\mathbf{b}_1 + T_i^{\,2}\mathbf{b}_2 + \cdots + T_i^{\,n}\mathbf{b}_n = T_i^{\,k}\mathbf{b}_k, \tag{10.7}$$

where $T_i^{\,k}$ are scalars (compare with (10.2)), which constitute components of a tensor $\mathbf{T}$.

The last thing is to substitute (10.7) into (10.6) to obtain

$$f(\mathbf{x}) = x^i \left(T_i^{\,k}\mathbf{b}_k\right). \tag{10.8}$$

This tells us that to find a vector, being the result of a linear operator f on a vector $\mathbf{x}$, all we need to do is sum up products of the components of the vector $\mathbf{x}$, the base vectors $\mathbf{b}_i$ and the values of tensor $\mathbf{T}$. Thus, to completely define f we need to know its associated tensor $\mathbf{T}$, as in (10.7). In the next sections we will see what happens if the base $\mathbf{b}_i$ is changed into $\mathbf{b}'_i$.

Finally, let us notice also that the notion of a linear function (10.3) can be extended into a more general concept of a linear operator which maps a vector $\mathbf{x}$ into vector $\mathbf{y}$:

$$f: \quad \mathbf{x} \to \mathbf{y}. \tag{10.9}$$

This way we can generalize the space of linear functionals on to such mathematical operations as differentiation or integration:

$$f \equiv \frac{\mathrm{d}(\cdot)}{\mathrm{d}x}, \quad f \equiv \int (\cdot)\mathrm{d}x, \text{ etc.} \tag{10.10}$$

However, because of the concept of a tensor, an application of the above linear operator can be represented as a simple multiplication, so instead of (10.3) we write simply

$$\mathbf{y} = \mathbf{T}\mathbf{x}, \tag{10.11}$$

where f and $\mathbf{T}$ are related by (10.7).

10.2.2 Change of Coordinate Systems: Jacobians

Equation (10.1) denotes a vector $\mathbf{x}$ with coordinates x^i given in a certain coordinate system U with base $\mathbf{b}_i$. Let us now connect x^i with the curvilinear coordinates x'^j of the system U' by the continuously differentiable and bijective functions, as follows:

$$x^i = x^i(x'^j), \quad \mathbf{x} = \mathbf{x}(x'^j), \tag{10.12}$$

$$x'^j = x'^j(x^i), \quad \mathbf{x}' = \mathbf{x}'(x^i), \tag{10.13}$$

i.e. x^i are functions of x^j and x^j are functions of x^i. Figure 10.1 depicts two 3D coordinate spaces. The first one has centre $\mathbf{O}$; the second has centre in a point $\mathbf{P}$ which traverses a curve C.

Now it is possible to define new vectors, called the local basis:

$$\mathbf{b}'_i = \frac{\partial \mathbf{x}}{\partial x'^i} = \frac{\partial (x^j \mathbf{b}_j)}{\partial x'^i} = \frac{\partial x^j}{\partial x'^i}\mathbf{b}_j = \alpha_i^j \mathbf{b}_j. \tag{10.14}$$

Thus for $U \to U'$ we have

$$\alpha_i^j = \frac{\partial x^j}{\partial x'^i}. \tag{10.15}$$

The ordered table of α_i^j is called a coordinate systems transformation matrix, known also a Jacobian matrix $\mathbf{J} = [\alpha_i^j]$. Since the transformation functions are assumed to be continuously

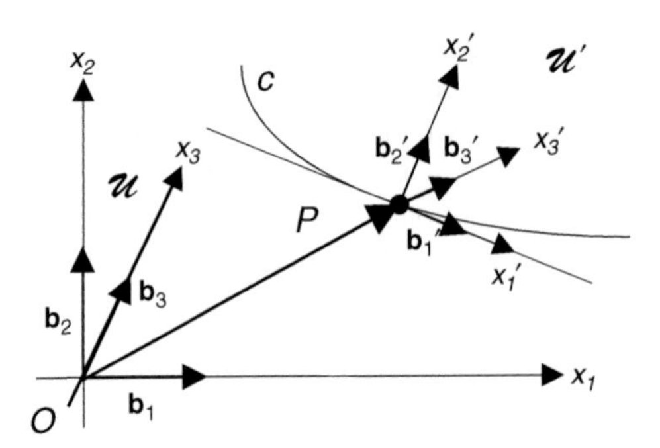

Figure 10.1 Two coordinate systems U and U' with base vectors

differentiable and bijective, then there is also an inverse matrix $\mathbf{J}^{-1}$ with elements

$$\beta_i^j = \frac{\partial x'^j}{\partial x^i} \tag{10.16}$$

for the transformation $U' \rightarrow U$.

10.3 Change of a Base

Let us assume now that the basis vectors $\mathbf{b}_i$ are transformed by a certain invertible matrix $\mathbf{A}$ and let us check the corresponding change of components of the vector $\mathbf{x}$. New basis vectors can be represented as

$$\mathbf{b}'_j = \mathrm{A}^i_j \mathbf{b}_i, \tag{10.17}$$

where A^i_j represents (i, j) elements of the matrix $\mathbf{A}$ transforming the base. The above can be written in short as

$$\mathbf{b}' = \mathbf{A}\mathbf{b}. \tag{10.18}$$

Since $\mathbf{A}$ is invertible then the original base can be found as

$$\mathbf{b} = \mathbf{A}^{-1}\mathbf{b}'. \tag{10.19}$$

From (10.19) and (10.2) we obtain vector representation in a new base as follows:

$$\mathbf{x}' = \mathbf{A}^{-1}\mathbf{x}. \tag{10.20}$$

This way we come to the important conclusion that although the base vectors transform in accordance with the matrix $\mathbf{A}$, coordinates of a vector $\mathbf{x}$ transform with an inverse matrix, i.e. $\mathbf{A}^{-1}$.

Let us assume that the vector **x** considered so far represents a point on a line **l**. Substituting (10.20) into (9.8) we obtain

$$\mathbf{x}'^{\mathrm{T}}(\mathbf{A}^{\mathrm{T}}\mathbf{l}) = 0. \tag{10.21}$$

However, the above can be interpreted as a line **l** in *a new base*, given as follows:

$$\mathbf{l}' = \mathbf{A}^{\mathrm{T}}\mathbf{l}. \tag{10.22}$$

In this case, however, a change of base vectors in accordance with (10.18) entails the *same* change of the coordinates of **l**, i.e. in accordance with the matrix **A**.

Based on the above analysis we can express a change of the coordinates of **x** and **l** in respect to the change of the basis given by a matrix **A**. These are as follows:

$$x'^{i} = (\mathbf{A}^{-1})^{i}_{j} x^{j} \tag{10.23}$$

and

$$l'_{i} = \mathbf{A}^{j}_{i} l_{j}. \tag{10.24}$$

Notice that coordinates of a tensor can transform in accordance with **A** or $\mathbf{A}^{-1}$, where – let us remember – **A** is an invertible matrix which transforms a base of the space under consideration. This turns out to be a feature characteristic to all tensors. The former transformation corresponds to the *covariant* tensors (lower indices), whereas the latter corresponds to the *contravariant* tensors (upper indices). There are also mixed tensors. The number of indices of a tensor is called its valence.

An example of a tensor of a valence (1, 1) is a tensor which is obtained for a given matrix **H** in a projective space, after simultaneous change of the base in input and output spaces in accordance with matrices **A** and **B**, respectively. Thus, the new matrix $\mathbf{H}'$ takes the form

$$\mathbf{H}' = \mathbf{B}^{-1}\mathbf{H}\mathbf{A}. \tag{10.25}$$

Coordinates of $\mathbf{H}'$ can be written now in the following form:

$$H'^{i}_{j} = (\mathbf{B}^{-1})^{i}_{p} A^{q}_{j} H^{p}_{q}. \tag{10.26}$$

An important conclusion is that the way of transformation of a certain value, caused by a change of the base of the space, determines whether this value is, or is not, a tensor, and if it is, then what the valence of that tensor is.

A scalar, a vector or a matrix are special cases of a more general m-th dimensional tensor of an n-th order, which contains m^n coordinates. In the 3D case, i.e. $m = 3$, scalars can be treated as tensors of zero order, for which 3^0 values need to be provided. Vectors, however, are tensors of first order and therefore they require 3^1 values, and so on [49].

10.4 Laws of Tensor Transformations

One of the most important features of tensors is the way in which they transform on a change of basis of a space. This is governed by a principle of tensor transformations, which in this section is discussed for first- and second-order tensors. However, the rules can be analogously generalized to tensors of higher rank. Let us note also that to check whether a given mathematical object is a tensor it is sufficient to check whether it transforms in accordance with the tensor transformation rule, discussed below.

In a change of a basis of a coordinate system U into U' covariant components of a first-order tensor **T** transform in accordance with the following rule:

$$T_i' = \alpha_{i'}^{k} T_k, \tag{10.27}$$

whereas the contravariant ones undergo the following transformation:

$$T'^i = \alpha_k^{i'} T^k, \tag{10.28}$$

where $\alpha_{i'}^{k}$ and $\alpha_k^{i'}$ are components of the direct and the inverse transformations of the basis of this system (10.15), i.e. from U into U', respectively. In the special case of linear systems, these are elements of the already introduced system transformation matrix **A**, as well as its inverse, respectively. In the case of orthogonal systems, $\alpha_{i'}^{k}$ can be treated as a cosine value of the angle between the i-th axis of the U' system and k-th axis of the axis in U.

On the other hand, there is also a connection between the covariant *and* contravariant components of a tensor (in one and the same coordinate system). These are given as follows:

$$T_i = g_{ik} T^k, \tag{10.29}$$

$$T^i = g^{ik} T_k, \tag{10.30}$$

where

$$g_{ik} = \mathbf{b}_i \cdot \mathbf{b}_k, \qquad g^{ik} = \mathbf{b}^i \cdot \mathbf{b}^k. \tag{10.31}$$

In the case of 2D tensors, apart from the pure covariant or contravariant components, the *mixed* components are also allowed. In this case the transformation law is given as

$$T_{ik}' = \alpha_{i'}^{l} \alpha_{k'}^{m} T_{lm}, \tag{10.32}$$

$$T'^{ik} = \alpha_l^{i'} \alpha_m^{k'} T^{lm}, \tag{10.33}$$

$$T_i'^k = \alpha_{i'}^{l} \alpha_m^{k'} T_l^m. \tag{10.34}$$

The relations (10.29) and (10.30) as well as (10.32)–(10.34) are used then to *check* whether a given entity is a tensor or not.

10.5 The Metric Tensor

Let us now observe that in accordance with (10.31) the following holds:

$$g'_{ik} = \mathbf{b}'_i \cdot \mathbf{b}'_k = \alpha^l_{i'}\mathbf{b}_l \cdot \alpha^m_{k'}\mathbf{b}'_m = \alpha^l_{i'}\alpha^m_{k'}\mathbf{b}_l \cdot \mathbf{b}'_m = \alpha^l_{i'}\alpha^m_{k'}g_{lm}, \tag{10.35}$$

which stays in agreement with (10.32). This means that g_{ik} is a second-order tensor which is called a *metric tensor*.

Due to the assumption of a bijective transformation of coordinates, the determinant of the Jacobian matrix is always different from zero. Therefore it is straightforward to define the inverse metric

$$g^{ik} = (g_{ik})^{-1}, \tag{10.36}$$

$$g^{ik}g_{kj} = \delta^i_j, \tag{10.37}$$

where δ^i_j and δ_{ij} are called Kronecker symbols, defined as follows:

$$\delta^i_j = \begin{cases} 1, & i = j \\ 0, & i \neq j \end{cases} \qquad \delta_{ij} = \begin{cases} 1, & i = j \\ 0, & i \neq j \end{cases}. \tag{10.38}$$

From this it follows that the inverse metric can be expressed as

$$g^{ik} = \beta^i_m \beta^k_n \delta^{mn}. \tag{10.39}$$

With g^{ik} we can define the dual local basis $\mathbf{b}^i$ as follows:

$$\mathbf{b}^i = g^{ik}\mathbf{b}_k. \tag{10.40}$$

In the Euclidean space the base vectors $\mathbf{b}_i = \mathbf{e}_i$ are orthonormal, therefore the following holds:

$$g_{ik}|_E = \mathbf{e}_i \cdot \mathbf{e}_j = \delta_{ij}. \tag{10.41}$$

10.5.1 Covariant and Contravariant Components in a Curvilinear Coordinate System

Figure 10.2 depicts a 2D coordinate system with a vector $\vec{\mathbf{q}}$. Its covariant coordinates and contravariant coordinates are obtained by two different projections on to the base axes $\mathbf{b}_1$ and $\mathbf{b}_2$. Namely, the contravariant coordinates q^1 and q^2 are obtained as a result of the parallel projection, whereas the covariant coordinates q_1 and q_2 are obtained from the orthogonal projection. Notice that in the case of an orthogonal system, i.e. $\phi_1 + \phi_2 = \frac{1}{2}\pi$, the corresponding coordinates would be equal.

Let us find values of q^1 and q^2 and also of q_1 and q_2. Starting from (10.2) we can write

$$\mathbf{q} = q'^i\mathbf{b}_i = q'^1\mathbf{b}_1 + q'^2\mathbf{b}_2. \tag{10.42}$$

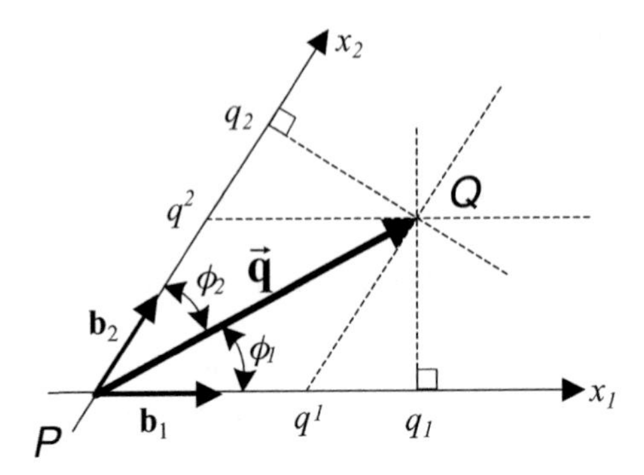

Figure 10.2 Covariant q_1, q_2 and contravariant q^1, q^2 coordinates of a vector $\vec{\mathbf{q}}$ in the nonorthogonal 2D coordinate system

However, we have to remember that in general $\mathbf{b}_i$ are not normalized, so the above can be transformed as follows:

$$\mathbf{q} = q'^1\mathbf{b}_1\frac{\|\mathbf{b}_1\|}{\|\mathbf{b}_1\|} + q'^2\mathbf{b}_2\frac{\|\mathbf{b}_2\|}{\|\mathbf{b}_2\|} = q'^1\sqrt{\mathbf{b}_1 \cdot \mathbf{b}_1}\frac{\mathbf{b}_1}{\|\mathbf{b}_1\|} + q'^2\sqrt{\mathbf{b}_2 \cdot \mathbf{b}_2}\frac{\mathbf{b}_2}{\|\mathbf{b}_2\|}, \tag{10.43}$$

which after (10.31) takes the form

$$\mathbf{q} = q'^1\sqrt{g_{11}}\frac{\mathbf{b}_1}{\|\mathbf{b}_1\|} + q'^2\sqrt{g_{22}}\frac{\mathbf{b}_2}{\|\mathbf{b}_2\|}. \tag{10.44}$$

Thus

$$q^1 = q'^1\sqrt{g_{11}} \quad \text{and} \quad q^2 = q'^2\sqrt{g_{22}}. \tag{10.45}$$

To find q_1 and q_2, notice that

$$q_i = \|\mathbf{q}\|\cos\phi_i = \frac{\|\mathbf{b}_i\|}{\|\mathbf{b}_i\|}\|\mathbf{q}\|\cos\phi_i = \frac{\mathbf{b}_i \cdot \mathbf{q}}{\|\mathbf{b}_i\|} = \frac{\mathbf{b}_i \cdot \mathbf{q}}{\sqrt{\mathbf{b}_i \cdot \mathbf{b}_i}} = \frac{\mathbf{b}_i \cdot \mathbf{q}}{\sqrt{g_{ii}}}, \tag{10.46}$$

which after (10.31) and defining

$$q'_i = \mathbf{b}_i \cdot \mathbf{q}, \tag{10.47}$$

takes the following form:

$$q_1 = \frac{q'_1}{\sqrt{g_{11}}} \quad \text{and} \quad q_2 = \frac{q'_2}{\sqrt{g_{22}}}. \tag{10.48}$$

Finally, from (10.47) and (10.42) we obtain easily that

$$q'_k = \mathbf{b}_k \cdot \mathbf{q} = \mathbf{b}_k \cdot \left(q'^i \mathbf{b}_i\right). \tag{10.49}$$

However, taking into consideration (10.31) we see that

$$q'_k = g_{ki} q'^i. \tag{10.50}$$

10.5.2 The First Fundamental Form

Metric can be interpreted as a certain entity that describes basic geometrical properties of its space with a coordinate system given by w^1 and w^2. Now, an expression on an infinitesimal arc length in that space can be stated as follows:

$$(ds)^2 = g_{ik} dw^i dw^k. \tag{10.51}$$

The *quadratic* equation above is also called *the first fundamental form*. Since g_{ik} in the above denotes a metric tensor (10.35), which transforms in accordance with (10.32), then ds^2 is invariant with respect to any allowable transformation of coordinates.

The first fundamental form (10.51) can be interpreted as an infinitesimal element of an arc of a curve, defined parametrically by $w1(t)$ and $w2(t)$, where t is a parameter [259, 263, 397]. It is a very important concept of differential geometry since it allows measurements of angles, lengths of arcs or areas on a surface. Thus, it defines a metric on a surface. Moreover, a metric defined by a quadratic differential form like (10.51) is called a Riemannian metric and the corresponding geometry a Riemannian geometry.

10.6 Simple Tensor Algebra

In this section we summarize the most important facts on basic algebraic operations of tensors, such as tensor summation or product, but also tensor contraction, inner product and finally reduction of a tensor to its principal axes. The definitions are provided for second-order tensors, although they can easily be extended to other dimensions.

10.6.1 Tensor Summation

The sum of the second-order tensors **A** and **B**, with components A_{ij} and B_{ij}, respectively, is a *second*-order tensor **C** with components given as follows:

$$C_{ij} = A_{ij} + B_{ij}. \tag{10.52}$$

A required condition for this operation is that the added tensors have the same structure, i.e. the same valence.

10.6.2 Tensor Product

A product of the second-order tensors **A** and **B**, with components A_{ij} and B_{ij}, respectively, is a *fourth*-order tensor **C** with components given as follows:

$$C_{ijkl} = A_{ij} B_{kl}. \tag{10.53}$$

This is called *an outer product* of tensors. It is easy to verify that this product is not commutative. However, contrary to addition (10.52), the outer product (10.53) can be computed for tensors with *different* valence.

For higher order tensors this operation is defined analogously to (10.53). A result is a tensor of a higher rank which is a sum of the ranks of tensors multiplied in this way.

10.6.3 Contraction and Tensor Inner Product

A contraction of a tensor **N** of rank $n \geq 2$ is a summation operation with respect to its *two* indexes. As a result, a tensor is obtained of rank $n-1$. For instance, if $n = 3$, then there are three ways of the tensor contraction, namely N_{iij}, N_{iji} and N_{ijj}, where e.g. $N_{iij} = N_{11j} + N_{22j} + N_{33j}$ for $j = 1, 2, 3$, and so on.

A tensor multiplication connected with tensor contraction with respect of the indices belonging to different components is called *an inner product* of a tensor.

For the special case of second-order tensors, their inner product is given as follows:

$$\mathbf{N} \cdot \mathbf{M} = \sum_{kl} N_{kl} M_{kl}, \tag{10.54}$$

where N_{kl} and M_{kl} are components of the tensors **N** and **M**, respectively.

10.6.4 Reduction to Principal Axes

Let us consider multiplication of a second-order tensor **T** with a first-order tensor **N** (i.e. a vector). As a result a tensor **M** is obtained which is of first order with coordinates

$$T_{ij} N_j = M_j. \tag{10.55}$$

The resulting tensor **M** is generally different from **N**. However, an important special case is when the above operation does not lead to tensor (vector) rotation, and only to a change of its modulus. Thus, we search for a solution to the following problem:

$$T_{ij} N_j = \lambda N_j, \tag{10.56}$$

where λ is a certain scalar value, called an eigenvalue.

Vectors that fulfil the above equation are called eigenvectors, whereas directions determined by them are called principal directions or characteristics for the tensor **T**.

10.6.5 Tensor Invariants

Tensor invariants are expressions composed of some tensor components, which do *not* change when transforming the coordinate systems from U to U'.

One of the commonly known and extensively used invariants is the Frobenius norm. For a second-order tensor **N** it is determined as follows:

$$\|\mathbf{N}\|^2 = \mathbf{N} \cdot \mathbf{N} = \sum_{kl} N_{kl}^2 = \sum_{\mathrm{i}} \lambda_i^2, \tag{10.57}$$

where N_{kl} represents components of **N**.

10.7 Closure

In this chapter we present a very concise introduction to the tensor calculus. Tensors, which can be seen as a generalization of the concept of linear operators, find application in computer vision to capture multilinear relations among geometrical objects, such as points, lines, planes, etc. It is interesting to observe that such commonly used concepts as scalars, vectors and matrices are also examples of tensors, so even while not naming them explicitly we use tensors in everyday life.

10.7.1 Further Reading

The tensor calculus has found many applications in computer vision and image processing areas. It offers a more compact representation of many quantities, such as relations among corresponding points in many views of the same scene. Tensors are mostly used in physics, especially in mechanics and theory of relativity. In this chapter we give a very brief introduction to this area. A very good (and inexpensive) introduction into the realm of tensors is the book by Borisenko and Tarapov [49]. For more formal treatment of tensors on manifolds one can refer to the classic book by Bishop and Goldberg [45]. A brief introduction with many exercises and examples is provided in the textbook by Kay [241]. More recent books on this subject are the seminal work by Penrose [344] and the monograph by Dimitrienko [103] which provides a unified geometric representation of a tensor and tensor operations. Both are excellent sources for self-study of the notions of manifolds, differential geometry and tensors.

Indirect sources of information on tensor calculus are books devoted to the differential geometry. Recommended reading in this area are books by Kreyszig [263] and Guggenheimer [167], as well as the works by Spivak [397] and the recent book by Kühnel [264]. Although most of them are not the most recent publications in the field, they are very intuitive and still can be obtained at very affordable prices from their publishers. A nice introduction to differential geometry, especially suited for self-study, is the frequently cited book by do Carmo [110].

Finally, tensors have found a profound place in works devoted to computer vision, such as the ones by Triggs [428, 429] or the excellent book by Hartley and Zisserman [180].

11

Distortions and Noise in Images

11.1 Abstract

Noise is an additional, usually unwanted, component that interferes with a pure signal. Its source comes from certain physical phenomena encountered during signal acquisition and transmission. There are many types of noise that can contaminate a 'pure' 2D signal of an image. In this chapter we discuss various types of noise that can be encountered in digital images, as well as different models of noise. However, sometimes we wish to generate an image with predefined type and level of noise. This is usually done for testing the tolerance of noise in feature detectors or matching modules. In this chapter we also discuss some simple algorithms for these tasks.

11.2 Types and Models of Noise

The presence of noise in a signal is usually modelled by either addition

$$\hat{f}(\mathbf{x}) = f(\mathbf{x}) + \eta \tag{11.1}$$

or multiplication

$$\hat{f}(\mathbf{x}) = \eta \cdot f(\mathbf{x}) \tag{11.2}$$

of the original signal by a random variable. In the above formulas $\hat{f}(\mathbf{x})$ stands for observable signal, $f(x)$ is a 'pure' signal and η is a random variable that models a noise which is characterized by the distribution function.

Table 11.1 describes the most common types of noise encountered in digital images. We do not provide information on many other types of noise, e.g. speckle noise and photographic grain noise, etc. Further information on the subject can be found, for instance, in Chan and Jianhong Shen [65] and Starck *et al.* [399, 400].

An Introduction to 3D Computer Vision Techniques and Algorithms Bogusław Cyganek and J. Paul Siebert

Table 11.1 Types of noise encountered in images.

Noise	Description
Gaussian noise	Gaussian-type noise is used to model such physical phenomena as thermal noise and sometimes photon counting and film grain noise. For this type of noise we use the additive model, in accordance with (11.1), in which the random variable η has the Gaussian density function $p_\eta(x)$, in the following form: $$p_\eta(x) = \frac{1}{\sqrt{2\pi}\sigma} e^{-\frac{(x-\mu)^2}{2\sigma^2}}, \qquad (11.3)$$ where μ is the mean and σ is the variance. If $\mu = 0, \sigma = 1$ then we have a normal Gaussian distribution. When applying directly (11.1) we have to consider the limited precision of the data representation used for pixels. Usually one has to fulfil the condition that $f(x) + \eta$ is positive and does not exceed the maximum range of a pixel. It is interesting to note that values x which exceed $\pm 3\sigma$ are 'highly improbable'.
Salt and pepper noise	The name of this type of noise comes from the visual effect which manifests as white and black dots in images – the same as scattering salt and pepper over a sheet of paper. One source of this phenomenon is transmitting lines of digital images. Assuming that B bits are used to code a value I of a pixel x we have $$I(x) = \sum_{k=0}^{B-1} b_k 2^k. \qquad (11.4)$$ Assuming further that each bit transferred over the channel under consideration can be flipped with probability α then the probability of the received value $J(x)$ fulfils the following condition: $$P\left(\lvert I - J \rvert = 2^k\right) = \alpha, \qquad (11.5)$$ where $k \in (0, \ldots, B-1)$. For the most significant bit (MSB) the mean square error $(I - J)^2$ follows $\alpha(2^{B-1})^2$. Usually salt and pepper noise is a result of a random change of the MSB in pixel representation, so white pixel becomes black or vice versa. This type of noise can be modelled as follows [65]: $$\hat{f}(\mathbf{x}) = (1-a) f(\mathbf{x}) + ab, \qquad (11.6)$$ where a is a random variable (of some distribution) characteristic of the probability $p = \Pr(a = 1)$ and b is a random variable characterized by $\Pr(b = f_{\text{MAX}}) = \Pr(b = f_{\text{MIN}}) = 0.5$. So the process of generating salt and pepper noise can be viewed as double drawing process: at first we generate a random variable a with probability p of the event $a = 1$. Then, if it happened that $a = 1$ with 50% probability, we draw for b to be f_{MAX} or f_{MIN}.

Table 11.1 (*Continued*)

Quantization noise	Quantization noise is a result of the change of a continuous signal into a digital representation which, of course, is of finite precision. It arises also in a change from one digital representation into another with smaller precision (fewer bits). Thus, we can say that this type of noise is a result of an introduced error into the data representation. Quantization noise is usually modelled as a random variable with uniform distribution [341]. An exception is the case for a small number of quantization levels, in which the quantization noise is signal dependent and cannot be modelled as uniform.
Photon counting noise	Photon counting noise arises from the physical properties of image acquisition systems that rely on photon counting. For instance, the speed of a shutter in a camera influences the number of photons that can reach the sensor and as a result adds to the photon counting noise. This type of noise is best modelled as a discrete random variable with Poisson distribution [341], as follows: $$\Pr(n = k) = \frac{e^{-\lambda}\lambda^k}{k!}, \quad (11.7)$$ where n denotes the number of counted photons in a certain (but constant) time interval, k is the number of actually counted photons in a single experiment (observation), thus $k = 0, 1, 2, 3$, and so on, and $\lambda > 0$ is a parameter. The expected value and variance of (11.7) is the same and equal to λ. It is interesting to note that the best estimation of the (usually unknown) parameter λ is given as a mean value from the population X [341]: $$\lambda \approx \overline{X} = \frac{1}{N}\sum_{k=1}^{N} X_k, \quad (11.8)$$ where N denotes number of elements in the population X. In our case $$\lambda \approx \frac{1}{N}\sum_{k=1}^{N} k n_k, \quad (11.9)$$ where n_k is the number of observations which resulted in exactly k photons and N denotes the total number of observations. One can conclude that the higher the number of counted photons k then the higher the value of λ. Therefore the brighter areas in an image have higher λ and therefore a higher noise variance (which is also λ).

11.3 Generating Noisy Test Images

Usually noise is an unwanted signal that contaminates the 'pure' signal and we try to filter it out. However, for some experiments it is useful to generate images with *a priori* given noise of known parameters. Such images can be used to test the behaviour of image processing algorithms, e.g. their resistance to noise. The most common practice is to add Gaussian or salt and pepper noise to the original image.

Gaussian noise can be modelled as an additive noise (11.1) in which η is a random variable with Gaussian distribution $N(\mu, \sigma)$ where μ is the mean and σ the variance. Usually however, a noise approximated by a random variable with normal Gaussian distribution $N(0,1)$ multiplied by a certain constant is added to the source image f, as follows:

$$\hat{f}(x) = f(x) + a\eta, \tag{11.10}$$

where a is a constant value.

To measure the difference between two images it is common to compute the mean-square error (MSE) between the two, as follows:

$$\mathrm{MSE}\left(\hat{f}, f\right) = E\left[\left(\hat{f} - f\right)^2\right]. \tag{11.11}$$

From the MSE one can compute the peak signal-to-noise ratio (PSNR), as follows:

$$\mathrm{PSNR}\left(\hat{f}, f\right) = 10\log_{10}\left(\frac{m^2}{\mathrm{MSE}\left(\hat{f}, f\right)}\right) \quad [\mathrm{dB}], \tag{11.12}$$

where m is the maximum pixel value (e.g. $m = 255$ for 8-bit images).

Introducing (11.10) and (11.11) into (11.12) we obtain

$$\begin{aligned} \mathrm{PSNR}\left(\hat{f}, f\right) &= 10\log_{10}\left(\frac{m^2}{E\left[\left((f + a\eta) - f\right)^2\right]}\right) \\ &= 10\log_{10}\left(\frac{m^2}{a^2 E\left[\eta^2\right]}\right) = 20\log_{10}\left(\frac{m}{a\sigma}\right). \end{aligned} \tag{11.13}$$

For 8-bit scalar-valued images (i.e. $m = 255$) and normal distribution (i.e. $\sigma = 1$), the above can be written in a simpler form:

$$\mathrm{PSNR}\left(\hat{f}, f\right) = 48.13 - 20\log_{10}(a), \tag{11.14}$$

where a is a constant that allows us to control the parameters of the added Gaussian noise in the same way as changing σ but in a more convenient way, since for all the time we can use the same random generator with the normal Gaussian distribution $N(0,1)$. Observe also that in this particular case if $a = m$ then $\mathrm{PSNR} = 0$ which stays in agreement with our intuition – if noise has potentially the same amplitude as signal then we cannot distinguish them. Thus, we can express a in relation to m and express it as a percentage. By this token 0% (i.e. the pure signal) stands for $a = 0$, whereas 100% (i.e. $\mathrm{PSNR} = 0$) stands for $a = m$.

Sometimes it is convenient to generate noise with a given PSNR value – for this case we compute easily the value a as follows:

$$a = 10^{\frac{48.13 - \mathrm{PSNR}\left(\hat{f}, f\right)}{20}}, \tag{11.15}$$

where the PSNR value is given in dB.

11.4 Generating Random Numbers with Normal Distributions

Generation of random numbers with a given distribution is a nontrivial task. Usually one starts from the pseudo-random generators of the uniform distribution. Most of the common software libraries for numerical computations are equipped with a version of such a generator. Then, based on uniform distribution, it is possible to build up generators of pseudo-random numbers which approximate other distributions [249].

To generate random variables with normal distribution based on a generator of uniform distribution, the method of the ratio-of-uniforms, proposed by Kinderman and Monahan [244], can be used. Algorithm 11.1 presents the C++ procedure *GenerateNormalValue( )* realizing this idea.

```
/////////////////////////////////////////////////////////////////
// This function returns a random variable featuring
// normal gaussian distribution.
/////////////////////////////////////////////////////////////////
//
// INPUT:
//                      none
//
// OUTPUT:
//                      random variable with a normal gaussian
//                              distribution
//
// REMARKS:
//                      The used method is based on uniform random
//                      values and the ratio-of-uniform theorem
//
//                      Remember to initialize the random generator
//                      before calling this function.
//
double GenerateNormalValue( void )
{
        const double kSqrE = 0.857763884960706796480189641278 77; // sqrt(2/e)

        double U, V, X, XX;
        bool accept = false;
        const double d_RandMax = (double)RAND_MAX;

        do
        {
                U = rand();                             // generate a uniform random value

                if( U == 0 )
                        U = 1e-10;                      // since we later divide by U

                else
                        U /= d_RandMax;                 // Let's normalize U, so U( 0, 1 )

                V = rand();
                V /= d_RandMax; // V( 0, 1 )
                V = kSqrE * ( 2.0 * V - 1.0 ); // V(-sqrt(2/e), sqrt(2/e))

                X = V / U;

                XX = X * X;
                if( XX <= 2.0 * ( 3.0 - U * ( 4.0 + U ) ) )
                        accept = true;
                else
                        if( XX <= 2.0/U - 2.0*U && XX <= - 4.0*log( U ) )
                                accept = true;


        } while( ! accept );

        return X;
}
```

Algorithm 11.1 Procedure to generate random variables with normal distribution $N(0, 1)$ based on the random generators of uniform distributions

11.5 Closure

This chapter presents a short overview on types and models of noise encountered in digital images. Also discussed are methods of generating noise of certain parameters. These can then be used in testing tolerance to noise of some other image algorithms.

11.5.1 Further Reading

In the literature on image processing, noise is usually discussed in terms of image restoration or signal filtering. From this category are books by Jähne [224] and Gonzalez and Woods [157]. Literature on filtering of digital signals is very ample. Recommended readings in this area are the books by Haykin [183], Oppenheim and Shafer [336], Mitra [312] and Mitra and Sicuranza [313], to name a few. For additional information on noise in digital images one can refer to the books by Chan and Jianhong Shen [65] or Starck *et al.* [399, 400].

12

Image Warping Procedures

12.1 Abstract

Image warping is a process of changing the appearance of an image as a result of changing pixel positions of an original image. The simplest change is a horizontal shift, to the left or to the right, by one, two or *n* pixel positions. Things change however if such a shift has to be done by a fractional displacement rather than integer positions and in two directions simultaneously. In such a case, the new pixel position will not fall into an integer grid of image samples. Thus, its new value has to be interpolated somehow to accommodate an arbitrary new position. Such situations arise usually when a geometrical transformation is applied to the coordinate system of an input image, such as rotation, translation, scaling or the already discussed projective transformation. Thus, the image warping system consists of the coordinate transformation and pixel interpolation stages, augmented by the warp module which sets the forward or backward (inverse) warping scheme. It is easy to imagine image warping employed in some artworks. However, it has broad applications in computer vision as well, such as in the already discussed log-polar transformation (section 6.3.8) or elastic area-based matching (section 6.7). This chapter presents details of software modules for image warping.

12.2 Architecture of the Warping System

Figure 12.1 depicts the architecture of the image warping system. It consists of the basic warp module, as well as the coordinate transformation and the pixel interpolation modules.

The coordinate transformation module is responsible for computation of the positions of pixels in the destination coordinate system, based on positions they had in the source system. For forward warping it converts coordinates from the input image to the output (warped) image, and for inverse warping the process is just reversed.

The pixel interpolation block is responsible for the computation of a value of a pixel in the new coordinate space. This process requires information on neighbouring pixels and their values.

Each of the modules of the warping system is implemented as separate class hierarchies. Thus, adding new coordinate transformations or pixel interpolation algorithms is straightforward and consists of deriving a new class for a particular task.

An Introduction to 3D Computer Vision Techniques and Algorithms Bogusław Cyganek and J. Paul Siebert

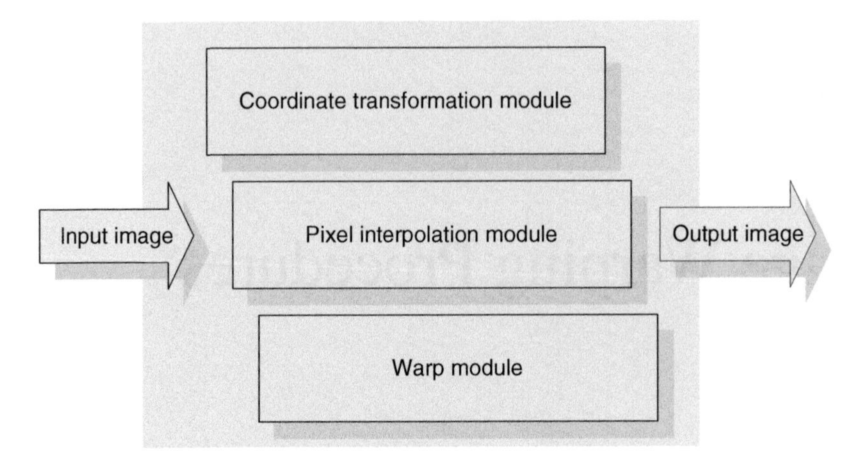

Figure 12.1 Architecture of the image warping system. It consists of three modules: warp engine, coordinate transformation and pixel interpolation modules

12.3 Coordinate Transformation Module

The coordinate transformation module is responsible for mapping of pixel coordinates between two spaces of the source and destination coordinate systems. Since in this solution we prefer the inverse warping scheme, then the coordinate transformation always does an inverse coordinate mapping, i.e. from the output (warped) space to the input (original) one. This means that we need to know parameters of an inverse transformation. However, usually the forward mapping is provided. In such a case the inverse mapping has to be found.

In this section the three types of coordinate transformations are discussed. We start with the projective and affine plane transformations, then present polynomial and generic approaches. The latter can be used in elastic stereo matching in which the reference image is warped in accordance with the current disparity map and then compared with the second image. This proceeds iteratively in the scale-space (section 6.7).

12.3.1 Projective and Affine Transformations of a Plane

An affine space $\Re^n$ is isomorphically transformed into the projective space $\wp^n$, as discussed in section 9.2. In the following discussion the homogeneous coordinates are employed, if not stated otherwise.

The projective homography is defined as a nonsingular matrix (9.41). In the case of planar homography we can rewrite this expression as

$$\mathbf{Hx} = \hat{\mathbf{x}}, \tag{12.1}$$

$$\begin{bmatrix} h_{11} & h_{12} & h_{13} \\ h_{21} & h_{22} & h_{23} \\ h_{31} & h_{32} & h_{33} \end{bmatrix} \begin{bmatrix} x_1 \\ x_2 \\ x_3 \end{bmatrix} = \begin{bmatrix} \hat{x}_1 \\ \hat{x}_2 \\ \hat{x}_3 \end{bmatrix}, \tag{12.2}$$

where $\mathbf{H}$ is the coordinate transformation (warping) matrix, $\mathbf{x}$ denotes pixel coordinates in the homogeneous coordinates and $\hat{\mathbf{x}}$ is a new position of a pixel in the wrapped output image. The projective homography of a plane requires nine parameters.

In many applications the affine transformation which corresponds to rotation, translation and scaling is used. It is defined by only six parameters. It can be written as

$$\mathbf{A}\mathbf{x} = \hat{\mathbf{x}},$$

$$\begin{bmatrix} a_{11} & a_{12} & a_{13} \\ a_{21} & a_{22} & a_{23} \\ 0 & 0 & 1 \end{bmatrix} \begin{bmatrix} x_1 \\ x_2 \\ x_3 \end{bmatrix} = \begin{bmatrix} \hat{x}_1 \\ \hat{x}_2 \\ 1 \end{bmatrix}. \tag{12.3}$$

A choice of the third coordinate in the homogeneous system is somehow arbitrary, since the projective transformations are defined up to a scaling factor; therefore in the case of affine transformation given by the above equation it is convenient to assume that $x_3 = \hat{x}_3 = 1$. This has an additional advantage of avoiding division when computing Cartesian coordinates of $\hat{\mathbf{x}}$:

$$\begin{bmatrix} x_1 \\ x_2 \end{bmatrix} = \begin{bmatrix} \hat{x}_1/1 \\ \hat{x}_2/1 \end{bmatrix} = \begin{bmatrix} \hat{x}_1 \\ \hat{x}_2 \end{bmatrix}. \tag{12.4}$$

12.3.2 Polynomial Transformations

The polynomial transformation of the point coordinates belongs to the class of nonlinear mappings. Thus, it can approximate wider group of transformations than for example the affine mappings.

A polynomial transformation can be defined as follows [351]:

$$\mathbf{y} = \mathbf{W} \cdot \mathbf{P}(\mathbf{x}), \tag{12.5}$$

where $\mathbf{x}$ denotes the input vector (a point), $\mathbf{y}$ is an output vector, $\mathbf{W}$ is a transformation matrix and $\mathbf{P}$ denotes a polynomial on $\mathbf{x}$.

Further, we will focus on the second-order polynomial transformation for which $\mathbf{W}$ and $\mathbf{P}$ in (12.5) can be stated as follows:

$$\mathbf{W} = \begin{bmatrix} w_{11} & w_{12} & w_{13} & w_{14} & w_{15} & w_{16} \\ w_{21} & w_{22} & w_{23} & w_{24} & w_{25} & w_{26} \end{bmatrix}_{2\times 6} \tag{12.6}$$

and

$$\mathbf{P}(\mathbf{x}) = \begin{bmatrix} 1 \\ x_1 \\ x_2 \\ x_1^2 \\ x_1 x_2 \\ x_2^2 \end{bmatrix}_{6\times 1} \quad \text{where} \quad \mathbf{x} = \begin{bmatrix} x_1 \\ x_2 \end{bmatrix} \quad \text{and} \quad \mathbf{y} = \begin{bmatrix} y_1 \\ y_2 \end{bmatrix}. \tag{12.7}$$

In contrast to the affine transformations – where elements of the transformation matrix can easily be found from given intuitive parameters, such as a rotation angle or scale – there are 12 parameters w_{ij} in (12.6) that need to be determined for a mapping with usually unknown analytical form. This can be achieved after manual (empirical) choice of the number of control points (at least six, since each point gives two equations) and their corresponding positions in the output image. Then (12.5) is solved for **W**. For the more general case of more than six corresponding points this can be achieved by the least-squares method (section 12.8).

12.3.3 Generic Coordinates Mapping

The group of generic transformations of point coordinates can be described by the following expression:

$$\mathbf{y} = \Theta(\mathbf{x}), \tag{12.8}$$

where Θ is a general (usually nonlinear) transformation function. This group is a generalization to the previously mentioned affine (12.3) and polynomial (12.6) versions. However, usually it requires determination of an unknown transformation function which can be in the form of a closed formula, fuzzy rules or look-up tables (i.e. look-up images). In the case of warping stereo images, Θ is just a disparity map.

It is easy to observe that the generic coordinate mapping Θ in the form of a look-up table can also be created by a single run of a coordinate transformation engine, such as affine or polynomial which were discussed in the previous sections. This is a very useful technique since further access to the look-up table is usually much faster than repetitive computation even of a linear transformation. Such an approach was undertaken when computing log-polar transformations for the selected image points – matching is then performed in the log-polar domain (section 6.3.8).

12.4 Interpolation of Pixel Values

Computed by the coordinate transformation module, positions of output pixels usually do not fall into the regular sampling grid of the input image (in the case of the inverse warping which is assumed further on). Therefore a value of each pixel has to be determined with some accuracy based on its neighbouring pixels. This is called value interpolation and is discussed in the next sections.

12.4.1 Bilinear Interpolation

Interpolation is a process of finding unknown values of data from some other, but known, values. It is often assumed that it is possible to determine a continuous function (e.g. a polynomial of certain order, etc.) that passes through the known data points (see also section 4.4.2). Then, the unknown value is simply a value of the interpolating function at the point of interest. However, there are many possible functions to be placed in this role and they require different amounts of known data points. For image interpolation we need to operate with 2D (or higher) interpolation functions.

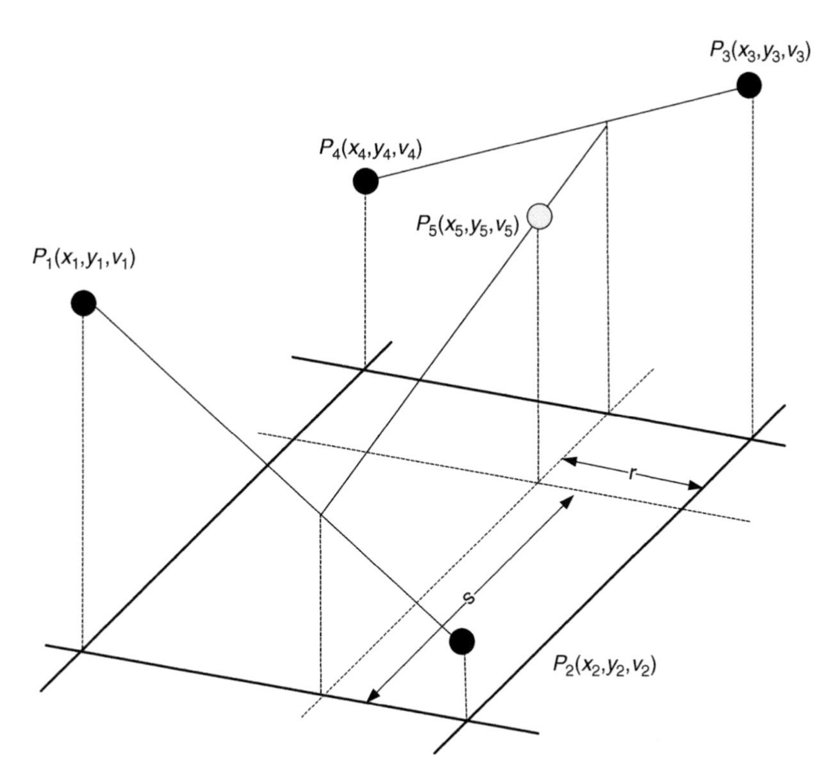

Figure 12.2 Scheme of bilinear interpolation. The points P_1–P_4 lie on a regular grid of an image coordinate system. The value v_5 of point P_5 is unknown and has to be determined

The bilinear interpolation of pixels relies on four nearest neighbours of a pixel whose value is unknown and has to be found. This scheme is depicted in Figure 12.2.

The four nearest neighbour points P_1–P_4 are acquired at a regular grid of the image and their intensity values v_1–v_4 are given. The interpolated point P_5 lies at some fractional distance from its neighbours. However, its value v_5 is unknown and has to be determined. In bilinear interpolation v_5 is linearly related to v_1–v_4 in terms of the local distances r and s (Figure 12.2), given as follows:

$$r = \frac{x_5 - x_1}{x_2 - x_1} \quad \text{and} \quad s = \frac{y_5 - y_3}{y_2 - y_3}. \tag{12.9}$$

For the neighbouring grid pixels, denominators in the above are equal to 1. The value v_5 can be computed as the linear combination of the values v_1–v_4 weighted by r and s, as follows:

$$v_5 = (1-r)(1-s)v_4 + r(1-s)v_3 + (1-r)sv_1 + rsv_2. \tag{12.10}$$

The fractionals (12.9), as well as the linear combination (12.10), can be easily and accurately computed with fixed-point arithmetic or even simpler binary arithmetic. The latter can be accomplished finding the nearest two fractionals for r and s in (12.9), then accordingly shifting the values v_1–v_4.

The bilinear interpolation method is very simple and therefore frequently used. It appears that such an approach is sufficient in many practical applications – introduction of interpolation functions of order higher than linear does not necessarily lead to better results since the type of the local relation among pixel values is usually not known *a priori*.

12.4.2 Interpolation of Nonscalar-Valued Pixels

Colour images belong to the class of nonscalar-valued images. For such a group the interpolation scheme (12.10) is applied to each colour component separately. In other words, each colour channel is treated as a separate mono-valued image and interpolation takes place on scalar values of that channel. However, the coordinate transformation (12.2) is computed only once for a given pixel position, so for this stage there is exactly the same number of computations regardless of the number of channels in the input image.

12.5 The Warp Engine

There are two common methods of image warping [449].

1. Forward transform.
2. Backward transform.

In forward warping (depicted in Figure 12.3) the input image is scanned line by line and the pixels are transformed to the output image. Their positions are given by the result of linear transformation. However this technique is troublesome since it results in images with holes due to nonoverlapping regions of the mapping.

It can happen for some transformations that different points from the input are mapped to the same point in the output image. However, all of them can have different values. Therefore for this method we need to store those values in accumulators for further interpolation stage.

Because of the aforementioned problems with forward warping, backward mapping is of interest, which is shown in Figure 12.4. This time, however, the *output* image is scanned pixel

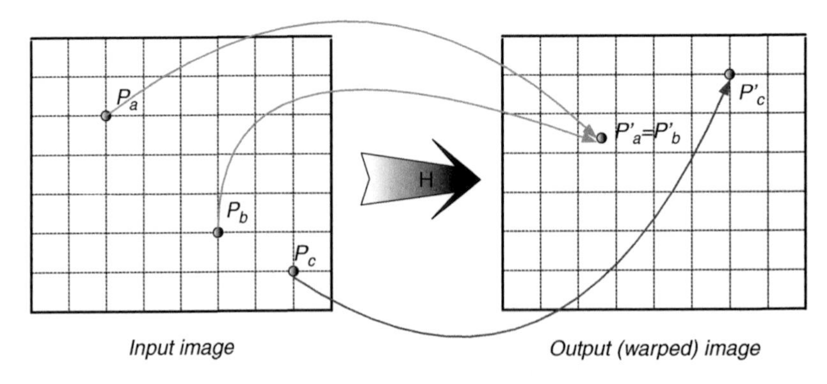

Figure 12.3 Forward warping scheme. Two different points P_a and P_b from the input space are mapped to the same point $P'_a = P'_b$ in the output space. P_c is mapped to P'_c

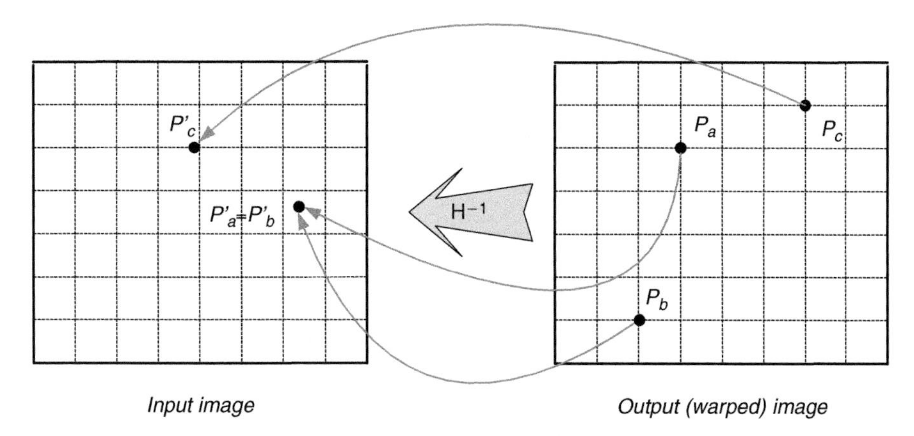

Figure 12.4 Backward (inverse) warping scheme. The points from the output space are mapped into the input space. Then, based on the nearest neighbours of a mapped point, its value is determined by pixel-value interpolation. Even if the two points are mapped to the same input position, it does not pose a problem

by pixel and the corresponding pixel position in the input image is computed. Once again, the new positions more often than not do not lie on the integer grid of the input image.

Thus to determine the value of a pixel we need to resample the original image. However, this time it is usually easier to find the closest neighbouring pixels of the input image that are necessary for interpolation.

The inverse warping scheme assumes knowledge of an inverse transformation $\mathbf{H}^{-1}$, i.e. the method of pixel mapping when going from the output image space to the input which usually does not pose much of a problem.

12.6 Software Model of the Warping Schemes

The presented warping modules were implemented in C++ in the form of class hierarchies: the coordinate transformation hierarchy, the interpolation hierarchy and the image warp hierarchy, respectively. These are briefly presented in the next sections. Complete source code can be accessed from the website of the book [216].

12.6.1 Coordinate Transformation Hierarchy

Figure 12.5 shows the *TCoordTransformEngine* class hierarchy for different groups of coordinate transformations. There are three main branches of derived classes.

1. The linear transformations, implemented by *TLinearTransformEngine*.
2. The nonlinear transformations, implemented by *TNonLinearTransformEngine*.
3. The generic transformations, from *TGenericTransformEngine*.

They reflect the transformation methods described in section 12.3.

Algorithm 12.1 presents the interface of the base *TCoordTransformEngine* class. The key method is a functional operator whose role is transformation of input coordinates passed by

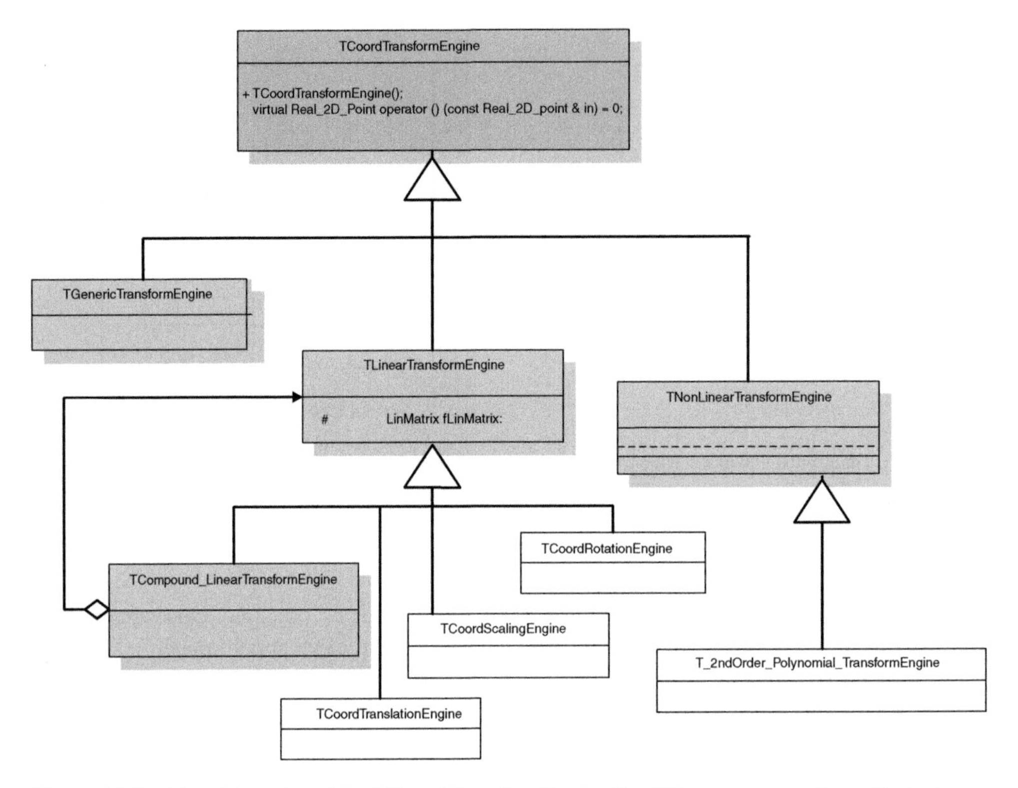

Figure 12.5 Class hierarchy of the *TCoordTransformEngine* for different groups of coordinate transformations

reference to the *Real_2D_Point*. Particular implementations are left to the derived classes, however.

If such transformation is not possible for some reason, then the *TCoordTransform Engine::kImpossiblePoint* is returned. This can happen if a transformation (or its inverse) is not defined for a given point.

12.6.2 Interpolation Hierarchy

Figure 12.6 presents the class hierarchy for pixel interpolation with the *TpixelInterpolation* base class.

Algorithm 12.2 contains the definition of the base *TPixelInterpolation* class. It provides a pure virtual functional operator whose role is to define a common interface for value interpolation. There are two input parameters, the reference to the input image and point coordinates, and one output value which is an interpolated value of a pixel.

12.6.3 Image Warp Hierarchy

Figure 12.7 presents the *TImageWarp* class hierarchy for different warp algorithms.

```
///////////////////////////////////////////////////////////////
// This class implements the hierarchy of the coordinate
// transformation classes.
///////////////////////////////////////////////////////////////
class TCoordTransformEngine
{
	public:

		// use it as an output whenever it is not
		// possible to determine the transformation
		static const Real_2D_Point kImpossiblePoint;

	public:

		// ====================================================
		TCoordTransformEngine( void ) {}
		// class virtual destructor
		virtual ~TCoordTransformEngine() {}
		// ====================================================

		///////////////////////////////////////////////////////////////
		// This function converts a given point based on some
		// external parameters (set in derived classes)
		///////////////////////////////////////////////////////////////
		//
		// INPUT:
		//		in - the input point
		//
		// OUTPUT:
		//		the output (converted) point
		//
		// REMARKS:
		//
		//
		virtual Real_2D_Point operator () ( const Real_2D_Point & in ) = 0;

};
```

Algorithm 12.1 Interface of the base *TCoordTransformEngine* class

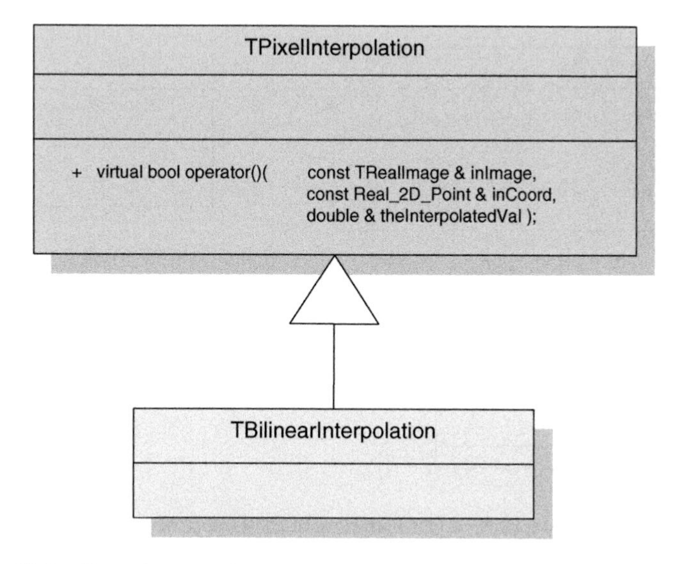

Figure 12.6 Class hierarchy for pixel interpolation with the base *TpixelInterpolation*

```
class TPixelInterpolation
{

	public:

		TPixelInterpolation( void ) {}
		virtual ~TPixelInterpolation() {}

	public:

		/////////////////////////////////////////////////////////////
		// This function interpolates a pixel value
		/////////////////////////////////////////////////////////////
		//
		// INPUT:
		//		inImage - the image that serves pixel values
		//			for interpolation
		//		pointCoords - a point to be interpolated (it
		//			can be further processed in derived classes
		//			e.g. for the inverse warping)
		//		theInterpolatedValue - the output interpolated
		//			value of a pixel
		//
		// OUTPUT:
		//		true if operation successful,
		//		false otherwise
		//
		// REMARKS:
		//
		//
		virtual bool operator()( const TRealImage & inImage,
						const Real_2D_Point & inCoord,
						double & theInterpolatedVal ) = 0;

};
```

Algorithm 12.2 Interface of the base *TPixelInterpolation* class

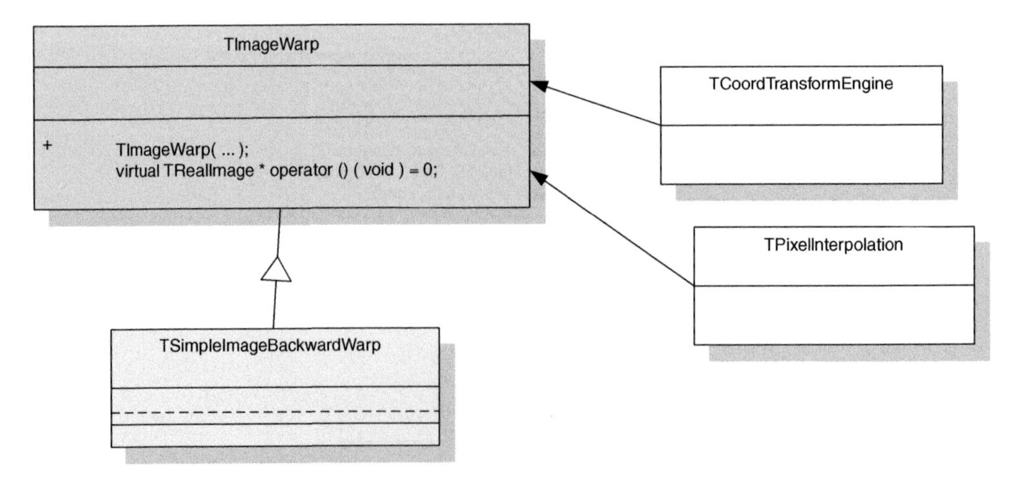

Figure 12.7 Class hierarchy of the *TImageWarp*

```
class TImageWarp
{
   protected:

        const TRealImage & fInImage;

        TCoordTransformEngine & fCoordTransformEngine;

        TPixelInterpolation & fPixelInterpolation;

   public:

        // ===================================================
        TImageWarp(    const TRealImage & inImage,
                       TCoordTransformEngine & coordTransformEngine,
                       TPixelInterpolation & pixelInterpolation );

        virtual ~TImageWarp() {}            // class virtual destructor
        // ===================================================


        ///////////////////////////////////////////////////////////////
        // This function performs the image warping.
        ///////////////////////////////////////////////////////////////
        //
        // INPUT:
        //          none
        //
        // OUTPUT:
        //          none
        //
        // REMARKS:
        //
        //
        virtual TRealImage * operator () ( void ) = 0;
};
```

Algorithm 12.3 Interface of the base *TImageWarp* class

Algorithm 12.3 presents the interface of the base *TImageWarp* class. The pure virtual functional operator defines a common interface for this class hierarchy. It does not take any input parameters which are supplied to the constructors. This technique allows the same interface which is required for the virtual functions. At the same time the necessary parameters can be changed from class to class by definition of new constructors. The warped image is returned.

12.7 Warp Examples

In this section we present some experimental results obtained from the software model described in the previous section.

Figure 12.8 presents the original test grey-valued image 'Airplane'. Its version rotated by -22° is depicted in Figure 12.8(b). It should be remembered that in the case of inverse (or backward) warping the supplied warp parameters should reflect the inverse transformation, i.e. from the output image space to the input space. It is also evident in Figure 12.8(b) that not all pixels from the output image can be mapped with this transformation to the valid places in the input image; thus they have to be filled with black.

Figure 12.9 presents another version of the affine transformations. The version is rotated by 22° around the central pixel, rotated -11° around the centre and translated by vector

(a)

(b)

Figure 12.8 (a) 'Airplane' test image (Source: USC-SIPI Image Database) and (b) its affine transformed version (rotated -22° around a point [33, 33])

[33, −17], rotated 11° and scaled by [0.77, 1.89] and a compound transformation consisting of rotation, translation and scaling.

Figure 12.10 depicts the 'Airplane' image transformed by polynomial inverse warping. The parameters in (12.5) are as follows: $\mathbf{W}$ = [0, 1, 0, 0.001, −0.001, 0.001] [0, 0, 1, 0.001,−0.001, 0.001] (Figure 12.10a), and $\mathbf{W}$ = [0,1, 0, 0, −0.001, 0] [0, 0, 1, 0.001,−0.005, 0.001] (Figure 12.10b).

Figure 12.11 presents the 'Airplane' image warped with the generic transformations. The horizontal variable was sinus modulated (Figure 12.11a), and both the horizontal and vertical variables were sinus modulated (Figure 12.11b).

The drawback of the generic transformations is the requirement of a look-up table of size equal to the size of the input image. However, they allow easy tiling and combination of all possible transformations. The other advantages come from easy implementation and fast execution.

Figure 12.12 depicts 'Kamil' colour image in the RGB space (see Plate 7), and the output image after the affine transformation consisting of a −43° rotation around a centre point, scaling by [0.7, 0.8] and translation of the [155, 0] vector. The interpolation was applied in each channel separately, as described in section 12.4.2.

12.8 Finding the Linear Transformation from Point Correspondences

In this section we discuss the problem of finding parameters of a coordinate transformation from point correspondences. This method can be used to assess linear parameters of a transformation, i.e. if the transformation can be written in the form of a set of linear equations. The number of equations can be equal to or greater than the number of unknown parameters.

Let us start from the affine transformation (12.3). However, because we perform inverse warping, instead of finding the matrix $\mathbf{A}$ we look for its inverse $\mathbf{B} = \mathbf{A}^{-1}$, assuming it exists.

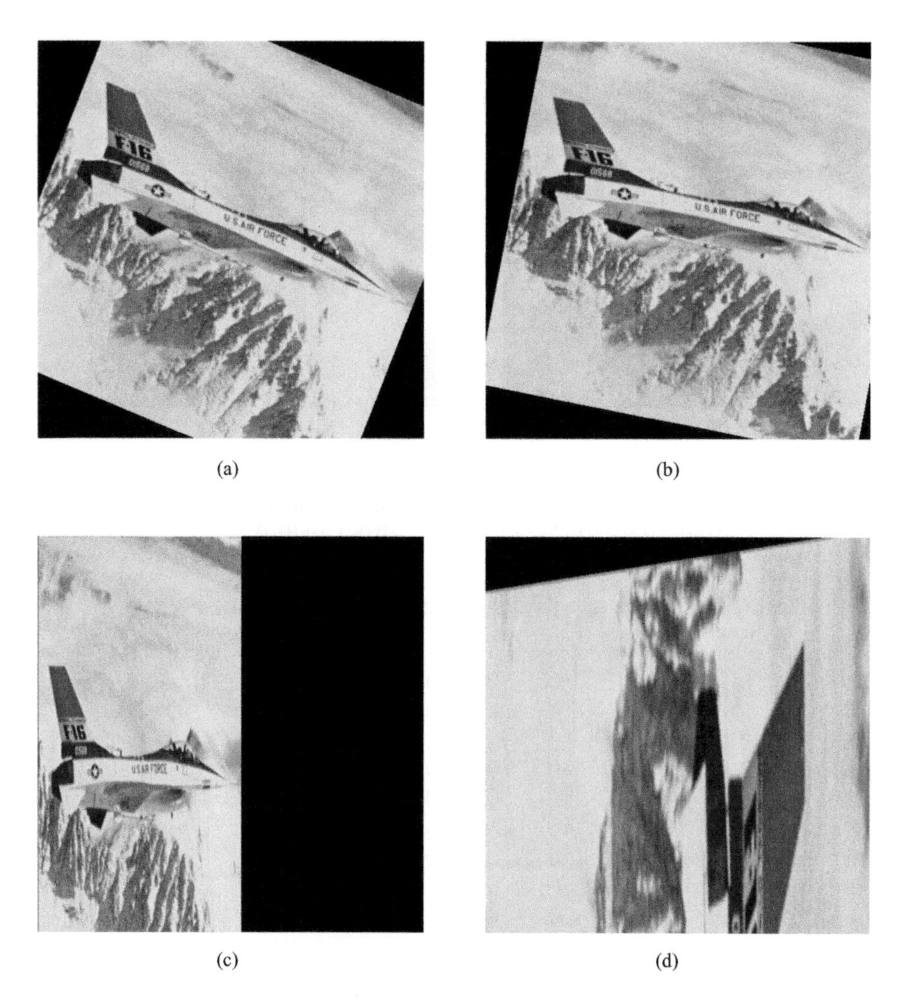

Figure 12.9 Affine transformed versions of 'Airplane': (a) rotated 22° around the central pixel; (Source: USC-SIPI Image Database) (b) rotated 11° around the centre and translated by [33, −17]; (c) rotated 11° and scaled by [0.77, 1.89]; (d) rotated (88°), translated (by [4, 5]) and scaled (by [0.2, 0.9])

B describes mapping from the output to the original (i.e. unwrapped) image space. Since we have six unknown parameters and each 2D point adds two equations, then at least three different points are necessary to determine the parameters of matrix **B**. Moreover the points should not all lie on a single line, since otherwise we end up with dependent equations which lead to a singularity. Such basic mapping can be written as follows:

$$\begin{cases} \mathbf{P}_1 = \mathbf{B}\tilde{\mathbf{P}}_1 \\ \mathbf{P}_2 = \mathbf{B}\tilde{\mathbf{P}}_2 \\ \mathbf{P}_3 = \mathbf{B}\tilde{\mathbf{P}}_3 \end{cases}, \tag{12.11}$$

where $\mathbf{P}_1 - \mathbf{P}_3$ and $\tilde{\mathbf{P}}_1 - \tilde{\mathbf{P}}_3$ are points in the original and warped spaces, respectively. To find

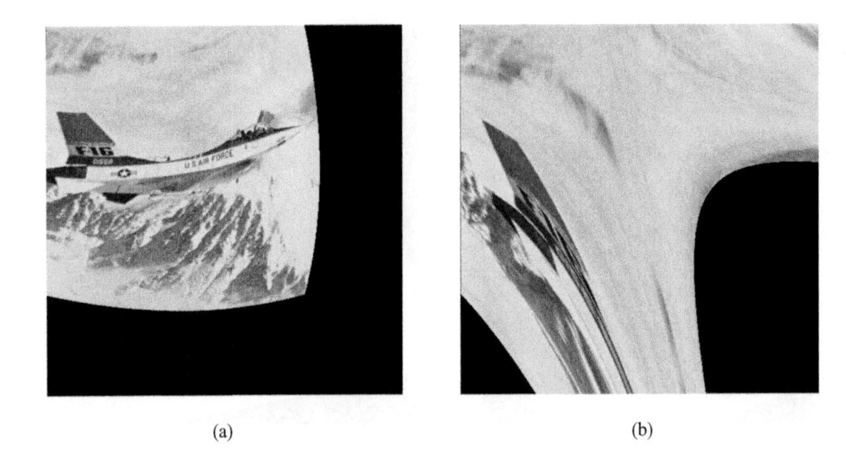

(a) (b)

Figure 12.10 'Airplane' image after polynomial inverse warping: (a) $\mathbf{W} = [0, 1, 0, 0.001, -0.001, 0.001]$ $[0, 0, 1, 0.001, -0.001, 0.001]$; (b) $\mathbf{W} = [0, 1, 0, 0, -0.001, 0]$ $[0, 0, 1, 0.001, -0.005, 0.001]$

B, we rewrite the above equation into the following representation:

$$\begin{bmatrix} \tilde{p}_{11} & \tilde{p}_{12} & \tilde{p}_{13} & 0 & 0 & 0 & 0 & 0 & 0 \\ 0 & 0 & 0 & \tilde{p}_{11} & \tilde{p}_{12} & \tilde{p}_{13} & 0 & 0 & 0 \\ 0 & 0 & 0 & 0 & 0 & 0 & \tilde{p}_{11} & \tilde{p}_{12} & \tilde{p}_{13} \\ \tilde{p}_{21} & \tilde{p}_{22} & \tilde{p}_{23} & 0 & 0 & 0 & 0 & 0 & 0 \\ 0 & 0 & 0 & \tilde{p}_{21} & \tilde{p}_{22} & \tilde{p}_{23} & 0 & 0 & 0 \\ 0 & 0 & 0 & 0 & 0 & 0 & \tilde{p}_{21} & \tilde{p}_{22} & \tilde{p}_{23} \\ \tilde{p}_{31} & \tilde{p}_{32} & \tilde{p}_{33} & 0 & 0 & 0 & 0 & 0 & 0 \\ 0 & 0 & 0 & \tilde{p}_{31} & \tilde{p}_{32} & \tilde{p}_{33} & 0 & 0 & 0 \\ 0 & 0 & 0 & 0 & 0 & 0 & \tilde{p}_{31} & \tilde{p}_{32} & \tilde{p}_{33} \end{bmatrix} \begin{bmatrix} b_{11} \\ b_{12} \\ b_{13} \\ b_{21} \\ b_{22} \\ b_{23} \\ b_{31} \\ b_{32} \\ b_{33} \end{bmatrix} = \begin{bmatrix} p_{11} \\ p_{12} \\ p_{13} \\ p_{21} \\ p_{22} \\ p_{23} \\ p_{31} \\ p_{32} \\ p_{33} \end{bmatrix}, \tag{12.12}$$

(a) (b)

Figure 12.11 'Airplane' warped with the generic transformations: (a) sinus modulated horizontal variable; (b) sinus modulated horizontal and vertical variables. (Source: USC-SIPI Image Database)

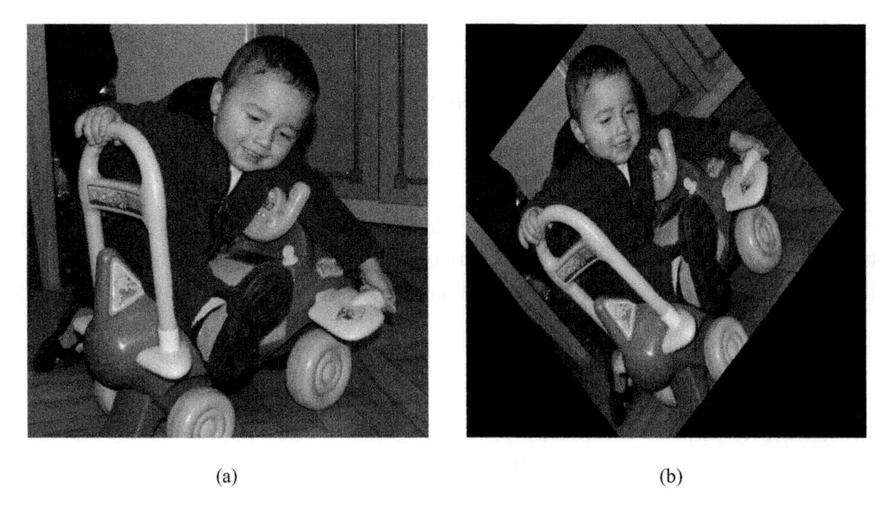

(a) (b)

Figure 12.12 'Kamil' image warped with affine transformations: (a) the original RGB image (colour version in Plate 7); (b) the output image after the affine transformation consisting of a -43° rotation around a centre point, scaling by [0.7, 0.8] and translation by the [155, 0] vector

where p_{ij} are components of points and b_{ij} are elements of **B**. This can be written also in a shorter form:

$$\tilde{\mathbf{P}}\mathbf{b} = \mathbf{P}, \tag{12.13}$$

where $\mathbf{b}_{9\times1}$ is a vector containing aligned elements of **B**, and $\tilde{\mathbf{P}}_{9\times9}$ and $\mathbf{P}_{9\times1}$ are given in (12.12).

In this section we briefly recall how to solve a linear system of equations, especially if the number of equations is greater than the number of unknowns, i.e. for the so-called overdetermined system of linear equations [259, 352]. In image warping this happens, for example, for the polynomial transformation with more control points specified than the number of polynomial coefficients.

The linear system of equations is given as

$$\mathbf{Ax} = \mathbf{B}, \tag{12.14}$$

where $\mathbf{A}_{\mathrm{M}\times\mathrm{N}}$ is a matrix of coefficients, $\mathbf{x}_{\mathrm{N}\times1}$ is a matrix (vector) of unknowns and $\mathbf{B}_{\mathrm{M}\times1}$ is a matrix (vector) of coefficients. The system (12.14) can be uniquely solved if M = N and **A** is not singular. In the case M > N (more equations than unknowns), the system (12.14) is overdetermined, and, in general, it has no solution since **B** is an M size vector and as such does not belong to the range(A), which is at most of N-th order. Therefore the above problem has to be reformulated, and instead of (12.14) we seek a solution to the equation

$$\mathbf{R} = \mathbf{B} - \mathbf{Ax} \tag{12.15}$$

that minimizes coefficients of $\mathbf{R}_{M\times 1}$. For the purpose of minimization we need to choose a certain norm on the vector. The most natural (or intuitive) choice is the L_2 norm (i.e. the Euclidean distance), for which the minimization problem (12.15) is expressed as follows:

$$\min_{\mathbf{x}} \|\mathbf{B} - \mathbf{Ax}\|_{L_2} . \tag{12.16}$$

A vector $\mathbf{x}$ which solves (12.16) under L_2 is the closest vector in the range($\mathbf{A}$) to the vector $\mathbf{B}$.

It can be proved that the solution to (12.16) is given by such a vector $\mathbf{x}_0$ for which the residual $\mathbf{R}$ is orthogonal to the rank($\mathbf{A}$). To find $\mathbf{x}_0$ let us build the following functional $E(\mathbf{x})$:

$$E(\mathbf{x}) = \|\mathbf{B} - \mathbf{Ax}\|_{L_2} , \tag{12.17}$$

which for the norm L_2 can be expressed as

$$E(\mathbf{x}) = (\mathbf{B} - \mathbf{Ax})^T (\mathbf{B} - \mathbf{Ax}) . \tag{12.18}$$

Differentiating the above equation with respect to $\mathbf{x}$ we obtain the so-called normal equation:

$$\begin{aligned} \frac{\mathrm{d}}{\mathrm{d}\mathbf{x}} E(\mathbf{x}) &= \frac{\mathrm{d}}{\mathrm{d}\mathbf{x}} \left(\mathbf{B}^{\mathrm{T}}\mathbf{B} - \mathbf{x}^{\mathrm{T}}\mathbf{A}^{\mathrm{T}}\mathbf{B} - \mathbf{B}^{\mathrm{T}}\mathbf{Ax} + \mathbf{x}^{\mathrm{T}}\mathbf{A}^{\mathrm{T}}\mathbf{Ax}\right) \\ &= -\mathbf{A}^{\mathrm{T}}\mathbf{B} - \mathbf{A}^{\mathrm{T}}\mathbf{B} + 2\mathbf{AA}^{\mathrm{T}}\mathbf{x} = -2\mathbf{A}^{\mathrm{T}} (\mathbf{B} - \mathbf{Ax}) . \end{aligned} \tag{12.19}$$

Then equating to zero to find its extreme point $\mathbf{x}_0$, we obtain

$$\mathbf{A}^{\mathrm{T}} (\mathbf{B} - \mathbf{Ax}_0) = 0. \tag{12.20}$$

Assuming that $\mathbf{A}^{\mathrm{T}}\mathbf{A}$ is nonsingular, the unique solution to the above is given by

$$\mathbf{x}_0 = \left(\mathbf{A}^*\mathbf{A}\right)^{-1} \mathbf{A}^*\mathbf{B}, \tag{12.21}$$

where $\mathbf{A}^*$ is a conjugate matrix to the matrix $\mathbf{A}$, and $\mathbf{A}^+ = (\mathbf{A}^*\mathbf{A})^{-1}\mathbf{A}^*$ is called a pseudoinverse of $\mathbf{A}$.

12.8.1 Linear Algebra on Images

A simple image is represented in a computer memory as a 2D array of discrete values, pixels. The same type of structure is used in mathematics to represent matrices, arrays of data. Thus, having defined a structure type for image representation (see section 3.7.1.2) we are granted a matrix representation as well, and vice versa. We have even more thanks to the recursive properties added by the C++ templates mechanism which in the case of images are used mostly to provide the type of pixel. With this technique, however, there is no obstacle to provide *any* pixel type which will be stored in an image, even another image, which has pixels of a certain type, and so on. This way we define multiply indexed structures which can be used to represent tensors (Chapter 10).

The accompanying software package has been endowed with a set of template functions for basic matrix operations such as multiplication, inverse and pseudo inverse, as well as a

```
/////////////////////////////////////////////////////////////
// This function returns a new matrix which is the
// result of multiplication of the operands a and b.
/////////////////////////////////////////////////////////////
//
// INPUT:
//      a - reference to the first matrix
//      b - reference to the second matrix
//
// OUTPUT:
//      a * b if possible
//      0 otherwise
//
// REMARKS:
//      Template parameter T stands for pixel type;
//      Template parameter D stands for intermediate
//         results accumulator;
//
//      The caller is responsible for disposing of
//      the returned object!!!
//
template< typename T, typename D/* = double*/ >
TImageFor< T > * Orphan_Mult_Matrix(      const TImageFor< T > & a,
                                          const TImageFor< T > & b);
```

Algorithm 12.4 *Orphan_Mult_Matrix* template function for multiplication of a matrix **a** times **b**. The matrices are represented as images with pixels of type T. All intermediate results are stored in variables of type D which can be the same or different from T. (Reproduced by permission of Pandora Int. Inc., London)

solution of the set of linear equations. Their declarators are presented in Algorithms 12.4–12.8. The functions can be used for instance to find point transformation matrices given by equations (12.14) and (12.21), discussed in the previous section.

The input and output parameters are in the form of image template classes *TImageFor<T>* where T denotes the type of pixel, i.e. type of element of the matrix.

```
/////////////////////////////////////////////////////////////
// This function returns a new matrix which is the
// inverted matrix of the input one (if possible).
/////////////////////////////////////////////////////////////
//
// INPUT:
//       in_data - reference to the first matrix
//
// OUTPUT:
//       inverse matrix of in_data, if possible
//       0 otherwise
//
// REMARKS:
//       Template parameter T stands for pixel type;
//       Template parameter D stands for intermediate
//          results accumulator;
//
//       The caller is responsible for disposing of
//       the returned object!!!
//
template < class T, class D /*= double*/ >
TImageFor< T > * Orphan_Inv_Matrix( const TImageFor< T > & in_data );
```

Algorithm 12.5 *Orphan_Inv_Matrix* template function which returns an inverse of the input matrix if such exists. The matrices are represented as images with pixels of type T. All intermediate results are stored in variables of type D which can be the same as or different from T. (Reproduced by permission of Pandora Int. Inc., London)

```
///////////////////////////////////////////////////////////////
// This function returns the conjugate (i.e. transposed)
// matrix.
///////////////////////////////////////////////////////////////
//
// INPUT:
//        in_matrix - the input matrix (= an image);
//           its value is only read
//
// OUTPUT:
//        a pointer to the orphaned transposed matrix if possible,
//        0 otherwise
//
// REMARKS:
//        The returned object is to be disposed of
//        by a caller (i.e. it is orphaned)
//
//        Actually it is not a proper conjugate, only transposed.
//
template < class T, class D /*= double*/ >
TImageFor<T> * Orphan_Conjugate_Matrix( const TImageFor< T > & in_data );
```

Algorithm 12.6 *Orphan_Conjugate_Matrix* template function which returns a conjugate of the input matrix. The matrices are represented as images with pixels of type T. All intermediate results are stored in variables of type D which can be the same as or different from T. (Reproduced by permission of Pandora Int. Inc., London)

There is also a second template parameter D which defines a type used for all intermediate results, such as summation of products, etc. This one should be chosen to allow desirable precision of the computations. In practice, the simplest choice is a built-in type *double* or *float* which represents numbers with the floating point format [258]. However, a fixed format can also be employed. Unfortunately, most of the modern programming languages do not provide a suitable type for such format. For this purpose the *FixedFor<>* template class has been added. More often than not it allows compact representation (smaller memory usage) of

```
///////////////////////////////////////////////////////////////
// This function returns the Penrose-Moore pseudo-inverse
// matrix (a matrix is an image):
//                       -1
//    A~ = ( A* x A )     x A*
//
///////////////////////////////////////////////////////////////
//
// INPUT:
//        in_matrix - the input matrix (= an image);
//           its value is only read
//
// OUTPUT:
//        a pointer to the orphaned pseudo-inverse matrix if possible,
//        0 otherwise
//
// REMARKS:
//        The returned object is to be disposed of
//        by a caller (i.e. it is orphaned)
//
template < class T, class D /*= double*/ >
TImageFor<T> * Orphan_PseudoInv_Matrix( const TImageFor<T> & in_matrix );
```

Algorithm 12.7 *Orphan_PseudoInv_Matrix* template function which returns a Moore–Penrose pseudo inverse if it exists. The matrices are represented as images with pixels of type T. All intermediate results are stored in variables of type D which can be the same as or different from T. (Reproduced by permission of Pandora Int. Inc., London)

```
//////////////////////////////////////////////////////////////
// This function returns the solution to the linear
// system of linear equations in the form:
//
//            Ax=B
//
// if such exists.
//////////////////////////////////////////////////////////////
//
// INPUT:
//        A - the input matrix
//        B - the input matrix (a vector)
//
// OUTPUT:
//        a pointer to the orphaned solution vector x,
//        0 otherwise
//
// REMARKS:
//        The function implements partial pivoting.
//
//        The returned object is to be disposed of
//        by a caller (i.e. it is orphaned)
//
template < class T, class D /*= double*/ >
TImageFor< T > * Orphan_Linear_Solution( const TImageFor< T > & A, const
TImageFor< T > & B );
```

Algorithm 12.8 *Orphan_Linear_Solution* template function which solves the set of linear equations $\mathbf{Ax} = \mathbf{B}$, if solution exists. The matrices are represented as images with pixels of type T. All intermediate results are stored in variables of type D which can be the same as or different from T. (Reproduced by permission of Pandora Int. Inc., London)

pixels or intermediate results compared to the floating point representation. It also allows more precise computations in a predefined dynamic range, however. Thus its application should be preceded by an analysis of the required dynamics of data to avoid overflow problems.

The template functions in C++ do not allow default template parameters. Therefore D has to be provided explicitly. An option is to put a template function into a template class which does not possess such restrictions (e.g. see definitions of the *TImageFor<>* or *_Convolve<>* for convolution).

The result of a matrix computation is returned as a pointer to a separate image object or 0 if computations cannot be finished for some reason (for instance a matrix was singular when trying to find its inverse, etc.). It should be remembered that this object is 'orphaned' which means that a caller is responsible for its disposal when the object is not used any more. If failing to do so, memory leaks will occur. A possible modification of this behaviour would be to employ the *std::auto_ptr<>* passed by value.

Complete implementations of the above are included in the accompanying software library [216].

12.9 Closure

This chapter is devoted to the problem of changing image geometry, called image warping. For this task, point mapping as well as image resampling are required. For the former, the methods of finding an affine transformation from point correspondences are presented. For the latter, the interpolation scheme is discussed. Finally, the object-oriented implementation of a simple warping software is also presented.

12.9.1 Further Reading

More information on the warping methods can be found in the book by Wolberg [449] or in many scientific papers such as the one by Zokai and Wolberg [465].

13

Programming Techniques for Image Processing and Computer Vision

13.1 Abstract

Image processing and vision by a computer are very demanding areas of computational science. The obvious initial observation is the amount of data which has to be processed in a limited time. But not only is the size of the input a problem here. Development of image processing methods and algorithms as well as their efficient implementations are the real challenges here.

When programming for engineering systems it can be useful to know the basic constructions, idioms and design patterns, so in the design stage we can catch most of them emerging in our project. It is the strength and beauty of the patterns that they appear to be ubiquitous in almost every system, regardless of its particular destination and application. Knowing their features we can make a more conscious choice. Such a strategy leads usually to more modular – or object aware – designs which are easier to comprehend and then to implement and maintain.

In-depth understanding of the subject being implemented and then tracing the execution steps in the run time is a must for proper building of any software system. Even educated guesses, vague assumptions or clutter in the code almost always lead to faulty designs. Unfortunately, the reverse does not unconditionally guarantee success. Nevertheless, it can move us closer to a desired solution.

In this chapter we provide a number of programming concepts explained by simple examples rather than by formal definitions. This learning-by-example method has been shown to be very efficient in practice, since all we need to do is understand and remember those examples, which is much easier. However, we assume a basic knowledge of C++ and object-oriented concepts, such as classes, objects, relations "has-a" and "is-a", basics behind class templates, etc.

An Introduction to 3D Computer Vision Techniques and Algorithms Bogusław Cyganek and J. Paul Siebert

13.2 Useful Techniques and Methodology

In this section we discuss a number of useful programming tips and techniques that, if properly used, can enhance the clarity of a design, as well as lead to improvement of the code quality and help achieve a correct solution.

13.2.1 Design and Implementation

When developing software we have to remember that programs are written primarily for people, then for computers. Computers do not need structural or object-oriented methods. For a computer a series of numbers denoting machine operations and data is what it needs to run a program. The better the quality of design and implementation, the better are the results of a program and the more the possibilities of reuse of it as a whole or its parts.

In this section we briefly discuss some practical techniques for making design and code more understandable to the creator and other programmers, such as commenting, naming conventions, as well as modelling conventions for better expression of concepts.

13.2.1.1 Comments and Descriptions of 'Ideas'

Each programming language is endowed with means of expressing some information which are not commands for execution by a computer. Instead, their role is to facilitate description of the ideas and concepts behind the code and they are intended for people. Interestingly enough, even a person creating a code finds his or her comments placed along the code useful after a week, a month or maybe a couple of years.

From a practical point of view, the most useful are brief descriptions of the main or basic ideas behind part of a code, a function or a class. Writing what the code does in each of its lines is sometimes useful, but most important are descriptions of the general ideas, algorithms and concepts.

Each programmer should develop his or her own style of writing code and comments on it. Algorithm 13.1 presents an example of a simple comment tag that can be placed at the beginning of each function in the code. The same can be used for classes, namespaces, components, etc.

```
//////////////////////////////////////////////////////////////////
// This function [put here main purpose of a function]
//////////////////////////////////////////////////////////////////
//
// INPUT:
//          [put description of the input parameters]
//
// OUTPUT:
//          [put description of the output parameters]
//
// REMARKS:
//          [place additional information on the function
//           or its calling conventions, etc. ]
//
```

Algorithm 13.1 A way of commenting on a function

Table 13.1 Naming conventions that increase code readability

Prefix	Application	Examples
A	Template parameters	*ADoubleTrait*
E	*enum* data in classes	*EErrorCode*
f	Data members of a class (fields)	*fErrorCode*
g	Static data that are not constant	*gRegularizationSpan*
k	Constant data (also *enum*)	*kPolyMatrixCols*
M	Mixin classes	*MPointMarker*
T	Base and standard classes	*TPixelInterpolation*
V	Virtual base classes	*VInOut*

13.2.1.2 Naming Conventions

After proper comments, naming conventions can greatly help to understand code. There are a number of rules which regulate how different groups of commands, data types, etc., are named. If used systematically, they allow understanding of the roles of particular identifiers only from their names, not referring to their context. This speeds up code analysis and adds to code readability.

Everybody can develop such naming conventions. However, it is a good idea to use some which are used by other people or groups. By this we will understand each other better. In the procedures presented in this book we adopt the notation conventions developed by Taligent® [414]. The most useful are presented in Table 13.1.

A useful hint when developing names for identifiers (variables, constants, functions, classes, etc.) is to put a concise but *informative* name. It can be composed of a number of words connected by the underscore or each starting with a capital letter. Use abbreviations sparingly, however. For instance, implementing a counter for rows in an image, instead of writing

```
int     tmp1;
```

place

```
int     row_counter;
```

or

```
int     rowCounter;
```

Almost all classes contain a number of members to set or get its private or protected data. These are commonly called setters and getters. Hence, it is common practice to start their names with *Set*... and *Get*... prefix, respectively.

Another group constitute the methods that create, orphan, copy or adopt some objects. It is recommended to start their names using these prefixes, i.e. *Create*..., *Orphan*..., *Copy*... and *Adopt*..., respectively. Semantics of these functions is discussed in section 13.4.

13.2.1.3 Unified Modelling Language (UML)

Unified Modelling Language (UML) is a set of rules for visualization of different facets of engineering design. These are described for instance in the book by Booch *et al.* [48] which has about 500 pages. In this section we briefly outline only the very basic concepts of UML, especially in the context of design of computer vision systems.

The basic concept of UML is a diagram. The main diagrams are named and briefly explained in the following list. More information on the subject can be found in Booch *et al.* [48].

1. Use case diagram – shows relationships between so-called *actors* and *use cases* in a system. These are used to model the behaviour of a system, subsystem or a class. Each presents use cases with actors, i.e. participating entities, as well as relations among them. More frequently than not they are used to present:
   - an environment of the system;
   - requirements of the system.

   The following depicts an exemplary case diagram of a vision system for face recognition. It models the requirements of that system. There are two actors: administrator (ADMIN) and user (USER). Use cases are placed in the adjacent rectangle. These are: system setup, image acquisition, filtering, face detection and face recognition.

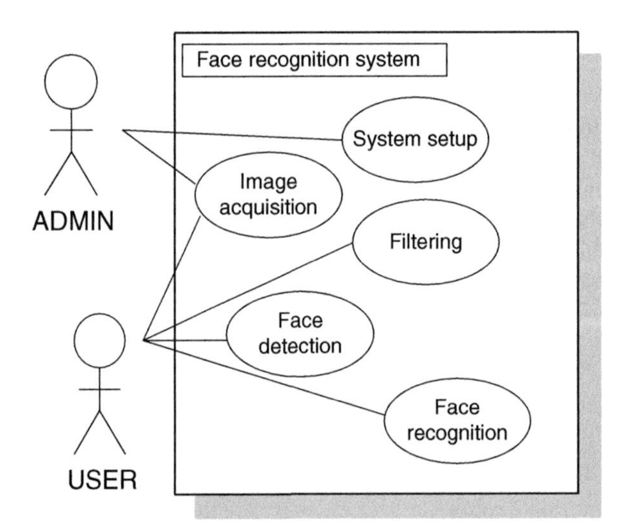

2. Activity diagram – models a *procedural flow* (from an action to an action) with *behaviour* in a system. Activity diagrams are used to model *dynamics* of a system. Usually it is a flow chart of sequential, but sometimes also parallel, computational steps required to fulfil a given task.

   Below, an activity diagram is depicted of a simple system for road sign recognition. The oval boxes denote single action steps. Synchronization or split of actions is depicted with thick bars. There are two parallel recognition stages. The first one (left branch) does figure detection followed by sign recognition. The second one (right branch) does template matching in the log-polar space. Thus it can directly recognize a sign from an input image. If the two branches give unanimous answers, then a recognized sign is reported to a user. Otherwise the process is started again.

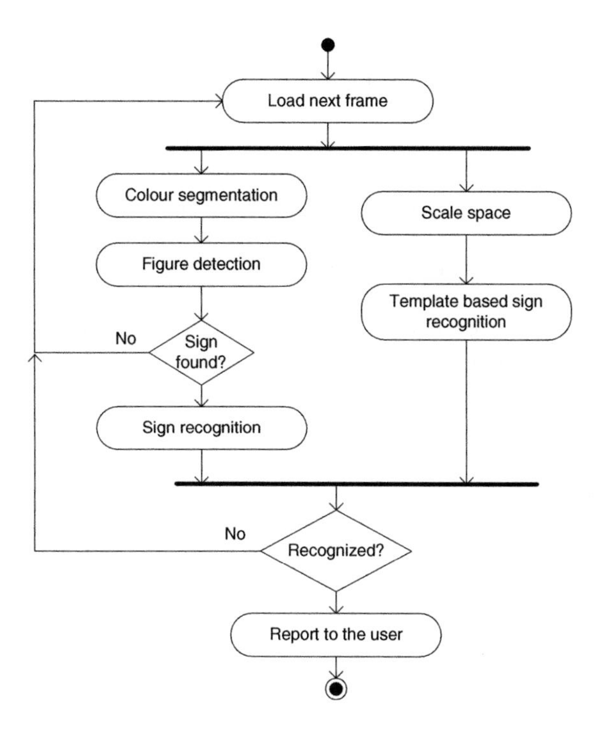

3. Interaction diagram – depicts patterns of *interaction* between objects in a time sequence. It also models *dynamic* aspects of a system. The key factors are: participating actors (objects, components, etc.), message sequence and time. Thus, this type of diagram should be used to model time dependencies among messages sent by participating objects.

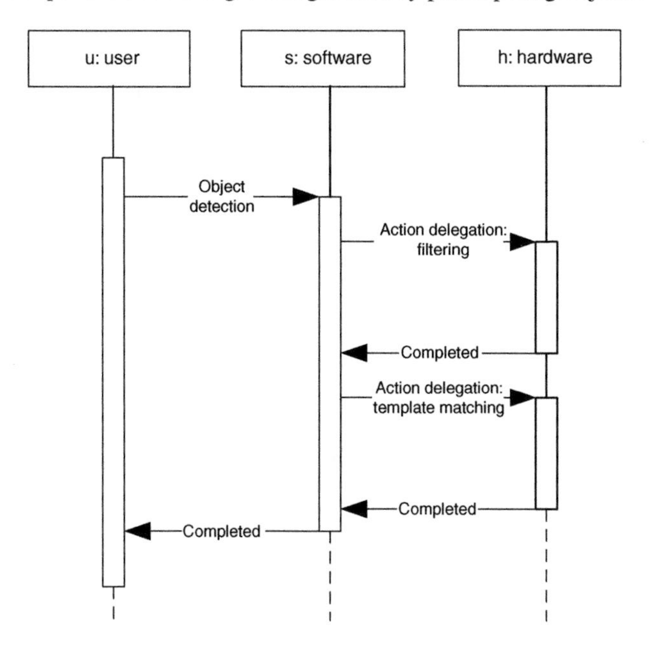

In the example there are three participants: user (u), software (s) and hardware (h). A message sent from a user results in a series of messages passed from software to hardware and vice versa. Upon completion of all of them, a final message is given back to the user.

4. State transition diagram – models *states* and *transitions* that show the response of a system to some excitations. State diagram models dynamic aspects of a system with states and transitions. These are elements of the state machines which can be of Mealy or Moore type. In the former, each next action is governed by the current state *and* the values of the current input. In the latter, actions depend only on the current state. There can be state machines that mix the two approaches as well.

   There are two specific states (shown below): a start (a single dot) and a stop (a double dot). The next state is determined from the current state and transition.

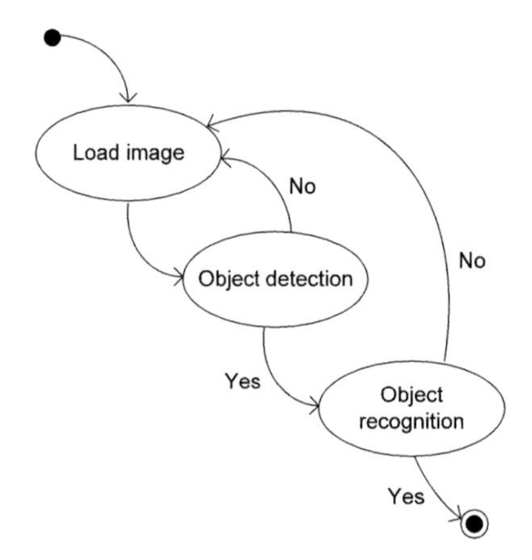

5. Component diagram – catches *relations* between software and external components.

   In the diagram below a relation among three components is visualized. The components are encapsulated self-contained programming entities, such as classes, packages, small programs, etc. In the provided example these are: user application, system libraries and image library. All have to be connected in some way on behalf of the user's application.

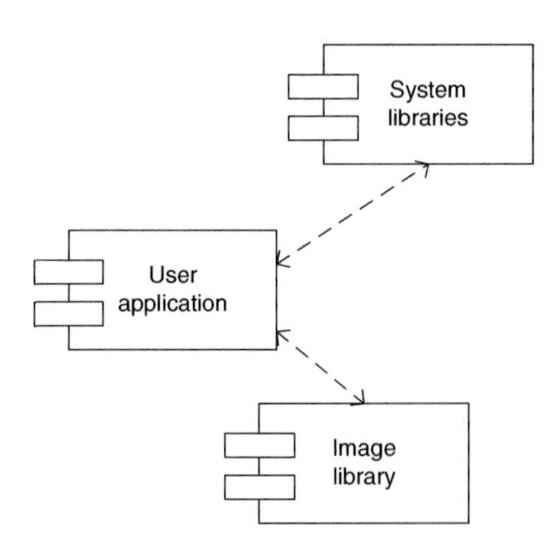

6. Deployment diagram – shows *deployment* and/or set up of components on processors and devices. Apart from visualizing components and their connections, the deployment diagram takes into consideration devices, such as microprocessors, computers, etc., in which components are installed. These devices are called nodes.

   In the diagram below there are two nodes, each with its own components. The nodes in this example are connected by the PCI Express connection.

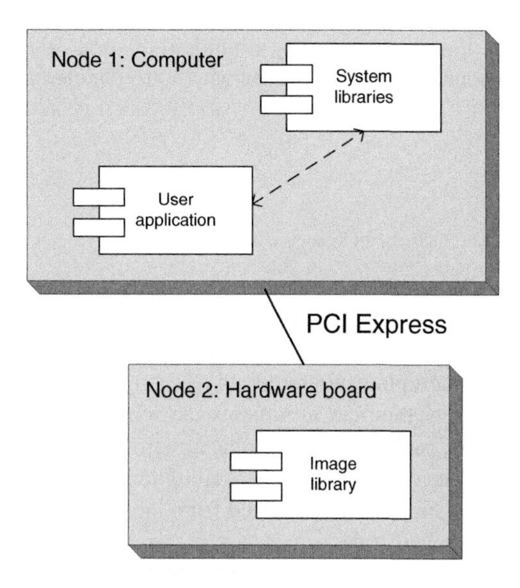

7. Class diagram – depicts the static *structure*, *relationships* and internal structure of objects. These are the most frequently used diagrams in this book. Many examples of real class hierarchies have been presented in previous chapters. The most basic is a concept of a class with its components. A class can contain attributes (data members) and operations

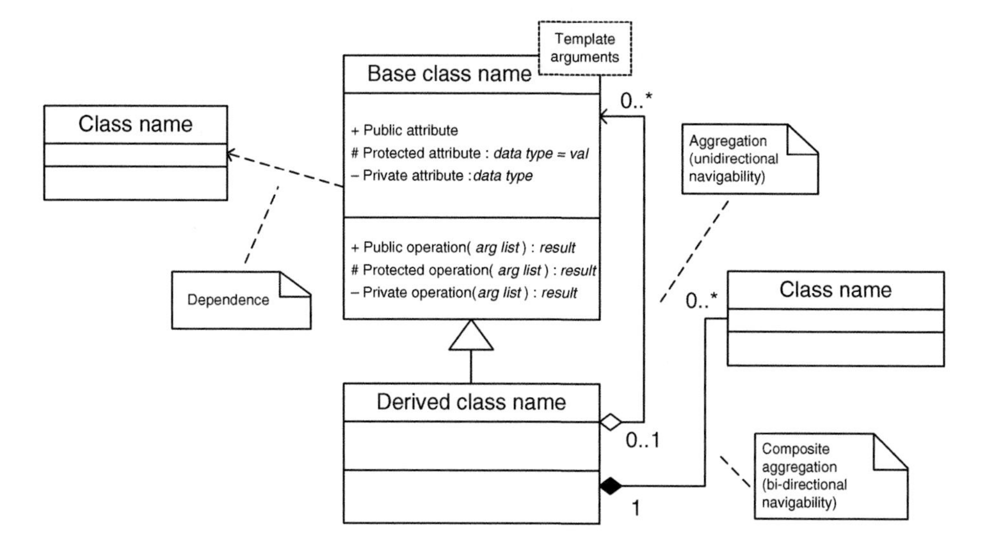

(function members). Each of them can belong to the public, protected or private section of a class. Public members are accessible to all users of an object of that class, whereas usage of protected members is restricted exclusively to this class and all publicly or protected derived classes from it. These restrictions are modelled by the +, # and – signs, respectively. A small triangle depicts a class derived from its base. The empty diamond denotes an aggregation which means that a class owns some other objects which are not necessarily members of that class. The filled diamond means a composite aggregation which indicates that an object of a class is actually composed of some other objects. There can be also some loose dependencies among classes which are denoted by a dashed line with an arrow.

13.2.2 Template Classes

Templates in C++ allow writing classes, for which some parameters are provided later, during template instantialization. This is called generic programming, that is, writing universal code for types which we might not even know. An example is the *TImageFor*< > class (Algorithm 13.8) which accepts pixel type as its template parameter. Pixels can be any objects which have a default constructor. Apart from types, template arguments can convey constant values which can be used in class instantialization, such as size of a static array or an initialization value for a variable. It is not our purpose to discuss all aspects of writing template classes – for such a discussion see [401, 434]. However, much can be learned from the examples provided in this chapter and from analysis of the attached code.

However, there is one particular construction related to template classes which is of special importance. This is template specialization which allows a kind of a break from a general definition of a template class and its implementation for specific type which for some reason should be treated differently. Algorithm 13.2 presents an example of a specialized template class. It is a *TImageFor*< *bool* >, i.e. a definition of an image for which pixels are bit values. Certainly, the general template *TImageFor*< > could be easily instantiated with the bool argument, but in this case we would sacrifice a whole byte for a single bit which in the case of images means significant waste of space. However, to access single bits in a computer word we need to change the implementation. Thus, *TImageFor*< *bool* > has different constructors and other members from the general class *TImageFor*< >, e.g. *SetPixel*( ) and *GetPixel*( ).

```
/////////////////////////////////////////////////////////////////
// Specialization for binary images.
// Its internal structures are organized in such a way as to
// save on space and access time.
/////////////////////////////////////////////////////////////////
template<> class TImageFor< bool >
{
   // Definitions specific to images with bit pixels
   // ...
};
```

Algorithm 13.2 Specialization of the template class *TImageFor*<> for pixels being bit values '*true*' or '*false*'

13.2.2.1 Expression Templates

The expression template is a technique of expressions encoded into template arguments [434]. Evaluation of an expression is postponed until the whole expression is created in a form of a compound template parameter. Such an approach allows for precise expression parsing and more efficient execution. Instead of single operations at a time, the most efficient way can be chosen. For example, if **a** and **b** are arrays of hundreds of real values (a double type), then for an expression[1]

$$\mathbf{a} = 11.3^*\mathbf{a} + \mathbf{b}^*\mathbf{a},$$

the naive solution would compute each component separately. This would result in

$$\begin{aligned}\mathbf{tmp\_1} &= 11.3^*\mathbf{a}\\ \mathbf{tmp\_2} &= \mathbf{b}^*\mathbf{a}\\ \mathbf{tmp\_3} &= \mathbf{tmp\_1} + \mathbf{tmp\_2}\end{aligned}$$

Then the final assignment is performed:

$$\mathbf{a} = \mathbf{tmp\_3}$$

However, a more efficient solution is to rewrite the above expression as follows:

$$\mathbf{a}^{*}= (11.3 + \mathbf{b})$$

and then perform computations starting from this representation.

The expression templates allow efficient evaluation of expressions. This method has many similarities to the other template technique, called mataprogramming. In the latter, some parameters are evaluated already at compilation time, during the template instantiation phase.

However, there are certain inherent limitations of this technique. For example it does not work for matrix vector multiplications, such as the following:

$$\mathbf{b} = \mathbf{a}^*\mathbf{b},$$

where $\mathbf{a}_{2\times 2}$ and $\mathbf{b}_{2\times 1}$. The problem comes from the fact that a temporary object should be created to store an intermediate result since the result which goes to **b** at the same time depends on each element of the input parameter which happened also to be **b**. Such situations promote creation of a run-time structure that represents the expression tree instead of encoding the tree in the type of expression template.

[1] We assume an element-by-element multiplication here.

13.2.3 Asserting Code Correctness

When designing programs one of the most important aspects is their correctness. This can be breached by many types of errors, however. The most common are due to simple programming bugs or code fragments unprotected against invalid input data. The situation gets even worse if the problems result from wrong design or simple misunderstanding of the subject.

However, countermeasures can be undertaken from the beginning of the design process to help write code which does its job, does not crash and does not cause memory leaks, at least.

The first thought is to analyse and *understand the problem* before we start writing a code. It is recommended to use *a top-down* approach or, in other words, *a divide and conquer strategy*. Each superior task should be divided into smaller tasks. Then work on each task separately having the same divide and conquer rule in mind, and so on. However, do not forget about *common interaction* among the modules.

13.2.3.1 Programming by Contract

To ensure the correctness of a program a good approach is to use the programming by contract technique [305]. This means treating a software procedure as a kind of a business contract which should have its pre- and post-conditions. Apart from this we have invariants, i.e. rules which should be true at whatever step of execution a program is. Thus, each software module, component, class, method or even code block can have its own pre- and post-conditions, as well as invariants. In the simplest, but very useful, approach all of them are called *code correctness assertions* or *requirements*. For proper operation of a software component all its assertions should be met during execution, in their *true* state, before and after execution of this component. Otherwise we say that an assertion was *fired*. What these are in a program are simply fragments of code, usually active only in its debug version, which check consistency of data or conditions which a programmer thinks should always be true. However, this is a different mechanism from, for example, checking whether input data is correct, although the two can be applied together.

A practical method to implement pre- and post-conditions with invariants is to implement a kind of a preprocessor macro command which checks a Boolean condition and if it evaluates to 'false' a message is displayed or other form of information issued to the programmer (e.g. it can be an entry in a log file, etc.). Algorithm 13.3 presents an exemplary implementation of the *REQUIRE* macro that accepts a Boolean expression. If the Boolean condition does not evaluate to 'true' then a message is launched, an example of which is presented in Figure 13.1. The *REQUIRE* macro can be used to insert assertions in a debug version of the code. In a 'release' version, *REQUIRE* is usually translated to an empty statement, however.

Finally, a version of the code is controlled by another flag, usually set by a tool used for program development. In the Microsoft Visual® C++ environment this is a *_DEBUG* flag which controls a multiplatform *DEBUGGING* flag (Algorithm 13.3).

Usually it is problematic as to what to do if *REQUIRE* fails. In Algorithm 13.3 a user is given three options which are unconditionally abort execution, stop and allow debugging at code fragment, or ignore and launch execution of further statements. Implementation provided in Algorithm 13.3 assumes the Windows® operating system. However, it can be easily changed to other platforms, by simply exchanging the *MessageBox*() function to its counterpart in another system. In some critical applications and if assertions do not slow the

```
// The _DEBUG flag is set by the Microsoft® Visual C++
// For other platforms change the flag to appropriate.
#if _DEBUG
   #define DEBUGGING 1
#else
   #define DEBUGGING 0
#endif

//////////////////////////////////////////////////////////////////////////

#if
      DEBUGGING
   #define _QUOTE(x) #x
   #define QUOTE(x) _QUOTE(x)

   // on REQUIRE violation exit if not IGNORE from the user
   #define REQUIRE(expr)       {                            \
      if( ! (expr) )                                        \
      {                                                     \
         int m;                                             \
         if( (m = MessageBox(NULL,                          \
            #expr "\n\n" "IN FILE: " __FILE__               \
            "\n" "IN LINE: " QUOTE(__LINE__),               \
            "HIL\'s REQUIRE doesn\'t hold",                 \
            MB_ICONSTOP | MB_ABORTRETRYIGNORE))==IDABORT )  \
               ExitProcess((UINT)-1);                       \
         else                                               \
            if( m == IDRETRY )                              \
               DebugBreak();                                \
      }                                                     \
   }

#else  // DEBUGGING

   #define REQUIRE(expr)    ;

#endif   // DEBUGGING
```

Algorithm 13.3 Definition of the *REQUIRE* macro for assertions in debug mode. Code version for Windows®

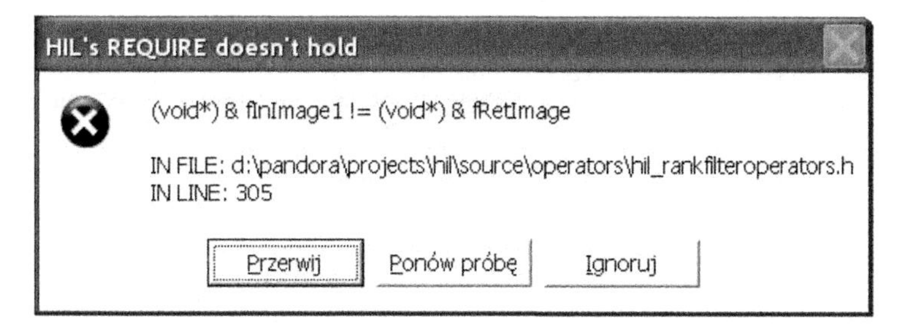

Figure 13.1 Windows® dialog launched after detecting false condition in a *REQUIRE*

```
//////////////////////////////////////////////////////////////
// This function initializes an image
//////////////////////////////////////////////////////////////
//
// INPUT:
//        col - number of columns
//        row - number of rows
//
// OUTPUT:
//        none
//
// REMARKS:
//        Memory for data is allocated but
//        data is NOT initialized.
//
void Create( Dimension col, Dimension row  )
{
    REQUIRE( col > 0 );              // pre-condtions
    REQUIRE( row > 0 );

    fRow = row;
    fCol = col;

    fElems = row * col;
    fData = (T*) new T[ fElems ];
    REQUIRE( fData != 0 );            // post-condition
}
```

Algorithm 13.4 Example of assertions with *REQUIRE* to check for input parameters (pre-conditions) and valid memory allocation (post-conditions) in a procedure creating a new image

computations too much, they can be left in the 'release' version of the program, however with no user dialog and with unconditional execution–abort exchanged into throwing an exception of proper type.

An even simpler solution would call a dummy function in which a breakpoint can be set. Thus, any false requirement would result in a stopped execution by this breakpoint.

Algorithm 13.4 presents an example of the *REQUIRE* macro used in a method creating a new image. It is placed at the beginning and at the end of the method, thus implementing pre- and post-conditions. The first one checks whether the declared size of an image is greater than 0, otherwise a logical error is encountered. The post-condition ensures that a memory has been really allocated for an image. If there is not enough memory space, then this value might result in a zero pointer for data allocation which means that an object has not been initialized as we expected.

More information on defensive programming and many other techniques improving code quality can be found, for instance, in the excellent books by McConnell [305] and by Stroustrup [401].

13.2.4 Debugging Issues

When developing a system a good strategy is not to postpone debugging until all components are finalized and connected together. Instead, their debugging should be performed parallel with the development of different components. Then the whole system has also to be checked when all its modules are assembled together. It is impossible to check all possible cases of execution or input data. However, some simple rules can save us from the majority of simple but sometimes very dangerous software malfunctions. One such rule postulates code debugging as soon as possible.

This concept can be facilitated by the object-oriented paradigm of a self-contained class and class encapsulation. Each class should be designed in a way which results in a clear implementation with a well-defined state and interface. Its dependence on the other objects should also be well specified. This creates a kind of a 'constraint space' which can be checked separately. Just after implementation a class should be debugged by its programmer. Such a strategy has good practical reasons since if debugging is postponed then some details have a tendency to become vague, which makes testing even more difficult. Then, if possible, a software component should also be checked by another person.

When testing computer vision systems the main problem is the size of the input data. For example, usually it is not possible to check processing of a procedure for each pixel in a video stream. Instead, a test pattern can be created for which a result can easily be predicted from an algorithm. Sometimes a very useful strategy is to create so-called 'border patterns', i.e. input data examples for the specific start or stop conditions, such as all pixels black, white or a fine grid, etc.

When dealing with iterative procedures it is also very important to check their stop conditions. If it is not possible to check all possible cases then an additional counter with a preset limit of iterations can be of help.

13.3 Design Patterns

Design patterns are specific engineering constructions which exhibit *similar* behaviour even when operating in different applications. They were first observed in architecture, and then were adopted to the realm of software development. There are many types of such software design patterns which can be further classified into specific categories such as creational, structural and behavioural patterns [144].

Design patterns are not a recipe for all problems encountered in software design. However, if used properly they can help greatly in this process by discovering characteristic and common system constructions which can be thought of and implemented in a unified way. Such a strategy usually leads to a more comprehensible design, reusable components and a more readable code. The following sections provide basic information on design patterns frequently encountered in computer vision software.

13.3.1 Template Function Objects

The function objects technique is a very powerful extension to the 'ordinary' functions concepts encountered in all programming languages. There are many varieties of the 'function object' (or functor) term, however [3, 401, 441]. Nevertheless, the main virtue of this technique comes from the fact that function objects can store a state. This is evident especially when compared to the simple function pointers or member pointers [3, 401]. They can also be easily extended in derived classes, as well as passed as arguments to other functors and template parameters. The basic concept of the function object can be explained based on the class in Algorithm 13.5.

```
1  template < class T >
2  class ExemplaryFunctor
3  {
4      protected:
5          T fState;
6      public:
7          ExemplaryFunctor( T & state ) : fState( state ) {}
8      public:
9          virtual void operator() ( void );   // uses fState
10 };
```

Algorithm 13.5 Exemplary functor class

The *ExemplaryFunctor<>* template class defines a family of classes that are differentiated by a template parameter T. There are two distinctive phases of its run-time behaviour.

1. Creation time – the *state* variable is provided and stored in the created object (line 5 in Algorithm 13.5). In such constructions it is always a question whether to store a copy of a *state* or only a reference or pointer to it. All three solutions have their implications which depend on their role and lifetime in a system. For 'simple' types, such as double the first option is most natural. For all others, it has to be decided taking into consideration what other parts of a program have access to *state* and what is its lifetime.
2. Call time – the *operator*( ) is invoked here, which can perform any implemented action, having access to the already set 'state' (line 9 in Algorithm 13.5). More often than not, this is a virtual function to allow polymorphic calls through references or pointers to the base class.

There are some limitations of this technique, however. The first is that if operators were to be implemented as function objects then they should be able to build class hierarchies and have a unified way of argument passing to the *operator*(). Moreover, this operator should be declared 'virtual', so the derived classes can be accessed by base pointers and references, and to allow for inheritance. This, in turn, poses a problem since different operations usually require different sets of parameters with different call policies. However, passing parameters can be solved by constructors which store necessary parameters in state members (variables) until *operator*() is invoked.

Nevertheless, the already mentioned functor techniques are very powerful and with some substantial modifications are used in many designs. The most important adjustment can be summarized as follows.

1. The input parameters of an operation are template arguments. This way we can build operation for any type.
2. The *operator*() is declared *virtual.*
3. The *operator*() takes *no* parameters in run-time. All parameters are passed (e.g. by reference) during *construction.*

Similar patterns are presented in the book by Gamma *et al.* [144], in which it is called the *Command* design pattern. Another approach, called *generalized functors*, is presented in the book by Alexandrescu [3].

13.3.2 Handle-body or Bridge

Systems are composed of modules. The modules communicate through interfaces. Thus, changing the specification of a single interface causes much variation in all affected cooperating parts of the system. Therefore interfaces should be well worked out before being put into the system. They should be changed only if necessary. However, behind interfaces are algorithms operating on some data structures. This part we call a body of a system. It undergoes much variation, not only during implementation but also when the system grows and changes through the years. Understanding the behaviour of these two realms is important to design complex computer systems. Therefore a question arises as to how we can join these two different parts.

An answer is to split a design into *two* separate lines of development. The first one is concerned with design and implementation of interfaces, the second with the body. In terms of design patterns the first is called *a handle*, the second *a body*. This is how a handle–body design pattern was developed.

Figure 13.2 depicts the relationship between a handle and its body. A primary role of a handle is to define an interface which is used to communicate with other components. The real execution of an action is delegated to the associated body part, however. The coupling is rather loose between the two, so the body can be easily changed. This leads naturally to the strategy design pattern, discussed in section 13.3.4.

In the image library the handle–body pattern is applied to separate implementation of the image operators from their representation (see Figure 3.29). This is the prime purpose of the library to allow two different ways of implementation of basic image operations: in software and in hardware. The latter allows much faster execution time at a cost of additional hardware connected to the computer. Then, a change of the implementation should not affect the application code which makes call to the library operators.

As an example let us analyse the code fragment presented in Algorithm 13.6. It lists a skeleton of the *_2D_Convolve_OperationFor* binary operator which is a handle, that is, it defines a common interface for the 2D convolution. The implementation is in a separate hierarchy of classes which tries to optimize this operation depending on such factors as mask separability (section 4.2.1), etc. Thus, a real action is delegated to the objects responsible for computation of the convolution with given parameters and on a given platform.

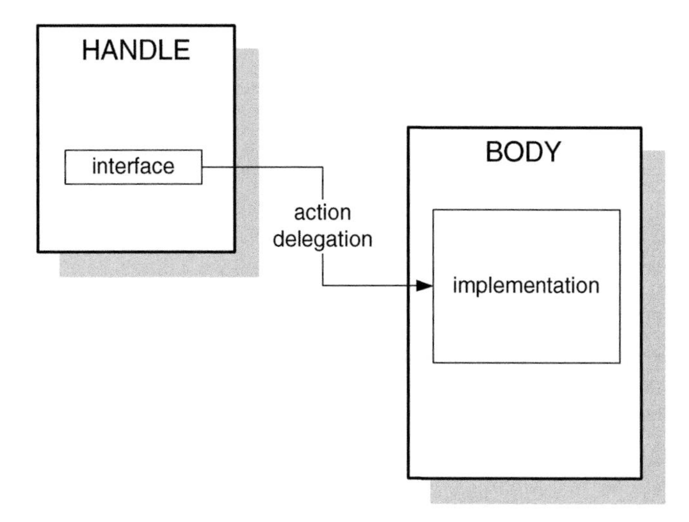

Figure 13.2 Handle–body design pattern

```
///////////////////////////////////////////////////////////////
//
// This class implements binary template image operation:
// the two dimensional (full) convolution of
// an image and template-image.
//
///////////////////////////////////////////////////////////////
template< typename RetIm_Type, typename InIm_Type1, typename InIm_Type2 >
class   _2D_Convolve_OperationFor   :   public   TImageTemplateOperationFor<
RetIm_Type,  InIm_Type1, InIm_Type2 >
{
   public:

      ///////////////////////////////////////////////////////////////
      // Class constructor
      ///////////////////////////////////////////////////////////////
      //
      // INPUT:
      //        retImage - reference to the output image
      //              of type RetIm_Type (specified by the
      //              first template parameter)
      //        inImage1 - constant reference to the first constant
      //              input image of type InIm_Type1 (specified
      //              by the second template parameter)
      //        inTemplateImage - constant reference to the constant
      //              input  template-image  of  type  InIm_Type2 (specified
      //              by the third template parameter). This
      //              is a 2D "mask" of the convolution.
      //        resourceAccessPolicy - optional reference to
      //              the thread security object (derivative
      //              of the TThreadSecurity class); by default
      //              the static kgThreadSecurity object is supplied
      //              which does nothing
      //        opCompCallback - optional reference to the callback
      //              object which is called upon completion of operation;
      //              by default the static kgOperationCompletionCallback
      //              object is supplied which does nothing
      //

      _2D_Convolve_OperationFor( RetType & retImage,
                              const InType_1 const & inImage1,
                              const InType_2 const & inTemplateImage,
                              TThreadSecurity & resourceAccessPolicy,
                        TOperationCompletionCallback & opCompCallback );


      ///////////////////////////////////////////////////////////////
      // The function operator which 2D convolves an image with
      // the supplied mask (it is either a one row image or
      // a vector<>) and puts a result to the output image.
      ///////////////////////////////////////////////////////////////
      //
      // INPUT:
      //        none
      //
      // OUTPUT:
      //        a pointer to the return image
      //
      // REMARKS:
```

Algorithm 13.6 Example of the handle – a 2D convolution operator. It defines only an interface for convolution. An action is delegated to the separate implementation in the form of *Convolve*() function. A change to implementation does not affect the interface. (Reproduced by permission of Pandora Int. Inc., London)

```
        //
        //        It is required that the input image and the mask,
        //        are different from the output object.
        //        Otherwise data will be corrupted!
        //
        virtual void * operator()( void )
        {
            MImageOperationRetinue theImageOperationRetinue( * this );
            Convolve( fInImage1, fImageTemplate, fRetImage ); // action
            return & fRetImage;

        }
};
```

Algorithm 13.6 (*Continued*)

If the computation method is changed now, due, for instance, to faster implementation (e.g. an optimized assembly code), or a computation platform is changed from software to hardware, then only the implementation part will be changed. In our example the *Convolve*() helper function will need to be implemented in a different way, depending on new circumstances.

As with most of the patterns there are some further questions on object behaviour and some special situations. These (or rather answers to them) can help further understand the pattern.

- What module is responsible for allocating/deallocating the body object(s)?
- What is the relation among handles/bodies; is it possible to assign multiple bodies to a single handle? A strategy pattern?
- What is the best way of developing hierarchies of handles and bodies?

Some hints on the above and further analysis of the handle/body pattern is provided for instance in [175]. In [144] handle–body is called *a bridge* pattern.

13.3.3 Composite

Composite belongs to one of the most common and interesting structural patterns [144, 441]. It allows composition of objects into tree-like structures which represent part/whole hierarchies. Thus, from an external point of view, a single leaf as well as a composition of leaves can be treated in the same way, i.e. they exhibit the same interface, although having different internal structures.

Figure 13.3 depicts a class hierarchy of a simple composite pattern extracted from the more complex hierarchy of the coordinate transformation engines in Figure 12.5. There are three types of objects involved: a component, a composite and a leaf. The first one defines a common interface and usually is implemented as a pure virtual class, i.e. one which does not serve to instantiate objects of its type. The composite and the leaf are children of the same level. However, the composite is able to aggregate one or more instances of such children, i.e. it can aggregate leaves and/or another composites, and so on. By this virtue recursive-like structures can be built. On the other hand, from a point of view of other modules, all children of the component (i.e. the *TLinearTransformEngine* in our example) share the same properties, such as a common interface (i.e. a set of public members) defined in the component.

In Figure 13.3 the *TLinearTransformEngine* is a base class and a component of the composite design pattern (actually it is also derived from the base *TCoordTransformEngine*).

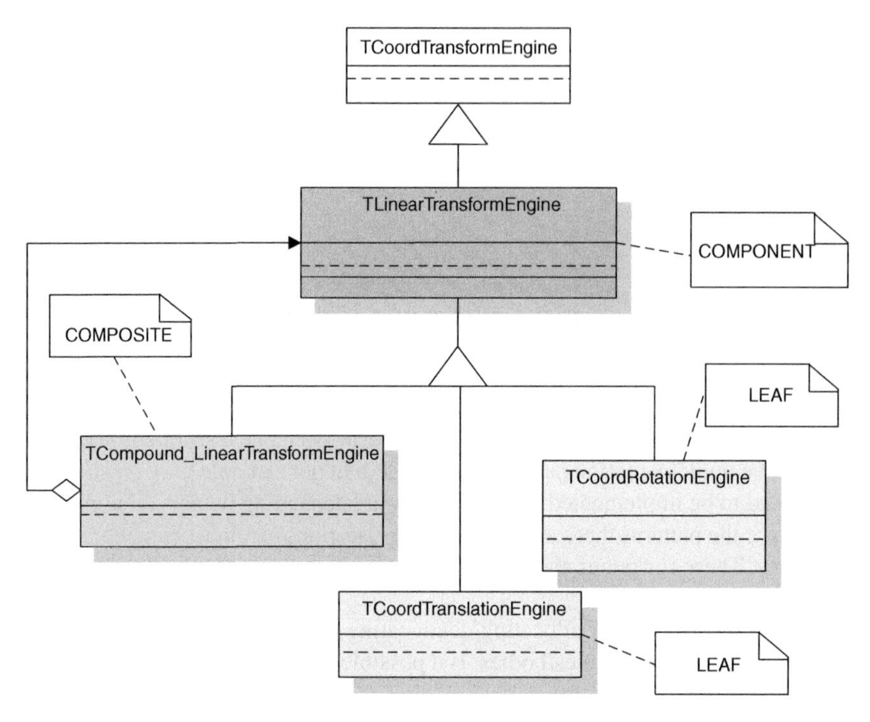

Figure 13.3 Example of a composite pattern. There are three types of objects: a component, a composite and a leaf

It defines a common interface for a group of linear transformations of image coordinates, such as rotation, translation and scaling. These are represented by the appropriate leaves, i.e. the classes *TCoordTranslationEngine* and *TCoordRotationEngine* in Figure 13.3. However, a compound linear transformation such as rotation and scaling can be represented by a combination of the appropriate leaves. Thus, any linear transformation can be represented by the *TCompound_LinearTransformEngine* a listing of which is presented in Algorithm 13.7.

The composite class is usually endowed with methods to add, remove and access its components, i.e. leaves. However, in the presented example only *add* was necessary. There are also subtleties in implementation of data storage for the components in the composite, as well as the iterators for their access. More often than not this is a vector or a set (STL is of great help in this place). Nevertheless, everything depends on the particular problem. For instance, the *TCompound_LinearTransformEngine* class inherits a 3×3 matrix from its base, i.e. from the *TLinearTransformEngine*, as all other children do. Then, each new component when added to the composite modifies entries of that matrix.

Another example of the composite pattern can be found in the image operations class hierarchy depicted in Figure 3.29. The base class *TImageOperation* defines a component. Then the *TComposedImageOperationFor<>* class implements the composite which can be composed of basic operations but also of other composed operations, and so on. The storage for operations constitutes the *vector* class from STL. In consequence, there is a linear access to the components of the composite object.

```
// A composite of linear transformations
class TCompound_LinearTransformEngine  : public TLinearTransformEngine
{
   public:

      // ====================================================
      TCompound_LinearTransformEngine( void );
      virtual ~TCompound_LinearTransformEngine() {}
      // ====================================================

      ///////////////////////////////////////////////////////////
      // This function adds a new linear transformation to the
      // one common linear transformation.
      ///////////////////////////////////////////////////////////
      //
      // INPUT:
      //          t - ref to the new lin transformation
      //
      // OUTPUT:
      //          none
      //
      // REMARKS:
      //
      //
      void AddNewTransformation( TLinearTransformEngine & t ) ;
};
```

Algorithm 13.7 Example of the composite class

The composite design pattern should be considered in the cases of the tree-like structures characteristic of the uniform interface to the external world.

13.3.4 Strategy

Strategy denotes a pattern that allows control over variability of algorithms. Its integral part constitutes an interface that allows uniform application of different algorithms. The algorithms can be defined in many ways, for instance as function objects (section 13.3.1). A particular algorithm is chosen based on some information on processed data.

Figure 13.4 depicts the structure of the strategy design pattern in the context of the image warping module. The base *TImageWarp* constitutes a context (refer to Algorithm 12.3). The strategy part starts in the *TCoordTransformEngine* (see also Figure 13.3). It actually realizes the strategy pattern which consists of assignment of a concrete strategy which in the present example can be one of the three subclasses *TLinearTransformEngine*, *TGenericTransformEngine* and *TNonLinearTransformEngine*.

In the strategy design pattern a context class contains a reference to a strategy object. All actions contained in the semantics of this strategy object are delegated from the context to strategy (see also the handle/body pattern in section 13.3.2). Depending on the particular transformation requested by an external module, actually one of the child concrete strategies is chosen in the run time and passed as a reference to the context (in our implementation it is passed directly to its constructor). Then, the chosen concrete strategy performs all actions on behalf of the context.

In the example given the *TImageWarp* class contains a reference to the *TCoordTransformEngine* base class. It is initialized to refer to the concrete object in the constructor of

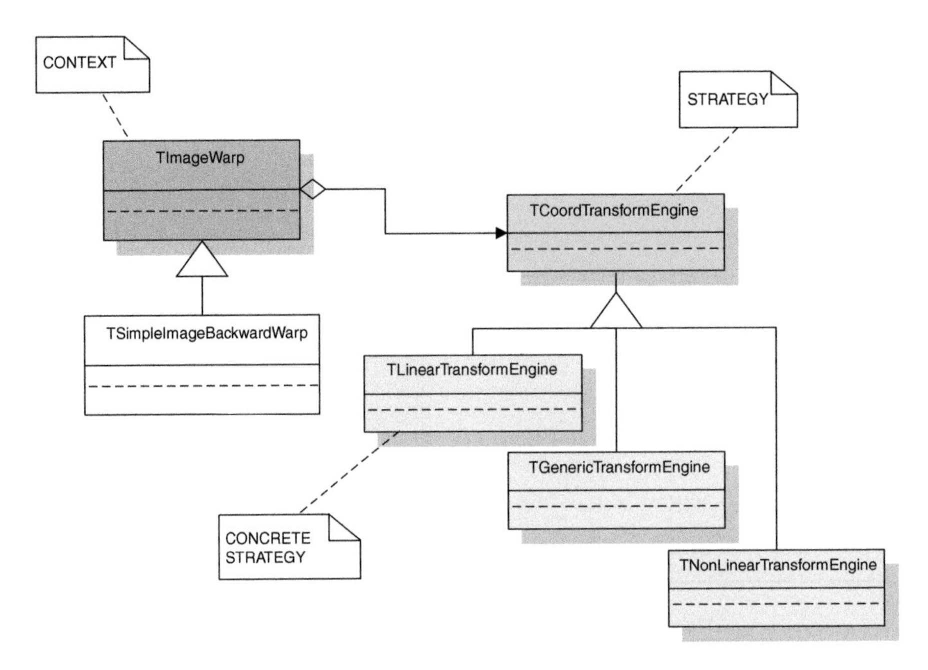

Figure 13.4 Structure of the strategy design pattern

TImageWarp. This concrete strategy object has to belong to the hierarchy of the *TCoordTransformEngine*. However, *TImageWarp* does not need to know which particular coordinate transformation object is actually chosen – the whole communication is obtained by an interface, common to the *TCoordTransformEngine* hierarchy (virtual members). Hence, any new (or future) object in this hierarchy will work as well. This is the way in which the library can be used and extended by users. Observe also that there is yet another strategy anchored within *TImageWarp*. It is the hierarchy of strategies for interpolation of pixels, with a reference *TImageWarp::fPixelInterpolation* (see Algorithm 12.3).

Strategy is a very ample behavioural pattern. It allows simple and uniform implementation of the families of related algorithms, as well as elimination of conditional statements for selection of different behaviours in run time. In the latter case it can speed up computations by elimination of a continuous condition check. Strategy nicely connects to other patterns, such as composite, functor or handle/body.

Finally, strategy can be implemented as a template parameter which is used to set up its class. In such a case, however, it is customary to name strategy a policy or a trait [3] as will be explained in the next section.

13.3.5 Class Policies and Traits

The strategy pattern allows choice and change of different behaviours in the run time, i.e. during code execution. For instance, depending on the requested precision, the bilinear or bicubic interpolation can be used in the warping module. However, a different behaviour can be obtained by adjusting an interface of a class before it is even compiled. This can be achieved with templates. Let us consider a simple template class:

```
template < typename T >
class TDerived : public T
{

};
```

Depending on a type *T* supplied to the *TDerived*<> it gets derived from quite different base class, i.e. from *T*. Deriving from a base means inheritance of its public and protected members. This is a very powerful technique which allows change of a class behaviour depending on its supplied template type. Depending on whether we wish to change behaviour or only some types, this technique is known as *a policy* or *a trait*, respectively [3]. The class *TDerived* is sometimes named a mixin. However, a policy or a trait should be used with caution. It should encapsulate only specific aspects of a behaviour of a class which can be changed in specific conditions rather than the whole interface (if this is the case, then a new class should be considered). Let us analyse an example from the vision library. The problem is that when defining the *TImageFor*<> class (see Algorithm 3.3 which lists a simplified version) for representation of an image, behaviour of particular members should be trimmed depending on the type of its pixels. For instance to access pixels, if these are modelled with built-in C++ types, such as *unsigned char* or *int*, their values should probably be passed by value. However, if the pixels are 'fat' objects or even other images, then these should be accessed by reference. This can be accomplished by special trait class supplied as a template parameter to the *TImageFor*<>. Thus, the fully fledged implementation of this class was endowed with the *PixelAccess_Trait*< *T* > in which *T* represents type of pixel of an image it is used in (Algorithm 13.8).

```
template< typename T, typename PAT = PixelAccess_Trait< T > > class
TImageFor
{
public:

   typedef typename T                          PixelType;

   typedef typename PAT::PixelAccessType       PixelAccessType;
   typedef typename PAT::ConstPixelAccessType  ConstPixelAccessType;

   // ...

public:
   void SetPixel(  Dimension xPixPosition, Dimension yPixPosition,
                   ConstPixelAccessType value ) const;

   PixelAccessType GetPixel ( Dimension xPixPosition,
                              Dimension yPixPosition ) const;

   // ...
};
```

Algorithm 13.8 Definition of the *TImageFor* template class with pixel access traits defined by template parameter class

```
template < typename T >
class PixelAccess_Trait
{
  public:
    // For all undefined access pixels by value
    typedef        T      PixelAccessType;
    typedef const  T      ConstPixelAccessType;
};


// Specific traits are implemented as specializations.
template <> class PixelAccess_Trait< double >
{
  public:
    // For double access pixels by reference
    typedef         double & PixelAccessType;
    typedef const   double & ConstPixelAccessType;
};

template <> class PixelAccess_Trait< int >
{
  public:
    // For double access pixels by value
    typedef         int PixelAccessType;
    typedef const   int ConstPixelAccessType;
};
```

Algorithm 13.9 Suite of pixel access traits. The traits for specific types of pixels are defined as template specializations for that type

The pixel access trait classes were designed with the help of the template and template specialization technique [434], as in Algorithm 13.9. It lists three template classes. The first one, *PixelAccess_Trait*, is the main template class. The other two are specializations of this class for the *double* and *int* types of pixels, respectively. We can see that the former are defined to be passed by reference, and the latter by value.

13.3.6 Singleton

One of the simplest is the singleton pattern. Its role is to ensure that a given class has only one instance (one object), but also to provide an access method to it [144]. More generally, we request a certain strictly controlled number of objects of a given type. In practice they represent some entities that should be unique or their number is restricted. For instance, the pattern can be used to represent a mirror-like interface to some hardware resources. Frequently objects responsible for management of other objects are implemented as singletons.

Although it is simple, its implementation is not trivial. It is difficult to ensure proper construction and disposal of the singleton object. This is an especially crucial problem in a multithreading environment. These issues are discussed by Vlissides [441] and by Alexandrescu [3], for instance. The latter, particularly, gives a thorough discussion and a policy-based template for users' singleton classes.

13.3.7 Proxy

The role of the proxy object is to become a placeholder or a surrogate in place of another object. An external caller cannot tell a proxy from its counterpart. However, to be useful the

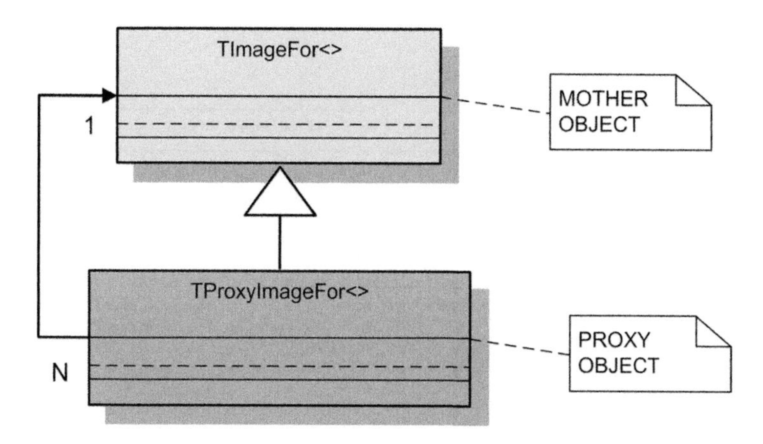

Figure 13.5 Structure of the proxy design pattern

proxy has to have some advantages over an object to which it is a surrogate. Usually, this is its smaller size or deferred implementation, etc.

In image processing it is very common to define an operation on a rectangular subregion of an original (mother) image. This subregion can be viewed as a separate image by itself, with its local coordinate system related to that region (i.e. a kind of manifold). However, it would be useful if the two concepts behaved analogously from an external caller point of view. If so, then we do not need a separate treatment of the two. Thus calling convolution or warping would be the same for a mother image and for its proxy. The idea of a proxy pattern in the context of the image library is depicted in Figure 13.5.

The proxy *TProxyImageFor<>* derives from the base *TImageFor<>*, so in accordance with the so-called Liskov substitution principle [401], it can be used in all places where its base can be used. However, contrary to a 'simple' image, a proxy image does not allocate any memory for its pixels. Instead it obtains a reference to its 'mother' object which contains all the pixels and in which it defines a rectangular region in the space of allowable pixel indices of the 'mother' object. Then, the proxy behaves as any other 'simple' image, although all operations are done in a predefined rectangular region and on the pixels of its 'mother'. The advantage of using a proxy is that to perform an operation on a subimage we do not need to create a separate image with its own storage to which a subregion would be copied. The main methods affected are *GetPixel*() and *SetPixel*() which in the proxy image have to recompute pixel positions from the local coordinates into the space of 'mother' coordinates.

13.3.8 Factory Method

In some situations we need an interface for creating objects of some hierarchy but a choice of the particular one is left to some classes in another hierarchy. In such a situation a factory method design pattern, also called a virtual constructor pattern, can be of help [144]. To understand the main idea behind this pattern let us analyse an example depicted in Figure 13.6. On the left we have a hierarchy of the Gaussian pyramids. It starts with the *TGaussianImagePyramids* base class and contains one derived class, the *TDOGImagePyra-*

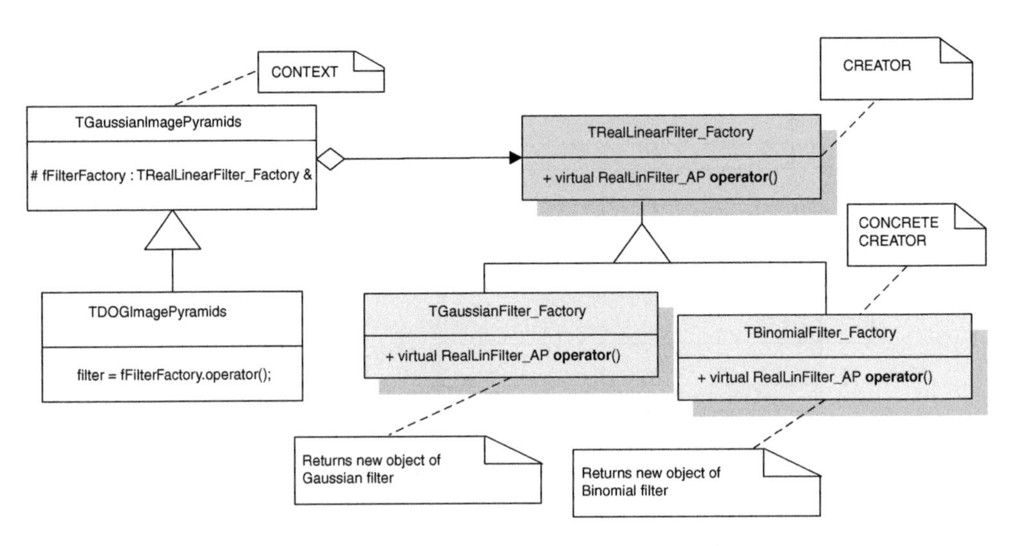

Figure 13.6 Structure of the factory method

mids (for difference of Gaussians, discussed in section 4.5.4). The classes in this hierarchy need to instantiate a concrete object representing one of the real filters, such as a Gaussian or a binomial filter. However, when the *TGaussianImagePyramids* hierarchy is created only the abstract framework for the hierarchy of filters is known. The latter is contained in the *TRealLinearFilter_Factory* base class which defines a functional operator returning an autopointer to the filter object. What we wish to achieve is to let derived classes from the pyramid hierarchy decide what type of filter to instantiate from the filter hierarchy. For this purpose we can use the factory method from Figure 13.6.

The base *TGaussianImagePyramids* keeps a reference to an object of type *TRealLinearFilter_Factory*. This reference has to be initialized with a concrete object of the filter hierarchy which can be done in the subclasses of *TGaussianImagePyramids*.

13.3.9 Prototype

Sometimes we are interested in creating a copy of an object accessed by a reference or pointer to its base class, however. Hence, we cannot easily tell which particular derivative of a base we have accessed. One of the ways to find out a real type of object is to use the C++ run-time-type-information (RTTI) mechanism. However, this results in not very friendly switch-case statements. The other way is to use a key mechanism of the *prototype* design pattern, namely the *virtual Clone*() method, defined for each class in a hierarchy. This method is responsible for creating and returning a new object of exactly the same type as its object and with exactly the same state as its object. In each *Clone*() this is implemented simply by calling the new operator and copy constructor of the class to which *Clone*() belongs (see Algorithm 13.10).

Figure 13.7 depicts the structure of the prototype design pattern. Central is the *TCoordTransform_Prototype* hierarchy which defines the *Clone*() operation. Each derived class implements its own version of this method. An exemplary implementation is listed in Algorithm 13.10.

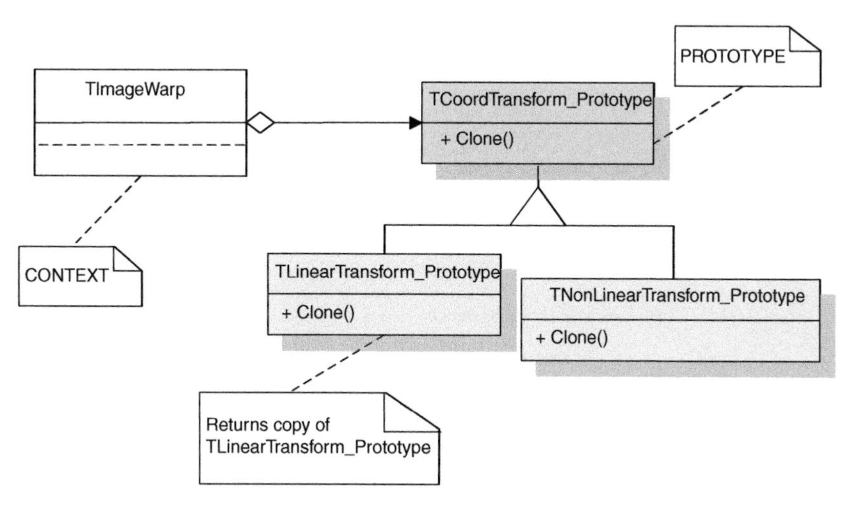

Figure 13.7 Structure of the prototype design pattern

It is also possible to combine the prototype technique with the factory method. A factory can contain a collection of prototypes, and then, on request (e.g. providing a tag or other means of object identification), it creates a new exemplar of a prototype from its collection.

13.4 Object Lifetime and Memory Management

Ensuring proper lifetime of objects gets quite complicated, even in relatively small-sized projects. In C++ there are three types of memory: automatic memory (stack allocations), static memory and free store memory (heap allocations) [401]. There is no memory management problem with the first two groups, although one should be aware of potential synchronization problems in a multithreading environment when using static memory, since static variables are shared by all software components. In the case of free store, allocations are done with the *new* operator, deallocations with *delete*. Each *new* should be matched by one *delete*, otherwise a program leaves allocated memory blocks until the next reboot of the system – a phenomenon called memory leaks.

However, we can use automatic allocation and deallocation of automatic data to control the lifetime of objects placed in the free store area. If an object is automatic, then when created its constructor is invoked. Then, when the object is automatically destroyed, its destructor is called. If such an object contains a pointer to some area on the free store, then it can delete this area in its destructor, which is called when the object is disposed of. The *auto_ptr<>*

```
TCoordTransform_Prototype * TLinearTransform_Prototype::Clone( void )
{
	return new TLinearTransform_Prototype( * this );	//	copy	constructor
}
```

Algorithm 13.10 Exemplary implementation of the *TLinearTransform_Prototype::Clone* method. It returns a new object which is an exact copy of itself

```
template< class T >
class auto_ptr
{
  public:

    // ====================================================
    explicit auto_ptr( T * w = 0 ) : fPointer( w ) {}

    template< class U >
      auto_ptr( auto_ptr<U> & ps ) : fPointer( ps.release() ) {}

    ~auto_ptr() { delete fPointer; }

    template< class U >
    auto_ptr<T> & operator = ( auto_ptr<U> & ps )
    {
      if( this != & ps ) reset( ps.release() );
      return *this;
    }
    // ====================================================

    T & operator  * () const { return * fPointer; }
    T * operator  -> () const { return fPointer; }
    T * get() const  { return fPointer; }

    T * release()
    {
      T * oldPointer = fPointer;
      fPointer = 0;
      return oldPointer;
    }

    void reset( T * w = 0 )
    {
      if( fPointer != w )
      {
        delete fPointer;
        fPointer = w;
      }
    }

  private:

    // Class inherent variables
    T * fPointer;
};
```

Algorithm 13.11 Exemplary implementation of the *auto_ptr* class

class embodies this idea. An exemplary implementation of the *auto_ptr<>* template class is presented in Algorithm 13.11.

The private member *fPointer* of type *T* holds a pointer to an object allocated on the heap. The constructor of the class accepts this pointer. The most interesting is the destructor which deletes *fPointer* thus releasing the memory. Apart from this, the *auto_ptr<>* behaves as an ordinary pointer due to overload dereferencing and field access operators. Thus, when allocating objects on the free store, a returned pointer should be controlled by *auto_ptr<>*.

For arrays we need a slightly different implementation of autopointers. Finally, the *std* namespace provides us with a more ample definition of the *auto_ptr<>*. Thus, an *auto_ptr* for monochrome images could look as follows:

```
typedef std::auto_ptr< MonochromeImage >          MIAP;
```

13.5 Image Processing Platforms

In this section we discuss some aspects of the software/hardware platforms for image processing and computer vision. Each such realization depends heavily on computer resources. However, thanks to some programming techniques they can be made more or less easily portable among evolving computer systems and development frameworks, which is a good feature in terms of software maintenance. Moreover, a layered architecture allows seamless hardware acceleration of some time-consuming operations, which is a desirable feature in terms of run-time speed.

13.5.1 Image Processing Libraries

There are many image processing libraries. The most popular are the following.

- Matlab Image Processing Toolbox by MathWorks [208].
- Open Image Library by Intel.
- Vision SDK by Microsoft.
- CImg, INRIA (Tschumperle), France.

However, there are many more. A choice of a particular one is not easy, though. Probably the most popular in the computer vision community is the OpenCV library by Intel.

13.5.2 Writing Software for Different Platforms

It is very desirable to write a code which works without a change on different computer platforms. Alas, more often than not this is not the case, except maybe for the simplest functions. Such a situation is caused by major differences among hardware and operating systems, as well as by diversity of environmental and compiler details even in the domain of the same programming language. What we can do is make the process of changing code when going from one computer platform to the other less cumbersome by proper design and some simple rules for implementation. Here we briefly outline the most important issues of systems development for multiple platforms.

- During design use object-oriented methods, design patterns, etc. They all help in code reusability and support multiplatform operations.
- Prefer layered designs with well-specified interfaces. When a layer needs to be changed, this can be done separately with a separate testing afterwards.
- When changing existing code try the methods of refactoring [243].
- Try to use pure language constructs, avoiding platform-specific settings, such as pragmas, precompiled headers, particular data representation (size, alignment), etc.

- If it is not possible to avoid specific features of a language or a system, then try to separate such a construction, e.g. by the handle–body pattern and action delegation paradigm (section 13.3.2).

13.6 Closure

In this chapter we discuss the basic methods and techniques of design and implementation of software systems. Their goal is to provide some methodology for creating modular, extensible and, as much as possible, bug-free systems. Although our main focus is vision processing, the methods can also be used in other areas of engineering.

13.6.1 Further Reading

The literature on the subject is extensive. A must in C++ is the book by Stroustrup [401]. It can be read from beginning to end, and then used as a reference. There are chapters on software development and design as well. It contains ample information on all aspects of C++, such as templates and STL. One of the best explanations and references on STL is the book by Josuttis [231]. Good design and programming practices can be found in the excellent book by McConnell [305]. In-depth treatment on C++ templates can be found in the book by Vandervoode and Josuttis [434]. Finally, design patterns and programming methodology are described in the books by Gamma *et al.* [144], Alexandrescu [3] and Vlissides [441], to name a few.

14

Image Processing Library

The main intention of the image processing library for hardware acceleration (HIL) was to develop a kind of bridge between software residing on a computer and an external hardware board accelerating some computations. Examples of procedures of the library, as well as the main class hierarchies, have been given in many places in this book. The software layers of the library are available for noncommercial applications from the web site of the book. The site contains also further description of the library, examples of its usage in different contexts and other interesting links [216].

An Introduction to 3D Computer Vision Techniques and Algorithms Bogusław Cyganek and J. Paul Siebert

14

Image Processing Library

References

[1] Adam, A., Rivlin, E. and Shimshoni, I. (2001) ROR: rejection of outliers by rotations. *IEEE Transactions on Pattern Analysis and Machine Intelligence*, **23** (1), pp. 78–84.

[2] Ahuja, N. (1993) *Motion and Structure from Image Sequences*, Springer-Verlag.

[3] Alexandrescu, A. (2001) *Modern C++ Design. Generic Programming and Design Patterns Applied*, Addison-Wesley.

[4] Alibhai, S. and Zucker, S.W. (2000) Contour-based correspondence for stereo. 6th European Conference on Computer Vision, ECCV 2000, Dublin, June/July 2000.

[5] Aksenov, P., Clark, I., Grant, D., Inman, A., Vartikovski, L. and Nebel, J.-C. (2003) 3D Thermography for the quantification of heat generation resulting from inflammation. In *Proceedings of the 8th 3D Modelling Symposium*. Paris, France.

[6] Alvarez, L., Deriche, R., Sanchez, J. and Weickert, J. (2002) Dense disparity map estimation respecting image discontinuities: a PDE and scale-space based approach. *Journal of Visual Communication and Image Representation*, **13** (1/2), 3–21.

[7] Anderson, B.L. (1994) The role of partial occlusion in stereopsis. *Nature*, **367**, 365–368.

[8] Antonini, G., Martinez, S.V., Bierlaire, M. and Thiran, J.P. (2006) Behavioral priors for detection and tracking of pedestrians in video sequences. *International Journal of Computer Vision*, **69** (2), 159–180.

[9] Antoniou, A. (2005) *Digital Signal Processing*, McGraw-Hill Professional.

[10] Armstrong, M.N. (1996) *Self-Calibration from Image Sequences*, Department of Engineering Science, University of Oxford.

[11] Arun, K.S., Huang, T.S. and Blostein, S.D. (1987) Least-squares fitting of two 3D point sets. *IEEE Transactions on Pattern Analysis and Machine Intelligence*, **9** (5), 698–700.

[12] Asada, H. and Brady, M. (1986) The curvature primal sketch. *IEEE Transactions on Pattern Analysis and Machine Intelligence*, **8** (1), 2–14.

[13] Aström, K., Heyden, A., Kahl, F. *et al.* (1998) *A Computer Vision Toolbox*, Department of Mathematics, Lund University.

[14] Aubert, G. and Kornprobst, P. (2002) *Mathematical Problems in Image Processing. Applied Mathematical Sciences*, Vol. 147, Springer.

[15] Avidan, S. and Shashua, A. (1996) *Tensorial Transfer: Representation of N>3 Views of 3D Scenes*, Hebrew University of Jerusalem, Institute of Computer Science.

[16] Avidan, S. and Shashua, A. (1998) *Tensor Embedding of the Fundamental Matrix*, Lecture Notes in Computer Science 1506, Springer, pp. 47–62.

[17] Ayache, N. and Hansen, C. (1988) Rectification of Images for Binocular and Trinocular Stereovision. *INRIA Technical Report 860*.

[18] Ayoub, A.F., Siebert, J.P., Wray, D. and Moos, K.F. (1996) A new three dimensional recording system of the face using a pair of stereo camera. *British Journal of Oral and Maxillofacial Surgery*.

[19] Ayoub, A.F., Siebert, J.P., Moos, K.F. *et al.* (1998) A vision-based three dimensional capture system for maxillofacial assessment and surgical planning. *British Journal of Oral and Maxillofacial Surgery*, **36** (5), 353–357.

An Introduction to 3D Computer Vision Techniques and Algorithms Bogusław Cyganek and J. Paul Siebert

[20] Ayoub, A.F., Xiao, Y., Khambay, B. *et al.* (2007) Towards building a photo-realistic virtual human face for craniomaxillofacial diagnosis and treatment planning. *International Journal of Oral and Maxillofacial Surgery*, **5**, 423–428.

[21] Baker, S., Nayar, S.K. and Murase, H. (1998) Parametric feature detection. *International Journal of Computer Vision*, **27** (1), 27–50.

[22] Baker, S., Scharstein, D., Lewis, J.P. *et al.* (2007) *A Database and Evaluation Methodology for Optical Flow*. IEEE International Conference on Computer Vision, October 2007.

[23] Baldock, R. and Graham, J. (2000) *Image Processing and Analysis. A Practical Approach*, Oxford University Press.

[24] Banks, J., Porter, R., Bennamoun, M. and Corke, P. (1997) A Generic Implementation Framework for Stereo Matching Algorithms. Technical Report, CSIRO Manufacturing Science and Technology, Australia.

[25] Banks, J. and Corke, P. (2001) Quantitative evaluation of matching methods and validity measures for stereo vision. *International Journal of Robotics Research*, **20** (7), 512–532.

[26] Banks, J., Bennamoun, M. and Corke, P. (1997) Fast and Robust Stereo Matching Algorithms for Mining Automation. Technical Report, CSIRO Manufacturing Science and Technology, Australia.

[27] Banks, J., Bennamoun, M. and Corke, P. (1997) Non-Parametric Techniques for Fast and Robust Stereo Matching. Technical Report, CSIRO Manufacturing Science and Technology, Australia.

[28] Banks, J., Bennamoun, M., Kubik, K. and Corke, P. (1998) Evaluation of New and Existing Confidence Measure for Stereo Matching. Technical Report, CSIRO Manufacturing Science and Technology, Australia.

[29] Banks, J., Bennamoun, M., Kubik, K. and Corke, P. (1999) A Constraint to Improve the Reliability of Stereo Matching Using the Rank Transform. Technical Report, CSIRO Manufacturing Science and Technology, Australia.

[30] Bazaraa, M.S., Jarvis, J.J. and Sherali, H.D. (2005) *Linear Programming and Network Flows*, 3rd edn, John Wiley & Sons, Ltd.

[31] Beaudet, P.R. (1978) Rotationally invariant image operators. In *Proceedings of the 4th International Joint Conference on Pattern Recognition*, Tokyo, pp. 579–583.

[32] Becker, S. and Hinton, G.E. (1992) Self-organizing neural network that discovers surfaces in random-dot stereograms. *Nature*, **355**, 161–163.

[33] Bellman, R. (2003) *Dynamic Programming*, Dover Publications.

[34] Benosman, R. and Kang, S.B. (eds) (2001) *Panoramic Vision. Sensors, Theory, and Applications*, Springer.

[35] Bertozzi, M., Broggi, A. and Fascioli, A. (1998) *Stereo Inverse Perspective Mapping: Theory and Applications*, University di Parma, Italy.

[36] Bertsekas, D.P. (1996) *Constraint Optimization and Lagrange Multiplier Methods*, Athena Scientific.

[37] Bertsekas, D.P. (2003) *Convex Analysis and Optimization*, Athena Scientific.

[38] Besl, P.J. and McKay, N.D. (1992) A method for registration of 3D shapes. *IEEE Transactions on Pattern Analysis and Machine Intelligence*, **14** (2), 239–256.

[39] Beucher, S. (1990) Segmentation d'images et morphologie mathématique. PhD thesis, Ecole des Mines de Paris.

[40] Bhat, D.N. and Nayarm, S.K. (1998) Ordinal measures for image correspondence. *IEEE Transactions on Pattern Analysis and Machine Intelligence*, **20** (4), 415–423.

[41] Bhat, D.N. and Nayar, S.K. (1997) Ordinal Measures for Image Correspondence. *Technical Report CUCS-009-96*, Department of Computer Science, Columbia University.

[42] Bigün, J., Granlund, G.H. and Wiklund, J. (1991) Multidimensional orientation estimation with applications to texture analysis and optical flow. *IEEE Transactions on Pattern Analysis and Machine Intelligence*, **13** (8), 775–790.

[43] Birchfield, S. (1999) Depth and motion discontinuities. PhD dissertation, Stanford University.

[44] Birchfield, S. and Tomasi, C. (1996) Depth Discontinuities by Pixel-to-Pixel Stereo. *Technical Report CS-TR-96-1573*, Computer Science Department, Stanford University.

[45] Bishop, R.L. and Goldberg, S.I. (1980) *Tensor Analysis on Manifolds*, Dover Publications.

[46] Black, M., Sapiro, G., Marimont, D.H. and Heeger, D. (1998) Robust anisotropic diffusion. *IEEE Transactions on Image Processing*, **7** (3), 421–432.

[47] Bobick, A.F. and Intille, S.S. (1995) Large Occlusion Stereo. Technical Report, Media Laboratory, MIT.

[48] Booch, G., Rumbaugh, J. and Jacobson, I. (1999) *The Unified Modeling Language. User Guide*, Addison-Wesley Longman.

[49] Borisenko, A.I. and Tarapov, I.E. (1979) *Vector and Tensor Analysis with Applications*, Dover Publications.

[50] Born, M. and Wolf, E. (2003) *Principles of Optics*, 7th edn, Cambridge University Press.

[51] Boykov, Y., Veksler, O. and Zabih, R. (2001) Fast approximate energy minimization via graph cuts. *IEEE Transactions on Pattern Analysis and Machine Intelligence*, **23** (11), 1222–1239.

[52] Boykov, Y. and Kolmogorov, V. (2004) An experimental comparison of min-cut/max-flow algorithms for energy minimization in vision. *IEEE Transactions on Pattern Analysis and Machine Intelligence*, **26** (9), 1124–1137.

[53] Bovik, A. (2000) *Handbook of Image Processing*, Academic Press.

[54] Böhm, M. (2000) Imagers using amorphous silicon thin film on application specific integrated circuit technology. *Journal of Non-Crystalline Solids*, **266–269**, 1145–1151.

[55] Brenner, E., Smeets, J.B.J. and Landy, M.S. (2001) How vertical disparities assist judgements of distance. *Vision Research*, **41**, 3455–3465.

[56] Brewster, D. (1856) *The Stereoscope. Its History, Theory, and Construction*. John Murray, London (available at: http://books.google.com/).

[57] Brown, M.Z., Burschka, D. and Hager, G.D. (2003) Advances in computational stereo. *IEEE Transactions on Pattern Analysis and Machine Intelligence*, **25** (8) 993–1008.

[58] Brox, T., Rousson, M., Derich, R. and Weickert, J. (2003) Unsupervised Segmentation Incorporating Colour, Texture, and Motion. *INRIA Technical Report 4760*.

[59] Brox, T., Bruhn, A., Papenberg, N. and Weickert J. (2004) High accuracy optical flow estimation based on a theory for warping. ECCV 2004, LNCS 3024, pp. 25–36.

[60] Canny, J. (1986) A computational approach to edge detection. *IEEE Transactions on Pattern Analysis and Machine Intelligence*, **8** (6), 679–698.

[61] Carlsson, B., Ahlén, A. and Sternad, M. (1989) *Optimal Differentiation Based on Stochastic Signal Models*, Department of Technology, Uppsala University.

[62] Carlsson, B., Sternad, M. and Ahlén, A. (1990) *Digital Differentiation of Noisy Data Measured through a Dynamic System*, Department of Technology, Uppsala University.

[63] Carlsson, S. (1994) *The Double Algebra: An Effective Tool for Computing Invariants in Computer Vision*, Computational Vision and Active Perception Laboratory, NADA-KTH, Stockholm.

[64] Casse, R. (2006) *Projective Geometry. An Introduction*, Oxford University Press.

[65] Chan, T.F. and Jianhong Shen, J. (2005) *Image Processing and Analysis: Variational, PDE, Wavelet, and Stochastic Methods*, Society for Industrial and Applied Mathematics (SIAM).

[66] Chaudhuri, S. and Rajagopalan, A.N. (1999) *Depth from Defocus: A Real Aperture Imaging Approach*, Springer Verlag.

[67] Chen, Y.-S., Hung, Y.-P. and Fuh, C.-S. (2000) *A Fast Block Matching Algorithm Based on the Winner-Update Strategy*. Proceedings of the Fourth Asian Conference on Computer Vision, Taipei, Taiwan, Vol. 2, pp. 977–982.

[68] Chen, T.-Y., Bovik, A.C. and Cormack, L. (1999) Stereoscopic ranging by matching image modulations. *IEEE Transactions on Image Processing*, **8** (6), 785–797.

[69] Chernyaev, E.V. (1995) Marching Cubes 33: Construction of Topologically Correct Isosurfaces. *CERN Technical Report CN 95-17*.

[70] Chesi, G., Garulli, A., Vicino, A. and Cipolla, R. (2000) *On the Estimation of the Fundamental Matrix: A Convex Approach to Constrained Least-Squares*. 6th European Conference on Computer Vision, ECCV 2000, Dublin, June–July 2000.

[71] Cohen, I. (1993) *Nonlinear Variational Method for Optical Flow Computation*. Proceedings of the Eighth Scandinavian Conference on Image Analysis, Tromsø, Norway, 1993, Vol. 1, pp. 523–530.

[72] Urquhart, C.W. (1997) *The Active Stereo Probe: The Design and Implementation of an Active Videometrics system*, PhD Dissertation, University of Glasgow, Glasgow, UK.

[73] Corke, P., Roberts, J. and Winstanley, G. (1997) 3D Perception for Mining Robotics. Technical Report, CSIRO Manufacturing Science and Technology, Australia.

[74] Cormen, T.H., Leiserson, C.E. and Rivest, R. (2003) *Introduction to Algorithms*, MIT Press.

[75] Cover, T.M. AND Thomas, J.A. (2006) *Elements of Information Theory*, 2nd edn, John Wiley & Sons, Ltd.

[76] Cox, I.J. (1994) *A Maximum Likelihood N-Camera Stereo Algorithm*. IEEE International Conference on Pattern Recognition, 1994, pp. 437–443.

[77] Cox, I.J. (1996) A maximum likelihood stereo algorithm. *Computer Vision and Image Understanding*, **63** (3), 542–567.

[78] Coxeter, H.S.M, (1998) *Projective Geometry*, 2nd edn, Springer-Verlag.

[79] Criminisi, A., Blake, A., Rother, C. *et al.* (2007) Efficient dense stereo with occlusions for new view-synthesis by four-state dynamic programming. *International Journal of Computer Vision*, **71** (1), 89–110.

[80] Cryer, J.E., Tsai, P.-S. and Shah, M. (1992) Combining Shape from Shading and Stereo Using Human Vision Model. *Technical Report CS-TR-92-25*, Department of Computer Science, University of Central Florida.

[81] Cumani, A. (1991) Edge detection in multispectral images. *Computer Vision, Graphics, and Image Processing: Graphical Models Image Processing*, **53** (1), 40–51.

[82] Curless, B. and Levoy, M. (1996) *A Volumetric Method for Building Complex Models from Range Images*. SIGGRAPH '96 Conference, New Orleans, LA, 1996, pp. 303–312.

[83] Cyganek, B. (2002) Novel stereo matching method that employs tensor representation of local neighborhood in binary images. *Machine Graphics & Vision*, **10**, 289–316.

[84] Cyganek, B. (2003) *Combined Detector of Locally Oriented Structures and Corners in Images*, Lecture Notes in Computer Science 2658, Springer, pp. 721–730.

[85] Cyganek, B. (2004) *Depth Recovery With an Area Based Version of the Stereo Matching Method With Scale-Space Tensor Representation*, Lecture Notes in Computer Science 3037, Springer, pp. 548–551.

[86] Cyganek, B. and Borgosz, J. (2004) *Fuzzy Nonparametric Measures for Image Matching*, Lecture Notes in Artificial Intelligence 3070, Springer, pp. 712–717.

[87] Cyganek, B. (2004) *Comparison of Nonparametric Transformations and Bit Vector Matching for Stereo Correlation*, Lecture Notes in Computer Science 3322, Springer, pp. 534–547.

[88] Cyganek, B. (2005) *Adaptive Window Growing Technique for Efficient Image Matching*, Lecture Notes in Computer Science 3522, Springer, pp. 308–315.

[89] Cyganek, B. (2005) *Object Detection in Multi-channel and Multi-scale Images Based on the Structural Tensor*, Lecture Notes in Computer Science 3691, Springer, pp. 570–578.

[90] Cyganek, B. (2006) *Matching of Multi-channel Images With Improved Nonparametric Transformations and Weighted Binary Distance Measures*, Lecture Notes in Computer Science 4040, Springer, pp. 74–88.

[91] Cyganek, B. and Socha, Ł. (2006) *Comparison of Matching Strategies for Colour Images*. First International Conference on Computer Vision Theory and Applications – VISAPP 2006, Setúbal, Portugal, pp. 364–369.

[92] Czajkowski, M. (2004) Overview of the corner detection methods in the low-level image processing. MSc thesis, AGH – University of Science and Technology, Poland.

[93] Da Vinci, L. (1270) *Trattato della Pittura* (available at: www.liberliber.it).

[94] Dalsa Corporation (2004) Image Sensor Architectures for Digital Cinematography, Technical Report (available at: http://www.dalsa.com/dc/documents/Image_Sensor_Architecture_Whitepaper_Digital_Cinema_00218-00_03-70.pdf).

[95] Davies, E.R. (1997) *Machine Vision. Theory, Algorithms, Practicalities*, 2nd edn, Academic Press.

[96] Demmel, J.W. (1997) *Applied Numerical Linear Algebra*, Society for Industrial and Applied Mathematics (SIAM).

[97] Deng, Y. and Lin, X. (2006) *A Fast Line Segment Based Dense Stereo Algorithm Using Tree Dynamic Programming, ECCV 2006*, Lecture Notes in Computer Science 3953, Springer, pp. 201–212.

[98] Deriche, R. (1987) Using Canny's criteria to derive a recursively implemented optimal edge detector. *International Journal of Computer Vision*, **1** (2), 167–187.

[99] Deriche, R. and Blaszka, T. (1993) *Recovering and Characterizing Image Features Using an Efficient Model Based Approach*. Proceedings of the Conference on Computer Vision and Pattern Recognition, New York, 1993, pp. 530–535.

[100] Devernay, F. and Faugeras, O. (1995) From Projective to Euclidean Reconstruction. *INRIA Technical Report 2725*.

[101] Dhond, U.R. and Aggarwal, J.K. (1995) Stereo matching in the presence of narrow occluding objects using dynamic disparity search. *IEEE Transactions on Pattern Analysis and Machine Intelligence*, **17** (7), 719–724.

[102] Di Zenzo, S. (1986) A note on the gradient of a multi-image. *Computer Vision, Graphics and Image Processing*, **33**, 116–125.

[103] Dimitrienko, I. (2003) *Tensor Analysis and Nonlinear Tensor Functions*, Springer.

[104] Dinic, E.A. (1970) Algorithm for solution of a problem of maximal flow in a network with power estimation. *Soviet Mathematics Doklady*, **11**, 1277–1280.

[105] Dixon, S.L. and Koehler, R.T. (1999) The hidden component of size in two-dimensional fragment descriptors: side effects on sampling in bioactive libraries. *Journal of Medicinal Chemistry*, **42**, 2887–2900.

[106] Downie, J., Mao, J., Lo, R. *et al.* (2008) A double-blind, clinical evaluation of facial augmentation treatments: a comparison of PRI 1, PRI 2, Zyplast® and Perlane®. *Journal of Plastic, Reconstructive & Aesthetic Surgery* (submitted).

[107] Dryden, I.L. and Mardia, K.V. (1998) *Statistical Shape Analysis*, John Wiley & Sons, Ltd.

[108] Duda, R.O., Hart, P.E. and Stork, D.G. (2001) *Pattern Classification*, John Wiley & Sons, Ltd.

[109] Dunn, P. and Corke, P. (1997) *Real-time Stereopsis using FPGAs*, CSIRO Division of Manufacturing Technology, Australia.

[110] do Carmo, M.P. (1976) *Differential Geometry of Curves and Surfaces*, Prentice-Hall.

[111] Eastman Kodak Company (2001) *CCD Primer*, MTD/PS-0218 (available at: http://www.kodak.com/ezpres/business/ccd/global/plugins/acrobat/en/supportdocs/chargeCoupledDevice.pdf).

[112] Edmonds, J. and Karp, R.M. (1972) Theoretical improvement in algorithmic efficiency for network flow problems. *JACM*, **19**, 248–264.

[113] Edmund Industrial Optics (2001) Optics and Optical Instruments Catalog.

[114] Egnal, G. and Wildes, R.P. (2002) Detecting binocular half-occlusions: empirical comparisons of five approaches. *IEEE Transactions on Pattern Analysis and Machine Intelligence*, **24** (8), 1127–1133.

[115] Ellis, M.A. and Stroustrup, B. (1990) *The Annotated C++ Reference Manual. ANSI Base Document*, Addison-Wesley.

[116] Euclid (1956) *The Thirteen Books of the Elements (Books I–XIII)*, Vols 1–3 (transl. T.L. Heath), Dover Publications.

[117] Farid, H. amd Simoncelli, E.P. (2004) Differentiation of discrete multidimensional signals. *IEEE Transactions on Image Processing*, **13** (4), 496–508.

[118] Faugeras, O., Hotz, B., Mathieu, H. *et al.* (1993) Real-Time Correlation-Based Stereo: Algorithm, Implementations and Applications. *INRIA Technical Report 2013.*

[119] Faugeras, O.D. and Luong, Q.-T. (2001) *The Geometry of Multiple Images*, MIT Press.

[120] Faugeras, O.D., Luong, Q.-T. and Maybank, S.J. (1992) Camera Self-Calibration: Theory and Experiment. *INRIA Technical Report.*

[121] Faugeras, O.D. and Luc, R. (1993) What Can Two Images Tell Us About a Third One? *INRIA Technical Report 2018.*

[122] Faugeras, O. (1993) *Three-Dimensional Computer Vision. A Geometric Viewpoint*, MIT Press.

[123] Faugera,s O. and Keriven, R. (1998) *Complete Dense Stereovision Using Level Set Methods.* ECCV '98, Proceedings of the 5th European Conference on Computer Vision, Vol. 1.

[124] Faugeras, O., Quan, L. and Strum, P. (2000) Self-calibration of a 1D projective camera and its application to the self-calibration of a 2D projective camera. *IEEE Transactions on Pattern Analysis and Machine Intelligence*, **22** (10), 1179–1185.

[125] Fineman, M. (1996) *The Nature of Visual Illusions*, Dover Publications.

[126] Fischler, M.A. and Bolles, R.C. (1981) Random sample consensus: a paradigm for model fitting with applications to image analysis and automated cartography. *Communications of the ACM*, **24** (6), 381–395.

[127] Fletcher, R. (2003) *Practical Methods of Optimization*, 2nd edn, John Wiley & Sons, Ltd.

[128] Fletcher, Y. and McAllister, D.F. (1989) A tension compatible patch for shape-preserving surface interpolation. *IEEE Computer Graphics*, **9** (3), 45–55.

[129] Fligner, M., Verducci, J., Bjoraker, J. and Blower, P. (2001) *A New Association Coefficient for Molecular Dissimilarity*. Conference on Chemoinformatics, Sheffield, UK, 2001.

[130] Florack, L., ter Haar Romeny, B., Koenderink, J. and Viergever, M. (1994) Linear scale-space. *Journal of Mathematical Imaging and Vision*, **4** (4), 325–351.

[131] Flynn, P.J. and Jain, A.K. (1994) Three-dimensional object recognition, in *Handbook of Pattern Recognition and Image Processing: Computer Vision*, Vol. 2 (ed. T.Y. Young), Academic Press.

[132] Foley, J.D., van Dam, A., Feiner, S.K. *et al.* (1994) *Introduction to Computer Graphics*, Addison-Wesley.

[133] Folsom, T.C. and Pinter, R.B. (1998) Primitive features by steering, quadrature, and scale. *IEEE Transactions on Pattern Analysis and Machine Intelligence*, **20** (11), 1161–1173.

[134] Ford, L. R. and Fulkerson, D.R. (1962) *Flows in Networks*, Princeton University Press.

[135] Forsyth, D.A. and Ponce J. (2003) *Computer Vision: A Modern Approach*, Prentice Hall.

[136] Förstner, W. (1994) *A Framework for Low Level Feature Extraction.* Proceedings of the 3rd European Conference on Computer Vision, Stockholm, Sweden, 1994, pp. 383–394.

[137] Freeman, W.T. and Roth, M. (1994) *Orientation Histograms for Hand Gesture Recognition*, Mitsubishi Electric Research Laboratories, TR-94-03a.

[138] Frisby, J.P. (1998) Stereo correspondence and neural networks, in *Handbook on Neural Networks*, MIT Press, pp. 937–941.

[139] Fua, P. (1991) A Parallel Stereo Algorithm that Produces Dense Depth Maps and Preserves Image Features. *INRIA Technical Report 1369.*

[140] Fusco, J. (2007) *The Linux Programmer's Toolbox*, Prentice Hall.

[141] Fusiello, A., Roberto, V. and Trucco, E. (1997) *Efficient Stereo With Multiple Windowing*. Proceedings of the IEEE Conference on Computer Vision and Pattern Recognition, 1997, pp. 858–863.

[142] Fusiello, A., Trucco, E. and Verri, A. (1998) *Rectification with Unconstrained Stereo Geometry*, Dipartimento di Matematica e Informatica, Universita di Udine.

[143] Fusiello, A., Trucco, E. and Verri, A. (2000) A compact algorithm for rectification of stereo pairs. *Machine Vision and Applications*, **12**, 16–22.

[144] Gamma, E., Helm, R., Johnson, R. and Vlissides, J. (1995) *Design Patterns. Elements of Reusable Object-Oriented Software*, Addison-Wesley.

[145] Gevers, T. and Smeulders, A.W.M. (2005) Content-based image retrieval: an overview, in *Emerging Topics in Computer Vision* (eds G. Medioni and S.B. Kang), IMSC Press, pp. 333–384.

[146] Gilat, A. (2007) *MATLAB: An Introduction with Applications*. 3rd edn, John Wiley & Sons, Ltd.

[147] Glassner, A.S. (1995) *Principles of Digital Image Synthesis*, Vols 1 and 2, Morgan Kaufmann Publishers.

[148] Gluckmann, J. and Nayar, S.K. (1997) *A Real-Time Catadioptric Stereo System Using Planar Mirrors*, Department of Computer Science, Columbia University.

[149] Gluckmann, J., Nayar, S.K. and Thoresz, K.J. (1999) *Real-Time Omnidirectional and Panoramic Stereo*, Department of Computer Science, Columbia University.

[150] Gluckmann, J. and Nayar, S.K. (1999) *Planar Catadioptric Stereo: Geometry and Calibration*, Department of Computer Science, Columbia University.

[151] Gluckmann, J. and Nayar, S.K. (2000) *Rectified Catadioptric Stereo Sensors*, Department of Computer Science, Columbia University.

[152] Gluckman, J. and Nayar, S.K. (2001) *Rectifying Transformations That Minimize Resampling Effects*. Proceedings of the IEEE Computer Society Conference on Computer Vision and Pattern Recognition, CVPR 2001, Vol. 1, pp. I-111–I-117.

[153] Golombek, M.P., Cook, R.A., Economou, T. *et al.* (1997) Overview of the Mars Pathfinder mission and assessment of landing site predictions. *Science*, **278**, 1743–1748.

[154] Golub, G.H. and Van Loan, C.F. (1996) *Matrix Computations*, 3rd edn, Johns Hopkins University Press.

[155] Gong, M. and Yang, Y.-H. (2002) Genetic-based stereo algorithm and disparity map evaluation. *International Journal of Computer Vision*, **47** (1–3), 63–77.

[156] Gong, M. and Yang, Y.-H. (2005) *Near Real-time Reliable Stereo Matching Using Programmable Graphics Hardware*. Proceedings of the IEEE Computer Society Conference on Computer Vision and Pattern Recognition, CVPR 2005, Vol. 1, pp. 924–931.

[157] Gonzalez, R.C. and Woods, R.E. (2002) *Digital Image Processing*, 2nd edn, Prentice Hall.

[158] Gonzalez, R.C., Woods, R.E. and Eddins, S.L. (2003) *Digital Image Processing Using Matlab*, 3rd edn, Prentice Hall.

[159] Goshtasby, A.A. (2005) *2-D and 3-D Image Registration*, Wiley Interscience.

[160] Granlund, G.H. and Knutsson, H. (1996) *Signal Processing for Computer Vision*, Kluwer Academic.

[161] Gregory, R.L. (1997) *Eye and Brain. The Psychology of Seeing*, Princeton University Press.

[162] Grewe, L.L. and Kak, A.C. (1994) Stereo vision, in *Handbook of Pattern Recognition and Image Processing: Computer Vision*, Vol. 2 (ed. T.Y. Young), Academic Press.

[163] Grimmet, G. and Stirzaker, D. (2001) *Probability and Random Processes*, 3rd edn, Oxford University Press.

[164] Grimson, W.E.L. (1993) Why Stereo Vision is Not Always About 3D Reconstruction. MIT, Artificial Intelligence Laboratory, Memo 1435.

[165] Grosso, E. and Tistarelli, M. (2000) *Log-Polar Stereo for Anthropomorphic Robots*. 6th European Conference on Computer Vision, ECCV 2000, Dublin, June–July 2000.

[166] Gruen, A. and Huang, T.S. (2001) *Calibration and Orientation of Cameras in Computer Vision*, Springer Series in Information Sciences, Springer.

[167] Guggenheimer, H.W. (1977) *Differential Geometry*, Dover Publications.

[168] Hajeer, M.Y., Mao, Z., Millett, D.T. *et al.* (2004) A new three-dimensional method of assessing facial volumetric changes following orthognathic treatment. *American Journal of Cleft & Craniofacial Surgery*, **42**, 113–120.

[169] Hall, E.L. (1994) Fundamental principles of robot vision, in *Handbook of Pattern Recognition and Image Processing: Computer Vision*, Vol. 2 (ed. T.Y. Young), Academic Press.

[170] Halliday, D., Resnick, R. and Walker, J. (2004) *Fundamentals of Physics*, John Wiley & Sons, Ltd.

[171] Hamanaka, M., Kenmochi, Y. and Sugimoto, A. (2005) *Discrete Epipolar Geometry*, Lecture Notes in Computer Science 3429, Springer, pp. 323–334.

[172] Haralick, R.M. and Shapiro, L.G. (1992) *Computer and Robot Vision*, Vol. 1, Addison-Wesley.

[173] Haralick, R.M. and Shapiro, L.G. (1993) *Computer and Robot Vision*, Vol. 2, Addison-Wesley.

[174] Harris, C. and Stephens, M. (1988) *A Combined Corner and Edge Detector*. Proceedings of the 4th Alvey Vision Conference, 1988, pp. 147–151.

[175] Harrison, N., Foote, B. and Rohnert, H. (2000) *Pattern Languages of Program Design 4*, Software Patterns Series, Addison-Wesley.

[176] Hartley, R.I. (1997) Kruppa's equations derived from the fundamental matrix. *IEEE Transactions on Pattern Analysis and Machine Intelligence*, **19** (2), 133–135.

[177] Hartley, R.I. (1997) In defense of the eight-point algorithm. *IEEE Transactions on Pattern Analysis and Machine Intelligence*, **19** (6), 1064–1075.

[178] Hartley, R.I. (1998) *Computation of the Quadrifocal Tensor*. Proceedings of the 5th European Conference on Computer Vision, ECCV '98, Vol. 1.

[179] Hartley, R.I. (1999) Theory and practice of projective rectification. *International Journal of Computer Vision*, **35** (2), 115–127.

[180] Hartley, R.I. and Zisserman, A. (2003) *Multiple View Geometry in Computer Vision*, 2nd edn, Cambridge University Press.

[181] Hauβecker, H. and Jähne, B. (1998) *A Tensor Approach for Local Structure Analysis in Multi-dimensional Images*, Interdisciplinary Centre for Scientific Computing, University of Heidelberg.

[182] Hauβecker, H. and Jähne, B. (1998) *A Tensor Approach for Precise Computation of Dense Displacement Vector Fields*, Interdisciplinary Centre for Scientific Computing, University of Heidelberg.

[183] Haykin, S. (1996) *Adaptive Filter Theory*, 3rd edn, Prentice-Hall.

[184] Haykin, S. (1999) *Neural Networks. A Comprehensive Foundation*, 2nd edn, Prentice-Hall.

[185] Hecht, E. (1998) *Optics*, 3rd edn, Addison-Wesley.

[186] Heikkilä, J. (2000) Geometric camera calibration using circular control points. *IEEE Transactions on Pattern Analysis and Machine Intelligence*, **22** (10), 1066–1077.

[187] Henkel, R.D. (1998) Fast Stereovision by Coherence Detection. Technical Report, Institute of Theoretical Neurophysics, University of Bremen (available at: henkel@theo.physik.uni-bremen.de).

[188] Hespanha, J.P., Dodds, Z., Hager, G.D. and Morse, A.S. (1998) *What Tasks Can Be Performed with an Uncalibrated Stereo Vision System?* Center for Computational Vision and Control, Yale University.

[189] Heyden, A. (1998) *A Common Framework for Multiple View Tensors*. Proceedings of the 5th European Conference on Computer Vision, ECCV '98, Vol. 1, pp. 3–19.

[190] Heyden, A. (1999) *Hand-Eye Calibration from Image Derivatives*, Department of Mathematics, Lund University.

[191] Heyden, A. (1995) Geometry and algebra of multiple projective transformations. PhD thesis, Department of Mathematics, Lund Institute of Technology.

[192] Heyden, A. and Åström, K. (1996) *Algebraic Varieties in Multiple View Geometry*, Department of Mathematics, Lund Institute of Technology.

[193] Hill, S. and Roberts, J.C. (1995) *Surface Models and the Resolution of N-Dimensional Cell Ambiguity, Graphics Gems V*, Academic Press, pp. 98–106.

[194] Hirschmüller, H. and Scharstein, D. (2007) *Evaluation of Cost Functions for Stereo Matching*. IEEE Computer Society Conference on Computer Vision and Pattern Recognition, CVPR 2007.

[195] Hirschmüller, H., Innocent, P. and Garibaldi, J. (2002) Real-time correlation-based stereo vision with reduced border errors. *International Journal of Computer Vision*, **47** (1–3), 229–246.

[196] Hodge, W.V.D. and Pedoe, D. (1947) *Methods of Algebraic Geometry*, Vol. 1, Cambridge University Press.

[197] van Hoff, A. (1992) *Efficient Computation of Gaussian Pyramids*, Turing Institute.

[198] van Hoff, A. (1992) *An Efficient Implementation of MSSM*, Turing Institute.

[199] Horn, B. and Schunck, B. (1981) Determining optical flow. *Artificial Intelligence*, **17**, 185–203.

[200] Hotca, S., Popa, C., Luncan, D. *et al.* (1997) *A Computer Application for Human Depth Perception Models: Computation of Stereo Disparities*. Proceedings of the 4th International Conference on Computers in Medicine, May 1997, pp. 151–156.

[201] Howard, I.P. and Rogers, B.J.R. (1995) *Binocular Vision and Stereopsis*, Oxford University Press.

[202] http://www.3dmd.com

[203] http://www.ai.sri.com/videos/

[204] http://www.brightbytes.com/

[205] http://www.cs.cmu.edu/~cil/vision.html

[206] http://www-dbv.cs.uni-bonn.de/stereo data/
[207] http://www.di3d.com
[208] http://www.mathworks.com/
[209] http://vision.middlebury.edu/stereo/
[210] http://vision.middlebury.edu/flow/
[211] http://www.precision3d.co.uk
[212] http://vasc.ri.cmu.edu/idb/html/stereo/ (test stereo pairs)
[213] http://www.vision.caltech.edu/bouguetj/calib_doc/
[214] http://www.wga.hu/
[215] http://www.wikipedia.org/
[216] http://www.wiley.com/go/cyganek3dcomputer
[217] http://www.wwl.co.uk/3dscanning-index.htm
[218] Huttenlocher, D. (1997) Computer vision, in *The Computer Science and Engineering Handbook* (ed. A. Tucker), CRC Press.
[219] Intille, S.S. and Bobick, A.F. (1996) Disparity-Space Images and Large Occlusion Stereo. *MIT Media Lab Perceptual Computing Group Technical Report 220*.
[220] Irani, M. and Anandan, P. (1999) *About Direct Methods*. International Workshop on Vision Algorithms, Corfu, Greece, September 1999.
[221] Ishikawa, H. and Geiger, D. (1998) *Occlusions, Discontinuities, and Epipolar Lines in Stereo*. Proceedings of the 5th European Conference on Computer Vision, ECCV '98, Vol. 1, pp. 233–248.
[222] Ivins, W.M. (1964) *Art and Geometry. A Study in Space Intuitions*, Dover Publications.
[223] Iwahori, Y., Woodham, R.J., Ozaki, M. *et al.* (1997) Neural network based photometric stereo with a nearby rotational moving light source. *IEICE Transactions on Information and Systems*, **E80-D** (9), 948–957.
[224] Jähne, B. (1997) *Digital Image Processing*, 4th edn, Springer-Verlag.
[225] Jähne, B. (1998) *Performance Characteristics of Low-Level Motion Estimation in Spatiotemporal Images*, Interdisciplinary Centre for Scientific Computing, University of Heidelberg.
[226] Jähne, B. (1997) *Practical Handbook on Image Processing for Scientific Applications*, CRC Press.
[227] Jähne, B. (1993) *Spatio-Temporal Image Processing*, Springer-Verlag.
[228] Janesick, J. (2002) Dueling detectors. CCD or CMOS? *SPIE OE Magazine*, February, 30–33 (available at: http://www.dalsa.com/shared/content/OE_Magazine_Dueling_Detectors_Janesick.pdf).
[229] Jin, Z. (1988) On the multi-scale iconic representations for low-level computer visions systems. PhD thesis, Turing Institute (University of Strathclyde).
[230] Jin, Z., Niblett, T.B. and Urquhart, C.W. (1999) Improved methods and apparatus for 3-D imaging. Patent application, PCT/GB99/03584.
[231] Josuttis, N.M. (2000) *The C++ Standard Library. A Tutorial and Reference*, Addison-Wesley.
[232] Ju, X., Boyling, T., Siebert, J.P. *et al.* (2004) *Integration of Range Images of Multi-View Stereo System*. 17th International Conference on Pattern Recognition, ICPR2004, Vol. 4. pp. 280–283. IEEE Computer Society Press.
[233] Ju, X., Mao, Z., Siebert, J. P. *et al.* (2004) *Applying Mesh Conformation on Shape Analysis with Missing Data*. 2nd International Symposium on 3D Data Processing, Visualization, and Transmission, 3DPVT 2004, pp. 696–702.
[234] Ju, X., Siebert, J.P., McFarlane, N. and Tillett, R. (2007) Ham volume measurement of live pigs. *Journal of Experimental Mechanics*, **22** (4).
[235] Julesz, B. (2006) *Foundations of Cyclopean Perception*, MIT Press.
[236] Kambhamettu, C., Goldgof, D.B., Terzopoulos, D. and Huang, T.S. (1994) Nonrigid motion analysis, in *Handbook of Pattern Recognition and Image Processing: Computer Vision*, Vol. 2 (ed. T.Y. Young), Academic Press.
[237] Kanade, T. and Okutomi, M. (1994) A stereo matching algorithm with an adaptive window: theory and experiment. *IEEE Transactions on Pattern Analysis and Machine Intelligence*, **16** (9), 920–932.
[238] Kanade, T. and Zitnick, C.L. (1999) A Cooperative Algorithm for Stereo Matching and Occlusion Detection. *Carnegie Mellon University Technical Report CMU-RI-TR-99-35*, Robotics Institute, Pittsburgh, PA.
[239] Kanatani, K. (2005) *Statistical Optimization for Geometric Computation. Theory and Practice*, Dover Publications.
[240] Kara, L.B. and Stahovich, T.F. (2005) An image-based, trainable symbol recognizer for hand-drawn sketches. *Computers & Graphics*, **29** (4), 501–517.

[241] Kay, D.C. (1988) *Theory and Problems of Tensor Calculus*, Schaum's Outlines Series, McGraw-Hill.

[242] Kecman, V. (2001) *Learning and Soft Computing: Support Vector Machines, Neural Networks, and Fuzzy Logic Models*, MIT Press.

[243] Kerievsky, J. (2005) *Refactoring to Patterns*, Addison-Wesley.

[244] Kinderman, A.J. and Monahan, J.F. (1977) Computer generation of random variables using the ratio of uniform deviates. *ACM Transactions on Mathematical Software*, **3**, 257–260.

[245] Kitchen, L. and Rosenfeld, A. (1982) Gray-level corner detection. *Pattern Recognition Letters*, **1** (2), 95–102.

[246] Klette, R., Schlüns, K. and Koschan, A. (1998) *Computer Vision. Three-Dimensional Data from Images*, Springer-Verlag.

[247] Klette, R. and Zamperoni, P. (1996) *Handbook of Image Processing Operators*, John Wiley & Sons, Ltd.

[248] Knuth, D. (1998) *The Art of Computer Programming: Vol. 1. Fundamental Algorithms*, 3rd edn, Addison-Wesley.

[249] Knuth, D. (1998) *The Art of Computer Programming: Vol. 2. Seminumerical Algorithms*, Addison-Wesley.

[250] Koenig, A. and Moo, B. (2000) Performance: myths, measurements, and morals: part 6. *Journal of Object-Oriented Programming*, **13** (2).

[251] Kolb, C., Mitchell, D. and Hanrahan, P. (1995) A Realistic Camera Model for Computer Graphics. Technical Report, Computer Science Department, Princeton University.

[252] Kolmogorov, V. (2004) Graph based algorithms for scene reconstruction from two or more views. PhD thesis, Cornell University.

[253] Kolmogorov, V. and Zabih, R. (2006) Graph cut algorithm for binocular stereo with occlusions, in *Handbook of Mathematical Models in Computer Vision* (eds N. Paragios, Y. Chen and O. Faugeras), Springer-Verlag, pp. 423–437.

[254] Kolmogorov, V. and Zabih, R. (2004) What energy functions can be minimized via graph cuts? *IEEE Transactions on Pattern Analysis and Machine Graphics*, **26** (2), 147–159.

[255] Konolige, K. (1999) *Small Vision Systems: Hardware and Implementation*, Artificial Intelligence Center, SRI International.

[256] Konolige, K. (1999) *Stereo Geometry*, Artificial Intelligence Center, SRI International.

[257] Kopparapu, S.K. and Corke, P. (1999) *The Effect of Measurement Noise of Intrinsic Camera Calibration Parameters*, CSIRO Manufacturing Science and Technology, Australia.

[258] Koren, I. (2002) *Computer Arithmetic Algorithms*, 2nd edn, A.K. Peters Ltd.

[259] Korn, G.A. and Korn, T.M. (2000) *Mathematical Handbook for Scientists and Engineers*, Dover Publications.

[260] Koschan, A. (1997) *Improving Robot Vision by Color Information*. Proceedings of the 7th International Conference on Artificial Intelligence and Information-Control Systems of Robots, Smolenice, Slovakia, 10–14 September 1997.

[261] Koschan, A., Rodehorst, V. and Spiller, K. (1996) *Color Stereo Vision Using Hierarchical Block Matching and Active Color Illumination*. Proceedings of the 13th International Conference on Pattern Recognition, ICPR '96, Vienna, Austria, 25–29 August 1996, Vol. I, pp. 835–839.

[262] Koschan, A. and Abidi, M. (2005) Detection and classification of edges in color images. A review of vector-valued techniques. *IEEE Signal Processing Magazine*, **22** (1), 64–73.

[263] Kreyszig, E. (1991) *Differential Geometry*, Dover Publications.

[264] Kühnel, W. (2003) *Differential Geometry. Curves, Surfaces, Manifolds*, 2nd edn, American Mathematical Society (AMS).

[265] Lan, Z.-D. and Mohr, R. (1995) Robust Matching by Partial Correlation. *INRIA Technical Report RR-2643*.

[266] Lasenby, J. and Lesenby, A. (1996) *Estimating Tensors for Matching over Multiple Views*, Engineering Department, Cambridge University.

[267] Laveau, S. and Faugeras, O. (1994) 3-D Scene Representation as a Collection of Images and Fundamental Matrices. *INRIA Technical Report 2205*.

[268] Lavest, J.-M., Viala, M. and Dhome, M. (1998) *Do We Really Need an Accurate Calibration Pattern to Achieve a Reliable Camera Calibration?* Proceedings of the 5th European Conference on Computer Vision, ECCV '98, June 1998, Vol. 1.

[269] Leclerc, Y.G., Luong, Q.-T. and Fua, P. (1999) *Self-Consistency: A Novel Approach to Characterizing the Accuracy and Reliability of Point Correspondence Algorithms*, Artificial Intelligence Center, SRI International.

[270] Lee, S.H., Park, J.-I., Inoue, S. and Lee, C.W. (1999) Disparity estimation based on Bayesian maximum a posteriori (MAP) algorithm. *IEICE Transactions on Fundamentals of Electronics, Communications and Computer Sciences*, **E82-A** (7), 1367–1376.

[271] Lee, S.H. and Kanatsugu, Y. (2002) MAP-based stochastic diffusion for stereo matching and line fields estimation. *International Journal of Computer Vision*, **47** (1–3), 195–218.

[272] Lee, H.-C. (2005) *Introduction to Color Imaging Science*, Cambridge University Press.

[273] Lerner, E.J. (2000) Laser microscopy opens a new dimension. *Laser Focus World*, **36** (12), 141–144.

[274] Levine, J. (1994) *Programming for Graphic Files in C and C++*, John Wiley & Sons, Ltd.

[275] Li, M. and Lavest, J.M. (1996) Some aspects of zoom lens calibration. *IEEE Transactions on Pattern Analysis and Machine Intelligence*, **18** (11), 1105–1110.

[276] Li, S.Z. (2001) *Markov Random Field Modeling in Image Analysis*, Springer.

[277] Lindeberg, T. (1994) Scale-space theory: a basic tool for analysis of structures at different scales. *Journal of Applied Statistics*, **21** (2), 224–270.

[278] Longuet-Higgins, H.C. (1981) A computer algorithm for reconstructing a scene from two projections. *Nature*, **293**, 133–135.

[279] Looney, C.G. (1997) *Pattern Recognition Using Neural Networks. Theory and Algorithms for Engineers and Scientists*, Oxford University Press.

[280] Lorensen, W.E. and Cline, H.E. (1987) *Marching Cubes: A High Resolution 3D Surface Construction Algorithm*. Computer Graphics, SIGGRAPH '87 Proceedings, July 1987, Vol. 21, pp. 163–169.

[281] Lotti, J.-L. and Giraudon, G. (1993) Adaptive Window Algorithm for Aerial Image Stereo. *INRIA Technical Report 2121*.

[282] Lourakis, M.I.A. and Deriche, R. (2000) Camera Self-Calibration Using the Kruppa Equations and the SVD of the Fundamental Matrix: The Case of Varying Intrinsic Parameters. *INRIA Technical Report 3911*.

[283] Lowe, D. (2004) Distinctive image features from scale-invariant keypoints. *International Journal of Computer Vision*, **60** (2), 91–110.

[284] Lucas, B.D. and Kanade, T. (1981) *An Iterative Image Registration Technique with an Application to Stereo Vision*. International Joint Conference on Artificial Intelligence, 1981, pp. 674–679.

[285] Luis-García, R., Deriche, R., Rousson, M. and Alberola-López, C. (2005) *Tensor Processing for Texture and Colour Segmentation*, Lecture Notes in Computer Science 3540, Springer, pp. 1117–1127.

[286] Luong, Q.-T. and Faugeras, O.D. (1993) Self-calibration of a Stereo Rig from Unknown Camera Motions and Point Correspondences. *INRIA Technical Report 2014*.

[287] Luong, Q.-T. and Faugeras, O.D. (1995) Camera Calibration, Scene Motion and Structure Recovery from Point Correspondences and Fundamental Matrices. INRIA Technical Report.

[288] Luong, Q.-T. and Faugeras, O.D. (1995) The Fundamental Matrix: Theory, Algorithms, and Stability Analysis. INRIA Technical Report.

[289] Luong, Q.-T., Deriche, R., Faugeras, O.D. and Papadopoulo, T. (1993) On Determining the Fundamental Matrix: Analysis of Different Methods and Experimental Results. *INRIA Technical Report 1894*.

[290] Ma, Y., Soatto, S., Košecká, J. and Sastry, S.S. (2004) *An Invitation to 3-D Vision. From Images to Geometrical Models*, Springer.

[291] Mallot, H.A. (1993) Computational psychophysics of stereoscopic depth perception, in *Grundlagen and Anwendungen der Kunstlichen Intelligenz*, Springer Verlag, pp. 60–73.

[292] Mallot, H.A. (2000) *Computational Vision. Information Processing in Perception and Visual Behavior*, MIT Press.

[293] Mao, Z., Ayoub, A. and Siebert, J.P. (2000) *Development of 3D Measuring Techniques for the Analysis of Facial Soft Tissue Change*. Proceedings of the Medical Image Computing and Computer-Assisted Intervention 2000 (MICCAI 2000), Pittsburgh, PA, pp. 1051–1061.

[294] Mao, Z., Siebert, J.P., Cockshott, W.P. and Ayoub, A.F. (2004) *A Coordinate-Free Method for the Analysis of 3D Facial Change*. Proceedings of SPIE on Medical Imaging, San Diego, CA, 14–19 February 2004.

[295] Mao, Z. (2005) Computer assisted methods to support the clinical assessment of human surface anatomy in 3D images. PhD dissertation, University of Glasgow.

[296] Mao, Z., Ju, X., Siebert, J.P. *et al.* (2006) Constructing dense correspondences for the analysis of 3D facial morphology. *Pattern Recognition Letters*, **27** (6), 597–608.

[297] Marchand-Maillet, S. and Sharaiha, Y.M. (2000) *Binary Digital Image Processing. A Discrete Approach*, Academic Press.

[298] Marr, D. and Poggio, T. (1976) Cooperative Computation of Stereo Disparity. AI Memo 364, Artificial Intelligence Laboratory, Massachusetts Institute of Technology.
[299] Marr, D. and Poggio, T. (1977) A Theory of Human Stereo Vision. AI Memo 451, Artificial Intelligence Laboratory, Massachusetts Institute of Technology.
[300] Matthies, L. and Xiong, Y. (1997) Error Analysis of a Real-Time Stereo System. Technical Report, Jet Propulsion Laboratory, Pasadena, CA.
[301] Matthies, L., Litwin, T., Owens, K. *et al.* (1998) *Performance Evaluation of UGV Obstacle Detection with CCD/FLIR Stereo Vision and LADAR 1*, Jet Propulsion Laboratory and National Institute of Standards and Technology.
[302] Mayhew, J. and Frisby, J. (1981) Psychophysical and computational studies towards a theory of human stereopsis. *Artificial Intelligence*, **17**, 349–385.
[303] Mayhew, J. and Longuet-Higgins, H.C. (1982) A computational model of binocular depth perception. *Nature*, **297**, 376–379.
[304] McCane, B., Novins, K., Crannitch, D. and Galvin, B. (2001) On benchmarking optical flow. *Computer Vision and Image Understanding*, **84** (1), 126–143.
[305] McConnell, S. (2004) *Code Complete*, 2nd edn, Microsoft Press.
[306] Mellor, J.P., Teller, S. and Lozano-Pérez, T. (1996) Dense Depth Maps from Epipolar Images. MIT, Artificial Intelligence Laboratory, Memo 1593.
[307] Meerbergen, V.G., Vergauwen, M., Pollefeys, M. and Van Gool, L. (2002) A hierarchical symmetric stereo algorithm using dynamic programming. *International Journal of Computer Vision*, **47** (1–3), 275–285.
[308] Meyer, C.D. (2000) *Matrix Analysis and Applied Linear Algebra*, Society for Industrial and Applied Mathematics (SIAM).
[309] Migdal, J. (2000) Depth Perception Using a Trinocluar Camera Setup and Sub-Pixel Image Correlation Algorithm. *Mitsubishi Electric Research Laboratories, Technical Report TR2000-20.*
[310] Mikołajczyk, K. (2002) Detection of local features invariant to affine transformations. PhD thesis, Institut National Polytechnique de Grenoble.
[311] Mikolajczyk, K. and Schmid, C. (2004) Scale and affine invariant interest point detectors. *International Journal of Computer Vision*, **60** (1), 63–86.
[312] Mitra, S.K. (2000) *Digital Signal Processing*, McGraw-Hill.
[313] Mitra, S.K. and Sicuranza, G.L. (2000) *Nonlinear Image Processing*, Academic Press.
[314] Mohr, R. and Triggs, B. (1996) Projective Geometry for Image Analysis. A tutorial given at ISPRS, Vienna, July 1996.
[315] Molton, N., Se, S., Brady, J.M. *et al.* (1998) A stereo vision-based aid for the visually impaired. *Image and Vision Computing*, **16** (4), 251–263.
[316] Molton, N.D. (1998) Computer vision as an aid for the visually impaired. PhD thesis, University of Oxford.
[317] Moon, T.K. and Stirling, W.C. (2000) *Mathematical Methods and Algorithms for Signal Processing*, Prentice-Hall.
[318] Moons, T., Frore, D., Vandekerckhove, J. and Gool, L.V. (1998) *Automatic Modeling and 3D Reconstruction of Urban House Roofs from High Resolution Aerial Imagery*. Proceedings of the 5th European Conference on Computer Vision, ECCV '98, June 1998, Vol. 1.
[319] Mordohai, P. and Medioni, G. (2006) Stereo using monocular cues within the tensor voting framework. *IEEE Transactions on Pattern Analysis and Machine Intelligence*, **28** (6), 968–982.
[320] Mordohai, P. and Medioni, G. (2007) *Tensor Voting. A Perceptual Organization Approach to Computer Vision and Machine Learning*, Morgan & Claypool Publishers.
[321] Moritsu, T. and Kato, M. (2000) Disparity mapping technique and fast rendering technique for image morphing. *IEICE Transactions on Information and Systems*, **E83-D** (2), 275–282.
[322] Mundy, J.L. and Zisserman, A. (1992) *Geometric Invariance in Computer Vision*, MIT Press.
[323] Mühlmann, K., Maier, D., Hesser, J. and Manner, R. (2002) Calculating dense disparity maps from color stereo images, and efficient implementation. *International Journal of Computer Vision*, **47** (1–3), 79–88.
[324] Müller, H. and Stark, M. (1993) Adaptive generation of surfaces in volume data. *Visual Computer*, **9** (4), 182–199.
[325] Myler, H.R. and Weeks, A. (1993) *The Pocket Handbook of Image Processing Algorithms in C*, Prentice-Hall.
[326] Nagel, H.-H. and Enkelmann, W. (1986) An investigation of smoothness constraints for the estimation of displacement vector fields from image sequences. *IEEE Transactions on Pattern Analysis and Machine Intelligence*, **8**, 565–593.

[327] Nebel, J.C., Cockshott, W.P., Yarmolenko, V. *et al.* (2005) Pre-commercial 3-D digital TV studio. *IEE Proceedings: Vision, Image, and Signal Processing*, **152** (6), 665–667.

[328] Nene, S.A. and Nayar, S.K. (1998) *Stereo Using Mirrors*, Department of Computer Science, Columbia University.

[329] Newton, I. (2000) *Opticks*, Dover Publications.

[330] Nishihara, H.K. (1993) *Real-Time Stereo- and Motion-Based Figure Ground Discrimination and Tracking Using LOG Sign Correlation.* Signals, Systems and Computers, 1993 Conference Record of the Twenty-Seventh Asilomar Conference on Volume, Vol. 1, Issue 1–3, pp. 95–100.

[331] Nocedal, J. and Wright, S.J. (1999) *Numerical Optimization*, Springer.

[332] Oda, K., Tanaka, M., Yoshida, A. *et al.* (1999) *A Video-Rate Stereo Machine and its Application to Virtual Reality*, Robotics Institute, Carnegie Mellon University 1999.

[333] Oehler, S.B., Siebert, J.P., Mao, Z. *et al.* (2007) *The Role of Geodesics in Human–Computer Interfaces for 3D Surface Anatomy Assessment.* 10th International Conference on Medical Image Computing and Computer Assisted Intervention, Brisbane, Australia, 2 November 2007.

[334] Ohta, Y. and Kanade, T. (1985) Stereo by intra- and inter-scanline search using dynamic programming. *IEEE Transactions on Pattern Analysis and Machine Intelligence*, **7** (2), 139–154.

[335] Okutomi, M. and Kanade, T. (1993) A multiple-baseline stereo. *IEEE Transactions on Pattern Recognition and Machine Intelligence*, **15** (4), 353–363.

[336] Oppenheim, A.V. and Schafer, R.W. (1989) *Discrete-Time Signal Processing*, Prentice-Hall.

[337] Pajares, G. and de la Cruz, J.M. (2003) Stereovision matching through support vector machines. *Pattern Recognition Letters*, **24**, 2575–2583.

[338] Paler, K., Föglein, J., Illingworth, J. and Kittler, J. (1984) Local ordered greylevels as an aid to corner detection. *Pattern Recognition*, **17** (5), 535–543.

[339] Pankanti, S. and Jain, A.K. (1995) Integrating vision modules: stereo, shading, grouping, and line labeling. *IEEE Transactions on Pattern Analysis*, **17** (9), 831–842.

[340] Papadimitriou, D.V. and Dennis, T.J. (1996) Epipolar line estimation and rectification for stereo image pairs. *IEEE Transactions on Image Processing*, **5** (4), 672–676.

[341] Papoulis, A. (1991) *Probability, Random Variables, and Stochastic Processes*, 3rd edn, McGraw-Hill.

[342] Parker, P. (1999) *Practical Image Algorithms*, John Wiley & Sons, Ltd.

[343] Pedrotti, L.S. and Pedrotti, F.L. (1998) *Optics and Vision*, Prentice-Hall.

[344] Penrose, R. (2005) *The Road to Reality. A Complete Guide to the Laws of the Universe*, Alfred A. Knopf.

[345] Perona, P. and Malik, J. (1990) Scale-space and edge detection using anisotropic diffusion. *IEEE Transactions on Pattern Analysis and Machine Intelligence*, **12** (7), 629–639.

[346] Pitas, I. and Venetsanopoulos, A.N. (1990) *Nonlinear Digital Filters. Principles and Applications*, Kluwer Academic.

[347] *Plastic & Reconstructive Surgery Journal*, 2008 (in press).

[348] Point Grey (2000) *TRICLOPS Stereo Vision System, Version 2.1, User's Guide and Command Reference*, Point Grey Research Inc, (www.ptgrey.com).

[349] Point Grey (2000) *TriclopsDemo Application 2.0. User's Manual*, Point Grey Research Inc, (www.ptgrey.com).

[350] Porikli, F. (2005) Integral Histogram: A Fast Way to Extract Histograms in Cartesian Spaces. *Mitsubishi Technical Report TR2005-057*.

[351] Pratt, W.K. (2001) *Digital Image Processing*, 3rd edn, John Wiley & Sons, Ltd.

[352] Press, W.H., Teukolsky, S.A., Vetterling, W.T. and Flannery, B.P. (2007) *Numerical Recipes in C. The Art of Scientific Computing*, 3rd edn, Cambridge University Press.

[353] Quan, L. and Triggs, B. (2000) *A Unification of Autocalibration Methods.* Asian Conference on Computer Vision, ACCV, 2000.

[354] Richter, J. (1999) *Advanced Windows. The Developer's Guide to the Win32® API for Windows NT™*, Microsoft Press.

[355] Riley, K.F., Hobson, M.P. and Bence, S.J. (2000) *Mathematical Methods for Physics and Engineering*, Cambridge University Press.

[356] Ritter, G. and Wilson, J. (2001) *Handbook of Computer Vision Algorithms in Image Algebra*, CRC Press.

[357] Rivest, J.-F., Soille, P. and Beucher, S. (1993) Morphological gradients. *Journal of Electronic Imaging*, **2** (4), 326–336.

[358] Robert, L., Zeller, C. and Faugeras, O. (1995) Application of Non-metric Vision to Some Visually Guided Robotics Tasks. *INRIA Technical Report 2584.*

[359] Robert, L. and Deriche, R. (1996) *Dense Depth Map Reconstruction: A Minimization and Regularization Approach Which Preserves Discontinuities*, Lecture Notes in Computer Science 1064, Springer, pp. 439–451.

[360] Robinson, J.O. (1998) *The Psychology of Visual Illusions*, Dover Publications.

[361] Rockett, P.I. (2003) Performance assessment of feature detection algorithms: a methodology and case study on corner detectors. *IEEE Transactions on Image Processing*, **12** (12), 1668–1676.

[362] Rohr, K. (1992) Recognizing corners by fitting parametric models. *International Journal of Computer Vision*, **9** (3), 213–230.

[363] Rothwell, C., Csurka, G. and Faugeras, O. (1995) A Comparison of Projective Reconstruction Methods for Pairs of Views. *INRIA Technical Report 2538.*

[364] Roy, S., Meunier, J. and Cox, I.J. (1997) *Cylindrical Rectification to Minimize Epipolar Distortion.* Proceedings of the IEEE International Conference on Computer Vision and Pattern Recognition, Puerto Rico, 1997, pp. 393–399.

[365] Russakoff, D.B., Tomasi, C., Rohlfing, T. and Maurer Jr, C.R. (2004) *Image Similarity Using Mutual Information of Regions*, Lecture Notes in Compute Science 3023, Springer, pp. 596–607.

[366] Santini, S. and Jain, R. (1999) Similarity measures. *IEEE Transactions on Pattern Analysis and Machine Intelligence*, **21** (9), 871–883.

[367] Schaffalitzky, F., Zisserman, A., Hartley, R.I. and Torr, P.H.S. (2000) *A Six Point Solution for Structure and Motion*, Department of Engineering Science, University of Oxford.

[368] Scharstein, D. and Szeliski, R. (1996) *Stereo Matching with Nonlinear Diffusion.* IEEE Computer Society Conference on Computer Vision and Pattern Recognition, CVPR '96, San Francisco, CA, June 1996, pp. 343–350.

[369] Scharstein, D. (1999) *View Synthesis Using Stereo Vision*, Lecture Notes in Computer Science 1582, Springer-Verlag.

[370] Scharstein, D. and Szeliski, R. (2002) A taxonomy and evaluation of dense two-frame stereo correspondence algorithms. *International Journal of Computer Vision*, **47** (1), pp. 7–42.

[371] Scharstein, D. and Szeliski, R. (2003) *High-Accuracy Stereo Depth Maps Using Structured Light.* IEEE Computer Society Conference on Computer Vision and Pattern Recognition, CVPR 2003, Vol. 1, pp. 195–202.

[372] Scharstein, D. and Pal C. (2007) *Learning Conditional Random Fields for Stereo.* IEEE Computer Society Conference on Computer Vision and Pattern Recognition, CVPR 2007.

[373] Schmid, C., Mohr, R. and Bauckhage, C. (2000) Evaluation of interest point detectors. *International Journal of Computer Vision*, **37** (2), 151–172.

[374] Schreer, O. (1998) *Stereo Vision-Based Navigation in Unknown Indoor Environment.* Proceedings of the 5th European Conference on Computer Vision, ECCV '98, June 1998, Vol. 1.

[375] Se, S. and Brady, M. (1998) *Stereo Vision-Based Obstacle Detection for Partially Sighted People.* Third Asian Conference on Computer Vision, ACCV '98, Vol. I, pp. 152–159.

[376] Se, S. and Brady, M. (2000) *Vision-Based Detection of Stair-Cases.* Fourth Asian Conference on Computer Vision, ACCV 2000, Vol. I, pp. 535–540.

[377] Se, S. and Brady, M. (2000) Zebra Crossing Detection for the Partially Sighted. Technical Report, University of Oxford.

[378] Sebe, N., Lew, M.S. and Huijsmans, D. (2000) Toward improved ranking metrics. *IEEE Transactions on Pattern Analysis and Machine Intelligence*, **22** (10), 1132–1143.

[379] Seetharaman, G.S. (1994) Image sequence analysis for three-dimensional perception of dynamic scenes, in *Handbook of Pattern Recognition and Image Processing: Computer Vision*, Vol. 2 (ed. T.Y. Young), Academic Press.

[380] Semple, J.G. and Kneebone, G.T. (1998) *Algebraic Projective Geometry*, 3rd edn, Oxford Classic Texts in the Physical Sciences, Oxford University Press.

[381] Seul, M., O'Gorman, L. and Sammon, M.J. (2000) *Practical Algorithms for Image Analysis. Description, Examples, and Code*, Cambridge University Press.

[382] Shannon, R.R. (1997) *The Art and Science of Optical Design*, Cambridge University Press.

[383] Shashua, A. (1995) Algebraic functions for recognition. *IEEE Transactions on Pattern Analysis and Machine Intelligence*, **17** (8), 779–789.

[384] Shashua, A. and Werman, M. (1995) *Fundamental Tensor: On the Geometry of Three Perspective Views*, Hebrew University of Jerusalem, Institute of Computer Science.

[385] Shirai, Y. (1987) *Three-dimensional Computer Vision*, Springer.

[386] Siebert, J.P. and Urquhart, C.W. (1990) Active Stereo: Texture Enhanced Reconstruction, *Electronics Letters*, **26** (7), 427–430.

[387] Siebert, J.P. and Urquhart, C.W. (1994) *C3D: a Novel Vision-Based 3-D Data Acquisition System*. Proceedings of the Mona Lisa European Workshop, Combined Real and Synthetic Image Processing for Broadcast and Video Production, Hamburg, Germany, 23–24 August 1994.

[388] Siebert, J.P. and Patterson, J.W. (1998) *Captivating Models*. Proceedings of the IEE Colloquium on Computer Vision for Virtual Human Modelling, London, UK, 1998.

[389] Siebert, J.P. and Marshall, S.J. (2000) Human body 3D imaging by speckle texture projection photogrammetry. *Sensor Review*, **20** (3), 218–226.

[390] Simoncelli, E.P. (1993) Distributed representation and analysis of visual motion. PhD thesis, MIT.

[391] Simoncelli, E.P. (1994) *Design of Multi-Dimensional Derivative Filters*. IEEE International Conference on Image Processing, November 1994.

[392] Sinha, S.S. and Jain, R. (1994) Range image analysis, in *Handbook of Pattern Recognition and Image Processing: Computer Vision*, Vol. 2 (ed. T.Y. Young), Academic Press.

[393] Slesareva, N., Bruhn, A. and Weickert, J. (2005) *Optic Flow Goes Stereo: A Variational Method for Estimating Discontinuity-Preserving Dense Disparity Maps*, Lecture Notes in Computer Science 3663, Springer, pp. 33–40.

[394] Smith, S. and Brady, J. (1997) Susan: a new approach to low level image processing. *International Journal of Computer Vision*, **23** (1), 45–78.

[395] Sochen, N., Kimmel, R. and Malladi, R. (1998) A general framework for low level vision. *IEEE Transactions on Image Processing*, **7** (3), 310–318.

[396] Soille, P. (2003) *Morphological Image Analysis. Principles and Applications*, Springer.

[397] Spivak, M. (1999) *A Comprehensive Introduction to Differential Geometry*, Vol. I, Publish or Perish Inc.

[398] Sporring, J., Nielsen, M., Florack, L. and Johansen, P. (1997) *Gaussian Scale-Space Theory*, Kluwer Academic.

[399] Starck, J.-L., Murtagh, F. and Bijaoui, A. (2000) *Image Processing and Data Analysis. The Multiscale Approach*, Cambridge University Press.

[400] Starck, J.-L. and Murtagh, F. (2002) *Astronomical Image and Data Analysis*, Springer.

[401] Stroustrup, B. (1998) *C++ Programming Language*, 3rd edn, Addison-Wesley.

[402] Sturm, P. (2000) A case against Kruppa's equations for camera self-calibration. *IEEE Transactions on Pattern Analysis and Machine Intelligence*, **22** (10), 1199–1204.

[403] Subbarao, M. and Choi, T. (1995) Accurate recovery of three-dimensional shape from image focus. *IEEE Transactions on Pattern Analysis and Machine Intelligence*, **17** (3), 266–274.

[404] Sudderth, E., Ihler, A., Freeman, W. and Willsky, A. (2002) Nonparametric Belief Propagation. *MIT LIDS Technical Report 2551*.

[405] Sun, J., Shum, H.-Y. and Zheng, N.-N. (2002) *Stereo Matching Using Belief Propagation*, ECCV 2002, Lecture Notes in Computer Science 2351, Springer, pp. 510–524.

[406] Sun, S. (2003) Uncalibrated three-view image rectification. *Image and Vision Computing*, **21**, 259–269.

[407] Sun, W. and Cooperstock, J.R. (2006) An empirical evaluation of factors influencing camera calibration accuracy using three publicly available techniques. *Machine Vision and Applications*, **17** (1), 51–67.

[408] Svoboda, T., Pajdla, T. and Hlavac, V. (1998) *Epipolar Geometry for Panoramic Cameras*. Proceedings of the 5th European Conference on Computer Vision, ECCV '98, June 1998, Vol. 1.

[409] Swaminathan, R. and Nayar, S.K. (2000) Nonmetric calibration of wide-angle lenses and polycameras. *IEEE Transactions on Pattern Analysis and Machine Intelligence*, **22** (10), 1172–1178.

[410] Synge, J.L. and Schild, A. (1978) *Tensor Calculus*, Dover Publications.

[411] Szeliski, R. and Coughlan, J. (1997) Spline-based image registration. *International Journal of Computer Vision*, **22** (3), 199–218.

[412] Szeliski, R. and Golland P. (1998) *Stereo Matching with Transparency and Matting*. Proceedings of the Sixth International Conference on Computer Vision, ICCV '98, 4–7 January 1998.

[413] Szeliski, R. and Zabih, R. (2000) An Experimental Comparison of Stereo Algorithms. Microsoft Technical Report (available at: www.research.microsoft.com/~szeliski).

[414] Taligent (1994) *Taligent's Guide to Designing Programs: Well-Mannered Object-Oriented Design in C++*, Addison-Wesley.

[415] Tanaka, S. and Kak, A.C. (1990) A rule-based approach to binocular stereopsis, in *Analysis and Interpretation of Range Images* (eds R.C. Jain and A.K Jain), Springer-Verlag.

[416] Tappen, M.F. and Freeman, W.T. (2003) *Comparison of Graph Cuts With Belief Propagation for Stereo, Using Identical MRF Parameters*. IEEE Computer Society Conference on Computer Vision and Pattern Recognition, 13–16 October 2003, Vol. 2, pp. 900–906.

[417] Tillett, R.D., McFarlane, N.J.B., Wu, J. *et al.* (2004) *Extracting Morphological Data from 3D Images of Pigs*, Agricultural Engineering, Leuven, pp. 203–222.

[418] Tokarczyk, R. and Mazur, T. (2006) Photogrammetry: principles of operation and application in rehabilitation. *Medical Rehabilitation*, **10** (4), 30–39.

[419] Tomasi, C. and Manduchi, R. (1998) *Bilateral Filtering for Gray and Color Images*. Proceedings of the 1998 IEEE International Conference on Computer Vision, Bombay, India.

[420] Torr, P.H.S. and Murray, D.W. (1997) The development and comparison of robust methods for estimating the fundamental matrix. *International Journal of Computer Vision*, **24** (3), 271–300.

[421] Torr, P.H.S. and Zisserman, A. (1998) *Robust Parametrization and Computation of the Trifocal Tensor*, Department of Engineering Science, University of Oxford.

[422] Torr, P.H.S. and Zisserman, A. (1999) *Feature Based Methods for Structure and Motion Estimation*. International Workshop on Vision Algorithms, Corfu, Greece, September 1999.

[423] Torr, P.H.S. and Fitzgibbon, A.W. (2004) Invariant fitting of two view geometry. *IEEE Transactions on Pattern Analysis and Machine Intelligence*, **26** (5), 648–650.

[424] Trapp, R., Drüe, S. and Hartmann, G. (1998) *Stereo Matching with Implicit Detection of Occlusions*, Lecture Notes in Computer Science 1407, Springer.

[425] Trefethen, L.N. and Bau, D. (1997) *Numerical Linear Algebra*, Society for Industrial and Applied Mathematics (SIAM).

[426] Triggs, B. (1997) *Autocalibration and the Absolute Quadric*. IEEE Computer Society Conference on Computer Vision and Pattern Recognition, CVPR 1997.

[427] Triggs, B. (1999) *Camera Pose and Calibration from 4 or 5 Known 3D Points*. International Conference on Computer Vision, 1999.

[428] Triggs, B. (1995) *The Geometry of Projective Reconstruction: I. Matching Constraints and the Joint Image*. Proceedings of the 5th International Conference on Computer Vision, 20–23 June 1995, pp. 338–343.

[429] Triggs, B. (2000) Plane + Parallax, Tensors and Factorization. INRIA Technical Report.

[430] Trucco, E. and Verri, A. (1998) *Introductory Techniques for 3-D Computer Vision*, Prentice-Hall.

[431] Tsai, R. and Huang, T. (1984) Uniqueness and estimation of three-dimensional motion parameters of rigid objects with curved surfaces. *IEEE Transactions of Pattern Analysis and Machine Intelligence*, **6** (1), 13–26.

[432] Ullman, S. (2000) *High-Level Vision. Object Recognition and Visual Cognition*. MIT Press.

[433] Urquhart, C.W. (1990) An investigation into active and passive methods for improving the performance of scale-space stereo. MEng dissertation, Heriot-Watt University and BBN System and Technologies Limited.

[434] Vandervoorde, D. and Josuttis, N.M. (2003) *C++ Templates. The Complete Guide*, Addison-Wesley.

[435] Veksler, O. (2003) *Fast Variable Window for Stereo Correspondence Using Integral Images*. International Conference on Computer Vision and Pattern Recognition, Vol. I, pp. 556–561.

[436] Veksler, O. (2005) *Stereo Correspondence by Dynamic Programming on a Tree*. IEEE Computer Society Conference on Computer Vision and Pattern Recognition, CVPR 2005, Vol. 2, pp. 384–390.

[437] Videre Design (1998) *STH-V1 Stereo Head User's Manual*, Videre Design Corp (www.videredesign.com).

[438] Videre Design (1999) *STH-MD1 Megapixel Digital Stereo Head User's Manual*, Videre Design Corp (www.videredesign.com).

[439] Viola, P. and Wells III, W.M. (1997) Alignment by maximization of mutual information. *International Journal of Computer Vision*, **24** (2), 137–154.

[440] Viola, P. and Jones, M. (2001) *Robust Real-Time Face Detection*. Proceedings of the International Conference on Computer Vision, pp. II. 747–755.

[441] Vlissides, J. (1998) *Pattern Hatching. Design Patterns Applied*, Addison-Wesley.

[442] Wandell, B.A. (1995) *Foundations of Vision*. Sinauer Associates.

[443] Wang, L., Liao, M., Gong, M. *et al.* (2006) *High-Quality Real-Time Stereo Using Adaptive Cost Aggregation and Dynamic Programming*. Third International Symposium on 3D Data Processing, Visualization, and Transmission, pp. 798–805.

[444] Wei, G.-Q., Brauer, W. and Hirzinger, G. (1998) Intensity- and gradient-based stereo matching using hierarchical gaussian basis functions. *IEEE Transactions on Pattern Analysis and Machine Intelligence*, **20** (11), 1143–1160.

[445] Wei, G.-Q. and Hirzinger, G. (1997) Parametric shape-from-shading by radial basis functions. *IEEE Transactions on Pattern Analysis and Machine Intelligence*, **19** (4), 353–365.

[446] Weickert, J. and Hagen, H. (ed) (2006) *Visualization and Processing of Tensor Fields*, Springer.

[447] Windyga, P.S. (2001) Fast impulsive noise removal. *IEEE Transactions on Image Processing*, **10** (1), 173–179.

[448] Witkin, A.P. (1983) *Scale-Space Filtering*. Proceedings of the International Joint Conference on Artificial Intelligence. ACM Inc., pp. 1019–1021.

[449] Wolberg, G. (1990) *Digital Image Warping*, John Wiley & Sons, Inc./IEEE Computer Society.

[450] Woodfill, J. and Von Herzen, B. (1997) *Real-Time Stereo Vision on the PARTS Reconfigurable Computer*. IEEE Symposium on FPGAs for Custom Computing Machines, April 1997.

[451] Wu, J., Tillett, T., McFarlane, N. *et al.* (2004) Extracting the three-dimensional shape of live pigs using stereo photogrammetry. *Computers and Electronics in Agriculture*, **44** (3), 203–222.

[452] Yang, R. and Pollefeys, M. (2005) A versatile stereo implementation on commodity graphics hardware. *Real-Time Imaging*, **11**, 7–18.

[453] Yokoya, N., Shakunaga, T. and Kanbara, M. (1999) Passive range sensing techniques: depth from images. *IEICE Transactions on Information and Systems*, **E82-D** (3), 523–533.

[454] Young T.Y. (ed.) (1994) *Handbook of Pattern Recognition and Image Processing: Computer Vision*, Vol. 2, Academic Press.

[455] Zabih, R. and Woodfill, J. (1998) *Non-parametric Local Transforms for Computing Visual Correspondence*, Computer Science Department, Cornell University.

[456] Zadeh, L.A. (1965) Fuzzy sets. *Information and Control*, **8**, 338–353.

[457] Zhang, Z. (1999) A Flexible New Technique for Camera Calibration. *Technical Report MSR-TR-98-71*, Microsoft Research, Microsoft Corporation (www.microsoft.com).

[458] Zhang, Z. (1996) Determining the Epipolar Geometry and its Uncertainty: A Review. *INRIA Technical Report 2927*.

[459] Zhang, Z. (2004) Camera calibration with one-dimensional objects. *IEEE Transactions of Pattern Analysis and Machine Intelligence*, **26** (7), 892–899.

[460] Zhang, Z., Deriche, R., Faugeras, O. and Luong, T.Q. (1994) A Robust Technique for Matching Two Uncalibrated Images Through the Recovery of the Unknown Epipolar Geometry. *INRIA Technical Report 2273*.

[461] Zheng, Z., Wang, H. and Teoh, E.K. (1999) Analysis of gray level corner detection. *Pattern Recognition Letters*, **20**, 149–162.

[462] Zhengping, J. (1988) On the multi-scale iconic representation for low-level computer vision. PhD thesis, Turing Institute and University of Strathclyde.

[463] Zitnick, C.L. and Kanade, T. (1998) A Volumetric Iterative Approach to Stereo Matching and Occlusion Detection. *Technical Report CMU-RI-TR-98-30*, Robotics Institute, Carnegie Mellon University.

[464] Zitnick, C.L. and Kanade, T. (1999) A Cooperative Algorithm for Stereo Matching and Occlusion Detection. *Technical Report CMU-RI-TR-99-35*, Robotics Institute, Carnegie Mellon University.

[465] Zokai, S. and Wolberg, G. (2005) Image registration using log-polar mappings for recovery of large-scale similarity and projective transformations. *IEEE Transactions on Image Processing*, **14** (10), 1422–1434.

[466] Cockshott, W.P., Hoff, S. and Nebel, J.-C. (2003) An experimental 3D digital TV studio. *IEE Proceedings: Vision, Image & Signal Processing Institute of Electrical Engineers*.

[467] Nebel, J.-C., Rodriguez-Miguel F.J. and Cockshott, W.P. (2001) Stroboscopic stereo rangefinder. In *Proceeding of Third International: 3-D Digital Imaging and Modeling, 2001*. Québec City, Canada.

[468] Hajeer, M.Y., Millett, D.T., Ayoub, A.F. and Siebert, J.P. (2004) Applications of 3D imaging in Orthodontics – Part I. *Journal of Orthodontics*, **31** (1), 62–70.

[469] Ju, X., Nebel J.C. and Siebert, J.P. (2004) 3D thermography imaging standardization technique for inflammation diagnosis. In *Proceedings of SPIE, Photonics Asia*, Beijing, China, 8–12 November 2004, Vols. 5640–46, pp. 5640–46.

Index

Plate 1 *Perspective* by Antonio Canal (1765, oil on canvas, Gallerie dell' Accademia, Venice). (See page 10)

Plate 2 Painting by Bernardo Bellotto *View of Warsaw from the Royal Palace* (1773, Oil on canvas, National Museum, Warsaw). (See page 11)

(a) (b)

Plate 3 Examples of the morphological gradient computed from the colour image (a, b). (See page 128)

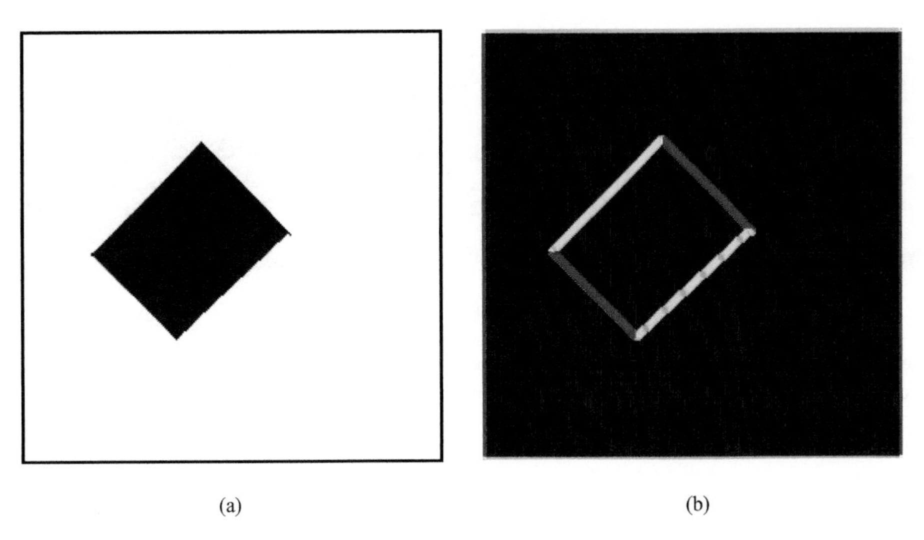

Plate 4 (a) Binary image of a skewed rectangle and (b) colour visualization of its structural tensor – hue H denotes a phase of local orientations, saturation S the coherence, and intensity I conveys trace of T. (See page 142)

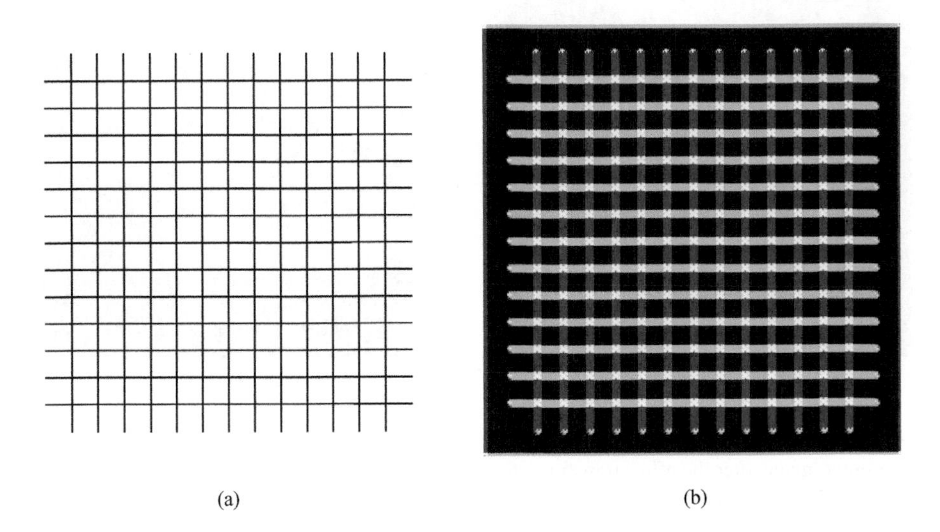

Plate 5 (a) Monochrome image of a grid and (b) the colour visualization of its structural tensor. (See page 142)

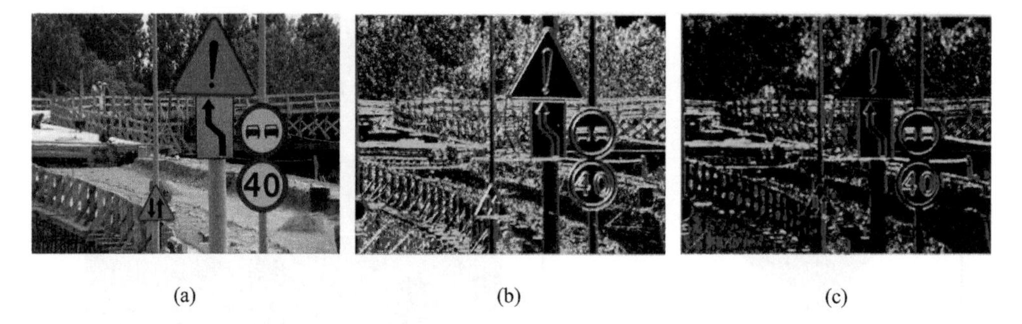

(a) (b) (c)

Plate 6 (a) Examples of the structural tensor operating on an RGB colour image. (b) Visualization of the structural tensor computed with the 3-tap Simoncelli filter. (c) Version with the 5-tap Simoncelli filter. (See page 145)

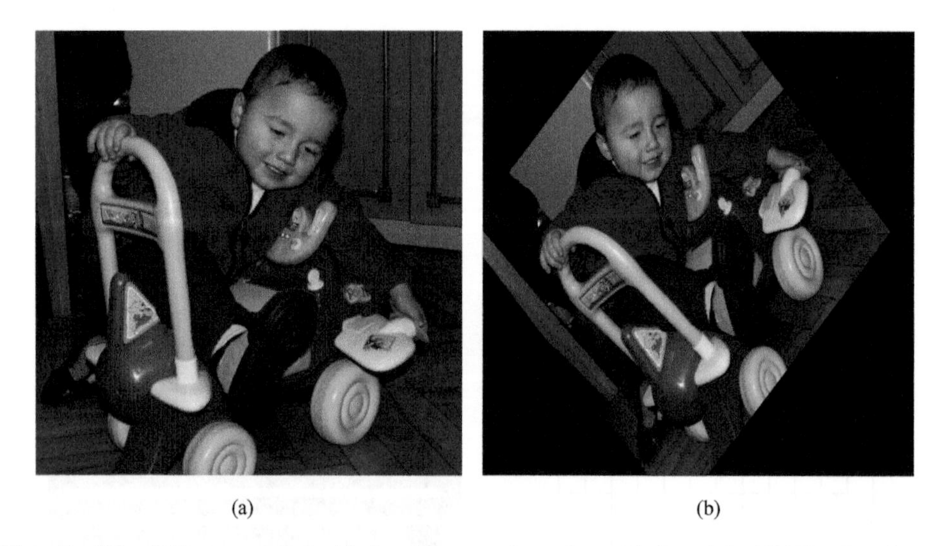

(a) (b)

Plate 7 "Kamil" image warped with the affine transformations: (a) the original RGB colour image, (b) the output image after the affine transformation consisting of the -43° rotation around a centre point, scaling by [0.7, 0.8] and translation by the [155, 0] vector. (See page 423)

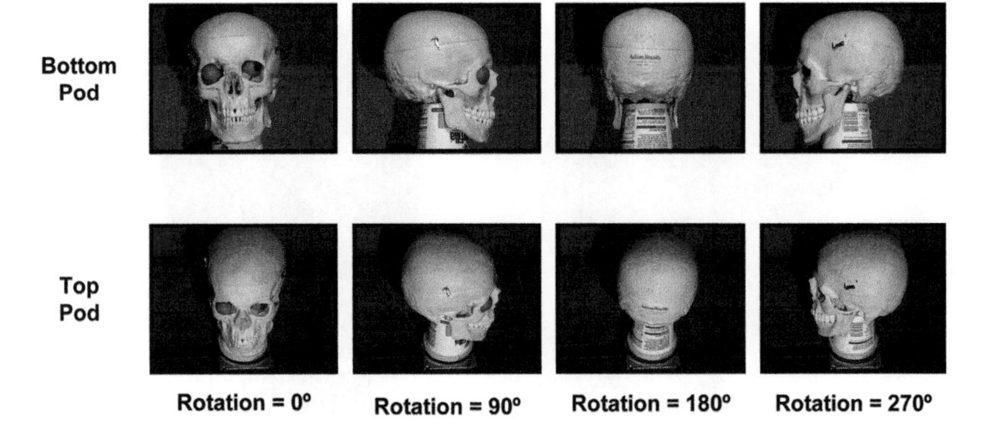

Plate 8 Eight dominant camera views of a skull. (See page 336)

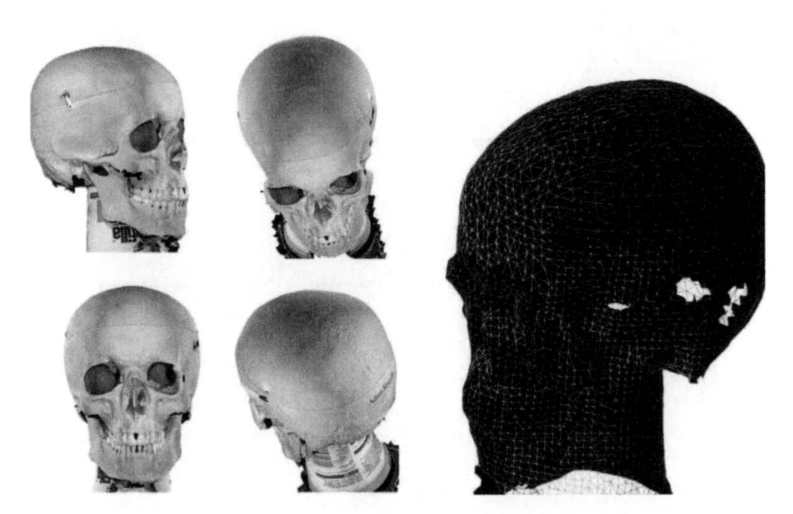

Plate 9 Five views (four of these have been texture-pasted) of a single complete 3D skull model computed by marching cubes integration of eight range surfaces. (See page 337)

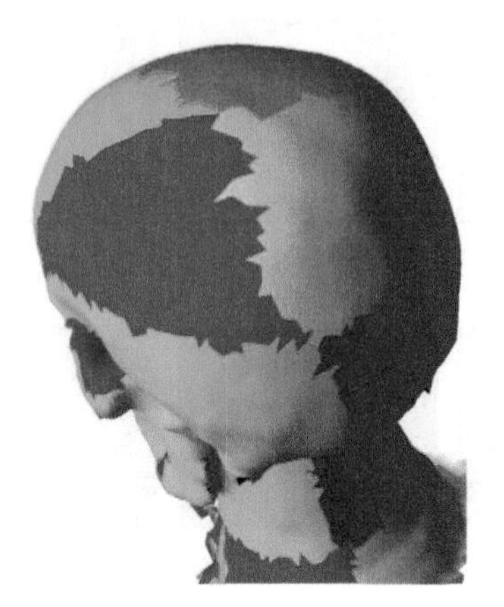
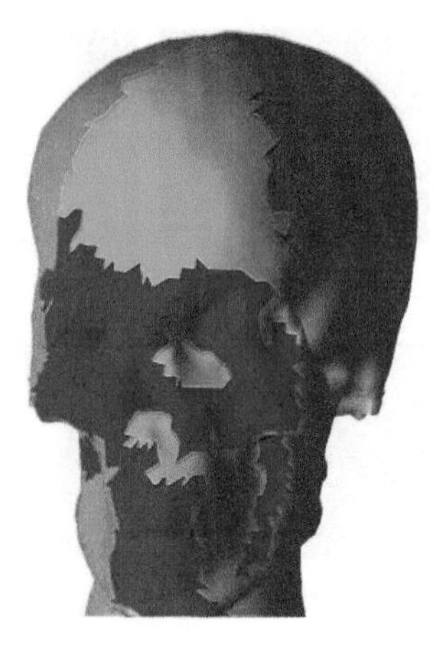

Plate 10 Two views of the integrated skull model showing the colour-coded contributions from different range maps. (See page 337)

Plate 11 Four rendered views of a 3D model captured by an experimental five-pod head scanner. (Subject: His Excellency The Honourable Richard Alston, Australian High Commissioner to the United Kingdom, 2005–2008). (See page 348)

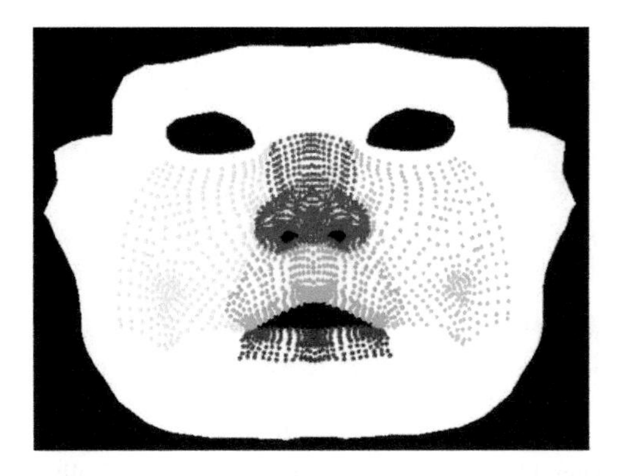

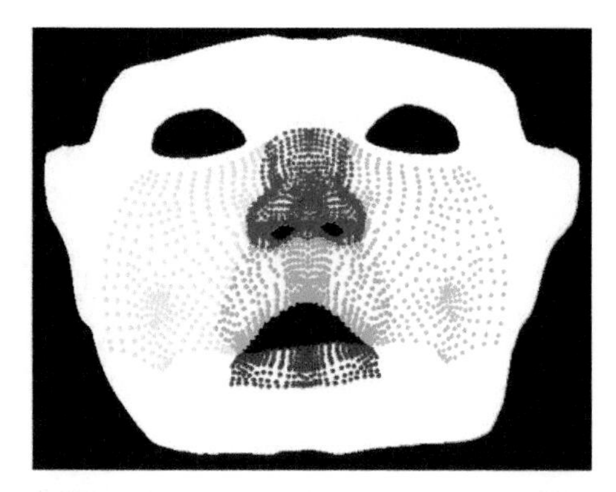

Plate 12 Left: a generic mesh colour coded to label different anatomic regions of the face. Right: the generic mesh conformed into the shape of a captured 3D face mesh, reproduced from [295] (see page 359)

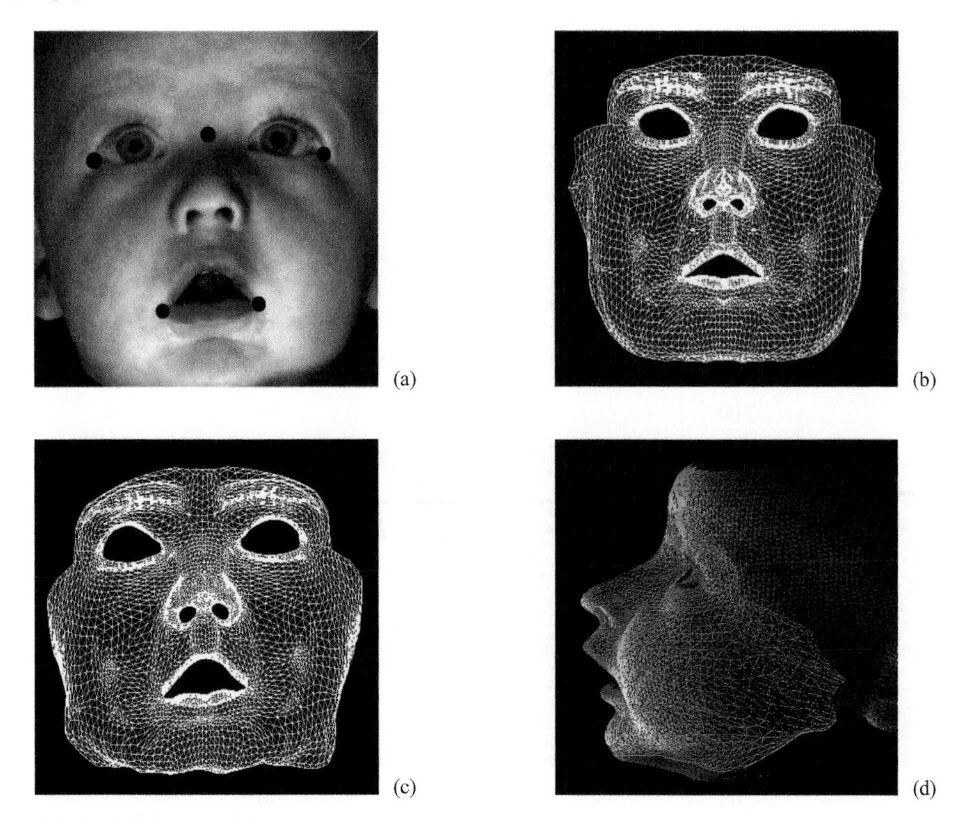

Plate 13 The result of the conformation process, using Mao's basic method, reproduced from [296]. **(a)** The scanned model with 5 landmarks placed for the global mapping; (b) the generic model; (c) the conformed generic model; reproduced from [295] (d) the scanned model aligned to the conformed generic model: the red mesh is the conformed generic model, the yellow mesh is the scanned model. (See page 358)

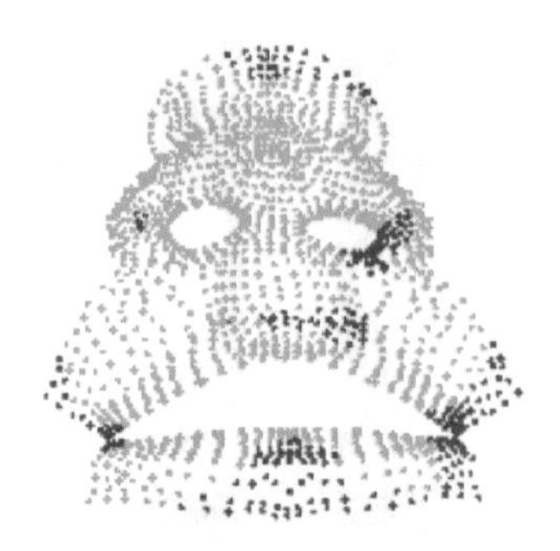

Plate 14 A comparison of corresponding vertices between the mean shapes for 3D face models of 1 & 2 year old children in a surgically managed group (unilateral facial cleft): green indicates no statistically significant difference, while the red indicates a significant difference between the models captured at the two different ages (0.05 significance), reproduced from [295]. (See page 361)

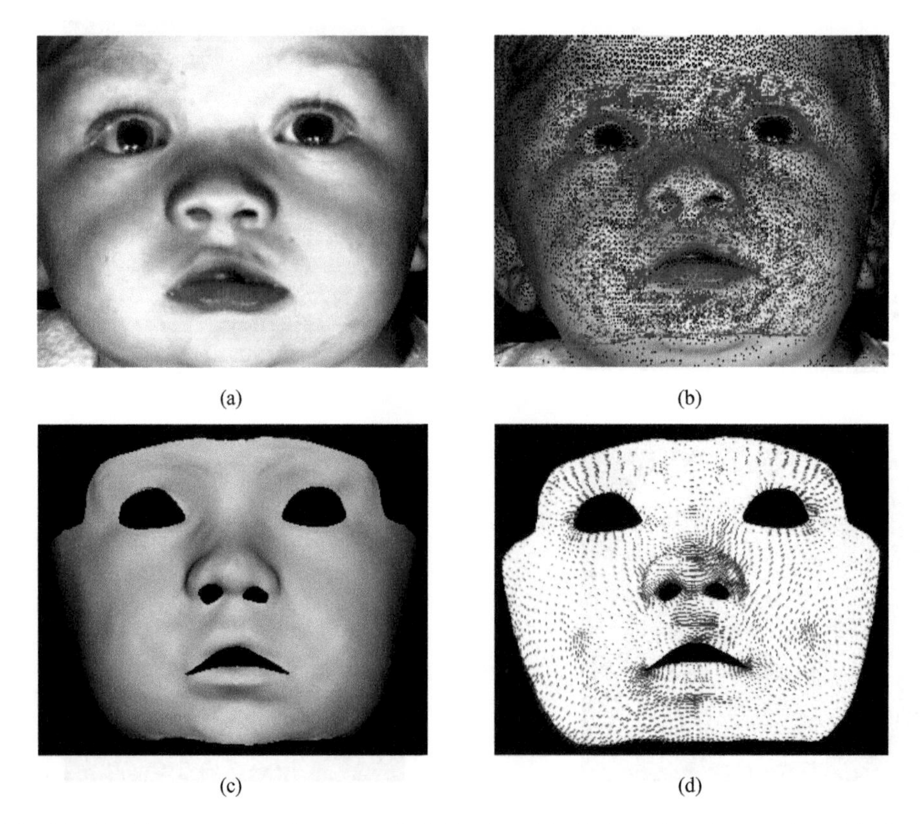

Plate 15 Facial symmetry analysis of an individual model: (a) the original scanned model, (b) the corresponding conformed model, (c) the original scanned model (the yellow mesh) aligned to the conformed model (the red mesh), (d) the calculated symmetry vector field, reproduced from [295]. (See page 362)

Printed and bound by CPI Group (UK) Ltd, Croydon, CR0 4YY

27/10/2024

14580290-0001